HYDRAULICS in Civil and Environmental Engineering

Also available from Spon Press

Coastal Engineering: Processes, Theory and Design Practice
D. Reeve, A. Chadwick and C. Fleming

Coastal, Estuarial and Harbour Engineer's Reference Book
M.B. Abbott and W.A. Price

Hydraulic Structures
2nd edition
P. Novak, A. Moffat, C. Nalluri and R. Naryanan

The Piping Guide
2nd edition
D. Sherwood

Public Water Supply: Models, Data and Operational Management
D. Obradovic and P.B. Lonsdale

Soil Mechanics
6th edition
R.F. Craig

Spon's Civil Engineering and Highway Works Price Book
Davis, Langdon and Everest

Streamflow Measurement
2nd edition
R.W. Herschy

Structural Analysis
4th edition
A. Ghali and A.M. Neville

Water Pollution Control
R. Helmer and I. Hespanhol

Water Quality Assessments
2nd edition
D. Chapman

Water Quality Monitoring
J. Bartram and R. Ballance

For more information about these and other titles please contact:
The Marketing Department, Spon Press, 11 New Fetter Lane, London, EC4P 4EE. Tel: (020) 7583 9855

HYDRAULICS in Civil and Environmental Engineering

Fourth edition

Andrew Chadwick
Professor of Coastal Engineering
School of Engineering, University of Plymouth

John Morfett
Formerly Principal Lecturer
Department of Civil Engineering, University of Brighton

Martin Borthwick
Senior Lecturer
School of Engineering, University of Plymouth

LONDON AND NEW YORK

First published 1986
by HarperCollins Academic
Reprinted 1989, 1991, 1992

Second edition published 1993
by E & FN Spon, an imprint of Chapman & Hall
Reprinted 1994 (twice), 1995, 1996, 1997

Third edition published 1998
by E & FN Spon, an imprint of Routledge
Reprinted 1999

Fourth edition published 2004
by Spon Press
11 New Fetter Lane, London EC4P 4EE

Simultaneously published in the USA and Canada
by Spon Press
29 West 35th Street, New York, NY 10001

Spon Press is an imprint of the Taylor & Francis Group

Typeset in Sabon by
Integra Software Services Pvt. Ltd, Pondicherry, India
Printed and bound in Great Britain by St Edmundsbury Press,
Bury St Edmunds, Suffolk

British Library Cataloguing in Publication Data
A catalogue record for this book is available from the British Library

Library of Congress Cataloging in Publication Data
A catalog record for this book has been requested

ISBN 0–415–30609–4

Contents

Preface

Dear readers, welcome to the fourth edition of *Hydraulics in Civil and Environmental Engineering*. To those of you about to read these pages for the first time, we would simply say that we hope you enjoy them. Our aims remain to provide a comprehensive coverage of civil engineering hydraulics in all its aspects and to provide an introduction to the principles of environmentally sound hydraulic engineering practice. You should find sufficient material to cover most first degree courses and much useful information for a taught higher degree and for professional practice. The references and further reading lists are comprehensive and point the way to further study.

To those of you who have previously used either the first, second or third edition, we commend the fourth edition. This edition contains much of the material from the previous editions. A completely revised chapter on flood hydrology has been written by our new third author, Martin Borthwick. He has preserved the original ordering of the material, but has completely updated the methods in line with the *Flood Estimation Handbook*, which largely supercedes the *Flood Studies Report*. Some additional material has been written for Chapters 8, 9 and 16 and minor revisions and updating of material and references have been undertaken throughout, where appropriate. Finally we now have a website (www.sponpress.com/supportmaterial/0415306094.html), which contains the solutions manual to the problems, some computer programs and spreadsheets and links to other websites of interest.

Acknowledgements

In addition to those from previous editions, we gratefully acknowledge the following, who have given permission for the use of hydrological data:

Environment Agency, South West, UK
Centre for Ecology and Hydrology, Wallingford, UK

Andrew Chadwick, John Morfett and Martin Borthwick, September 2003

Preface to the third edition

Dear readers, welcome to the third edition of *Hydraulics in Civil and Environmental Engineering*. To those of you about to read these pages for the first time, we would simply say that we hope you enjoy them. Our aims remain to provide a comprehensive coverage of civil engineering hydraulics in all its aspects and to provide an introduction to the principles of environmentally sound hydraulic engineering practice. You should find sufficient material to cover most first degree courses and much useful information for a taught higher degree and for professional practice. The references and further reading lists are comprehensive and point the way to further study.

To those of you who have previously used either the first or second edition, we commend the third edition. This edition contains much of the material from the previous editions, to which has been added a completely new chapter on computational hydraulics, several other chapters have been comprehensively revised and many more minor revisions and updating of material and references have been undertaken. All of the more significant changes are summarized below.

Chapter 3 Material on the Navier–Stokes equations removed and added to the new Chapter 14 in revised and expanded form. Revised description of the Prandtl eddy model and addition of the $\kappa-\varepsilon$ turbulence model.

Chapter 5 Material on gradually varied unsteady flow removed and added to the new Chapter 14 in revised form.

Chapter 6 Material on the method of characteristics removed and added to the new Chapter 14 in revised form.

Chapter 8 A completely revised and updated chapter containing new material in most of the previous sections and new sections concerning wave–current interaction, wave conservation equations, infra-gravity waves, refraction and diffraction of directional spectra, long-term wave statistics, effects of climate change and prediction of extreme still water levels.

Chapter 9 New material added, including Van Rijn's equations for the threshold of motion, the updated Ackers and White equations for total sediment transport load, revised description of suspended sediment transport and a new example on suspended sediment transport.

Chapter 11 A reduced description of the matrix method of dimensional analysis. New material added to the section on hydraulic models.

Chapter 12 Material on the numerical solution to the surge tower equations removed and added to the new Chapter 14 in revised form.

Chapter 14 A completely new chapter concerning computational hydraulics. The first part of the chapter introduces the basic concepts of mathematical and numerical models, derivation of the conservation equations, the properties of typical partial differential equations and their numerical solution. The second part discusses the application of computational hydraulics to a range of problems, including surge towers, unsteady river flow and compressible surge in pipelines. Two new computer programs are included for explicit and implicit solution of the equations for unsteady river flow.

Chapter 15 A revised and updated version of the previous Chapter 14, including the latest research results for flow in compound channels, numerical modelling of river morphology and environmentally sound river engineering practice.

Chapter 16 A completely revised and updated version of the previous Chapter 15, containing new material in most of the previous sections and new sections concerning equilibrium profiles and beach erosion, shoreline evolution modelling, shoreline management planning, coastal defence principles, coastal defence techniques and computational wave modelling.

Chapter 17 An updated version of the previous Chapter 16.

Acknowledgements

We are again extremely grateful to a number of colleagues, fellow academics and practitioners who have contributed in various ways during the drafting of the manuscript. We would particularly like to express our thanks to Dr K. Anastasiou of Imperial College, London, UK and Professor M.B. Abbott of the International Institute for Hydraulic and Environmental Engineering, Delft, The Netherlands, for reviewing Chapter 14, to Professor C.A. Fleming, Director of Maritime Engineering, Sir William Halcrow and Partners, Swindon, UK for reviewing Chapters 8 and 16, and to Professor D.A. Huntley of Plymouth University, UK for reviewing Chapter 8. We also wish to thank Colin Prior for producing all the new diagrams and Susan Gardner and Lynne Saunders for word processing most of the new text.

Finally, we gratefully acknowledge the following, who have given permission for the reproduction of illustrative and tabular material:

Professor C.A. Fleming for Figures 16.7 and 16.8.
MAFF for Tables 16.1, 16.2 and 16.3 and Figures 16.2, 16.4, 16.10.
Arun District Council for Figure 16.5.

Preface to the second edition

Since the publication of the first edition, the theory and practice of civil engineering hydraulics has seen sustained and often dramatic advances. This is particularly so in the application of computational modelling techniques to problems in river and coastal engineering and in water quality modelling. There has also been a growing public and governmental awareness of, and concern for, global environmental issues and local environmental issues associated with construction and development projects. The civil engineering profession has always been intimately involved in resolving the conflicts of development pressures and environmental protection, but in recent years there has been a mushrooming of professional activity with regard to environmental assessment, and in evolving civil engineering scheme designs which may be termed environmentally sound.

In this edition, therefore, we have tried to incorporate new information and explanations of the theoretical advances and the current state of the art in the computational modelling as applied to river and coastal waters. In addition, we have drawn upon some of the concepts (or design philosophies) of environmentally sound river and coastal engineering which are increasingly evident in current thinking and practice. Perhaps most significantly, we have written a completely new chapter on water quality modelling. Our aim in this chapter is to introduce the reader to the hydraulic aspects of the dispersion of water- borne pollutants in rivers and estuaries and along coastlines.

In preparing this edition, we have been fortunate in that our publishers have permitted us to retain the original contents whilst adding the (not inconsiderable) new material. The scope of the book has thus been widened and we feel that the new title is both appropriate and justified.

Finally, our overall aim is similar to that of the first edition, namely to provide a comprehensive coverage of hydraulics that is appropriate to a degree course in civil engineering and useful reference for practising engineers. To this is now added a second aim, namely to provide an introduction to environmental aspects of hydraulic engineering. We hope you like it and wish you happy reading.

Acknowledgements

Again, we are extremely grateful to a number of colleagues, fellow academics and practitioners who have contributed in various ways during the drafting of the manuscript. We would particularly wish to express our thanks to Dr G. Pender of Glasgow University, Dr M. Garcia of the University of Illinois and Mr B. G. S. Ellet of Oxford Polytechnic for reviewing our original proposals for the second edition. We are also indebted to Dr R. G. Matthew of Bradford University for comments on Chapter 16, Professor D. M. McDowell of University of Brighton for comments on Chapters 8 and 16, Dr D. W. Knight of the University of Birmingham for providing much material concerning the SERC Flood Channel Facility and for his comments on Chapters 5 and 14, Dr R. Bettess of Hydraulics Research Limited for comments on Chapter 14 and again to Dr G. Pender for comments on Chapters 5, 10 and 14. Their many constructive comments have undoubtedly improved the final text. Finally we wish to record our appreciation of the loving labours of Colin Prior who produced all the new diagrams and Mrs Susan Gardner who word-processed most of the new text.

Andrew Chadwick and John Morfett

Preface to the first edition

Our aim in writing this book has been to provide a comprehensive coverage of hydraulics that is appropriate to a degree course in civil engineering. It should also be suitable for practising civil engineers as a guide and source of reference to some of the more specialist aspects of hydraulic engineering.

The material falls into two distinct sections. Part I presents the fundamental theoretical concepts. Part II exemplifies some of the ways in which basic concepts may be applied to the design of hydraulic systems. A substantial number of worked examples are included in the text, since it is our experience that these are essential if the student is to gain confidence in applying theory to typical engineering problems. In certain respects, the content of this book differs from many of its predecessors. The coverage has been broadened to include Flood Hydrology, Sediment Transport and the cognate areas of River and Coastal Engineering. The presentation also takes account of the widespread availability of the microcomputer. There are examples of various numerical techniques, together with some flow charts and an appendix (Appendix B) containing some programs.

All engineering is a synthesis, and a book such as this owes much to the many engineers and researchers in the field of hydraulics. We have tried to give due acknowledgement to all our principal sources, and to trace ownership of copyright where required. If any attribution is inaccurate, we would be glad to be informed so that future editions may be corrected. If any errors are discovered by readers we would also be grateful to be informed.

Acknowledgements

We are extremely grateful to a number of colleagues who have contributed in various ways during the drafting of the manuscript:

F. W. Matthews wrote Appendix A, and also assisted with some of the computational work.
Colin Prior produced all of the line illustrations.

Hanna Chadwick undertook the considerable task of word-processing the manuscript.
Dr B. O. Hilson (now professor), Professor D. M. McDowell (formerly of Manchester University) and Mr M. Rees, all of Brighton Polytechnic (now the University of Brighton), read various sections of the draft manuscript and gave us the benefit of their constructive criticism.

We are also indebted to Mr J. H. Loveless of King's College London, Mr R. D. Faulkner of Loughborough University of Technology and Dr G. Fleming (now professor) of Strathclyde University for reviewing the complete typescript. Their many constructive criticisms have undoubtedly improved the final text.

Any remaining errors are, of course, the sole responsibility of the authors.

Finally, we gratefully acknowledge the following individuals and organizations who have given permission for the reproduction of illustrative and tabular material:

Figure 4.5 reproduced from *Charts for the hydraulic design of channels and pipes*, 5th edn, courtesy of Hydraulics Research Limited, Wallingford, UK. Figures 8.13 and 8.14 reproduced from H. Darbyshire and L. Draper, 'Fore-casting wind-generated sea waves', *Engineering* (15 April 1963), by permission of L. Draper and the Design Council, London. Figures 10.3 and 10.4 reproduced from *Flood studies report* vol. I, and Tables 10.3 and 10.4 from *Flood studies report* vol. IV, by permission of the Institute of Hydrology, Wallingford, UK. Figure 10.21 reproduced from *The Wallingford procedure* vol. I, courtesy of Hydraulics Research Limited, Wallingford, UK. Figures 13.1, 13.4, 13.5 and 13.8 are extracted from *BS3680* and are reproduced by permission of the British Standards Institution: complete copies of the document can be obtained from BSI, Linford Wood, Milton Keynes MK14 6LE, UK. Figure 13.13 reproduced from D. A. Ervine, *Proc. Instn Civ. Engrs* Part 2 **61**, 383–400 (June) by permission of the author and the Institution of Civil Engineers.

Figure 3.2 courtesy of Armfield Technical Education Co. Ltd, Ringwood, Hampshire, UK. Figures 4.1, 8.8, 9.2, 11.3, 13.9 and 16.5 courtesy of HR Wallingford, UK. Figure 12.11 courtesy of the Central Electricity Generating Board (photography by John Mills, Liverpool). Figure 15.4 courtesy of Shephard, Hill and Co. Ltd., Hillingdon, Middlesex.

Andrew Chadwick and John Morfett

Principal symbols

a	amplitude
A	area (m^2)
A	catchment area (km^2)
A_F	amplification factor
$A(h)$	plan area of reservoir at stage h (m^2)
A_p	areal packing of grains
A_R	ratio of partially open value area to fully open area
A_S	cross-sectional area of sediment particle (m^2)
A_{ST}	cross-sectional area of surge tower (m^2)
A_V	cross-sectional area of flow at value (m^2)
$\mathbf{B}_*$	sediment transport parameter
B	centre of buoyancy
B	surface width of channel (m)
b	width (m)
b	amplitude
b	probability weighted moment
c	celerity (m/s)
c	concentration (g/m^3 or parts per million)
C	Chézy coefficient ($m^{1/2}/s$)
C	concentration (mg/l)
C	concentration (g/m^3 or parts per million)
C_c	contraction coefficient
C_d	discharge coefficient
C_D	drag coefficient
C_E	entrainment coefficient
C_f	friction coefficient
C_G	group celerity of waves (m/s)
C_L	lift coefficient
C_0	celerity of waves in deep water (m/s)
Cr	courant number
C_v	velocity coefficient
d	water depth under a wave (m)
d	diameter (m)
D	diameter (m)

D	difference operator
D	duration (h)
d_B	water depth for a breaking wave (m)
D_e	energy dissipation
D(f, θ)	directional spreading function
D_m	hydraulic mean depth (m)
D_M	molecular diffusion coefficient (m^2/s or m^2/day)
D_S	sediment size (m)
D_T	turbulent diffusion coefficient (m^2/s or m^2/day)
e	pipe wall thickness (m)
e	sediment transport 'efficiency'
e_r	error
E	Young's modulus (N/m^2)
E	energy (J)
E_K	kinetic energy of wave (J)
E_P	potential energy of wave (J)
E_S	specific energy (m)
F	fetch length (km)
F	force (N)
F_D	drag force (N)
Fr	Froude Number
f	wave frequency (Hz)
f_W	wave friction factor
g	gravitational acceleration (m/s^2)
G	centre of gravity
H	head (m)
H	wave height (m)
H_0	wave height in deep water (m)
H_B	wave height at breaker line (m)
H_C	cavitation head (m)
h_f	frictional head loss (m)
h_L	local head loss (m)
H_P	pump head (m)
H_t	turbine head (m)
i	rainfall intensity (mm/h)
I	inflow (to reservoir) (m^3/s)
I	second moment of area (m^4)
K	bulk modulus (N/m^2)
K	channel conveyance (m^3/s)
k	wave number (m^{-1})
K	von Kármán constant
K_d	diffraction coefficient

k_L local loss coefficient
K_N dimensionless specific speed
K_R refraction coefficient for waves
k_S surface roughness (m)
K_S shoaling coefficient for waves
K_X longitudinal mixing coefficient (m^2/s or m^2/day)

l length (m)
L length (dimension)
L wave length (m)
l L-moment
$\bar{l}$ distance to centroid of area (m)
l' distance to centre of pressure (m)
L_{SW} salt wedge intrusion (m)

m beach slope
m mass (kg)
m spectral moment
M mass (dimension)
M metacentre
M momentum (kg/m s)

n Manning's roughness coefficient
N rotational speed (1/s)
N record length
N_S specific speed

O outflow (from reservoir) (m^3/s)

P power (kW)
P rainfall depth (mm)
p^* piezometric pressure (N/m^2)
p pressure (N/m^2)
P probability
P wetted perimeter (m)
p_s porosity
P_S sill or crest height (m)

Q discharge (m^3/s)
q discharge per unit channel width (m^3/ms)
q_{LS} sediment transport per unit width at a point (m^3/ms)
Q_{LS} longshore sediment transport by waves (m^3/s or $m^3/annum$)
Q_{MLS} longshore sediment transport by waves (kg/s)
Q_p peak runoff (m^3/s)
Q_p spectral peakedness
Q_S sediment discharge (m^3/s)
q_s sediment discharge per unit channel width (m^3/ms)
Q_{Tr} flood discharge of return period T_r (m^3/s)

r	radius (m)
R	radius (m)
R	hydraulic radius (m)
R	reaction force (N)
R	wave reflection parameter
R_p	reading on pressure gauge
Re	Reynolds' Number
$\mathrm{Re_W}$	wave Reynolds' Number
S	radiation stress (N/m)
S_c	slope of channel bed to give critical flow
S_{CR}	slope of channel bed to give critical shear stress for sediment transport
S(f)	spectral energy density (m^2 s)
S(f, θ)	directional spectral energy density (m^2/s)
S_f	slope of hydraulic gradient
S_0	slope of channel bed
S_s	slope of water surface
S_{XX}, S_{YY}	principal radiation stress (N/m)
S_{XX}, S_{YY}	radiation stresses in x, y directions (N/m)
St	Strouhal Number
T	temperature
T	tension force (N)
t	time (s)
t	L-moment ratio
T	wave period (s)
TB	time base of the unit hydrograph (h)
t_c	time of concentration (min)
t_e	time of entry into drainage system (min)
t_f	time of flow through a drainage pipe (min)
T_p	periodic time (s)
T_{p}	time to peak runoff of the unit hydrograph (h)
T	return period (years)
T_T	periodic time of tide (h)
T_S	surface tension (N/m)
u	velocity (m/s)
U	wind speed (m/s)
$\overline{u}$	average velocity (m/s)
$\mathrm{u_m}$	maximum nearbed orbital velocity (m/s)
u_0	initial velocity (m/s)
U_∞	'free stream' velocity (m/s)
u_*	friction velocity (m/s)
v	velocity (m/s)
v_l	period-averaged mean longshore velocity (m/s)
V	velocity (usually mean velocity of flow) (m/s)

V volume (m^3)
V_A absolute velocity (m/s)
V_F radial velocity in hydraulic machine (m/s)
V_{FS} fall velocity of a sediment particle (m/s)
V_I tangential velocity of pump impeller (m/s)
V_R relative velocity (m/s)
V_S volume of sediment particle (m^3)
V_{SR} volume of surface runoff
V_{ST} velocity of flow in a surge tower (m/s)
V_W tangential (whirl) velocity in hydraulic machine

w velocity (m/s)
W weight (N)
W' immersed weight of a particle (N)
We Weber Number

x growth curve factor
X body force (N)
X length (m)

Y body force (N)
y water depth (m)
y reduced variate
y_c critical depth (m)
y_n normal depth (m)

z height above datum (m)
Z length (m)
Z body force (N)

α angle (degree or rad)
α velocity coefficient
β momentum coefficient
β slope of seabed or reach
Γ circulation
γ wave height ratio
δ boundary layer thickness (m)
δ_* displacement thickness (m)
ε spectral width
ε_d energy dissipation (kg/s^3)
ζ vorticity
η efficiency
η excursion (m)
$\overline{\eta}_d$ set-down (m)
$\overline{\eta}_u$ set-up (m)
θ angle (degree or rad)
θ momentum thickness (m)
θ Preissmann scheme weighting factor

λ friction factor
λ scale factor
μ absolute viscosity (kg/m s)
μ_T eddy viscosity (kg/ms)
ν kinematic viscosity (m^2/s)
ξ_p Irribaren number
Π dimensionless group
ρ density of liquid (kg/m^3)
ρ_s density of sediment (kg/m^3)
σ hoop stress (N/m^2)
σ normal stress (N/m^2)
τ shear stress (N/m^2)
τ_b seabed shear stress (N/m^2)
τ_{CR} critical shear stress for sediment transport (N/m^2)
τ_0 shear stress at a solid boundary (N/m^2)
ϕ velocity potential
Φ sediment transport parameter
ψ stream function
Ψ sediment transport parameter
ω rotational speed (l/s)
ω angular frequency (rad/s)

A short history of hydraulics

Hydraulics is a very ancient science. The Egyptians and Babylonians constructed canals, both for irrigation and for defensive purposes. No attempts were made at that time to understand the laws of fluid motion. The first notable attempts to rationalize the nature of pressure and flow patterns were undertaken by the Greeks. The laws of hydrostatics and buoyancy were enunciated, Ctesibius and Hero designed hydraulic equipment such as the piston pump and water clock, and (of course) there was the Archimedes screw pump. The Romans appear, like the Egyptians, to have been more interested in the practical and constructional aspects of hydraulics than in theorizing. Thus, development continued slowly until the time of the Renaissance, when men such as Leonardo Da Vinci began to publish the results of their observations. Ideas which emerged then, respecting conservation of mass (continuity of flow), frictional resistance and the velocity of surface waves, are still in use, though sometimes in a more refined form. The Italian School became famous for their work. Toricelli *et al.* observed the behaviour of water jets. They compared the path traced by a free jet with projectile theory, and related the jet velocity to the square root of the pressure generating the flow. Guglielmini *et al.* published the results of observations on river flows. The Italians were hydraulicians in the original sense of the word, i.e. they were primarily empiricists. Up to this point, mathematics had played no significant part in this sort of scientific work. Indeed, at that time mathematics was largely confined to the principles of geometry, but this was about to change.

In the 17th century, several brilliant men emerged. Descartes, Pascal, Newton, Boyle, Hooke and Leibnitz laid the foundations of modern mathematics and physics. This enabled researchers to perceive a logical pattern in the various aspects of mechanics. On this basis, four great pioneers – Bernoulli, Euler, Clairaut and D'Alembert – developed the academic discipline of hydrodynamics. They combined a sound mathematical framework with an acute perception of the physical phenomena which they were attempting to represent. In the 18th century, further progress was made, both in experimentation and in analysis. In Italy, for example, Poleni investigated the concept of discharge coefficients. However, it was French and German thinkers who now led the way. Henri de Pitot constructed a device which could measure flow velocity. Antoine Chézy (1718–98), followed by Eytelwein and Woltmann, developed a rational equation to describe flow in streams. Men such as Borda, Bossut and du Buat not only extended knowledge, but took pains to see that the available knowledge was disseminated. Woltmann and Venturi used Bernoulli's work as a basis for developing the principles of flow measurement.

The 19th century was a period of further advance. Hagen (1797–1884) constructed experiments to investigate the effects of temperature on pipe flow. His understanding of the nature of fluid viscosity was limited to Newton's ideas, yet so careful was his work that the results were within 1% of modern measurements. He injected sawdust into the fluid for some of his experiments, in order to visualize the motion. He was probably the first person knowingly to observe turbulence, though he was unable to grasp its significance fully. At almost the same time, a French doctor (Poiseuille) was also making observations on flow in pipes (in an attempt to understand the flow of blood in blood vessels), which led to the development of equations for laminar flow in pipes. Further contributions were made by Weisbach, Bresse and Henri Darcy, who developed equations for frictional resistance in pipe and channel flows (the first attempts to grapple with this problem coincided with signs of an incipient awareness of the existence of the 'boundary layer'). During the later part of the century, important advances were made in experimentation. The first practical wind tunnel, the first towing tank for model testing of ships and the first realistic attempt to model a tidal estuary (by O. Reynolds) were all part of this flowering of knowledge. These techniques are still used today. Reynolds also succeeded in defining the different types of flow, observing cavitation and explaining Darcy's friction law in greater detail.

Even at this stage, studies of fluid flows were subdivided into 'Classical Hydrodynamics' (which was a purely mathematical approach with little interest in experimental work) and experimental 'Hydraulics'. Navier, Stokes, Schwarz, Christoffel and other hydrodynamicists all contributed to the development of a formidable array of mathematical equations and methods (including the conformal transformation). Their work agreed only sporadically with the practitioners (the hydraulicians) and, indeed, there were frequently yawning disparities between the results suggested by the two schools. The rapid growth of industry in the 19th and 20th centuries was by now producing a demand for a better understanding of fluid flow phenomena. The real break-through came with the work of Prandtl. He proposed (in 1901) that flow was 'divided into two interdependent parts. There is on the one hand the free fluid which can be treated as inviscid (i.e. which obeyed the laws of hydrodynamics)...and on the other hand the transition layer at the fixed boundaries' (the transition layer is the thin layer of fluid within which frictional forces dominate). With this brilliant insight, Prandtl effectively fused together the two disparate schools of thought and laid the foundation for the development of the unified science of Fluid Mechanics.

The 20th century has, in consequence, seen tremendous advances in the understanding and application of fluid mechanics in almost every branch of engineering. It is only possible to give the barest outline here. Prandtl and Th. von Kármán published a series of papers in the 1920s and 1930s, covering various aspects of boundary layer theory and turbulence. Their work was supplemented by increasingly sophisticated laboratory research (e.g. the work of Dryden and his colleagues at NACA in the USA). These efforts had an impact on every aspect of engineering fluid mechanics. In the 1930s, the efforts of Nikuradse (in Germany), Moody (in America), Colebrook (Great Britain) and others resulted in a clearer understanding of pipe flows and, in particular, of the factors affecting pipe friction. This led directly to the modern methods for estimating flows in pipes and channels.

Since 1945, the advent of the electronic computer, and advances in sensing and data logging equipment have revolutionized many aspects of hydraulics. Our understanding of the nature of turbulence, steady and unsteady flows in channels, sediment transport and maritime phenomena have developed rapidly. This has been matched by developments in software. Today's engineer, with a modern personal computer or workstation at his or her disposal, has the tools to achieve more effective and economic designs.

International research continues at an intensive pace, as our world faces many problems, social and environmental. On the threshold of the 21st century, there is every reason to suppose that engineers and scientists will face exciting and difficult challenges that will place the fullest demands upon their skills.

Introductory notes

Before formally embarking on a study of the engineering aspects of the behaviour of fluids, it is worth pausing to consider the area to be studied, the nature of the substance, and one or two other basic points. Students who already have some grounding in fluid mechanics may wish to proceed directly to the main body of the text.

Civil engineering hydraulics

'Fluid mechanics' is the general title given to the study of all aspects of the behaviour of fluids which are relevant to engineers. Within this very broad discipline, a number of subsections have developed. Of these subsections, hydraulics is the branch which concentrates on the study of liquids. Civil engineers are largely, though not exclusively, concerned with one liquid, namely water. The development of the industrial society rests largely on the ability of civil engineers to provide adequate water services, such as the supply of potable water, drainage, flood control, etc.

The nature of fluids

A **fluid** is a substance which can readily flow, i.e. in which there can be a continuous relative motion between one particle and another. A fluid is inelastic in shear, and therefore continuously deforms under application of a shear force without the possibility of a return to its original disposition.

Fluids are subdivided into the following:

Liquids which have a definite volume for a given mass, i.e. they cannot readily be altered (say, compressed) due to changes of temperature or pressure – if liquid is poured into a container, a clearly defined interface is established between the liquid and the atmosphere;
Gases generally exhibit no clear interface, will expand to fill any container, and are readily compressible.

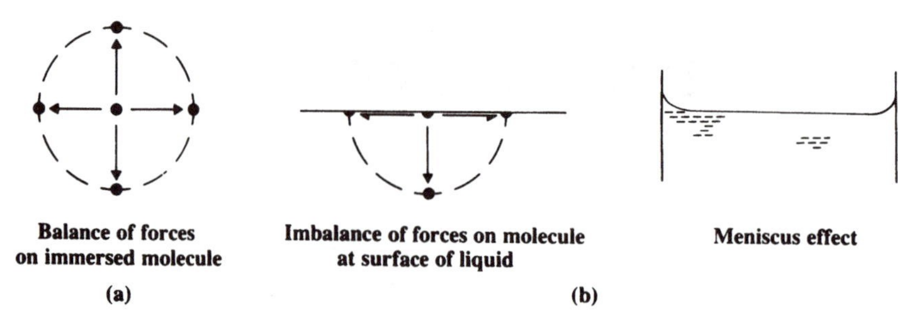

Figure 1 Surface tension.

Properties of fluids

Density is the ratio of mass of a given quantity of a substance to the volume occupied by that quantity.

Specific weight is the ratio of weight of a given quantity of a substance to the volume occupied by that quantity. An alternative definition is that specific weight equals the product of density and gravitational acceleration.

Relative density (otherwise called 'specific gravity') is the ratio of the density of a substance to some standard density (usually the standard density is that of water at 4°C).

Viscosity represents the susceptibility of a given fluid to shear deformation, and is defined by the ratio of the applied shear stress to rate of shear strain. This property is discussed in detail in Chapter 3.

Surface tension is the tensile force per unit length at the free surface of a liquid. The reason for the existence of this force arises from intermolecular attraction (Fig. 1a). In the body of the liquid, a molecule is surrounded by other molecules and intermolecular forces are symmetrical and in equilibrium. At the surface of the liquid (Fig. 1b), a molecule has this force acting only through 180°. This imbalance of forces means that the molecules at the surface tend to be drawn together, and they act rather like a very thin membrane under tension. This causes a slight deformation at the surface of a liquid (the meniscus effect).

Bulk modulus is a measure of the compressibility of a liquid, and is the ratio of the change in pressure to the volumetric strain caused by the pressure change. This property is discussed in Chapter 6.

Units

In order to define the magnitudes of physical quantities (such as mass or velocity) or properties (density, etc.), it is necessary to set up a standardized framework of units. In this book, the Système Internationale d'Unités (SI system) will generally be used. There are six primary units from which

Table 1 The six primary units in the SI system.

Quantity	Symbol of quantity	SI unit	Symbol of unit
length	L	metre	m
mass	M	kilogramme	kg
time	T	second	s
electric current	i	ampere	A
luminous intensity	I	candela	Cd
temperature	t	kelvin	K

all others are derived, and these are listed in Table 1. For a complete specification of a magnitude, a number and a unit are required (e.g. 5 m, 3 kg).

The principal derived units are as follows:

Velocity is the distance travelled in unit time and therefore has the dimensions of LT^{-1} and units m/s;
Acceleration is the change of velocity in unit time and has dimensions LT^{-2} (= velocity per second) and units m/s^2;
Force is derived by the application of Newton's second law, which may be summarized as 'force = mass × acceleration' – it is measured in newtons, where 1 newton (N) is that force which will accelerate a mass of 1 kg at 1 m/s^2, and has dimensions MLT^{-2} and units kg m/s^2;
Energy (or 'work', or heat) is measured in joules, 1 joule being the work equivalent of a force of 1 N acting through a distance of 1 m, and has dimensions ML^2T^{-2} and units N m;
Power, the rate of energy expenditure, is measured in watts (= joules per second), and has dimensions ML^2T^{-3} and units N m/s.

The dimensions and units of the properties of fluids are the following:

Density, which by definition is mass/volume = ML^{-3}, has units of kg/m^3;
Specific weight is the force applied by a body of given mass in a gravitational field – its dimensions are $ML^{-2}T^{-2}$ and its units are N/m^3;
Relative density is a ratio, and is therefore dimensionless;
Viscosity $= \dfrac{\text{shear stress}}{\text{rate of shear strain}} = \dfrac{\text{(force/area)}}{\text{(velocity/distance)}}$ has dimensions $ML^{-1}T^{-1}$ and its units may be expressed as either N s/m^2 or kg/m s;
Surface tension is defined as force per unit length, so its dimensions are MT^{-2} and its units are N/m.

Dimensional homogeneity

If any equation is to represent some physical situation accurately, then it must achieve numerical equality and dimensional equality (or equality of units).

It is strongly recommended that when constructing equations the numerical value and the units of each term are written down. It is then possible to check that both of the required equalities are, in fact, achieved. Thus, if at the end of a calculation one has

$$55\,\mathrm{kg\,m} = 55\,\mathrm{s}$$

then a mistake has been made, even though there is numerical equality. This point is especially important when equations incorporate coefficients. Unless the coefficients are known to be dimensionless, a check should be made to ascertain precisely what the units relevant to a given numerical value are. If an engineer applies a particular numerical value for a coefficient without checking units, it is perfectly possible to design a whole system incorrectly and waste large sums of money. Subsequent excuses ('I didn't know it was in centimetres') are not usually popular!

Part One

Principles and Basic Applications

1

Hydrostatics

1.1 Pressure

Hydrostatics is the study of fluids at rest, and is therefore the simplest aspect of hydraulics. The main characteristic of a stationary fluid is the force which it brings to bear on its surroundings. A fluid force is frequently specified as a pressure, p, which is the force exerted on a unit area. Pressure is measured in N/m^2 or in 'bar' ($1\,\text{bar} = 10^5\ N/m^2$).

Pressure is not constant everywhere in a body of fluid. In fact, if pressure is measured at a series of different depths below the upper surface of the fluid, it will be found that the pressure reading increases with increasing depth. An exact relationship can be developed between pressure, p, and depth, y, as follows. Suppose there is a large body of liquid (a lake or swimming pool, for example), then take any imaginary vertical column of liquid within that main body (Fig. 1.1). The column of fluid is at rest, therefore all of the forces acting on the column are in equilibrium. If this statement is to be true for any point on the boundary surfaces of the column, the action and reaction forces must be perpendicular to the boundary surface. If any forces were not perpendicular to the boundary, then a shear force component would exist; this condition arises only for fluids in motion. It follows that the only force which is supporting the column of fluid is the force acting upwards due to the pressure on the base of the column. For the column to be in equilibrium, the upward force must exactly equal the weight force acting downward.

The weight of the column is $\rho g V$ where V is the volume of the column. The volume is the product of the horizontal cross-sectional area and height, $V = Ay$, and thus the weight is given by $\rho g A y$.

The force acting upwards is the product of pressure and horizontal cross-sectional area, i.e. pA. Therefore

$$pA = \rho g A y$$

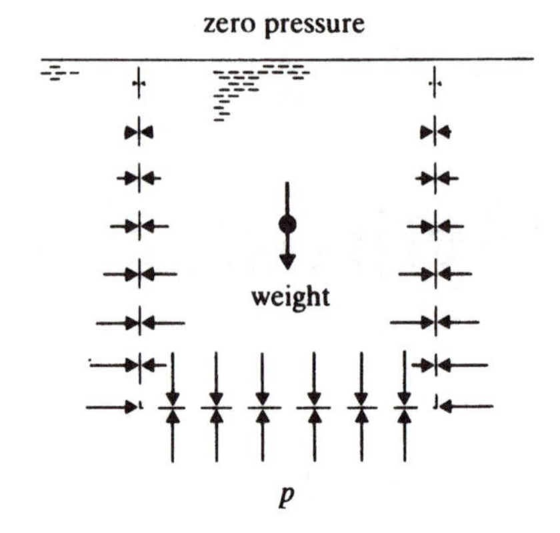

Figure 1.1 Pressure distribution around a column of liquid.

and so

$$p = \rho g y \tag{1.1a}$$

This is the basic hydrostatic equation or 'law'. By way of example, in fresh water (which has a density of $1000\,\text{kg/m}^3$) the pressure at a depth of 10 m is

$$\begin{aligned} p\,(\text{N/m}^2) &= 1000\,(\text{kg/m}^3) \times 9.81\,(\text{m/s}^2) \times 10\,(\text{m}) \\ &= 98\,100\,\text{N/m}^2 \\ &= 98.1\,\text{kN/m}^2 \end{aligned}$$

The equation is correct both numerically and in terms of its units. For all practical purposes, the value of $g\,(= 9.81\,\text{m/s}^2)$ is constant on the earth's surface. The product ρg will therefore also be constant for any homogeneous incompressible fluid, and (1.1a) then indicates that pressure varies linearly with the depth y (Fig. 1.2).

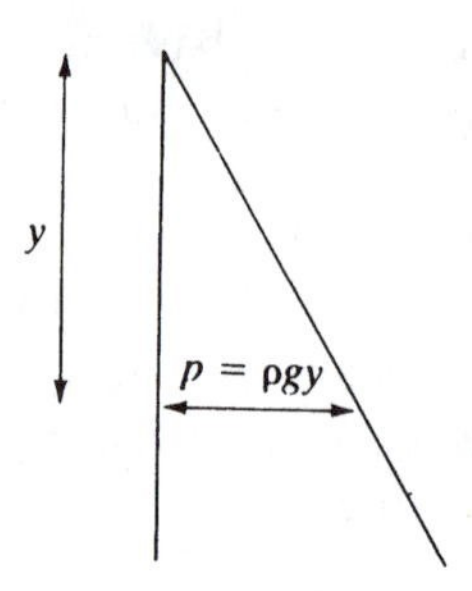

Figure 1.2 Pressure variation with depth.

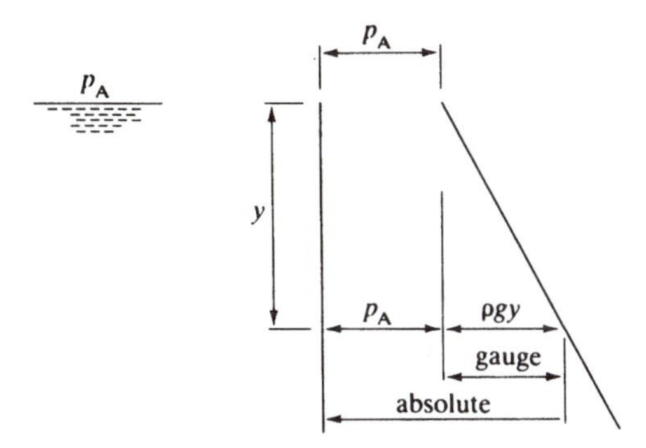

Figure 1.3 Gauge and absolute pressure.

Gauge pressure and absolute pressure

An important case of pressure variation is that of a liquid with a gaseous atmosphere above its free surface. The pressure of the gaseous atmosphere immediately above the free surface is p_A (Fig. 1.3). For equilibrium, the pressure in the liquid at the free surface is p_A, and therefore at any depth y below the free surface the absolute pressure p_{ABS} (i.e. the pressure with respect to absolute zero) must be

$$p_{ABS} = p_A + \rho g y$$

The gauge pressure is the pressure with respect to p_A (i.e. p_A is treated as the pressure 'datum'):

$$p = \rho g y = \text{gauge pressure}$$

It is possible for gauge pressure to be positive (above p_A) or negative (below p_A). Negative gauge pressures are usually termed vacuum pressures. Virtually every civil engineering project is constructed on the earth's surface, so it is customary to take atmospheric pressure as the datum. Most pressure gauges read zero at atmospheric pressure.

1.2 Pressure measurement

The argument so far has centred upon variation of pressure with depth. However, suppose that a pipeline is filled with liquid under pressure (Fig. 1.4(a)). At one point the pipe has been pierced and a vertical transparent tube has been attached. The liquid level would rise to a height y, and since (1.1a) may be rearranged to read

$$p/\rho g = y \tag{1.1b}$$

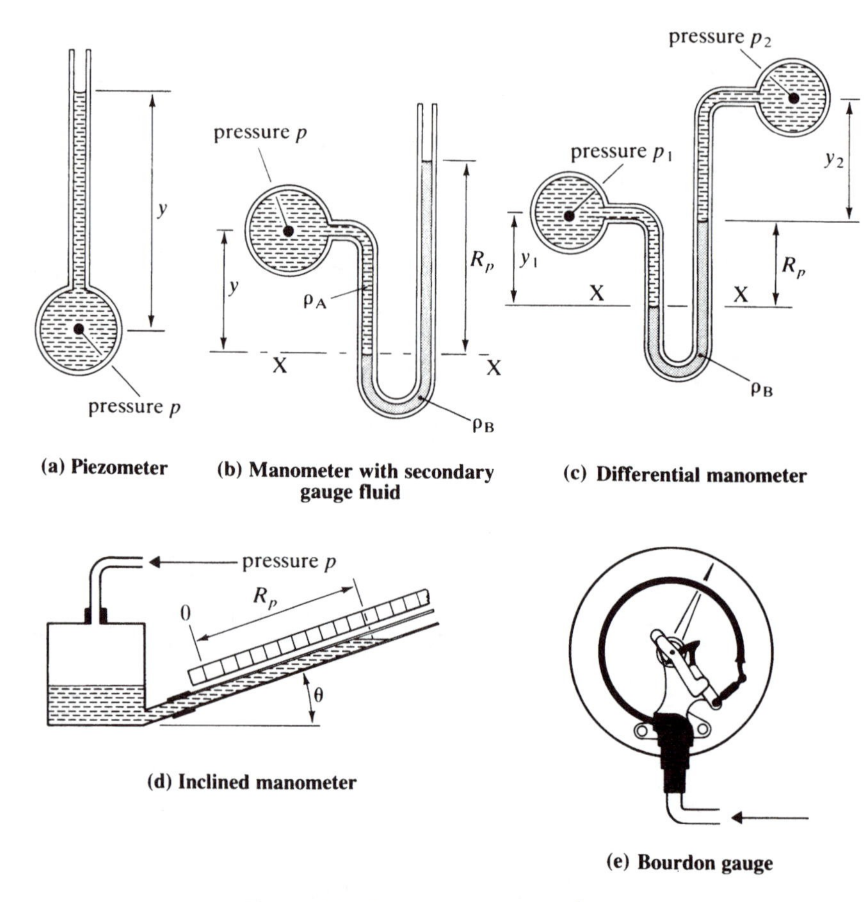

Figure 1.4 Pressure measuring devices.

this height will indicate the pressure. The term $p/\rho g$ is often called 'pressure head' or just 'head'. A vertical tube pressure indicator is known as a piezometer. The piezometer is of only limited use. Even to record quite moderate water pressures, the height of a piezometer becomes impracticably large. For example, a pressure of water of $100\,\mathrm{kN/m^2}$ corresponds to a height of roughly 10 m, and civil engineers commonly design systems for higher pressure than this. The piezometer is obviously useless for gas pipelines.

To overcome this problem, engineers have developed a range of devices. From a purely hydrostatic viewpoint, perhaps the most significant of these is the manometer (Fig. 1.4(b)), which is a U-shaped transparent tube. This permits the use of a second gauge fluid having a density (ρ_B) which differs from the density (ρ_A) of the primary fluid in the pipeline. The two fluids must be immiscible, and must not react chemically with each other. When the pressure in the pipeline is p, the gauge fluid will be displaced so that there is a difference of R_p metres between the gauge fluid levels in the left- and right-hand vertical sections. To evaluate the pressure p, we proceed as follows.

Set a horizontal datum X–X at the lower gauge fluid level, i.e. at the interface between fluid A and fluid B in the left-hand vertical section. The pressure in the left-hand section at X–X is the sum of the pressure p at the centre of the pipe and the pressure due to the height y of fluid A, i.e.

$$p_{\mathrm{X-X}} = p + \rho_{\mathrm{A}} g y$$

In the right-hand section the pressure at X–X is due to the height R_p of fluid B (atmospheric pressure is ignored, so the final answer will be a gauge pressure), i.e.

$$p_{\mathrm{X-X}} = \rho_{\mathrm{B}} g R_p$$

In any continuous homogeneous fluid, pressure is constant along any horizontal datum. Since X–X passes through the interface on the left-hand side and through fluid B on the right-hand side, $p_{\mathrm{X-X}}$ must be the same on both sides. Therefore

$$p_{\mathrm{X-X}} = p + \rho_{\mathrm{A}} g y = \rho_{\mathrm{B}} g R_p$$

so

$$p = \rho_{\mathrm{B}} g R_p - \rho_{\mathrm{A}} g y \tag{1.2a}$$

Thus, if ρ_{A}, ρ_{B} and y are known, the pressure p may be calculated for any reading R_p on the manometer. Equation (1.2a) is often written in terms of pressure head:

$$\frac{p}{\rho_{\mathrm{A}} g} = \frac{\rho_{\mathrm{B}}}{\rho_{\mathrm{A}}} R_p - y \tag{1.2b}$$

The manometer may also be connected up to record the pressure difference between two points (Fig. 1.4(c)).

Other devices include the following:

1. The sloping tube manometer (Fig. 1.4(d)), which works on the same essential principle as the U-tube. However, the right-hand limb comprises a transparent tube sloping at an angle θ while the left-hand limb comprises a tank whose horizontal cross-section is much larger than that of the tube. The primary fluid extends from the pipeline through

a flexible connecting tube into the top of the tank. The pressure p forces fluid from the tank into the transparent sloping tube, giving the reading R_p. The level in the tank falls only by a relatively small amount, so the reading is usually taken from a fixed datum. The equation of pressure may be adapted from (1.2a) if it is realized that $y = R_p \sin\theta$.

2. The Bourdon gauge is a semi-mechanical device (Fig. 1.4(e)). It consists of a tube formed into the shape of a question mark, with its upper end sealed. The upper end is connected by a linkage to a pointer. An increase in pressure causes a slight deformation of the tube. This is transmitted to the pointer, which rotates through a corresponding angle. Bourdon gauges can be designed for a much wider range of pressures than can manometers. On the other hand, they are somewhat less precise and require a certain amount of maintenance.
3. Pressure transducers are devices which convert a pressure into a corresponding electrical output. They are particularly applicable for use in conjunction with automatic (computer-controlled) systems, or where variable or pulsating pressures need to be measured. For details of such devices, the manufacturer's literature is usually the best source of information.

1.3 Pressure forces on submerged bodies

The omnidirectional nature of pressure

In developing the equation of hydrostatic pressure distribution, it was emphasized that pressure forces were assumed to be perpendicular to the imaginary surfaces of the column of liquid. However, this, taken to its logical conclusion, must mean that at any point in the fluid the pressure acts equally in all directions. That this is true may be formally demonstrated as follows. Imagine a small sphere of fluid which is stationary within a larger body of the same fluid (Fig. 1.5). Action and reaction forces must be in equilibrium at all points round the sphere. If the sphere diameter is reduced, then even though $d \to 0$ (and therefore the weight of the sphere tends to zero) the equilibrium statement must still be true. Therefore, in an uninterrupted fluid continuum, pressure at any point acts equally in all directions.

Pressure on plane surfaces

If a body is immersed in a fluid, it must be subject to the external pressure of the fluid and must provide the reaction necessary for equilibrium. Once again, it is emphasized that in the absence of fluid motion the action and reaction forces must be perpendicular to the surface of the body at any point. Generally, in design calculations it is the total force of the fluid on

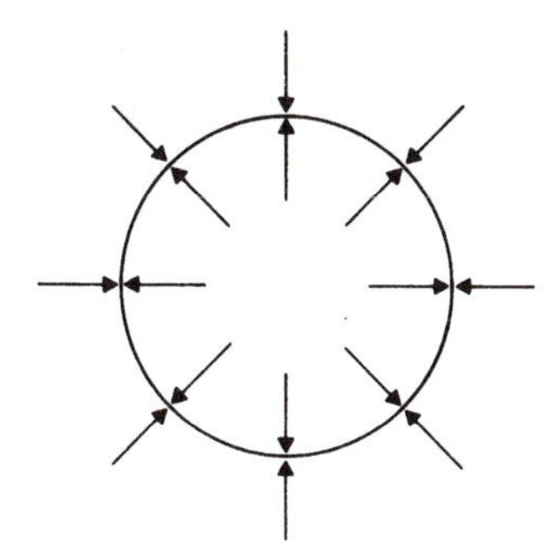

Figure 1.5 Pressure at a point in a liquid.

the body that is required, rather than the pressure at any given location. To show how this is evaluated, take an arbitrary plane surface which is immersed in a liquid (Fig. 1.6). Somewhere on the surface, take a small element of area δA. The force δF on that area, due to the pressure of the liquid, is $p\,\delta A = \delta F$. Obviously, the total force acting on the whole plane surface must be the sum of all the products $(p\,\delta A)$. However, p is not a constant. Thus if the element is a distance y below the free surface of the liquid, then $p = \rho g y$ and $p\delta A = \delta F = \rho g y\,\delta A$ and therefore the total force is $F = \rho g \int y\,\mathrm{d}A$ (ρ and g, being constants, are outside the integral). But

$$y = l \sin\theta$$

therefore

$$F = \rho g \int l \sin\theta\,\mathrm{d}A = \rho g(\sin\theta) \int l\,\mathrm{d}A$$

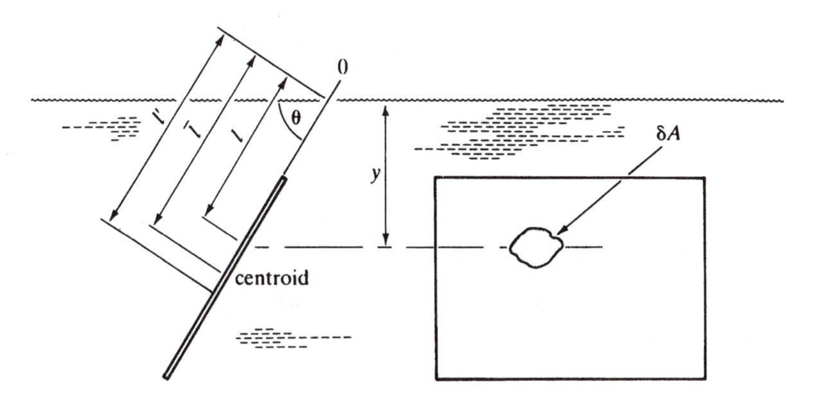

Figure 1.6 Pressure on a plane surface.

The quantity $\int l\,\mathrm{d}A$ is a geometrical characteristic of the shape, and is known as the first moment of area. This integral can be evaluated for any shape and equals the product $A\bar{l}$, where A is the area of the plane surface and $\bar{l}$ is the distance from the origin O to the centroid of the plane. Therefore

$$F = \rho g(\sin\theta)A\bar{l} \tag{1.3}$$

In addition to the magnitude of the force F, it is usually necessary to know the precise location and angle of its line of action. The pressure is, of course, distributed over the whole of the immersed body, but it is almost always more convenient to regard F as a concentrated point load.

Returning to the element of area, the force δF produces a moment $\delta F\,l$ about the origin. This may be written as

$$\begin{aligned}\text{moment} = \delta F\,l &= \rho g y\,\delta A l \\ &= \rho g l(\sin\theta)\delta A l \\ &= \rho g(\sin\theta)l^2\,\delta A\end{aligned}$$

In the limit, $\mathrm{d}F\,l = \rho g(\sin\theta)l^2\mathrm{d}A$, so for the whole surface the moment is $\rho g(\sin\theta)\int l^2\mathrm{d}A$.

The quantity $\int l^2\mathrm{d}A$ is another geometrical characteristic, known as the second moment of area, which has the symbol I (see Appendix A). Therefore

$$\text{moment about origin} = \rho g(\sin\theta)I$$

Therefore, the distance from the origin to the point of action of F is

$$l' = \frac{\text{moment}}{\text{force}} = \frac{\rho g(\sin\theta)I}{\rho g(\sin\theta)A\bar{l}} = \frac{I}{A\bar{l}} \tag{1.4}$$

I is usually evaluated by use of the Parallel Axes theorem.

Example 1.1 Pressure forces on a dam

A rockfill dam is to have the cross-section shown in Figure 1.7(a). The reservoir design depth is to be 10 m. Estimate:

(a) the force on the dam per unit width;
(b) the location of the centre of pressure.

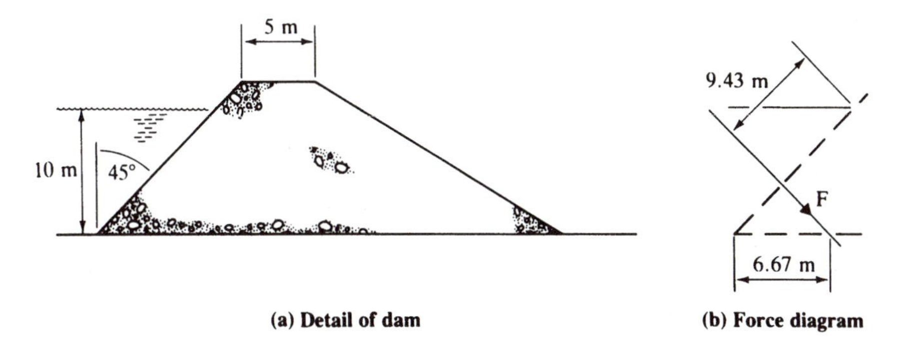

Figure 1.7 Dam section (Example 1.1).

Solution

Using trigonometry, the wetted slant height is 14.14 m.

$$\text{Force on dam} = \rho g(\sin\theta)\int l\,dA = \rho g(\sin\theta)A\bar{l}$$

$$= \rho g \sin\theta\, \text{area} \times \left(\frac{\text{wetted height}}{2}\right)$$

$$= 1000 \times 9.81 \times \sin 45°(14.14 \times 1) \times 14.14/2$$

$$= 693\,460\,\text{N/m width}$$

The moment of the force is $\rho g \sin\theta \times$ (2nd moment of area). Therefore

$$\text{M} = 1000 \times 9.81 \times \sin\theta\,[I_0 + A(\text{wetted height}/2)^2]$$

where I_0 = second moment of area about centroid. Hence,

$$M = 1000 \times 9.81 \times \sin 45° \left(\frac{1 \times 14.14^3}{12} + (1 \times 14.14)\left(\frac{14.14}{2}\right)^2\right)$$

$$= 6\,537\,040\,\text{N m}$$

So

$$l' = 6\,537\,040/693\,460 = 9.43\,\text{m}$$

(see Fig. 1.7b). Therefore the slant height from base to the centre of pressure is 4.71 m.

Pressure on curved surfaces

Not all immersed structures are flat. Some dams have an upstream face which may be curved in both the vertical and the horizontal planes. Certain types of adjustable hydraulic structures (used to control water levels), such as gates, may also be curved.

It has already been shown that pressure is perpendicular to an immersed surface. The pressure on a curved surface would be distributed as shown in Figure 1.8(a). To calculate the total force directly is inconvenient. At any small area δA, the force $\delta F (= p\,\delta A)$ will be perpendicular to the surface. This force can be resolved into two components (Fig. 1.8b), one vertical, δF_y, and one horizontal, δF_x. It can easily be shown that $\Sigma\,\delta F_y$ is the total weight, W, of the volume of liquid above the curved surface. The horizontal force, $F_x = \Sigma\,\delta F_x$, is equal to the pressure force on a vertical plane surface equal in height to the projected height of the curved surface. The resultant force may be obtained from the triangle of forces (Fig. 1.8c). The weight component of the volume will act through the centre of gravity, while the horizontal component will act at the distance l' below the free surface of

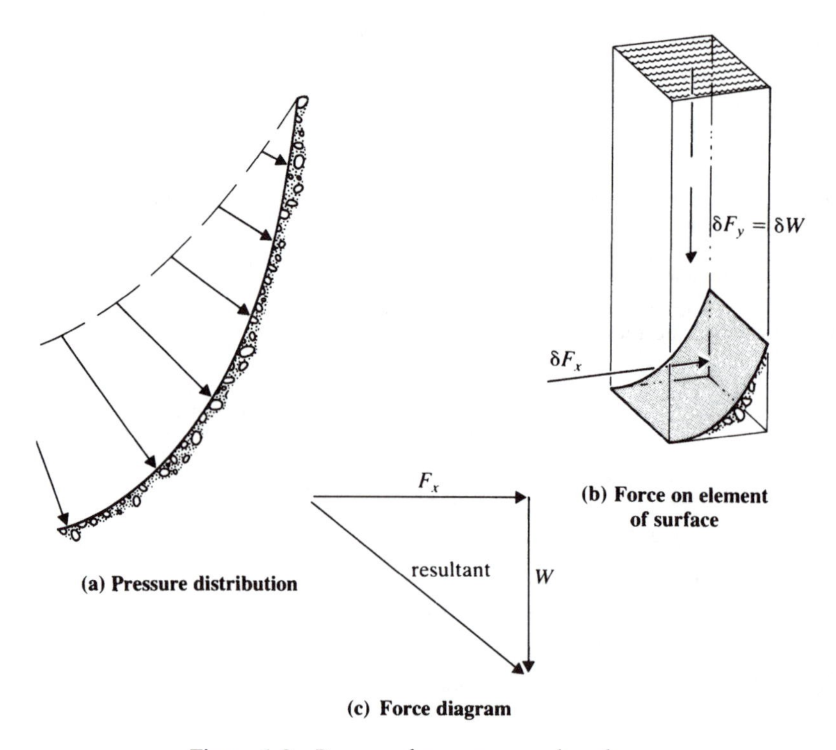

Figure 1.8 Pressure force on curved surface.

the liquid. The resultant force will act through the point of intersection of the lines of action of W and F_x.

Example 1.2 Hydrostatic force on a quadrant gate

Find the magnitude and direction of the resultant force of water on the quadrant gate shown in Figure 1.9(a). The principal dimensions of the gate are:

$$\text{radius of gate} = 1\,\text{m}$$

$$\text{width of gate} = 3\,\text{m}$$

$$\text{water density} = 1000\,\text{kg/m}^3$$

The position of the centre of gravity is, as shown, $4R/3\pi$ horizontally from the origin.

Solution

(a) For vertical component,

$$\text{area of sector} = \pi R^2/4 = \pi \times 1^2/4 = 0.785\,\text{m}^2$$

$$\text{volume of water} = 0.785 \times 3 = 2.355\,\text{m}^3$$

$$\text{weight of water} = 9.81 \times 1000 \times 2.355 = 23\,100\,\text{N}$$

$$\text{centre of gravity is } 4 \times 1/3\pi = 0.424\,\text{m from origin}$$

(b) For horizontal component, centroid of vertical projected area is 0.5 m below free surface of water. Therefore, from (1.3),

$$\text{force} = \rho g A\bar{l} = 1000 \times 9.81 \times 3 \times 1 \times 0.5 = 14\,715\,\text{N}$$

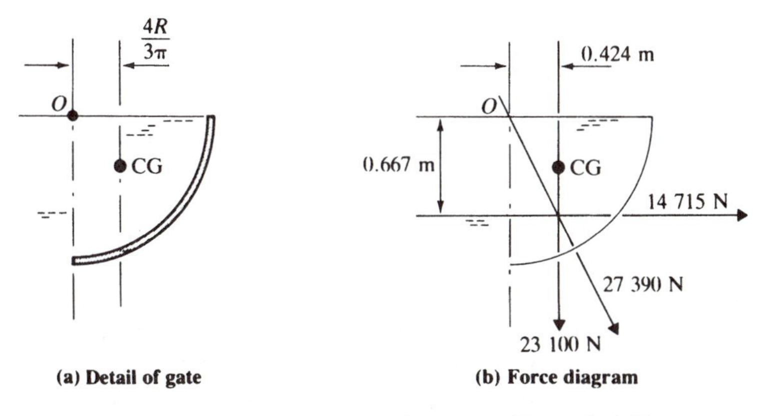

Figure 1.9 Pressure on quadrant gate (Example 1.2).

The point of action of this component lies l' below the free surface:

$$l' = I/A\bar{l}, \text{ where } I = A\bar{l}^2 + I_0 = (3\times 1)0.5^2 + (3\times 1^3)/12$$

$$l' = \frac{(3\times 1)\times 0.5^2 + (3\times 1^3)/12}{(3\times 1)\times 0.5} = 0.667\,\text{m}$$

Since this is for the vertical projected surface, l' is a vertical distance. The resultant force, F, is given by

$$F = \sqrt{23\,100^2 + 14\,715^2} = 27\,390\,\text{N}$$

This will act at an angle $\tan^{-1}(23\,100/14\,715)$, i.e. 57.5° below the horizontal. The force is therefore disposed as shown in Figure 1.9(b), with its line of action passing through 'O'.

1.4 Flotation

Buoyancy forces

Although civil engineers are not boat designers, they do have to deal with cases of buoyancy from time to time. Some typical examples are:

1. buried gas pipelines in waterlogged ground;
2. exploration rigs used by oil or gas corporations;
3. towing large steel dock/lock gates by sea or river (assuming that the structure can float, of course).

Figure 1.10 shows the vertical forces acting on an immersed cylinder of horizontal cross-sectional area A with its axis vertical.

The force acting downwards is due to pressure on the top surface i.e. $p_1 A = \rho g y A$. The force acting upwards is due to pressure on the bottom surface, and is given by $p_2 A = \rho g(y+L)A$. So

$$\text{total upthrust} = F_B = \rho g(y+L)A - \rho g y A = \rho g L A$$

Where LA is the volume of the cylinder. This leads to Archimedes' principle that 'the upthrust on a body is equal to the weight of the fluid displaced'. The upthrust acts through the centre of buoyancy B, which is the centre of gravity of the displaced fluid. For the case of a floating body, there must be equilibrium between F_B and the weight of the body.

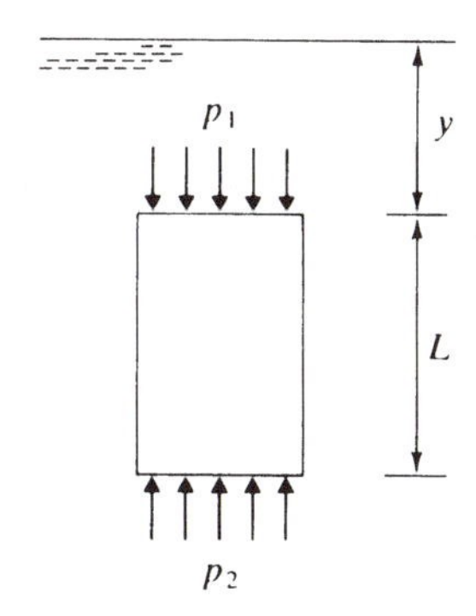

Figure 1.10 Pressure on immersed body.

Stable flotation

If a body is designed to float, it must do so in a stable fashion. This means that if the body suffers an angular displacement, it will automatically return to the original (correct) position. To appreciate how a floating body may exhibit this self-righting property, it must be realized that (with the exception of circular sections) the angular displacement of a body causes a lateral displacement in the position of B. Consider, for example, the pontoon of rectangular section shown in Figure 1.11(a). For upright flotation, the pressure distribution across the base is uniform. The centre of buoyancy lies on the vertical centreline. Since the pontoon is upright, the weight, W, must also be acting along the vertical centreline. If the vessel now rotates about its longitudinal axis ('heels over') through angle θ (Fig. 1.11b), the pressure distribution on the base becomes non-uniform, although still linear. This is, therefore, similar to the other pressure distribution patterns which have been examined. The shift from a uniform distribution (with the vessel upright) to a non-uniform distribution necessarily implies a corresponding shift in the position of the line of action of the resultant force. The buoyancy force now acts through B′ rather than through the original centre of buoyancy B. If a vertical line (representing the buoyancy force) is drawn through B′, it intercepts the centreline of the vessel at point M, which is called the 'metacentre'. Using trigonometry, the distance $\overline{\mathrm{BB'}}$ may be found:

$$\overline{\mathrm{BB'}} = \overline{\mathrm{MB'}} \sin\theta$$

Unfortunately, the distances $\overline{\mathrm{MB'}}$ and $\overline{\mathrm{BB'}}$ are both unknown. In order to evaluate them, use is made of the pressure distribution as follows.

The pressure, p, on the base at any distance a from the centreline is

$$p = \rho g(Y + a\sin\theta)$$

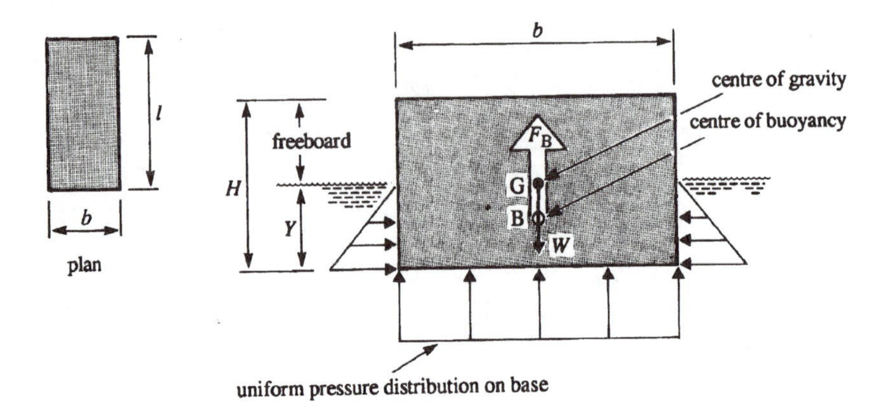

(a) Pontoon in upright position

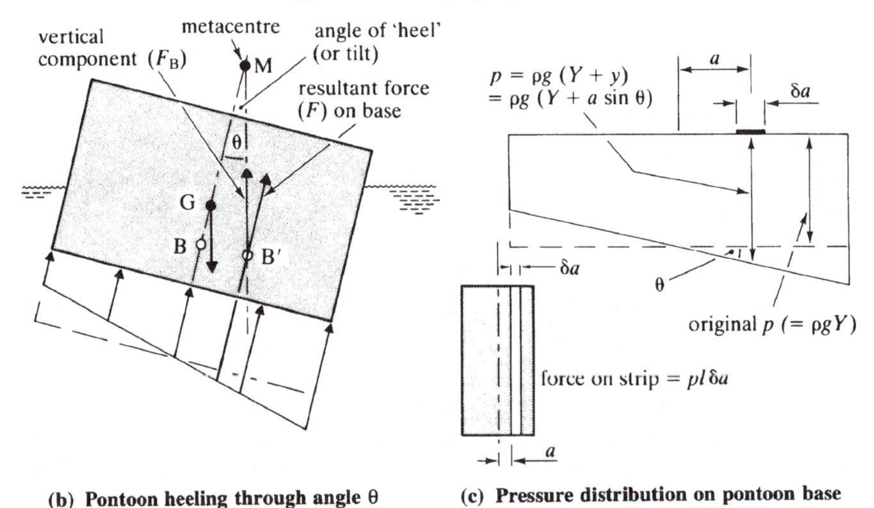

(b) Pontoon heeling through angle θ

(c) Pressure distribution on pontoon base

Figure 1.11 Flotation of pontoon.

The corresponding force on an element of width δa and length l is

$$\delta F = pl\,\delta a = \rho g(Y + a \sin\theta) l\,\delta a$$

The vertical component of δF is the buoyancy component, so that

$$\delta F_B = \rho g(Y + a \sin\theta) l\,\delta a \cos\theta$$

The moment of this component about **B** is

$$\rho g(Y + a \sin\theta) l\,\delta a(\cos\theta) a$$

So, for the whole vessel the moment due to the buoyancy force is given by

$$\int_{-b/2}^{+b/2} \rho g(Y + a \sin\theta) l(\cos\theta) a \, da$$

to which the solution is

$$\text{moment} = \rho g(\sin\theta)(\cos\theta)(b^3 l/12) = \rho g(\sin\theta)(\cos\theta) I$$

Remember, this is the moment due to the change in position of the centre of buoyancy. It has already been shown that buoyancy force $F_B = \rho g V$ and that $\overline{BB'} = \overline{MB'} \sin\theta$. Thus

$$\text{moment} = F_B \times \overline{BB'} = F_B \times \overline{MB'} \sin\theta$$
$$= \rho g(\sin\theta)(\cos\theta) I$$

Substituting for F_B and rearranging,

$$\overline{MB} = (I/V) \cos\theta$$

If θ is small, $\cos\theta \to 1$ and

$$\overline{MB} = I/V \tag{1.5}$$

The appearance of the second moment of area could have been anticipated in view of its association with linear non-uniform stress distributions.

The self-righting ability of a body is a function of the geometrical relationship between M and the centre of gravity G:

1. if M is above G, the body is stable (i.e. self-righting);
2. if M coincides with G, the body is neutrally stable and will neither capsize nor be self-righting;
3. if M is below G, the body is unstable and will capsize.

A glance at the disposition of the forces shows that the righting action is due to the moment of buoyancy force about the centre of gravity. This may be expressed as

$$\text{righting moment} = F_B \times \overline{MG}\,\theta$$

where θ is measured in radians. A number of simplifications are inherent in the derivation of these relationships. For example, the effect of the pressure distribution on each side has been ignored. However, as long as θ is small, the resulting errors are negligible.

Example 1.3 Buoyancy (uplift) force on quadrant gate

For a gate having the same dimensions as for Example 1.2, find the magnitude and direction of the force if the water is on the opposite side of the gate (Fig. 1.12).

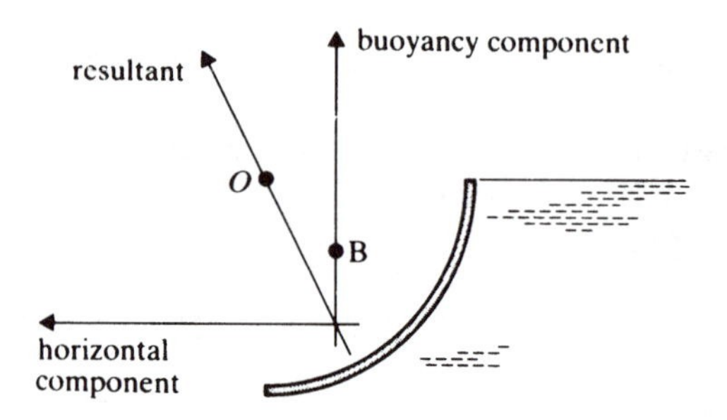

Figure 1.12 Pressure on quadrant gate (Example 1.3).

Solution

(a) For vertical component,

$$\text{area of sector (as before)} = 0.785\,\text{m}^2$$

$$\text{volume of water displaced} = 2.355\,\text{m}^3$$

$$\text{buoyancy force, } F_B = 9.81 \times 1000 \times 2.355 = 23\,100\,\text{N}$$

The centre of gravity of the sector is 0.424 m from the origin, so this will be the position of the centre of buoyancy.

(b) For horizontal component, centroid of vertical projected area is 0.5 m below the free surface of the water, so

$$\text{force} = \rho g A\bar{l} = 1000 \times 9.81 \times 3 \times 1 \times 0.5 = 14\,715\,\text{N}$$

The calculation of the position of the points of action is as before. Therefore, $l' = 0.667$ m in the vertical plane. The resultant force, F, is given by

$$F = \sqrt{23\,100^2 + 14\,715^2} = 27\,390\,\text{N}$$

The solution is numerically identical to that of Example 1.2. The only difference is that here each force is acting in the opposite direction to the corresponding force in Example 1.2.

Example 1.4 Pontoon stability

During serious river flooding, one span of a road bridge collapses. Government engineers decide to install a temporary pontoon bridge while repairs are in progress (Fig. 1.13). The river is 80 m wide and 7 m deep, and is not tidal. An outline specification of the temporary scheme is:

$$\text{clearance between pontoon base and river bed} = 5.5\,\text{m}$$

$$\text{pontoon freeboard (i.e. distance from water line to deck)} = 1.56\,\text{m}$$

$$\text{maximum pontoon self weight} = 220\,\text{t}$$

$$\text{width of highway} = 10\,\text{m}$$

$$\text{maximum side tilt due to 40 t vehicle load} = 4°$$

Assume that the centre of gravity of the vehicle is 3 m above the bridge deck and 2 m off-centre. Assume that the pontoon centre of gravity lies at the geometrical centre of the pontoon.

Estimate the principal pontoon dimensions.

Solution

$$\text{Pontoon draft (i.e. depth of immersion)} = 7 - 5.5 = 1.5\,\text{m}$$

$$\text{total height of pontoon} = \text{draft} + \text{freeboard} = 1.5 + 1.5 = 3\,\text{m}$$

$$\text{displacement due to pontoon} + \text{vehicle} = 220 + 40 = 260\,\text{t}$$

Therefore, the volume of water displaced is 260 m^3, since 1 m^3 of water has a mass of 1 t. So,

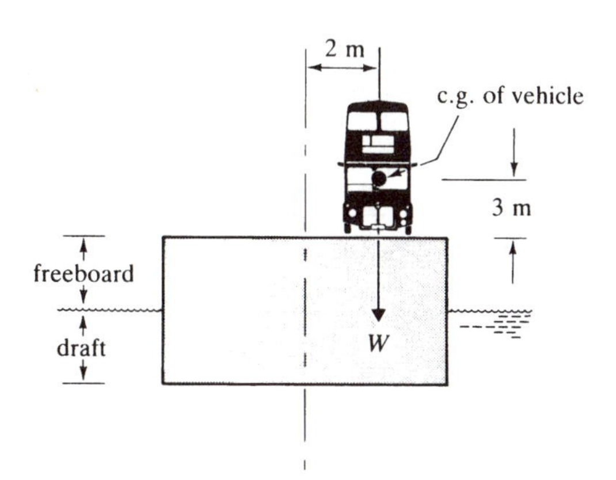

Figure 1.13 Pontoon bridge (Example 1.4).

$$\text{pontoon length} = \frac{\text{volume displaced}}{\text{width} \times \text{draft}}$$
$$= \frac{260}{10 \times 1.5} = 17.333\,\text{m}$$

This is not a whole submultiple of the river width, so say that the pontoon length is 20 m, with a corresponding draft of 1.3 m.

Check that the pontoon stability is within specified limits:

$$\overline{\text{MB}} = \frac{I}{V} = \frac{(20 \times 10^3)/12}{260} = 6.41\,\text{m}$$

(from (1.5)). The pontoon centre of gravity is at its geometrical centre, 1.5 m above the base. The centre of gravity of the vehicle is 3 m above the deck, and therefore 6 m above the base of the pontoon. Taking moments about the pontoon base, the combined centre of gravity position is found:

$$\text{height of combined c.g. above the base} = \frac{(220 \times 1.5) + (40 \times 6)}{260}$$
$$= 2.192\,\text{m}$$

The centre of buoyancy, B, is at the centre of gravity of the displaced fluid. This lies at the geometrical centre of the immersed section of the pontoon, and is therefore 0.65 m above base.

Therefore, the distance $\overline{\text{MG}}$ between the metacentre and the centre of gravity is

$$\overline{\text{MG}} = 6.41 + 0.65 - 2.192 = 4.868\,\text{m}$$

$$\text{overturning moment due to the vehicle} = 40 \times 2 = 80\,\text{tm}$$

$$\text{righting moment} = 260 \times 4.868 \times \theta^c$$

For equilibrium, righting moment = overturning moment, so

$$\theta^c = \frac{80}{260 \times 4.868} = 0.062\,\text{rad} = 3.55°$$

which is within specification.

2

Principles of fluid flow

2.1 Introduction

The problems encountered in Chapter 1 involved only a small number of quantities, basically p, g and y. In consequence, the equations developed were simple and precise.

Thus, if the results of hydrostatic experimentation and calculation are compared, they are found to agree within the limits of experimental accuracy. Unfortunately, this combination of simplicity and accuracy does not apply when we turn our attention to problems involving fluid flows. Engineering flows are mostly very complex, and it is not usually possible to evolve precise theoretical models.

However, a great deal can be learnt about flows by adopting the techniques and equations which were developed by the hydrodynamicists. These equations are reasonably straightforward because they are developed for the case of an 'ideal' fluid. An ideal fluid has no viscosity (i.e. it is inviscid), has no surface tension and is incompressible. Viscosity and compressibility are the major reasons for the complexity of real fluid flows. Of course, no such substance as an 'ideal' fluid actually exists. Nevertheless, for certain types of problem the equations for 'ideal' flows are remarkably accurate.

2.2 Classification of flows

Flows can be classified in a number of ways. The system generally adopted is to consider the flow as being characterized by two parameters – time and distance. The class into which any particular flow falls is usually a reliable guide to the appropriate method of solution.

The first major subdivision is based on consideration of the timescale. This categorizes all flows as either steady or unsteady.

A flow is steady if the parameters describing that flow do not vary with time. Typical parameters of a flow are velocity, discharge (volume per second passing a given point), pressure, or depth of flow (e.g. in a river or channel). Conversely, a flow is unsteady if these parameters do vary with time.

Because of the complex equations associated with unsteady flows, engineers often use steady flow equations even where a small degree of unsteadiness is present. A case in point is that of flows in rivers, where the discharge is rarely, if ever, absolutely steady. Nevertheless, if changes occur slowly, steady flow equations give quite accurate results.

The second subdivision relates to the scale of distance. This classifies flows as being uniform or non-uniform.

A flow is uniform if the parameters describing the flow do not vary with distance along the flow path. Conversely, for a non-uniform flow, the magnitude of the parameters varies from point to point along the flow path.

The existence of a uniform flow necessarily implies that the area of the cross-section perpendicular to the direction of flow is constant. A typical example is that of the flow in a pipeline of constant diameter. By contrast, one would have a non-uniform flow in a tapered pipe.

The two sets of subdivisions or classifications (steady–unsteady, uniform–non-uniform) are not mutually exclusive. Some flows exhibit changes with respect to both time and distance, while others change with respect to time or distance only. However, the majority of flows will fall into one of the classifications listed below.

Steady uniform flow. For such a flow the discharge is constant with time, and the cross-section through which the flow passes is of constant area. A typical example is that of constant flow through a long straight pipe of uniform diameter.

Steady non-uniform flow. The discharge is constant with time, but the cross-sectional area varies with distance. Examples are flow in a tapering pipe and flow with constant discharge in a river (the cross-section of a river usually varies from point to point).

Unsteady uniform flow. The cross-section through the flow is constant, but the discharge varies with time. This is a complex flow pattern. An example is that of pressure surge in a long straight pipe of uniform diameter.

Unsteady non-uniform flow. The cross-section and discharge vary with both time and distance. This is typified by the passage of a flood wave in a natural channel, and is the most complex flow to analyse.

2.3 Visualization of flow patterns

Streamlines

The fundamental method of visualizing a flow pattern is by means of 'streamlines'. A streamline is constructed by drawing a line which is tangential to the velocity vectors of a connected series of fluid particles (Fig. 2.1(a)).

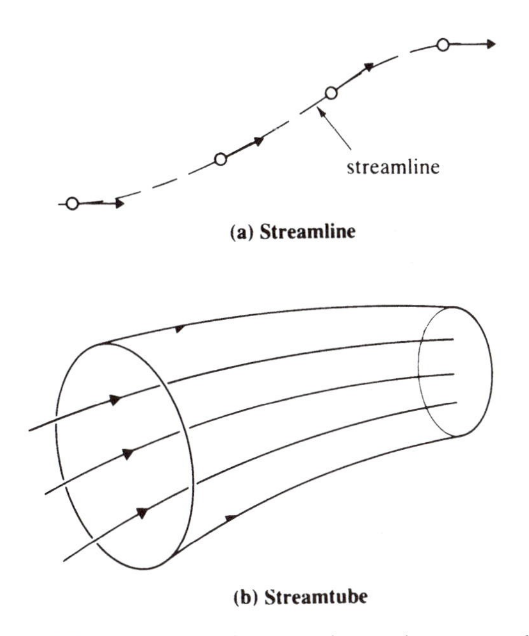

Figure 2.1 Definitions of a streamline and a streamtube.

The streamline is thus a line representing the direction of flow of the series of particles at a given instant. Because the streamline is always tangential to the flow, it follows that there is no flow across a streamline.

A set of streamlines may be arranged to form an imaginary pipe or tube. This is known as a 'streamtube' (Fig. 2.1(b)). Under certain specific circumstances, streamtubes can actually be identified. For example, the internal surface of a pipeline must also be a streamtube, since the vectors representing the flow adjacent to the surface must be parallel to that surface. The surface is therefore 'covered' with streamlines.

The hydrodynamicists developed a theoretical framework which makes it possible to construct streamlines for a variety of ideal flows, by using a graphical technique. The resulting diagrams are known as 'flow nets', and these are discussed in section 2.9.

Streaklines

Clearly, it will be difficult to construct a streamline for a real flow, since individual particles of fluid are not visible to the eye. A simple method of obtaining approximate information regarding streamline patterns is to inject a dye into the flow. The dye will trace out a path known as a 'streakline', which may be photographed. Usually the characteristics of the dye (its density, etc.) will not be identical to those of the fluid. The streakline, therefore, will not necessarily be absolutely identical to the streamline.

One-, two- and three-dimensional flow

Most real flows are three-dimensional, in that velocity, pressure and other parameters may vary in three directions (x, y, z). There may also be variation of the parameters with time.

In practice it is nearly always possible to consider the flow to be one- or two-dimensional. This greatly simplifies the equations of flow. Minor adjustments to these simplified equations can often be incorporated to allow for a two- or three-dimensional flow. For example, steady uniform flow in a pipe is considered to be one-dimensional, the flow being characterized by a streamline along the centreline of the pipe. The velocity and pressure variations across the pipe are ignored, or are considered separately as secondary effects.

2.4 The fundamental equations of fluid dynamics

Description and physical basis

In order to develop the equations which describe a flow, hydrodynamicists assumed that fluids are subject to certain fundamental laws of physics. The pertinent laws are:

1. conservation of matter (conservation of mass);
2. conservation of energy;
3. conservation of momentum.

These principles were initially developed for the case of a solid body, and it is worth expanding them a little before proceeding.

1. The law of conservation of matter stipulates that matter can be neither created nor destroyed, though it may be transformed (e.g. by a chemical process). Since this study of the mechanics of fluids excludes chemical activity from consideration, the law reduces to the principle of conservation of mass.
2. The law of conservation of energy states that energy may be neither created nor destroyed. Energy can be transformed from one guise to another (e.g. potential energy can be transformed into kinetic energy), but none is actually lost. Engineers sometimes loosely refer to 'energy losses' due to friction, but in fact the friction transforms some energy into heat, so none is really 'lost'. The basic equation may be derived from the First Law of Thermodynamics, though here it will be obtained in a slightly simplified manner.

3. The law of conservation of momentum states that a body in motion cannot gain or lose momentum unless some external force is applied. The classical statement of this law is Newton's Second Law of Motion, i.e.

$$\text{force} = \text{rate of change of momentum.}$$

Applying these laws to a solid body is relatively straightforward, since the body will be of measurable size and mass. For example, if the mass of a body is known, the force required to produce a given acceleration is easily calculated. However, for the case of a fluid the attempt to apply these laws presents a problem. A flowing fluid is a continuum – that is to say, it is not possible to subdivide the flow into separate small masses. How, then, can the three basic laws be applied? How can a suitable mass of fluid be identified so that we can investigate its momentum or its energy? The answer lies in the use of 'control volumes'.

Control volumes

A control volume is a purely imaginary region within a body of flowing fluid. The region is usually (though not always) at a fixed location and of fixed size. Inside the region, all of the dynamic forces cancel each other. Attention may therefore be focused on the forces acting externally on the control volume. The control volume may be of any shape. Therefore, a shape may be selected which is most convenient for any particular application. In the work which follows, a short streamtube (Fig. 2.1(b)) will be used as a control volume. This may be visualized as a short transparent pipe or tube. Fluid enters through one end of the tube and leaves through the other. Any forces act externally along the boundaries of the control volume.

2.5 Application of the conservation laws to fluid flows

The continuity equation (principle of conservation of mass)

During any time interval δt, the principle of conservation of mass implies that for any control volume the mass flow entering minus the mass flow leaving equals the change of mass within the control volume.

If the flow is steady, then the mass must be entering (or leaving) the volume at a constant rate. If we further restrict our attention to incompressible flow, then the mass of fluid within the control volume must remain fixed. In other words, the change of mass within the control volume is zero. Therefore, during time δt,

$$\text{mass flow entering} = \text{mass flow leaving}$$

Since the flow is incompressible, the density of the fluid is constant throughout the fluid continuum. Mass flow entering may be calculated by taking the product

$$\text{(density of fluid)} \times \text{(volume of fluid entering per second)}$$

Mass flow is therefore represented by the product ρQ, hence

$$\rho Q(\text{entering}) = \rho Q(\text{leaving})$$

But since flow is incompressible, the density is constant, so

$$Q(\text{entering}) = Q(\text{leaving}) \tag{2.1a}$$

This is the 'continuity equation' for steady incompressible flow.

The dimensions of Q are L^3T^{-1} (SI units m^3/s). This can be expressed alternatively as L^2LT^{-1}, which is the product of an area and a velocity. Supposing that measurements are taken of the velocity of flow across the entry to the control volume, and that the velocity is constant at u_1 m/s. Then, if the cross-sectional area of the streamtube at entry is A_1,

$$Q(\text{entering}) = u_1A_1$$

Note that the area must be that of a plane perpendicular to the direction of flow. Similar remarks apply to the flow leaving the volume. Thus, if the velocity of flow leaving the volume is u_2 and the area of the streamtube at exit is A_2, then

$$Q(\text{leaving}) = u_2A_2$$

Therefore, the continuity equation may also be written as

$$u_1A_1 = u_2A_2 \tag{2.1b}$$

The energy equation (principle of conservation of energy)

This may be simply derived by considering the forms of energy available to the fluid. Figure 2.2 shows the control volume used to develop the equation. The combination of a flow and a pressure implies that work is done. Thus, if pressure p acts on area A, the corresponding force is pA. If the fluid is flowing, then in travelling through a length L the work done equals the product of force and distance, i.e.

$$\text{'flow work' done} = pAL$$

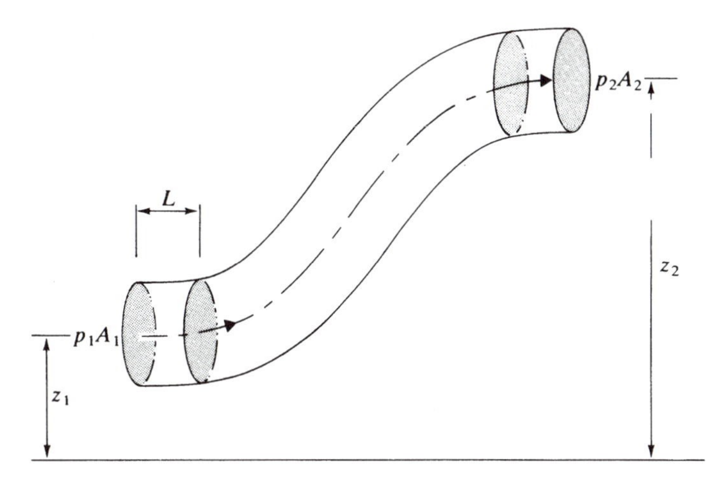

Figure 2.2 Streamtube used to derive the energy equation.

For the control volume under consideration, the fluid entering the system travels through distance L during time interval δt. The flow work done during this time is $p_1 A_1 L$. The mass, m, entering the system is $\rho_1 A_1 L$ during δt. Therefore the kinetic energy (KE) entering the system is

$$\mathrm{KE} = \frac{1}{2} m u^2 = \frac{1}{2} \rho_1 A_1 L u_1^2$$

The potential energy of a body is mgz, where z is the height of the body above some arbitrary datum. The potential energy of the mass $\rho_1 A_1 L$ entering during δt is $\rho_1 A_1 L g z$. The total energy entering is the sum of the flow work done and the potential and kinetic energies, i.e.

$$p_1 A_1 L + \frac{1}{2} \rho_1 A_1 L u_1^2 + \rho_1 A_1 l g z_1$$

It is more convenient to consider the energy per unit weight of fluid. Note that the weight of fluid entering the system is $\rho_1 g A_1 L$, so the energy per unit weight of fluid is

$$\frac{p_1 A_1 L}{\rho_1 g A_1 L} + \frac{1}{2} \frac{\rho_1 A_1 L u_1^2}{\rho_1 g A_1 L} + \frac{\rho_1 A_1 L g z_1}{\rho_1 g A_1 L} = \frac{p_1}{\rho_1 g} + \frac{u_1^2}{\rho_1 g} + z_1$$

Similarly, at the exit the total energy per unit weight of fluid leaving the system is

$$\frac{p_2}{\rho_2 g} + \frac{u_2^2}{2g} + z_2$$

If, during the passage from entry to exit, no energy is supplied to the fluid or extracted from the fluid, then clearly

$$\text{energy entering} = \text{energy leaving}$$

If the flow is incompressible, then $\rho_1 = \rho_2 = \rho$, hence

$$\frac{p_1}{\rho g} + \frac{u_1^2}{2g} + z_1 = \frac{p_2}{\rho g} + \frac{u_2^2}{2g} + z_2 = H = \text{constant} \tag{2.2}$$

This is the 'Bernoulli equation' named after Daniel Bernoulli (1700–82), who published one of the first books on fluid flow in 1738.

Several points should be made about this equation.

(a) The statement that 'no energy is extracted from the fluid' implies that the fluid is frictionless. If this were not so, frictional forces would transform some of the energy into heat, then

$$\text{energy entering at 1} > \text{energy leaving at 2}$$

i.e. there would be a 'loss' of energy between 1 and 2.

(b) The reader should check the dimensions of each separate component of the Bernoulli equation:

$$\frac{p}{\rho g} \equiv \frac{\text{N}}{\text{m}^2} \cdot \frac{\text{m}^3}{\text{kg}} \cdot \frac{\text{s}^2}{\text{m}} \equiv \frac{ML}{L^2T^2} \cdot \frac{L^3}{M} \cdot \frac{T^2}{L} = L\,(\text{m})$$

$$\frac{u^2}{2g} \equiv \frac{\text{m}^2}{\text{s}^2} \cdot \frac{\text{s}^2}{\text{m}} \equiv \frac{L^2}{T^2} \cdot \frac{T^2}{L} = L\,(\text{m})$$

$$z \equiv L\,(\text{m})$$

Thus, all the constituent parts of the equation have units of metres. For this reason, each term may be regarded as a 'head': $p/\rho g =$ pressure head, $u^2/2g =$ kinetic, or velocity head, $z =$ potential or elevation head.

(c) The equation was developed for a streamtube of finite area A. However, A must be 'small', otherwise the height (z) of a streamline at the bottom of the tube is significantly less than that of a streamline at the top of the tube. Strictly, therefore, Bernoulli's equation applies along a single streamline. However, if the diameter of the appropriate control volume is small compared with its length, engineers apply Bernoulli's equation between two points without specific reference to a particular streamline.

It is possible to derive Bernoulli's equation in a mathematically rigorous fashion, as is now demonstrated.

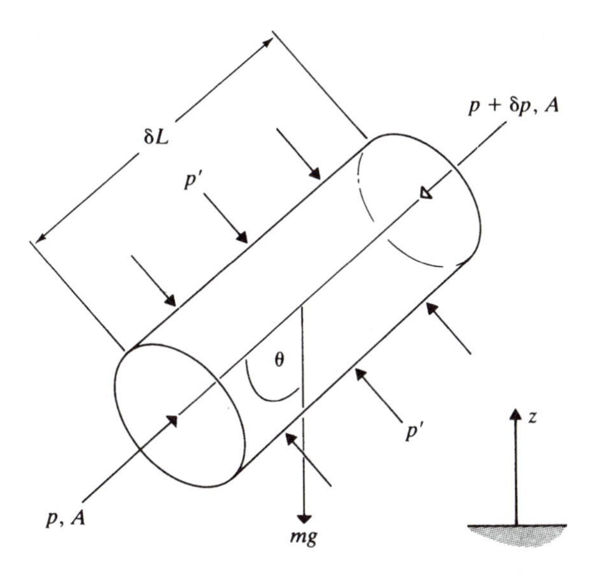

Figure 2.3 An elemental cylindrical streamtube.

Figure 2.3 shows a small (elemental) streamtube of cylindrical section. Taking this to be a control volume at an instant in time, the forces acting in the direction of flow along the streamtube are:

Pressure forces

upstream end	$+pA$
downstream end	$-(p+\delta p)A$
circumference	zero (pressure forces p' cancel)

Weight force

$$-mg\cos\theta = -\rho A\,\delta L g\cos\theta$$

Newton's Second Law (force = mass × acceleration) may be used to relate these forces:

$$pA-(p+\delta p)A-\rho A\,\delta L g\cos\theta = \rho A\,\delta L(\mathrm{d}u/\mathrm{d}t)$$

Setting $\delta L\cos\theta = \delta z$ and cancelling terms,

$$-A\,\delta p-\rho A g\,\delta z = \rho A\,\delta L(\mathrm{d}u/\mathrm{d}t)$$

Dividing by $\rho A\,\delta L$, and in the limit

$$\frac{1}{\rho}\frac{\mathrm{d}p}{\mathrm{d}L}+\frac{\mathrm{d}u}{\mathrm{d}t}+g\frac{\mathrm{d}z}{\mathrm{d}L}=0$$

For steady flow, velocity (u) only varies with distance (L). Therefore,

$$\frac{\mathrm{d}u}{\mathrm{d}t} = \frac{\mathrm{d}L}{\mathrm{d}t}\frac{\mathrm{d}u}{\mathrm{d}L} = u\frac{\mathrm{d}u}{\mathrm{d}L}$$

Hence, substituting for $\mathrm{d}u/\mathrm{d}t$,

$$\frac{1}{\rho}\frac{\mathrm{d}p}{\mathrm{d}L} + u\frac{\mathrm{d}u}{\mathrm{d}L} + g\frac{\mathrm{d}z}{\mathrm{d}L} = 0 \tag{2.3}$$

Equation (2.3) is known as Euler's equation (in honour of the Swiss mathematician Leonhard Euler (1707–83)).

For incompressible fluids, Euler's equation may be integrated to yield

$$\frac{p}{\rho} + \frac{u^2}{2} + gz = \text{constant}$$

or, dividing by g,

$$\frac{p}{\rho g} + \frac{u^2}{2g} + z = \text{constant}$$

This is, of course, Bernoulli's equation (cf. (2.2)).

The momentum equation (principle of conservation of momentum)

Again, consider the streamtube shown in Figure 2.2 and apply Newton's Second Law in the form force = rate of change of momentum (i.e. $F = \mathrm{d}(mu)/\mathrm{d}t$)). In a time interval δt,

$$\text{momentum entering} = \rho\,\delta Q_1 \delta t u_1$$

$$\text{momentum leaving} = \rho\,\delta Q_2 \delta t u_2$$

and $\delta Q_1 = \delta Q_2 = \delta Q$ by the continuity principle. Hence, the force required to produce this change of momentum is

$$\delta F = \frac{\rho\,\delta Q\,\delta t(u_2 - u_1)}{\delta t} = \rho\,\delta Q(u_2 - u_1)$$

As the velocities may have components in the x-, y- and z-directions, it is more convenient to consider each component separately. Hence,

$$\delta F_x = \rho\,\delta Q(u_{2x} - u_{1x})$$

and similarly for δF_y and δF_z. Integrating over a region,

$$F_x = \rho Q(V_{2x} - V_{1x}) \tag{2.4}$$

(similarly for F_y and F_z) provided that the velocity (V) is uniform across the collection of streamtubes.

Equation (2.4) is the momentum equation for steady flow for a region of uniform velocity. The momentum force is composed of the sum of all external forces acting on the streamtube (control volume) and may include pressure forces (F_P) and reaction forces (F_R), i.e.

$$F_x = \Sigma(F_P + F_R) \tag{2.5}$$

This is explained more fully in section 2.7, where the momentum equation is used to find the forces acting on pipe bends, etc.

Energy and momentum coefficients

In deriving the momentum equation, it was stressed that it could only be applied directly to a large region if the velocity (u) was constant across the region (set equal to V). The energy equation was stated to apply to a streamline with velocity u. In many real fluid flow problems, u varies across a section (refer, for example, to pipe flow in Chapter 4). The velocity distribution across a section is not always known, and the mean velocity V (defined as discharge divided by area) must be used instead of u. Both the energy and momentum equations may still be used by introducing the energy and momentum coefficients, α and β, respectively. These are defined as follows:

$$\alpha = \int u^3 \mathrm{d}A / V^3 A \tag{2.6}$$

$$\beta = \int u^2 \mathrm{d}A / V^2 A \tag{2.7}$$

The derivation of the equation for the energy coefficient rests on the principle that the total kinetic energy possessed by a bundle of streamtubes is equal to α multiplied by the kinetic energy represented by the mean velocity V. Stated mathematically,

$$\sum \frac{1}{2} m u^2 = \alpha \frac{1}{2} M V^2$$

or for unit time

$$\sum \frac{1}{2} \rho \delta Q u^2 = \alpha \frac{1}{2} \rho Q V^2$$

hence, as $\delta Q = u \delta A$ and $Q = VA$,

$$\sum \frac{1}{2} \rho u^3 \delta A = \alpha \frac{1}{2} \rho V^3 A$$

This leads to (2.6). The momentum coefficient β may be derived similarly by equating momentum for a bundle of streamtubes.

The Bernoulli energy equation may then be rewritten in terms of mean velocity V as

$$\frac{p}{\rho g}+\frac{\alpha V^2}{2g}+z=\text{constant} \tag{2.8}$$

and the momentum equation as

$$F_x=\rho Q\beta(V_{2x}-V_{1x}) \tag{2.9}$$

The values of α and β must be derived from the velocity distributions across a region. They always exceed unity, but usually by only a small margin, so they are frequently omitted. However, this is not always the case, and they should therefore not be forgotten.

Example calculations for these coefficients for open-channel flow are given in Chapter 5. In the case of turbulent pipe flow, $\alpha=1.06$ and $\beta=1.02$, and both may safely be ignored.

2.6 Application of the energy equation

Bernoulli's equation may be applied to any continuous flow system. The simplest system might be a single pipeline with a frictionless fluid discharging through it.

For such a system, (2.2) may be rewritten as

$$\frac{p_1}{\rho g}+\frac{u_1^2}{2g}+z_1=\frac{p_2}{\rho g}+\frac{u_2^2}{2g}+z_2=\text{constant}$$

where subscripts 1 and 2 refer to two points along any streamline. Hence Bernoulli's energy equation, in conjunction with the continuity equation, may be used to determine the variation of pressure and velocity along any streamline.

Example 2.1 Application of Bernoulli's equation

For the frictionless syphon shown in Figure 2.4, determine the discharge and the pressure heads at A and B, given that the pipe diameter is 200 mm and the nozzle exit diameter is 150 mm.

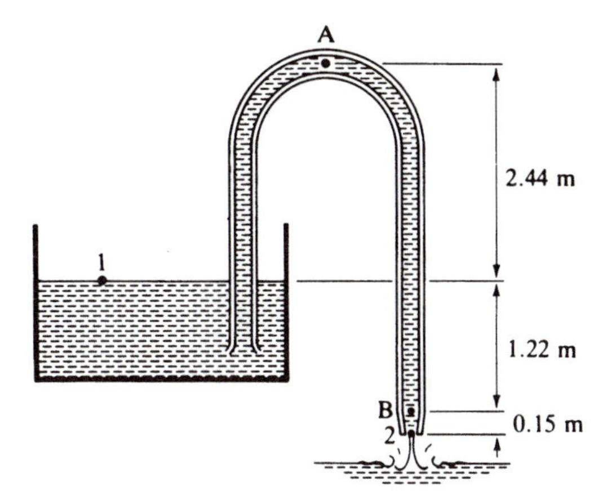

Figure 2.4 Syphon arrangement for Example 2.1.

Solution

To find the discharge, first apply Bernoulli's equation along the streamline between 1 and 2:

$$\frac{p_1}{\rho g} + \frac{u_1^2}{2g} + z_1 = \frac{p_2}{\rho g} + \frac{u_2^2}{2g} + z_2$$

At both 1 and 2, the pressure is atmospheric, and at 1 the velocity is negligible, hence

$$z_1 - z_2 = \frac{u_2^2}{2g}$$

From Figure 2.4,

$$z_1 - z_2 = 1.22 + 0.15 = 1.37\,\text{m}$$

hence

$$\frac{u_2^2}{2g} = 1.37$$

or

$$u_2 = 5.18\,\text{m/s}$$

Next apply the continuity equation to find the discharge:

$$\begin{aligned} Q &= uA \\ &= 5.18 \times \left(\frac{\pi \times 0.15^2}{4}\right) \end{aligned}$$

hence

$$Q = 0.092\,\mathrm{m}^3/\mathrm{s}$$

To find the pressure head at A, again apply Bernoulli's equation along the streamline from 1 to A

$$\overset{0}{\cancel{\frac{p_1}{\rho g}}} + \overset{0}{\cancel{\frac{u_1^2}{2g}}} + z_1 = \frac{p_\mathrm{A}}{\rho g} + \frac{u_\mathrm{A}^2}{2g} + z_\mathrm{A}$$

as $p_1 = 0$ and $u_1 = 0$ then

$$\frac{p_\mathrm{A}}{\rho g} = (z_1 - z_\mathrm{A}) - \frac{u_\mathrm{A}^2}{2g}$$

$$= -2.44 - \frac{u_\mathrm{A}^2}{2g}$$

and

$$u_\mathrm{A} A_\mathrm{A} = Q$$

$$u_\mathrm{A} = 0.092 \Big/ \left(\frac{\pi \times 0.2^2}{4}\right) = 2.93\,\mathrm{m/s}$$

$$\frac{u_\mathrm{A}^2}{2g} = 0.44\,\mathrm{m}$$

hence

$$\frac{p_\mathrm{A}}{\rho g} = -2.44 - 0.44 = -2.88\,\mathrm{m}$$

The pressure head at B is found similarly:

$$\overset{0}{\cancel{\frac{p_1}{\rho g}}} + \overset{0}{\cancel{\frac{u_1^2}{2g}}} + z_1 = \frac{p_\mathrm{B}}{\rho g} + \frac{u_\mathrm{B}^2}{2g} + z_\mathrm{B} \quad (p_1, u_1 \to 0)$$

$$\frac{p_\mathrm{B}}{\rho g} = (z_1 - z_\mathrm{B}) - \frac{u_\mathrm{B}^2}{2g}$$

$$(z_1 - z_\mathrm{B}) = 1.22\,\mathrm{m}$$

$$u_\mathrm{B} = u_\mathrm{A} = 2.93\,\mathrm{m/s}$$

$$\frac{p_\mathrm{B}}{\rho g} = 1.22 - 0.44 = 0.78\,\mathrm{m}$$

It may be observed that the pressure head at A is negative. This is because atmospheric pressure was taken as zero (i.e. pressures are gauge pressures) and therefore the pressure head at A is still positive in terms of absolute pressure, because atmospheric pressure is about 10.1 m head of water.

Modifications to Bernoulli's equation

In practice, the total energy of a streamline does not remain constant. Energy is 'lost' through friction, and external energy may be added by means of a pump or extracted by a turbine. These energy gains and losses are explained in later chapters, but it is worth introducing them into Bernoulli's equation at this point so that it can be extended to cover a wider range of practical cases. Consider a streamline between two points 1 and 2. If the energy head lost through friction is denoted by h_f and the external energy head added (say by a pump) is E, then Bernoulli's equation may be rewritten as

$$H_1 + E = h_2 + h_f$$

or

$$\frac{p_1}{\rho g} + \frac{u_1^2}{2g} + z_1 + E = \frac{p_2}{\rho g} + \frac{u_2^2}{2g} + z_2 + h_f$$

This equation is really a restatement of the First Law of Thermodynamics for an incompressible fluid.

Example 2.2 Application of Bernoulli's equation with energy gains and losses

A pump delivers water from a lower to a higher reservoir. The difference in elevation between the reservoirs is 10 m. The pump provides an energy head of 11 m and the frictional head losses are 0.7 m. If the pipe diameter is 300 mm, calculate the discharge.

Solution

Apply the modified Bernoulli's equation from the lower to the higher reservoir:

$$H_1 + 11 = H_2 + 0.7$$

or

$$\frac{p_1}{\rho g} + \frac{u_1^2}{2g} + z_1 + 11 = \frac{p_2}{\rho g} + \frac{u_2^2}{2g} + z_2 + 0.7$$

Taking $p_1 = p_2 = 0$ and $u_1 = 0$,

$$\begin{aligned}\frac{u_2^2}{2g} &= (z_1 - z_2) + 11 - 0.7 \\ &= -10 + 11 - 0.7 \\ &= 0.3\,\text{m} \\ u_2 &= 2.426\,\text{m/s}\end{aligned}$$

Applying the continuity equation

$$Q = u_2\left(\frac{\pi \times 0.3^2}{4}\right) = 0.171\,\text{m}^3/\text{s}$$

2.7 Application of the momentum equation

Range of applications

The momentum equation may be used directly to evaluate the force causing a change of momentum in a fluid. Such applications include determining forces on pipe bends and junctions, nozzles and hydraulic machines. Examples of some of these are given in this section.

In addition, the momentum equation is used to solve problems in which energy losses occur that cannot be evaluated directly, or when the flow is unsteady. Examples of such problems include local head losses in pipes, the hydraulic jump and unsteady flow in pipes and channels. These applications are discussed in later chapters.

Forces exerted on pipework

Whenever there is a change in geometry or direction, the fluid will exert a force on the pipework. These forces may be considerable, and must be resisted in order that the pipeline does not move. For underground pipes, the forces are normally resisted by thrust blocks which transfer the force to the surrounding earth. For exposed pipework, the forces are transmitted by supports at the pipe joints to the nearest structural member (e.g. a wall or beam). The calculation of these forces is now illustrated by three examples.

Example 2.3 Force exerted by a firehose

Calculate the force required to hold a firehose for a discharge of 5 l/s if the nozzle has an inlet diameter of 75 mm and an outlet diameter of 25 mm.

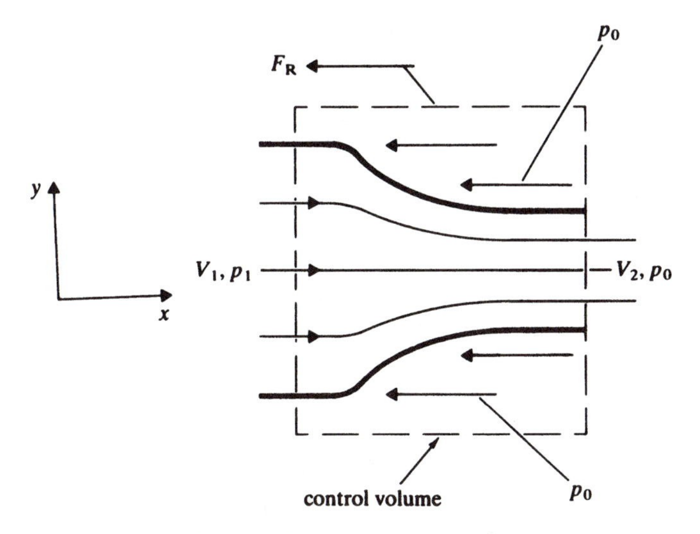

Figure 2.5 Forces on a nozzle.

Solution

The nozzle, surrounded by a suitable control volume, is shown in Figure 2.5. The forces acting in the x-direction on the control volume are the pressure forces (F_P) and the reaction force (F_R) of the nozzle on the fluid (here shown in the negative x-direction.) The sum of these must equal the momentum force (F_M). Hence (2.4) and (2.5) apply, i.e.

$$F_M = F_P + F_R$$

(where all terms are taken positive with x) and

$$F_M = \rho Q(V_2 - V_1)$$

V_1 and V_2 may be found using the continuity equation:

$$V_1 = Q/A_1 = 0.005 \Big/ \left(\frac{\pi \times 0.075^2}{4}\right)$$
$$= 1.13\,(\text{m/s})$$

and

$$V_2 = Q/A_2 = 0.005 \Big/ \left(\frac{\pi \times 0.025^2}{4}\right)$$
$$= 10.19\,(\text{m/s})$$

The pressure forces are

$$F_P = p_1 A_1 - p_0 A_2 - p_0(A_1 - A_2)$$

The pressure p_1 may be found using Bernoulli's equation:

$$\frac{p_1}{\rho g} + \frac{V_1^2}{2g} = \frac{p_0}{\rho g} + \frac{V_2^2}{2g}$$

Taking $p_0 = 0$ (*Note*: gauge pressures may be used because atmospheric pressure p_0 acts over the whole area A_1),

$$\frac{p_1}{\rho g} = \frac{V_2^2 - V_1^2}{2g}$$

or

$$p_1 = (\rho/2)(V_2^2 - V_1^2) = (1000/2)(10.19^2 - 1.13^2)$$
$$= 51.28\,\text{kN/m}^2$$

hence

$$F_P = 51.28 \times 10^3 \times \pi \times 0.075^2/4$$
$$= 0.226\,\text{kN}$$

The momentum force is

$$F_M = 10^3 \times 0.005(10.19 - 1.13)$$
$$= 0.0453\,\text{kN}$$

Hence, the reaction force is

$$F_R = F_M - F_P$$
$$= 0.0453 - 0.226\,\text{kN}$$
$$= -0.181\,\text{kN}$$

F_R is the force exerted on the fluid by the nozzle. The fireman must, of course, provide a force of equal magnitude but opposite direction to F_R.

Example 2.4 Force on a pipe bend

Calculate the magnitude and direction of the force exerted by the pipe bend shown in Figure 2.6 if the diameter is 600 mm, the discharge is 0.3 m^3/s and the upstream pressure head is 30 m.

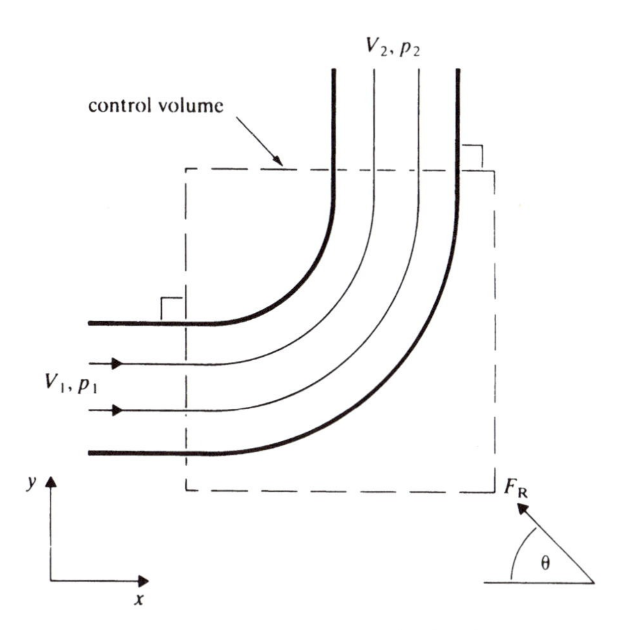

Figure 2.6 Forces on a pipe bend.

Solution

In this case, the force exerted is due to the change of direction and the x- and y-components of the force must be calculated separately. Hence

$$F_{Mx} = F_{Px} + F_{Rx}$$
$$F_{My} = F_{Py} + F_{Ry}$$

and

$$F_{Mx} = \rho Q(V_{2x} - V_{1x})$$
$$F_{My} = \rho Q(V_{2y} - V_{1y})$$

As the pipe is of constant diameter

$$V_2 = V_1 = 0.3 \Big/ \left(\frac{\pi \times 0.6^2}{4}\right)$$
$$= 1.06\,\text{m/s}$$

Neglecting the small energy loss around the pipe bend,

$$p_2 = p_1 = 30\rho g = 294.3\,\text{kN/m}^2$$

Pressure forces

$$F_{Px} = p_1 A_1 - 0 = 294.3 \times \pi \times 0.6^2/4$$
$$= 83.21\,\text{kN}$$

and

$$F_{Py} = 0 - p_2 A_2 = -83.2\,\text{kN}$$

Momentum forces

$$F_{Mx} = \rho Q(0 - V_1)$$
$$= 10^3 \times 0.3(-1.06)$$
$$= -0.318\,\text{kN}$$

and

$$F_{My} = \rho Q(V_2 - 0)$$
$$= +0.318\,\text{kN}$$

Reaction forces

$$F_{Rx} = F_{Mx} - F_{Px}$$
$$= -0.318 - 83.21$$
$$= -83.528\,\text{kN}$$

and

$$F_{Ry} = F_{My} - F_{Py}$$
$$= 0.318 - (-83.21)$$
$$= 83.528\,\text{kN}$$

hence

$$F_R = \sqrt{F_{Rx}^2 + F_{Ry}^2}$$
$$= 118.1\,\text{kN}$$

and

$$\theta = \tan^{-1}(F_{Ry}/F_{Rx})$$
$$= 45^\circ \text{ (from negative } x\text{-direction to positive } y\text{-direction)}$$

Example 2.5 Force on a T-junction

Calculate the magnitude and direction of the force exerted by the T-junction shown in Figure 2.7 if the discharges are $Q_1 = 0.3\,\text{m}^3/\text{s}$, $Q_2 = 0.15\,\text{m}^3/\text{s}$, $Q_3 = 0.15\,\text{m}^3/\text{s}$, the diameters are $D_1 = 450\,\text{mm}$, $D_2 = 300\,\text{mm}$, $D_3 = 200\,\text{mm}$ and the upstream pressure $p_1 = 500\,\text{kN/m}^2$.

Solution

In this case there are changes of direction and pressure and velocity. First find the three velocities by continuity, then apply Bernoulli's equation to find the pressures p_2 and p_3. Then apply the momentum equation.

Velocities

$$V_1 = Q_1/A_1 = 1.886\,\text{m/s}$$

$$V_2 = Q_2/A_2 = 2.122\,\text{m/s}$$

$$V_3 = Q_3/A_3 = 4.775\,\text{m/s}$$

Pressures

$$\frac{p_1}{\rho g} + \frac{V_1^2}{2g} = \frac{p_2}{\rho g} + \frac{V_2^2}{2g} = \frac{p_3}{\rho g} + \frac{V_3^2}{2g}$$

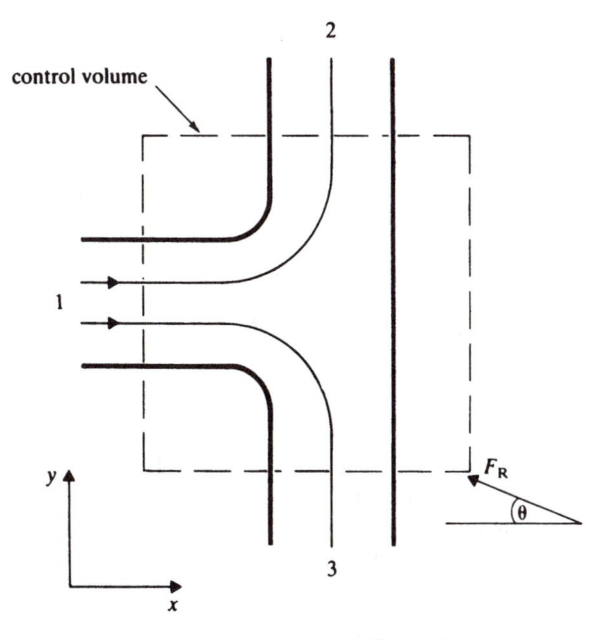

Figure 2.7 Forces on a T-junction.

hence

$$p_2 = p_1 + \rho(V_1^2 - V_2^2)/2 = 499.53\,\text{kN/m}^2$$
$$p_3 = p_1 + \rho(V_1^2 - V_3^2)/2 = 490.38\,\text{kN/m}^3$$

Pressure forces

$$F_{Px} = p_1 A_1 - 0 = 79.52\,\text{kN}$$
$$F_{Py} = p_3 A_3 - p_2 A_2 = -19.96\,\text{kN}$$

Momentum forces

$$F_{Mx} = 0 - \rho Q_1 V_1 = -0.566\,\text{kN}$$
$$F_{My} = \rho Q_2 V_2 + (-\rho Q_3 V_3) - 0 = -0.40\,\text{kN}$$

Reaction forces

$$F_{Rx} = F_{Mx} - F_{Px} = -80.09\,\text{kN}$$
$$F_{Ry} = F_{My} - F_{Py} = +19.50\,\text{kN}$$

hence

$$F_R = \sqrt{F_{Rx}^2 + F_{Ry}^2} = 82.43\,\text{kN}$$

and

$$\theta = \tan^{-1}(F_{Ry}/F_{Ry})$$
$$= 13.7^\circ \text{ (from negative } x\text{-direction to positive } y\text{-direction)}$$

2.8 Velocity and discharge measurement

Velocity measurement

The pitot tube, shown diagramatically in Figure 2.8(a), is used to measure velocity. At the nose of the pitot tube, the fluid is brought to rest and the height of the fluid in the pitot tube therefore corresponds to $p/\rho g + u^2/2g$. This is known as the stagnation pressure. The pressure head $(p/\rho g)$ is measured separately by a second tube, and hence

$$\frac{p}{\rho g} + \frac{u^2}{2g} = \frac{p}{\rho g} + h$$

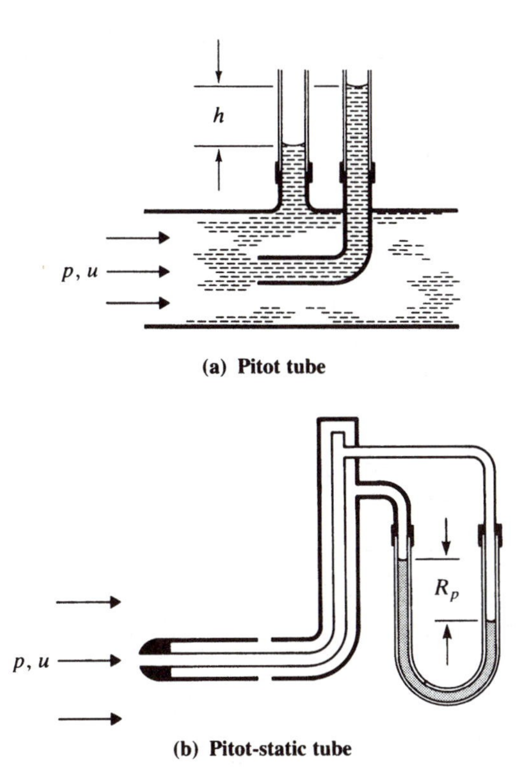

Figure 2.8 Velocity measurement.

or

$$u = \sqrt{2gh} \tag{2.10}$$

The two pressure heads are normally measured by a single integrated instrument called the pitot-static tube, as shown in Figure 2.8(b). This instrument, when connected to a suitable manometer, may be used to measure point velocities in pipes, channels and wind tunnels.

Discharge measurement in pipelines

The venturi meter, shown in Figure 2.9(a), is an instrument which may be used to measure discharge in pipelines. It essentially consists of a narrowed section tapering out to the pipe diameter at each end. In the throat section, the velocity is increased, and consequently the pressure is decreased. By measuring the difference in pressure, an estimate of discharge may be made as follows.

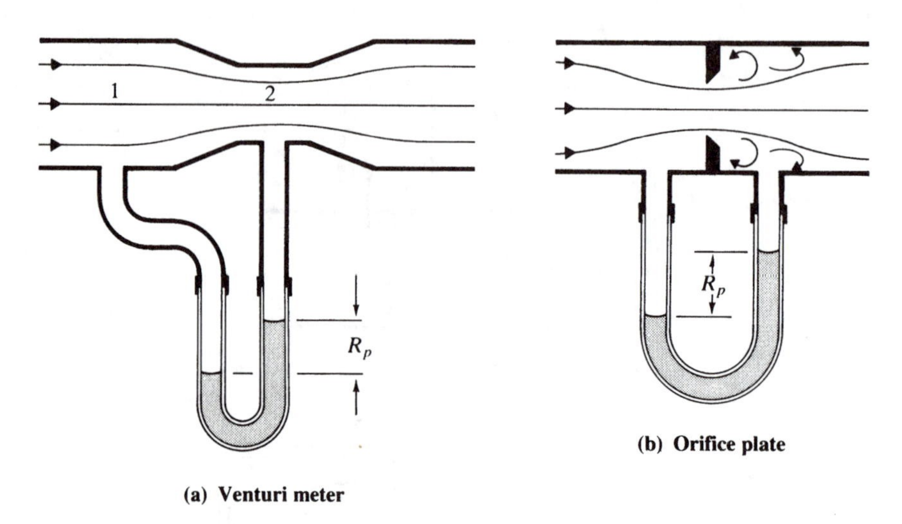

Figure 2.9 Discharge measurement.

Consider a streamline from the upstream position (1) to the throat position (2). Then

$$\frac{p_1}{\rho g} + \frac{V_1^2}{2g} = \frac{p_2}{\rho g} + \frac{V_2^2}{2g}$$

(assuming no energy losses and using mean velocities), or

$$V_2^2 - V_1^2 = 2g\left(\frac{p_1}{\rho g} - \frac{p_2}{\rho g}\right)$$

The pressure head difference is indicated by the manometer reading, R_p, hence

$$V_2^2 - V_1^2 = 2gR_p\left(\frac{\rho_g}{\rho} - 1\right)$$

where ρ_g is the density of the gauge fluid. Also, by continuity,

$$Q = V_1 A_1 = V_2 A_2$$

Substituting for V_2 in terms of V_1 yields

$$V_1^2[(A_1/A_2)^2 - 1] = 2gh^*$$

where

$$h^* = R_p\left(\frac{\rho_g}{\rho} - 1\right)$$

Solving for V_1,

$$V_1 = \frac{1}{\sqrt{(A_1/A_2)^2 - 1}}\sqrt{2gh^*}$$

hence

$$Q_{\text{ideal}} = A_1 V_1 = \frac{A_1}{\sqrt{(A_1/A_2)^2 - 1}}\sqrt{2gh^*}$$

Taking $m = D_1/D_2$,

$$Q_{\text{ideal}} = \frac{\pi D_1^2}{4}\left(\frac{1}{\sqrt{m^4 - 1}}\right)\sqrt{2gh^*} \tag{2.11}$$

The actual discharge will be slightly less than this, due to energy losses in the converging section. These energy losses are accounted for by introducing a coefficient of discharge C_d such that $Q_{\text{actual}} = C_d Q_{\text{ideal}}$. Energy losses may be minimized by careful design, as evidenced in British Standard 1042, Part I Section 1.1 (1981) or the *Water measurement manual* (USBR, 1967). Venturi meters designed in accordance with these standards have C_d values between 0.97 and 0.99.

It is useful to note that the indicated head (R_p), and consequently the discharge equation, is independent of the inclination of the meter. You may care to prove this for yourself, by deriving the expression for h^* for an inclined venturi meter.

A second device (shown in Fig. 2.9(b)) for measuring discharge in a pipeline is an orifice plate. Its function is similar to the venturi meter, in that a region of low pressure is created by a local constriction. The same discharge equation therefore applies. However, the major difference between the devices lies in the fact that, downstream of the orifice plate, the flow area expands instantaneously while the fluid is unable to expand at the same rate. This creates a 'separation zone' of turbulent eddies in which large energy losses occur. Consequently, the coefficient of discharge is considerably lower than that for the venturi meter (typically about 0.65). The advantages of the orifice plate is its lower cost and its compactness. So long as the increased energy losses are acceptable, it may be used instead of a venturi meter.

Discharge through a small orifice

Figure 2.10 shows a jet of water issuing from a large tank through a small orifice. At a small distance from the tank, the streamlines are straight and

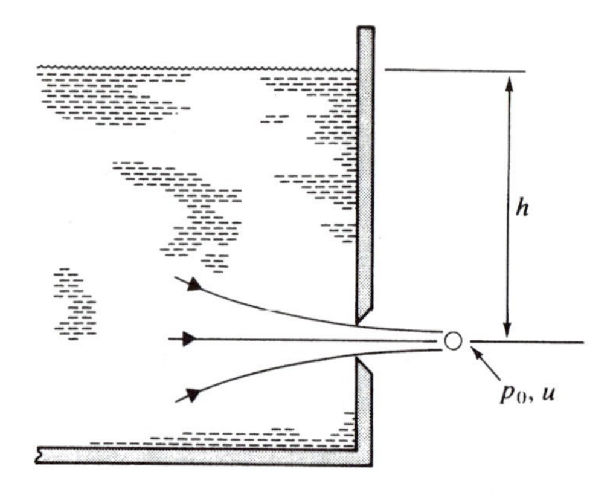

Figure 2.10 Discharge through a small orifice.

parallel, and the pressure is atmospheric. Application of Bernoulli's equation between this point and the water surface in the tank yields

$$\frac{p_0}{\rho g} + h = \frac{p_0}{\rho g} + \frac{u^2}{2g}$$

or

$$u = \sqrt{2gh}$$

This result is usually attributed to Torricelli, who demonstrated its validity experimentally in 1643. The discharge may be calculated by applying the continuity equation

$$Q = uA$$

or

$$Q_{\text{ideal}} = A\sqrt{2gh} \tag{2.12}$$

where A is the area of the jet.

The area of the jet, however, is smaller than that of the orifice, due to the convergence of the streamlines, as is shown in Figure 2.10. The contraction of the jet is called the vena contracta. Experiments have shown that the jet area (A) and orifice area (A_0) are related by

$$A = C_c A_0$$

where C_c is the coefficient of contraction, and is normally in the range 0.61–0.66.

In addition, energy losses are incurred at the orifice, and therefore a second coefficient C_v, the velocity coefficient, is introduced to account for these. C_v is normally in the range 0.97–0.99. Hence, the true discharge may be written as

$$Q_{\text{actual}} = C_v C_c A_0 \sqrt{2gh}$$

or

$$Q_{\text{actual}} = C_d A_0 \sqrt{2gh} \tag{2.13}$$

where C_d is the overall coefficient of discharge.

Discharge through a large orifice

If the orifice is large, then the small orifice equation will no longer be accurate. This is a result of the significant variation of h (and therefore u) across the orifice. In such cases, it is necessary to find the discharge as follows (see Fig. 2.11).

For any elemental strip of thickness δy, across a rectangular orifice,

$$\delta Q = b\,\delta y\sqrt{2gy}$$

and

$$\begin{aligned} Q &= b\sqrt{2g}\int_{y=h_1}^{y=h_2} \sqrt{y}\,\mathrm{d}y \\ &= b\sqrt{2g}\left(\frac{2}{3}y^{3/2}\right)_{h_1}^{h_2} \\ &= \frac{2}{3}b\sqrt{2g}(h_2^{3/2} - h_1^{3/2}) \end{aligned} \tag{2.14}$$

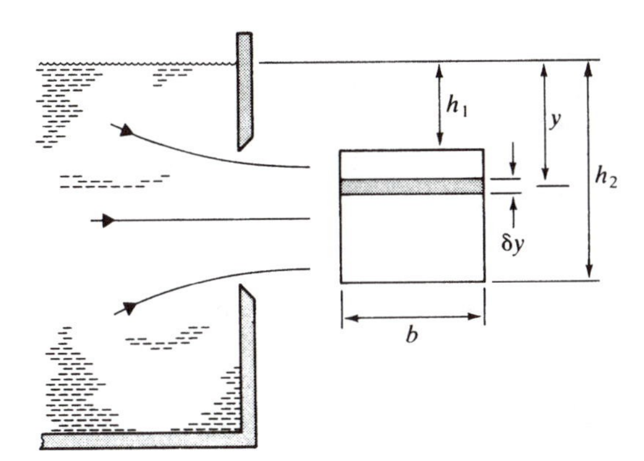

Figure 2.11 Discharge through a large orifice.

If h_1 tends to zero, the upper edge of the orifice no longer influences the flow. This corresponds to a thin plate weir. Consequently, (2.14) forms the basis for a discharge formula of thin plate weirs. This is developed more fully in Chapter 13.

2.9 Potential flows

In the problems studied so far, each flow has been considered as a continuum. However, a great deal can be learned by investigating the behaviour of streamlines in a flow. The laws governing their behaviour were investigated by the school of pure mathematicians (the 'hydrodynamicists') that studied flow problems. In order to develop mathematical relationships, the hydrodynamicists developed the concept of an 'ideal' fluid, the characteristics of which were outlined at the beginning of the chapter.

It has already been pointed out that streamlines are everywhere tangential to the flow, so that no flow ever crosses a streamline. Flows which can be represented by streamlines are often known as 'potential flows' or 'ideal flows'.

If the flow vectors are known (or can be guessed) for a number of successive points in a flow, then the streamline may be drawn. Therefore, the streamline gives a graphical representation of the flow. A series of streamlines may be used to construct a 'map' of the flow pattern, though it will emerge that there are mathematical conditions which govern the validity of such a map. For certain flow patterns involving real fluids, these conditions are not fulfilled. It has already been stated that any solid boundary (e.g. a pipe wall) which encloses a flow may be regarded as a surface covered by a continuous series of adjacent streamlines (i.e. a 'stream surface' or streamtube).

Properties of streamlines

The stream function. Let two streamlines, 1 and 2 (Fig. 2.12), enclose a two-dimensional flow. The streamlines are not necessarily parallel or straight. It follows, from the definition of a streamline, that the discharge crossing boundary AB must equal the discharge crossing CD. Another approach to this statement is to argue that if point A is taken as an arbitrary 'zero', there is a certain 'increment' of discharge, δQ, between A and B. If point C is now taken as a 'zero' position, then the increment of discharge from C to D must be the same as that between A and B. Therefore any streamline may be assigned a numerical value corresponding to the increment of discharge between that streamline and some arbitrary 'zero' streamline. This number is called the stream function, Ψ. The concept of a stream function may be expressed algebraically as follows.

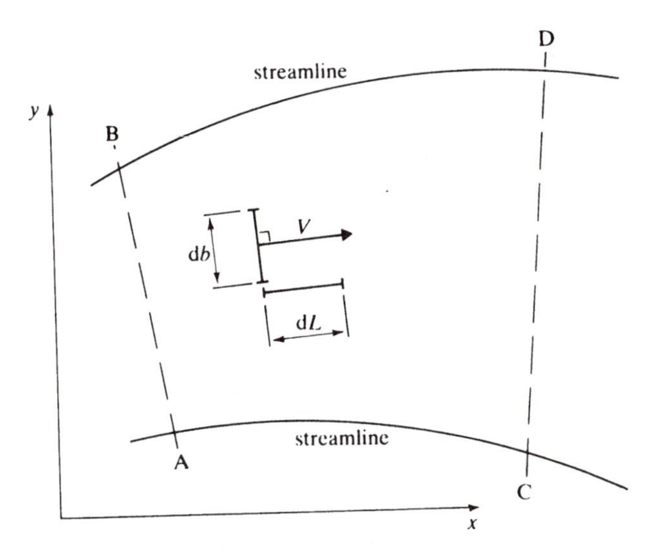

Figure 2.12 Two-dimensional flow.

Let the velocity of flow at any point be V (Fig. 2.12). Then for unit depth of flow (depth is assumed to lie in the z-direction in Cartesian co-ordinates) the increment of discharge between streamline 1 and streamline 2 is

$$\Psi = \int_1^2 d\Psi = \int_A^B V\,db = \int_C^D V\,db = \text{constant} \tag{2.15}$$

For convenience, in algebraic manipulations it is usual to use a sign convention as well as a magnitude to define Ψ. The convention adopted here is that if an observer faces downstream (i.e. with the flow moving away from the observer in the direction of sight), Ψ will increase positively to the left (see Fig. 2.13). Since Ψ is an arbitrary function of position, the position of the reference or zero streamline is also arbitrary, so it is perfectly possible to have positive and negative values of Ψ. The negative sign does not have any special significance (it does not indicate the existence of a 'negative flow', whatever that might be) and should be interpreted purely graphically.

The velocity potential function. Another function can also be used to characterize a flow. This is known as the 'potential function' or 'velocity potential', which is defined by the equation $d\phi/dL = V$. Hence,

$$\phi = \int V\,dL \tag{2.16}$$

Conventionally, ϕ increases in the direction of flow. Lines of constant ϕ are known as 'equipotential' lines, and are orthogonal to streamlines. A diagram which represents a flow in terms of its lines of Ψ and ϕ is a 'flow

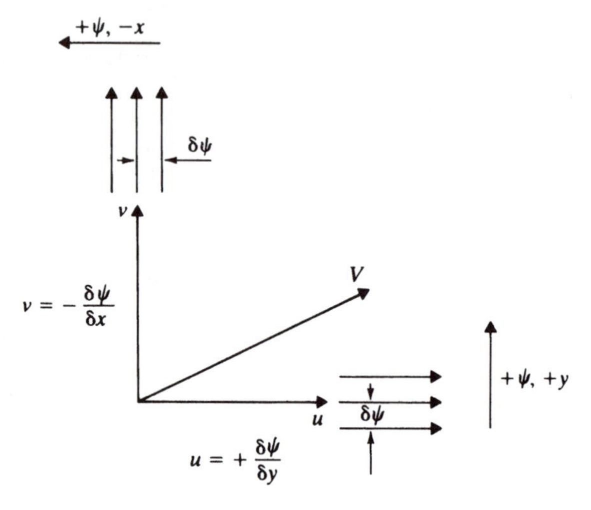

Figure 2.13 Components of flow in Cartesian co-ordinates.

net'. An accelerating flow (V increasing) is indicated by convergence of the streamlines and equipotential lines, whereas a decelerating flow will be represented by diverging streamlines and equipotential lines.

Conditions for the validity of a flow net

In order to preserve the condition of orthogonality of the lines representing the stream and potential functions, the flow must be 'irrotational', i.e. there must be no net rotation of a given element of fluid (Fig. 2.14). This may be expressed algebraically as follows.

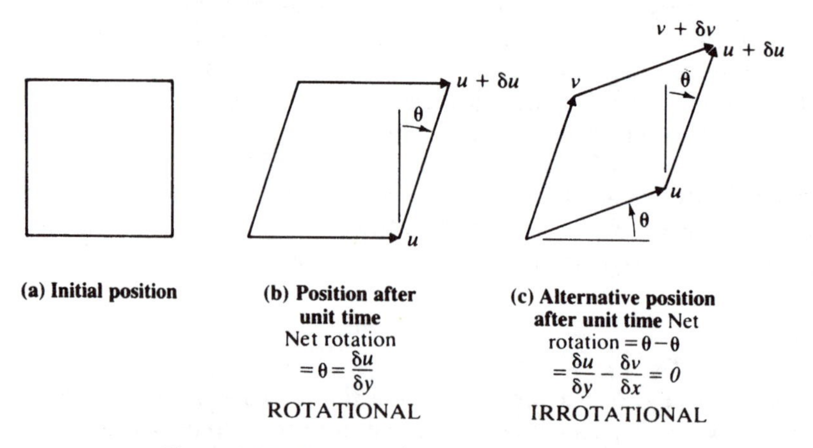

Figure 2.14 Rotational and irrotational flows.

The velocity, V, is resolved into its x- and y-components, u and v, respectively. If there is a velocity u at ordinate y, and a velocity $(u+\delta u)$ at $(y+\delta y)$, then the net rotation of an element (Fig. 2.14(b)) may be defined as $\delta u/\delta y$.

However, for an element with velocities u (at y), $(u+\delta u)$ (at $y+\delta y$) and v (at x), $v+\delta v$ (at $x+\delta x$), the net rotation is $\delta u/\delta y - \delta v/\delta x$ (see Fig. 2.14(c)). In an irrotational flow, and for the limit $\delta x \to 0, \delta y \to 0$,

$$\frac{\partial u}{\partial y} - \frac{\partial v}{\partial x} = 0 \tag{2.17}$$

Stream and potential functions in Cartesian co-ordinates

It is often useful to express the two flow functions in terms of their Cartesian co-ordinates, especially for some of the algebraic manipulations which arise (Fig. 2.13). The x-direction component of V is u, so from the definition of Ψ

$$u = \frac{\partial \Psi}{\partial y} \tag{2.18}$$

Similarly for the y-component,

$$v = -\frac{\partial \Psi}{\partial x} \tag{2.19}$$

By a similar process of reasoning,

$$u = \frac{\partial \phi}{\partial x} \quad v = \frac{\partial \phi}{\partial y} \tag{2.20}$$

If these equations are substituted into (2.17),

$$\frac{\partial u}{\partial y} = \frac{\partial \phi}{\partial x \partial y} \quad \text{and} \quad \frac{\partial v}{\partial x} = \frac{\partial \phi}{\partial x \partial y}$$

Therefore

$$\frac{\partial u}{\partial y} - \frac{\partial v}{\partial x} = \frac{\partial \phi}{\partial x \partial y} - \frac{\partial \phi}{\partial x \partial y} = 0$$

which demonstrates that a potential flow satisfies the condition of irrotationality. An alternative way of making the same statement is as follows:

$$u = \frac{\partial \Psi}{\partial y} \quad \text{so} \quad \frac{\partial u}{\partial y} = \frac{\partial^2 \Psi}{\partial y^2}$$

$$v = -\frac{\partial \Psi}{\partial x} \quad \text{so} \quad \frac{\partial v}{\partial x} = -\frac{\partial^2 \Psi}{\partial x^2}$$

Therefore

$$\frac{\partial u}{\partial y} - \frac{\partial v}{\partial x} = \frac{\partial^2 \Psi}{\partial y^2} + \frac{\partial^2 \Psi}{\partial x^2} = 0$$

This is the Laplace equation, which must be satisfied if the flow is to be a potential flow.

2.10 Some typical flow patterns

Uniform rectilinear flow

For the case of a positive uniform flow parallel to the x-axis (Fig. 2.15) (velocity $= u =$ constant), $u = \mathrm{d}\Psi/\mathrm{d}y$ (from (2.18)). Therefore

$$\Psi = \int \mathrm{d}\Psi = \int u\, \mathrm{d}y = uy \tag{2.21}$$

(if $\Psi = 0$ when $y = 0$, then the constant of integration $= 0$). Similarly, for a positive uniform flow parallel to the y-axis (from (2.19)),

$$\Psi = \int -v\, \mathrm{d}x = -vx \tag{2.22}$$

For a uniform flow with velocity V and parallel to neither axis,

$$V = \frac{\mathrm{d}\Psi}{\mathrm{d}b}$$

therefore

$$\delta\Psi = V\, \delta b$$

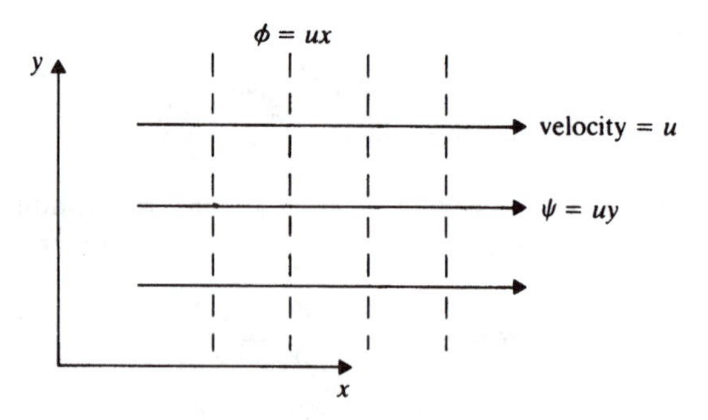

Figure 2.15 Rectilinear flow.

This may be expressed in terms of the x- and y-components:

$$\delta\Psi = V\,\delta b = u\,\delta y - v\,\delta x$$

Therefore

$$\Psi = uy - vx \tag{2.23}$$

Example 2.6 Drawing streamlines

Draw the streamlines for the following flows (up to a maximum Ψ value of 20):

(a) a uniform flow of 5 m/s parallel to the x-axis positive direction;
(b) a uniform flow of 5 m/s parallel to the y-axis negative direction;
(c) the flow resulting from the combination of (a) and (b).

Solution

(a) For a flow of 5 m/s the stream function is

$$\Psi_u = uy = +5y$$

The streamlines are drawn in Figure 2.16.

(b) For a flow of 5 m/s in the negative y-direction the stream function is

$$\Psi_v = -(-vx) = +5x$$

These streamlines are also shown in Figure 2.16.

(c) The combined stream function is

$$\Psi_c = 5y + 5x$$

This pattern of streamlines is obtained by simple graphical addition. Thus at the junction between the streamlines $\Psi_u = 5$ and $\Psi_v = 5$, the total is $\Psi_c = 5 + 5 = +10$, and so on. Points having the same total are then joined to form the streamline pattern for the total flow.

Radial flows

Radial flows may be flows whose velocity vectors point away from the centre ('source') or towards the centre ('sink') (see Fig. 2.17). For unit depth of flow, the discharge between any pair of radial lines, subtending angle θ at the centre, must be equal to the product of radial velocity and cross-sectional area between those lines, i.e.

$$\delta Q = V_r r\,\delta\theta$$

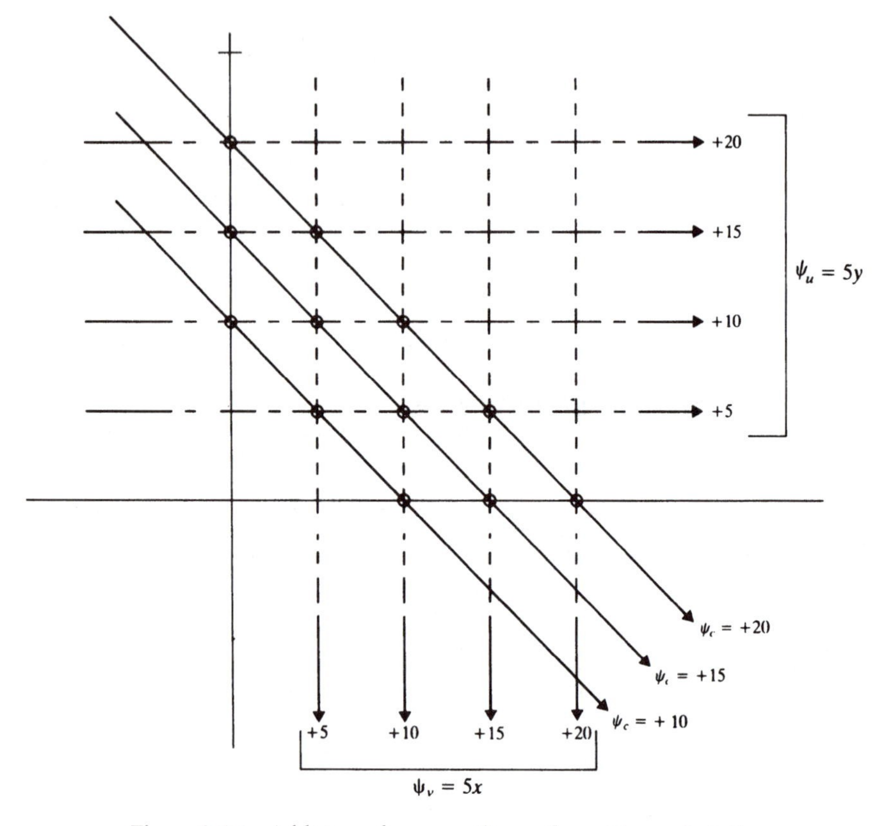

Figure 2.16 Addition of two rectilinear flows (Example 2.6).

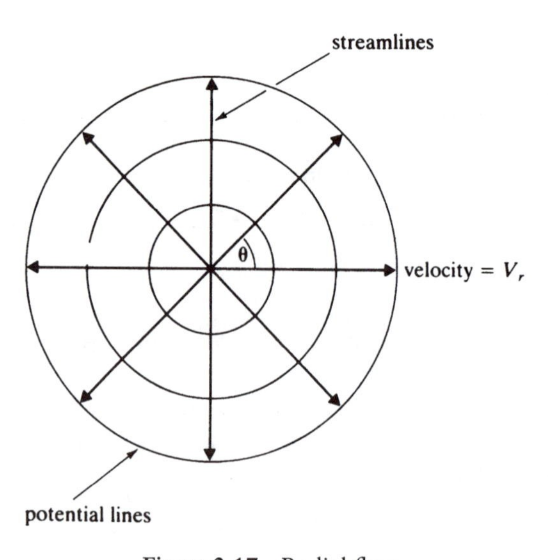

Figure 2.17 Radial flow.

for a source. Therefore, from the definition of Ψ,

$$\Psi = V_r r \theta \tag{2.24}$$

If the total discharge from the source is $Q_s = V_r r 2\pi$, then

$$\Psi = Q_s \theta / 2\pi \tag{2.25}$$

Since a sink is simply a 'negative source', the stream function for a sink is

$$\Psi = -Q_s \theta / 2\pi \tag{2.26}$$

Since potential lines are orthogonal to streamlines, it follows that an equipotential is in the form of a circle. From (2.24) and (2.25), $V_r = Q_s/2\pi r$ for a source. Therefore

$$\begin{aligned} \Phi &= \int V \mathrm{d}L = \int V_r \mathrm{d}r = \int \frac{Q_s}{2\pi} \frac{\mathrm{d}r}{r} \\ &= \frac{Q_s}{2\pi} \ln\left(\frac{r}{c}\right) \end{aligned} \tag{2.27}$$

Flows in a curved path

A streamline may be curved in any arbitrary fashion, depending on the pattern of forces to which it is being subjected. Therefore, from Newton's Second Law, if a flow is following a curved path, then there must be some lateral force acting upon the fluid. An equation to express this is now developed.

Consider an element of fluid constrained to travel in a path which is curved as shown in Figure 2.18. The mean velocity of the element is $(V + \frac{1}{2}\delta V)$ and its mass is $\rho a \delta r$. The radial force exerted by the element on its surrounds is given by

$$\text{mass} \times \frac{(\text{velocity})^2}{\text{radius}} = \rho a \delta r \frac{(V + \frac{1}{2}\delta V)^2}{(r + \frac{1}{2}\delta r)}$$

For the element to be in equilibrium, the surrounding fluid must exert a reaction force on the element. This reaction is the result of an increase in pressure with radius. Therefore, equating forces,

$$\delta p a = \rho a \delta r \frac{(V + \frac{1}{2}\delta V)^2}{(r + \frac{1}{2}\delta r)}$$

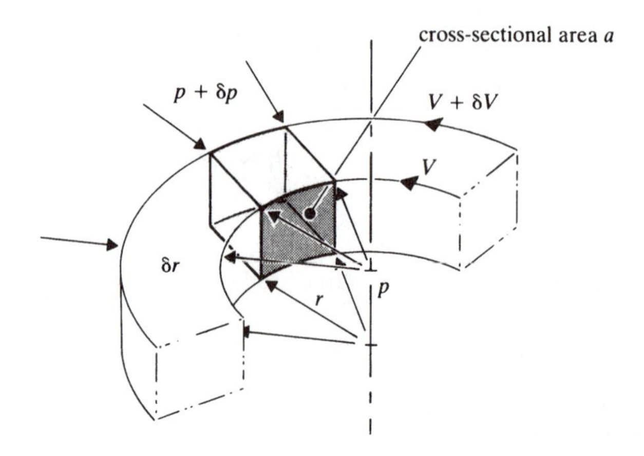

Figure 2.18 Flow in a curved path.

So, if $\delta V \ll V$ and $\delta r \ll r$, we obtain

$$\frac{\delta p}{\delta r} = \frac{\rho V^2}{r} \tag{2.28}$$

The Bernoulli equation may be applied along any streamline. If the streamlines are assumed to lie in a horizontal plane, then the potential energy is constant so the 'z' term may be dropped, and the Bernoulli equation becomes

$$\frac{V^2}{2g} + \frac{p}{\rho g} = \text{constant} = H$$

The rate of change of head between one streamline and another is therefore

$$\frac{\delta H}{\delta r} = \left(\frac{(V + \delta V)^2 - V^2}{2g\,\delta r} + \frac{(p + \delta p) - p}{\rho g\,\delta r} \right)$$

If products of small quantities are ignored,

$$\frac{\delta H}{\delta r} = \frac{2V\,\delta V}{2g\,\delta r} + \frac{\delta p}{\rho g\,\delta r}$$

Substituting for $\delta p/\delta r$ from (2.28),

$$\frac{\delta H}{\delta r} = \frac{V\,\delta V}{g\,\delta r} + \frac{V^2}{gr} = \frac{V}{g}\left(\frac{\delta V}{\delta r} + \frac{V}{r} \right)$$

or, in the limit ($\delta r \to 0$),

$$\frac{dH}{dr} = \frac{V}{g}\left(\frac{dV}{dr} + \frac{V}{r}\right) \tag{2.29}$$

Vortices

Equation (2.29) is a completely general equation for any flow with circumferential streamlines. Two cases of such flows are of particular interest. They are both members of the same 'family' of flows, which are given the general title of vortices. The two specific cases examined here are the free vortex and the forced vortex.

Forced vortex. A forced vortex is a circular motion approximating to the pattern generated by the action of a mechanical rotor on a fluid. The rotor 'forces' the fluid to rotate at uniform rotational speed ω rad/s, so $V = \omega r$. Therefore,

$$\frac{dV}{dr} = \frac{V}{r} = \omega$$

Substitution in (2.29) yields

$$\frac{dH}{dr} = \frac{\omega r}{g}(\omega + \omega) = \frac{2\omega^2 r}{g} \tag{2.30}$$

A forced vortex is a rotational flow, so it cannot be represented by a flow net.

Free vortex. A free vortex approximates to naturally occurring circular flows (e.g. the circumferential component of the flow down a drain hole or around a river bend) in which there is no external source of energy. It follows that there can be no difference of total head between one streamline and another, i.e.

$$\frac{dH}{dr} = 0 = \frac{V}{g}\left(\frac{dV}{dr} + \frac{V}{r}\right)$$

so

$$\frac{dV}{dr} = -\frac{V}{r}$$

This equation is satisfied by a flow such that

$$V = \frac{\text{constant}}{r} = \frac{K}{r}$$

since

$$\frac{\mathrm{d}V}{\mathrm{d}r} = -\frac{K}{r^2} = -\frac{V}{r}$$

A free vortex is irrotational. This may seem surprising, nevertheless the absence of a mechanical rotor does mean that for any fluid element at radius r the net rotation is zero, as defined by Figure 2.13. The element at the centre is an apparent exception to this (and is known as a 'singular point'), but as the statement $r \to 0$ also implies that $V \to \infty$ it is of no practical significance. The free vortex therefore conforms to the conditions for construction of a flow net. Since the flow is in a circumferential path, $\mathrm{d}\psi/\mathrm{d}r = V$. Therefore

$$\psi = \int V\,\mathrm{d}r = \int \frac{K}{r}\,\mathrm{d}r = K\ln\left(\frac{r}{c}\right) \tag{2.31}$$

As the potential lines are orthogonal to the streamlines, a line of constant potential must be a radial line. From the definition of ϕ,

$$\phi = \int \mathrm{d}\phi = \int V\,\mathrm{d}L = Vr\theta = K\theta \tag{2.32}$$

(The reader may like to compare the ψ and ϕ functions for a source or sink with those for the free vortex.)

Circulation and vorticity

The study of vortices forms a useful starting point for developing the concepts of circulation and vorticity (Fig. 2.19). Circulation is simply the

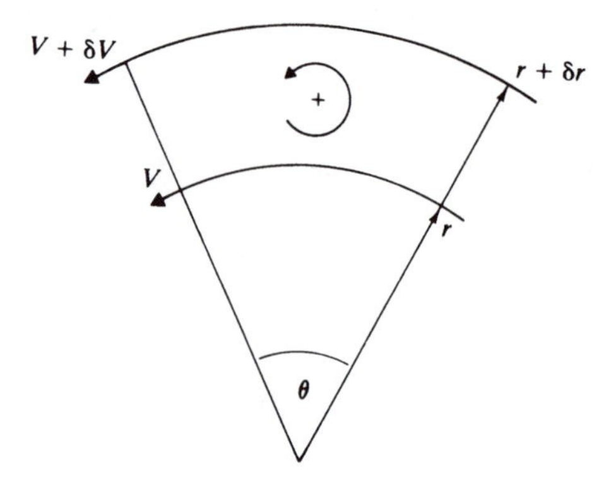

Figure 2.19 Definition diagram for vorticity and circulation.

product (velocity × length) around a circuit. For the element of fluid shown in the figure, the circulation Γ (assuming positive flows are anticlockwise) is

$$\Gamma = (V + \delta V)(r + \delta r)\theta - Vr\theta = V\,\delta r\,\theta + r\,\delta V\,\theta$$

The flow is assumed to be in a circular arc, so velocities in the radial direction are zero.

The vorticity ζ is simply the intensity of circulation, i.e.

$$\frac{\Gamma}{\text{area}} = \zeta = \frac{V\,\delta r\theta + r\,\delta V\,\theta}{r\theta\,\delta r} = \left(\frac{V}{r} + \frac{\delta V}{\delta r}\right) \tag{2.33}$$

which should be compared with (2.29).

Example 2.7 Vortex flow

A cylinder of diameter 350 mm is filled with water and is rotated at 130 rad/s. Assuming that there is zero gauge pressure on the centreline at the top of the cylinder, estimate:

(a) the pressure on the side wall of the cylinder;
(b) the loading on the upper end plate due to the pressure.

Solution

(a) As the cylinder is rotated mechanically, the flow induced in the fluid will be a forced vortex. Equation (2.30) is therefore appropriate:

$$\frac{\mathrm{d}H}{\mathrm{d}r} = \frac{2\omega^2 r}{g}$$

This is rearranged as

$$\int \mathrm{d}H = \int \frac{2\omega^2 r}{g}\,\mathrm{d}r$$

and hence

$$H = \frac{\omega^2 r^2}{g} + C$$

Given that pressure is zero at the centre, and knowing that velocity is also zero ($\omega r \to 0$ when $r \to 0$), the constant C must be zero.

Therefore, at $r = 0.175$ m,

$$H = \omega^2 r^2 / g$$

But

$$H = \frac{p}{\rho g} + \frac{V^2}{2g}$$

or, since $V = \omega r$,

$$H = \frac{p}{\rho g} + \frac{\omega^2 r^2}{2g}$$

Therefore

$$p = \frac{1}{2}\rho\omega^2 r^2 = \frac{1}{2} \times 1000 \times 130^2 \times 0.175^2$$
$$= 258.8\,\text{kN/m}^2$$

(b) At any radius r within the cylinder,

$$p = \frac{1}{2}\rho\omega^2 r^2$$

The force δF on a circular annulus of radius r and thickness δr is

$$\delta F = \frac{1}{2}\rho\omega^2 r^2 2\pi r\ \delta r = \pi\rho\omega^2 r^3 \delta r$$

Therefore

$$F = \int_0^{0.175} \mathrm{d}F = \int_0^{0.175} \pi\rho\omega^2 r^3\, \mathrm{d}r = \left[\frac{\pi\rho\omega^2 r^4}{4}\right]_0^{0.175}$$
$$= \frac{\pi \times 1000 \times 130^2 \times 0.175^4}{4} = 12.45\,\text{kN}$$

Note that hydrostatic forces have been ignored in this example, since they would be small compared with the forces due to the vortex.

Combinations of flow patterns

The real value and power of potential flow methods can only be appreciated when the various individual flow patterns are combined (or superposed). A large number of combined patterns have been developed, closely approximating to real flows. Some idea of the wide range of patterns available may be obtained by consulting Milne-Thompson (1968). There is space here only for a brief introduction.

It is possible to combine flows graphically or algebraically. An example of the first method has already been presented (Example 2.1) and another follows. For the algebraic approach it is sometimes best to apply polar, rather than Cartesian, co-ordinates.

Example 2.8 Streamlines

A linear flow, having a uniform negative velocity of 60 m/s parallel to the x-axis, is combined with a source with a discharge of 160 m^3/s. Draw the streamlines for the combined flow and locate the stagnation point.

Solution

The stream function for the linear flow is $-60y$, and that for the source is $+160\,\theta/2\pi$. The streamlines for the two flows are shown in Figure 2.20. The combined streamlines are obtained by using graphical addition, as outlined in Example 2.6.

A stagnation point (a point where the fluid comes to rest) occurs where the velocity due to the source is equal and opposite to the linear flow, i.e. where $V_r = +60$ m/s. So, using (2.24) and (2.25),

$$V_r r\,\theta = \frac{Q_S\theta}{2\pi} = \Psi$$

Hence $V_r = Q_S/2\pi r$, and for the stagnation point $60 = 160/2\pi r$, so $r = 0.424$ m. The velocities are equal and opposite only along the axis $y = 0$ (see Fig. 2.20).

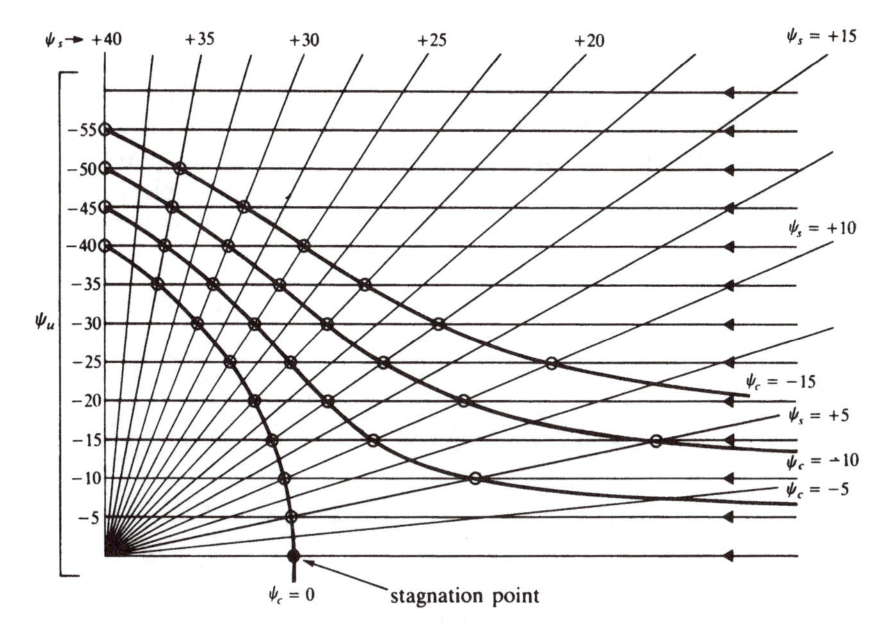

Figure 2.20 Streamline pattern – source, plus linear flow (Example 2.8). (*Note*: Heavy lines represent Ψ_c.)

The streamlines for the combined flow divide about the line $\Psi_c = 0$. This line can be treated as the external surface of a solid body, all the streamlines with a negative Ψ_c then represent the flow around such a shape.

Other important examples of combined flows are the following:

(a) Source plus sink plus linear flow. A source with discharge Q_S has its centre at $(-k, 0)$, and a sink with discharge $-Q_S$ has its centre at $(+k, 0)$. These are combined with a linear flow u parallel to the x-axis (Fig. 2.21). The streamline $\Psi_c = 0$ is oval in shape, and the streamlines with negative Ψ_c represent the flow around such a shape.

(b) If the combination of flows is as in (a) but $k \to 0$, the streamline $\Psi_c = 0$ becomes a circle. (Note that there are mathematical conditions for this, e.g. as $k \to 0$, $Q_S \times 2k$ must remain finite, otherwise the source and sink simply cancel each other out – see Milne-Thompson, 1968.)

(c) For a combination of flows as in (b), but with the addition of a free vortex surrounding the circular streamline $\Psi_c = 0$, the addition of the vortex displaces the position of the stagnation points. If the streamlines are plotted, it is found that the velocity pattern is symmetrical about the y-axis, but asymmetrical about the x-axis. This implies an asymmetrical pressure distribution, which in turn implies that there is a component of force perpendicular to the direction of the linear flow. This is known as a 'lift force'. This flow pattern leads to some important equations which are used in aerodynamics.

Applications of a number of the flow patterns mentioned in this section will be found scattered through the remainder of the book. Typical examples are flows around bends in siphon spillways or rivers (which approximate to free vortex flows), and flows around streamlined bridge piers (which approximate to the combination source + sink + linear flow mentioned above).

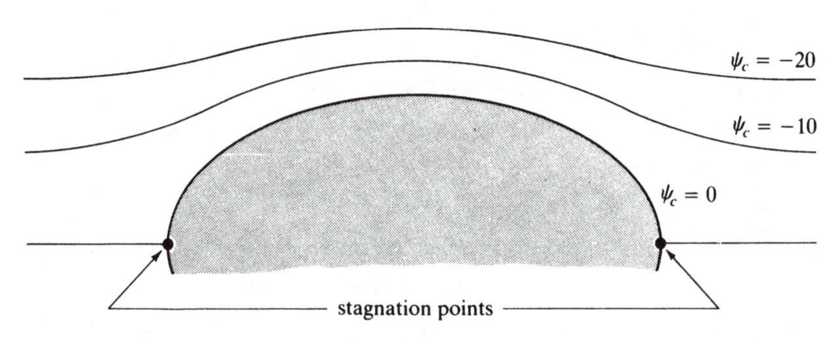

Figure 2.21 Combination of source, sink and linear flow.

References and further reading

British Standards Institution (1981) BS1042 Part 1 Sec 1.1 (1981). *Orifice plates, nozzles and venturi tubes inserted in circular cross section conduits running full.* BS1042. Part 1, Section 1.1. London: BSI.

Massey, B. S. (revised J. Ward-Smith) (1998) *Mechanics of Fluids*, 7th edn, Stanley Thornes, Cheltenham.

Milne-Thompson, L. M (1968) *Theoretical Hydrodynamics*, 5th edn, Macmillan, London.

US Bureau of Reclamation (1967) *Water Measurement Manual*, 2nd edn, US Department of the Interior, Washington, DC.

3

Behaviour of real fluids

3.1 Real and ideal fluids

In Chapter 2, equations describing fluid motion were developed. The equations were mathematically straightforward because the fluid was assumed to be ideal, i.e. it possessed the following characteristics:

1. it was inviscid;
2. it was incompressible;
3. it had no surface tension;
4. it always formed a continuum.

From a civil engineering standpoint, surface tension problems are encountered only under rather special circumstances (e.g. very low flows over weirs or in hydraulic models). Compressibility phenomena are particularly associated with the high speed gas flows which occur in chemical engineering or aerodynamics. The civil engineer may occasionally need to consider one particular case of compressibility associated with surge, and this will be investigated in Chapters 6 and 12.

In this chapter, it is the effect of viscosity which will dominate the discussion, since it is this characteristic which differentiates so many aspects of real flows from ideal flows. This difference is exemplified if the ideal and real flows around a bluff (i.e. non-streamlined) shape are compared (Fig. 3.1). The ideal fluid flows smoothly around the body with no loss of energy between the upstream and downstream sides of the body. The pressure distribution around the body is therefore symmetrical, and may be obtained by applying Bernoulli's equation. The real fluid flow approximates to the ideal flow only around the upstream portion of the body, where the streamlines are converging. Shortly after the streamlines start to diverge (as the fluid starts to negotiate the downstream part of the body), the streamlines fail to conform to the symmetrical pattern. A region of strongly eddying flow occurs (Fig. 3.2), in which there are substantial energy losses.

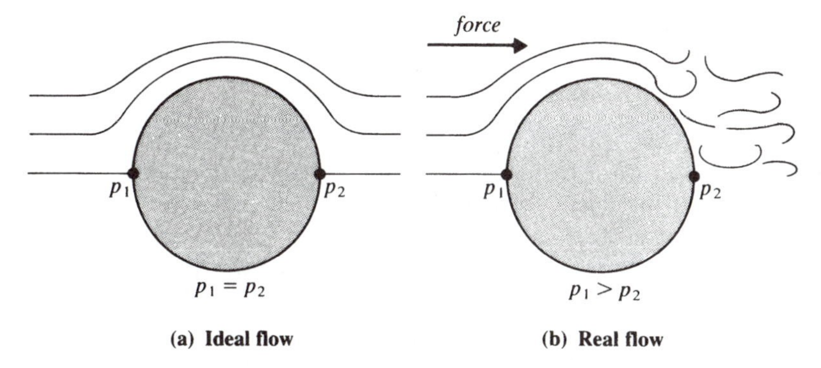

Figure 3.1 Ideal and real flow around a cylinder.

Figure 3.2 Flow around a cylinder showing eddy formation. (Courtesy of Armfield Technical Education.)

Consequently, there is now an asymmetrical pressure distribution, which in turn implies that there is now a force acting on the body. An explanation for the behaviour of the real fluid will emerge as the nature of viscosity is considered.

3.2 Viscous flow

An approach to viscosity

Perhaps the best way to approach viscosity is to contrast the effect of shearing action on a real fluid with that on an ideal fluid. Imagine, for example, that the fluids are both lying sandwiched between a fixed solid surface on one side and a movable belt (initially stationary) on the other (Fig. 3.3). Turning first to the ideal fluid (Fig. 3.3(a)), if the belt is set in motion, experimental measurements will indicate:

(a) that the force required to move the belt is negligible;
(b) that the movement of the belt has no effect whatsoever on the ideal fluid, which therefore remains stationary.

By contrast, if the other belt is slowly started and comparable measurements are taken on the real fluid, it will be found that:

(a) A considerable force is required to maintain belt motion, even at slow speed.
(b) The whole body of fluid is deforming and continues to deform as long as belt motion continues. Closer investigation will reveal that the deformation pattern consists in the shearing, or sliding, of one layer of fluid over another. The layer of fluid immediately adjacent to the solid surface will adhere to that surface and, similarly, the layer adjacent to the belt adheres to the belt. Between the solid surface and the belt the fluid velocity is assumed to vary linearly as shown in Figure 3.3(c). It is characteristic of a viscous fluid that it will deform continuously under a shear force. It is emphasized that this discussion is based on the assumption that the velocities are low, and that the flow is two-dimensional only.

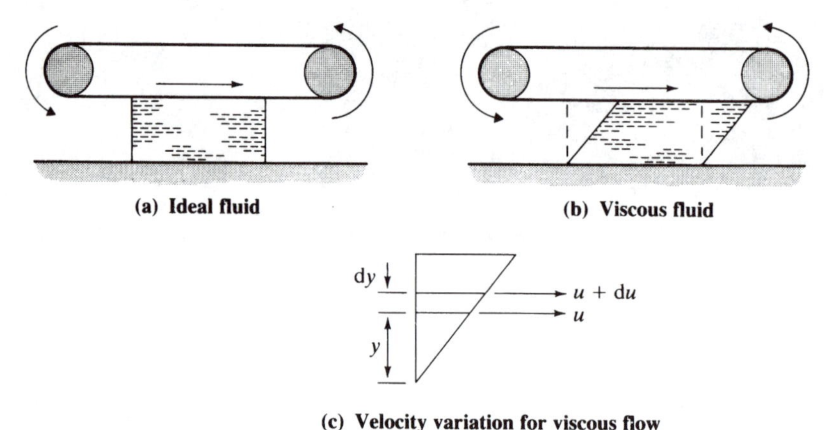

Figure 3.3 Effect of shear force on fluid.

A definition of viscosity

The pattern of events outlined in the previous section raises two questions. First, how must viscosity be defined? Secondly, based on this definition, can numerical values of viscosity be obtained for each fluid? Referring to the velocity diagram in Figure 3.3, it is evident that the shear force is being transmitted from one layer to another. If any pair of adjacent layers are taken, the lower one will have some velocity, u, whilst the upper one will be travelling with velocity $u + \mathrm{d}u$. The rate of shear strain is thus $\mathrm{d}u/\mathrm{d}y$. Newton postulated that the shear force applied and the rate of fluid shear were related by the equation

$$F = A \times \text{constant} \times (\mathrm{d}u/\mathrm{d}y)$$

where A is the area of the shear plane (i.e. the cross-sectional area of the fluid in the x–z plane). The equation may be rewritten as

$$\frac{F}{A} = \tau = \text{constant} \times \frac{\mathrm{d}u}{\mathrm{d}y} \tag{3.1}$$

(this is valid only for 'laminar' flows, see section 3.3). The constant is known as the absolute coefficient of viscosity of a fluid, and is given the symbol μ. Experiments have shown that:

1. μ is not a constant;
2. for the group of fluids known as 'Newtonian fluids', μ remains constant only at constant temperature – if the temperature rises, viscosity falls, and if the temperature falls, viscosity rises (air and water are Newtonian fluids);
3. there are some ('non-Newtonian') fluids in which μ is a function of both temperature and rate of shear.

Another form of the coefficient of viscosity is obtained if absolute viscosity is divided by fluid density to produce the coefficient of kinematic viscosity, ν:

$$\nu = \mu/\rho$$

Both coefficients have dimensions: μ is measured in $\mathrm{N\,s/m^2}$ (or kg/m s) and ν is in $\mathrm{m^2/s}$.

Example 3.1 Viscous flow

A moving belt system, such as that depicted in Figure 3.3, is to be used to transfer lubricant from a sump to the point of application. The working length of the belt is assumed to be straight and to be running parallel to a stationary metal surface.

The belt speed is 200 mm/s, the perpendicular distance between the belt and the metal surface is 5 mm, and the belt is 500 mm wide. The lubricant is an oil, having an absolute viscosity of $0.007\,\mathrm{N\,s/m^2}$. Estimate the force per unit length of belt, and the quantity of lubricant discharged per second.

Solution

$$\tau = \mu \frac{du}{dy} = 0.007 \times \frac{200 \times 10^{-3}}{5 \times 10^{-3}} = 0.28\,\mathrm{N/m^2}$$

$$\text{area } A \text{ of a 1 m length of belt} = 1 \times 0.5 = 0.5\,\mathrm{m^2}$$

$$\text{force per unit length of belt} = 0.28 \times 0.5 = 0.14\,\mathrm{N}$$

To estimate the discharge, note that the velocity is assumed to vary linearly with perpendicular distance from the belt. The discharge may therefore be obtained by taking the product of the mean velocity, V, and the area of the flow cross-section, by:

$$Q = Vby = \frac{200 \times 10^{-3}}{2} \times 0.5 \times 0.005 = 2.5 \times 10^{-4}\,\mathrm{m^3/s}$$

3.3 The stability of laminar flows and the onset of turbulence

Introduction

The flows examined so far have involved only low speeds (of the belt and fluid). However, what happens if, say, the speed of the belt, and therefore the rate of shear, is increased? The answer is that the pattern of linearly sheared flow will continue to exist only up to a certain belt velocity. Above that velocity a dramatic transformation takes place in the flow pattern. But why?

Effect of a disturbance in a sheared flow

Consider first a slowly moving sheared flow. What might be the effect of a small disturbance (due perhaps to a small local vibration) on the otherwise rectilinear flow pattern? The pathlines might be slightly deflected (Fig. 3.4(a)) bunching together more closely at A and opening out correspondingly at B. This implies that the local velocity at A, u_A, increases slightly compared with the upstream velocity u, while at the same time u_B reduces. Now, from Bernoulli's equation,

$$\frac{p}{\rho g} + \frac{u^2}{2g} = \frac{p_A}{\rho g} + \frac{u_A^2}{2g} = \frac{p_B}{\rho g} + \frac{u_B^2}{2g}$$

Therefore $p_B > p_A$. It can thus be reasoned that the disturbance will produce a small transverse resultant force acting from B towards A. The lateral

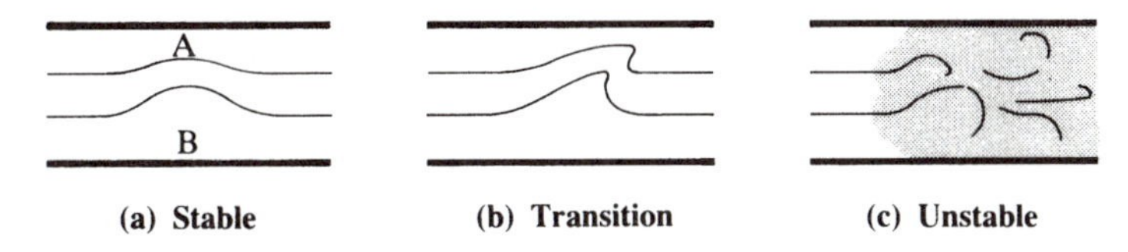

Figure 3.4 Effect of disturbance on a viscous flow.

component of velocity will also produce a corresponding component of viscous shear force, which acts in the opposite sense to the resultant disturbing force. As long as the fluid is moving slowly, the resultant disturbing force tends to be outweighed by the viscous force. Disturbances are therefore damped out. As the rate of shear increases, the effect of the disturbance becomes more pronounced:

1. The difference between u_A and u_B increases.
2. The pressure difference $(p_A - p_B)$ increases with $(u_A^2 - u_B^2)$, so the deflection of the pathline becomes more pronounced.
3. The greater shear results in a deformation of the crest of the pathline pattern (Fig. 3.4(b)). When the rate of shear is sufficiently great, the deformation is carried beyond the point at which the rectilinear pattern of pathlines can cohere (one can think of an analogy with a breaking wave). The flow pattern then disintegrates into a disorderly pattern of eddies in place of the orderly pattern of layers.

The above is (deliberately) a radical oversimplification of a complex physical pattern. It does, however, explain why the neat and mathematically tractable viscous flows only exist at low velocities. The other eddying flow is known as turbulent flow. The existence of these distinct patterns of flow was first investigated scientifically by Osborne Reynolds (1842–1912) at Manchester University.

Reynolds' experiment

Classical hydrodynamicists (most of them were primarily mathematicians) had long been puzzled by certain aspects of flow which did not conform to the known mathematical formulations. Towards the end of the 19th century, Reynolds designed an experiment in which a filament of dye was injected into a flow of water (Fig. 3.5). The discharge was carefully controlled, and passed through a glass tube so that observations could be made. Reynolds discovered that the dye filament would flow smoothly along the tube as long as the velocities remained very low. If the discharge was increased gradually, a point was reached at which the filament became wavy. A small further increase in discharge was then sufficient to trigger

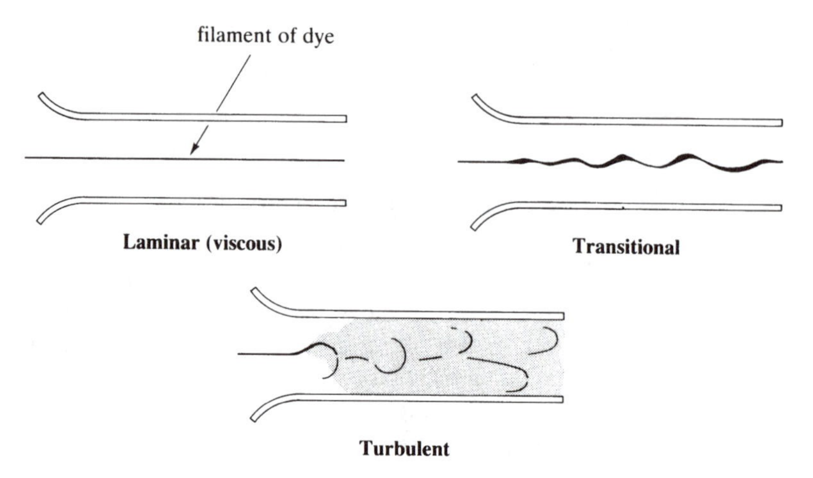

Figure 3.5 Reynolds' experiment.

a vigorous eddying motion, and the dye mixed completely with the water. Thus three distinct patterns of flow were revealed:

'Viscous' or 'laminar' – in which the fluid may be considered to flow in discrete layers with no mixing.
'Transitional' – in which some degree of unsteadiness becomes apparent (the wavy filament). Modern experimentation has demonstrated that this type of flow may comprise short 'bursts' of turbulence embedded in a laminar flow.
'Turbulent' – in which the flow incorporates an eddying or mixing action. The motion of a fluid particle within a turbulent flow is complex and irregular, involving fluctuations in velocity and direction.

Most of the flows which are encountered by civil engineers are turbulent flows.

The Reynolds Number

An obvious question which now arises is 'can we predict whether a flow will be viscous or turbulent?'. Reynolds' experiments revealed that the onset of turbulence was a function of fluid velocity, viscosity and a typical dimension. This led to the formation of the dimensionless Reynolds Number (symbol Re):

$$\mathrm{Re} = \frac{\rho u l}{\mu} = \frac{u l}{\nu} \tag{3.2}$$

It is possible to show that the Reynolds Number represents a ratio of forces (see Example 11.1):

$$\mathrm{Re} = \frac{\text{inertia force}}{\text{viscous force}}$$

For this reason, any two flows may be 'compared' by reference to their respective Reynolds Numbers. The onset of turbulence therefore tends to occur within a predictable range of values of Re. For example, flows in commercial pipelines normally conform to the following pattern:

for $\mathrm{Re} < 2000$, laminar flow exists;
for $2000 < \mathrm{Re} < 4000$, the flow is transitional;
for $\mathrm{Re} > 4000$, the flow is turbulent.

These values of Re should be regarded only as a rough guide (in some experiments laminar flows have been detected for $\mathrm{Re} \gg 4000$).

3.4 Shearing action in turbulent flows

General description

Shearing in laminar flows may be visualized as a purely frictional action between adjacent fluid layers. By contrast, shearing in turbulent flows is both difficult to visualize and less amenable to mathematical treatment. As a consequence, the solutions of problems involving turbulent flows tend to invoke experimental data.

In section 3.3, a (rather crude) model of flow instability was proposed. This led to a description of the way in which a streamline might be broken down into an eddy formation. An individual eddy may be considered to possess a certain 'size' or scale. This size will obviously be related to the local rate of shearing and to the corresponding instability. The passage of a succession of eddies leads to a measurable fluctuation in the velocity at a given point (Fig. 3.6). The eddies are generally irregular in size and shape, so the fluctuation of velocity with time is correspondingly irregular. For convenience, this fluctuating velocity is broken down into two components:

(a) The time-averaged velocity of flow at a point, u. For the present this is assumed to be parallel to the x-axis.
(b) Fluctuating components u' (in the x-direction), v' (y-direction) and w' (z-direction). The time average of u', v', or w' is obviously zero. The magnitudes of u', v' and w' and the rapidity with which they fluctuate give an indication of the structure of the eddy pattern.

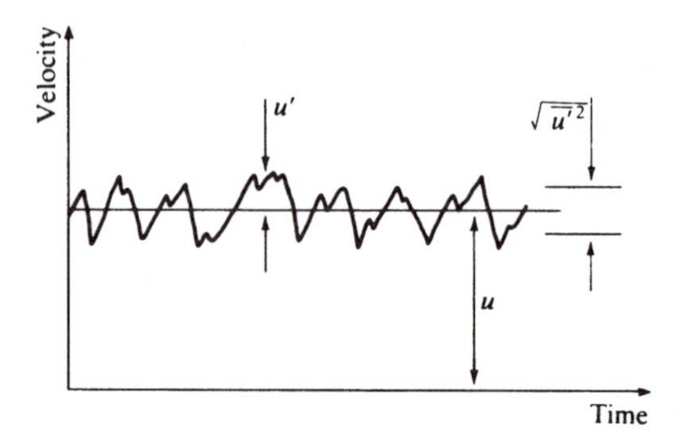

Figure 3.6 Variation of velocity with time in turbulent flow.

A measure of the magnitude of the fluctuations may be based upon the root-mean-square of the quantity u' (i.e. $\sqrt{\overline{u'}^2}$). This measure is usually known as the 'intensity of turbulence', and is the ratio $\sqrt{\overline{u'}^2}/u$.

Simple models of turbulent flows

Turbulence implies that within a fluid flow individual particles of fluid are 'jostling' (migrating transversely) as they are carried along with the flow. An attempt to plot the pathlines of a series of particles on a diagram would result in a tangled and confused pattern of lines. Since particles are continuously interchanging, it follows that their properties will similarly be continuously interchanging (hence the rapid mixing of the dye in Reynolds' observations of turbulence). Due to the complexity of the motion, it is impossible to produce an equation or numerical model of turbulence which is simple, complete and accurate. Nevertheless, some useful insights may be gained by investigating two elementary models of turbulence, even though both are known to have serious shortcomings.

The 'Reynolds' Stress' model. It has been stated above that the fluctuating components of velocity are u', v' and w'. If we restrict our attention to a two-dimensional flow (x- and y-components), then only u' and v' are present (Fig. 3.7(a)). Hence, during a time interval dt, the mass of fluid flowing in the y-direction through a small horizontal element of area δA is

$$\rho v' \, \delta A \, \delta t$$

This mass has an instantaneous horizontal velocity of $u + u'$. Its momentum, δM, is therefore

$$\delta M = \rho v' \, \delta A \, \delta t (u + u')$$

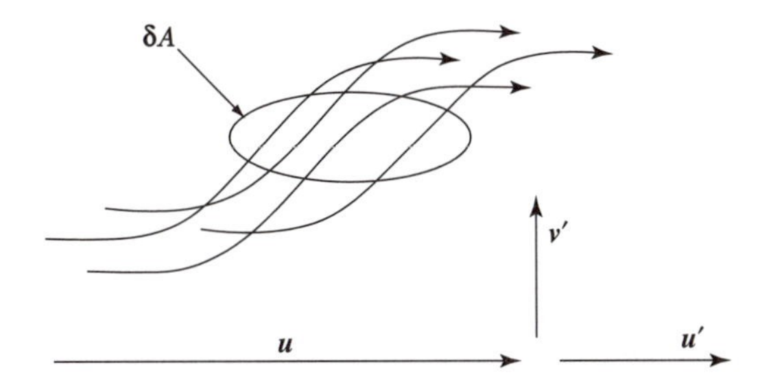

(a) Reynolds' eddy model

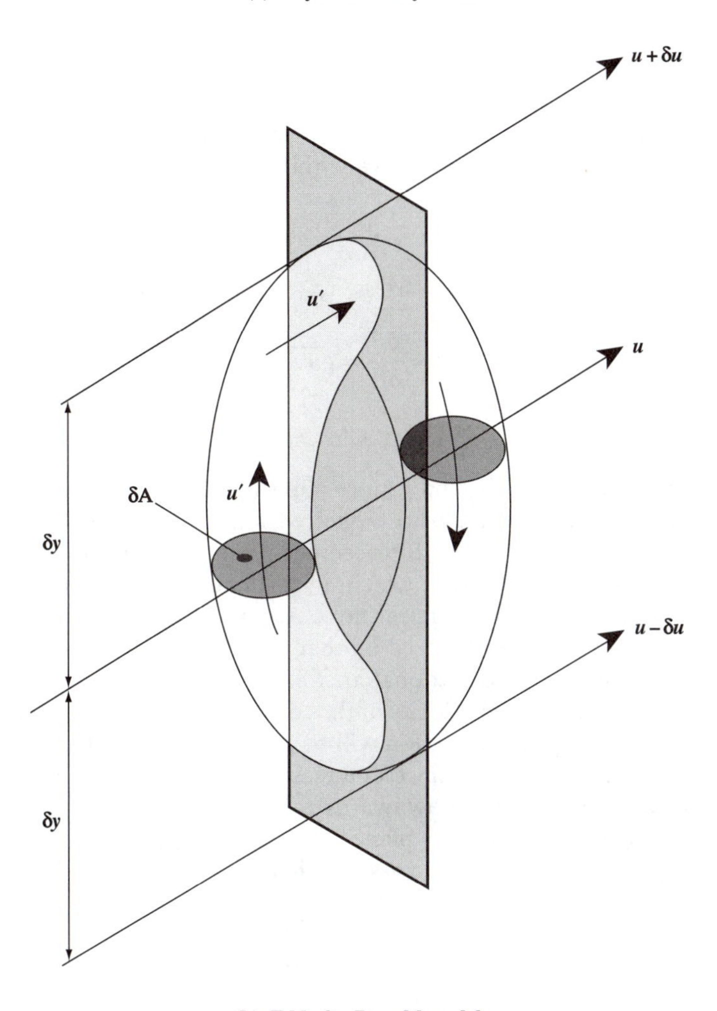

(b) Eddy for Prandtl model

Figure 3.7 Turbulent eddies.

Therefore, the rate of transport (or rate of interchange) of momentum during the particular instant is

$$\frac{\delta M}{\delta t} = \rho v' \,\delta A(u+u') = \rho \overline{v'}\,\delta A u + \rho \overline{v'u'}\,\delta A$$

The average rate of transport of momentum will be a function of the time-averaged velocities of the fluid particles. The magnitude of u remains constant. The averaged values of u' and v' are $\overline{u'}$ and $\overline{v'}$, but $\overline{u'}$ and $\overline{v'}$ must both be zero (see Fig. 3.6 and the associated text, above). Despite this fact, the product $\overline{u'v'}$ may not be zero, so

$$\text{average}\frac{\delta M}{\delta t} = \rho\overline{u'v'}\,\delta A$$

The existence of a 'rate of interchange' of momentum necessarily implies the existence of a corresponding force within the fluid:

$$\delta F = \rho\overline{u'v'}\,\delta A$$

Alternatively, since force/area = stress, τ,

$$\tau = \frac{\delta F}{\delta A} = \rho\overline{u'v'} \tag{3.3}$$

This is termed a Reynolds' Stress.

Prandtl eddy model. A slightly more sophisticated approach is due to Prandtl, a German engineer, who pioneered much of the work on shear flows in the early years of the 20th century. Prandtl sought to develop a simple model of an eddy, from which an analogue to viscosity might be derived. Consider, then, a rectilinear flow. At a point (x, y) in the flow, the velocity of flow is u. Superimposed on that flow is an eddy (Fig. 3.7(b)) in the form of a ring of cross-sectional area δA. The eddy is rotating, rather like a wheel. The tangential velocity of the eddy is u', and thus the mass flow transferred at any one cross-section will be $\rho\,\delta A u'$. The rate of interchange of mass is therefore $2\rho\,\delta A u'$ (since one mass of fluid migrates upwards as another equal mass migrates downwards).

Between the plane at y and the plane at $y+\delta y$, the velocity increases from u to $u+\delta u$, so it is reasonable to assume that

$$\delta u = u' = \delta y\frac{\mathrm{d}u}{\mathrm{d}y}$$

so

$$\text{rate of interchange} = 2\rho\,\delta A\,\delta y\,\mathrm{d}u/\mathrm{d}y$$

The rate of interchange of momentum is the product of the rate of mass interchange and the change in velocity in the y direction, i.e. δu. Therefore, the rate of change of momentum across the eddy is

$$\frac{dM}{dt} = 2\rho\delta A\left(\delta y\frac{du}{dy}\right)\left(\delta y\frac{du}{dy}\right)$$

Now, shear stress, τ = force/area of shear plane ($2\delta A$), and force $= dM/dt$, therefore

$$\tau = \rho\left(\delta y\frac{du}{dy}\right)^2 = \mu_T\frac{du}{dy} \tag{3.4}$$

where μ_T is the 'eddy viscosity' and is given by

$$\mu_T = \rho\delta y^2\left(\frac{du}{dy}\right)$$
$$= \rho l^2\left(\frac{du}{dy}\right) \tag{3.5}$$

μ_T is analogous to μ, but is a function of the rate of shear, and of eddy size (l).

Eddy viscosity is therefore too complex to be reducible (for example) to a convenient viscosity diagram. Prandtl's treatment is now rather dated. Nevertheless, because of its links with the physics of turbulence, it is a convenient starting point for the student. Furthermore, Prandtl's model formed the basis of much of the progress in the analysis of turbulence which continues to the present.

Velocity distribution in turbulent shear flows

It is possible to develop a relationship between y and u on the basis of Prandtl's work. In order to achieve this, an assumption must be made regarding eddy size. The simplest guess might be that $(\delta y) = \text{constant} \times y$, i.e. a linear relationship. If the constant is K, then (3.5) may be rewritten as

$$\tau = \rho(Ky)^2\left(\frac{du}{dy}\right)^2$$

or

$$\frac{\tau}{\rho} = (Ky)^2\left(\frac{du}{dy}\right)^2 \tag{3.6}$$

Now the ratio $\sqrt{\tau/\rho}$ is known as the 'friction velocity', u_*. No such velocity actually exists in the flow; u_* is just a 'reference' value. Also, from (3.3), $u_* = \sqrt{\overline{u'v'}}$.

Equation (3.6) may therefore be rewritten in terms of u_*:

$$u_* = Ky\frac{\mathrm{d}u}{\mathrm{d}y} \tag{3.6a}$$

Therefore

$$\mathrm{d}u = \frac{u_*}{K}\frac{\mathrm{d}y}{y} \tag{3.6b}$$

so

$$u = \frac{u_*}{K}\ln y + C$$

or

$$\frac{u}{u_*} = \frac{1}{K}\ln y + C \tag{3.7}$$

No further progress can now be made on a purely mathematical basis; it is necessary to use experimental data to evaluate K and C. It was originally thought that K was a constant (equal to 0.4), but this is now known to be an oversimplification. C is a function of surface roughness, k_S, and must be determined for different materials.

The k–ε turbulence model. Another technique, which has become quite widely used in computational modelling, is based on the use of two parameters, namely the turbulent kinetic energy $k(= 0.5(u'^2 + v'^2 + w'^2))$ and the rate of dissipation of k due to viscous damping, which is denoted by ε.

Computational models based on this technique usually incorporate two equations, to establish the rates of change of k and of ε. The eddy viscosity is then estimated as $\mu_T = \rho C_\mu k^2/\varepsilon$, where C_μ is a coefficient. Readers who wish to pursue this, and other techniques currently in use, should refer to Rodi (1984) or to the comprehensive book by McComb (1991).

3.5 The boundary layer

Description of a boundary layer

The concepts used in analysing shearing action in fluids have been introduced above. For illustrative purposes, it was assumed that the shearing action was occurring in a fluid sandwiched between a moving belt and a stationary solid surface. The fluid was thus bounded on two sides. It may have occurred to the reader that such a situation is not common in civil engineering. Some flows (e.g. the flow of air round a building) are bounded on one side only, while others (e.g. the flow through a pipe) are completely surrounded by a stationary solid surface. To develop the boundary layer concept, it is helpful to begin with a flow bounded on one side only. Consider, therefore, a rectilinear flow passing over a stationary flat plate which lies parallel to the flow (Fig. 3.8(a)). The incident flow (i.e. the flow just upstream of the plate) has a uniform velocity, U_∞. As the flow comes into contact with the plate, the layer of fluid immediately adjacent to the plate decelerates (due to viscous friction) and comes to rest. This follows from the postulate that in viscous fluids a thin layer of fluid actually 'adheres' to a solid surface. There is then a considerable shearing action between the layer of fluid on the plate surface and the second layer of fluid. The second layer is therefore forced to decelerate (though it is not quite brought to rest), creating a shearing action with the third layer of fluid, and so on. As the fluid passes further along the plate, the zone in which shearing action occurs tends to spread further outwards (Fig. 3.8(b)). This zone is known as a 'boundary layer'. Outside the boundary layer the flow remains effectively free of shear, so the fluid here is not subjected to viscosity-related forces. The fluid flow outside a boundary layer may therefore be assumed to act like an ideal fluid.

As with any other sheared flow, the flow within the boundary layer may be viscous or turbulent, depending on the value of the Reynolds Number. To evaluate Re we need a 'typical dimension' and in boundary layers this dimension is usually the distance in the x-plane from the leading edge of the solid boundary. The Reynolds Number thus becomes $\mathrm{Re}_x = \rho U_\infty x/\mu$.

A moment's reflection should convince the reader that if the solid surface is sufficiently long, a point will be reached at which the magnitude of Re indicates the onset of turbulence. This hypothesis accords with experimental observations. To add further to the complications, the very low velocities in the flow close to the plate imply a low local Re, and the consequent

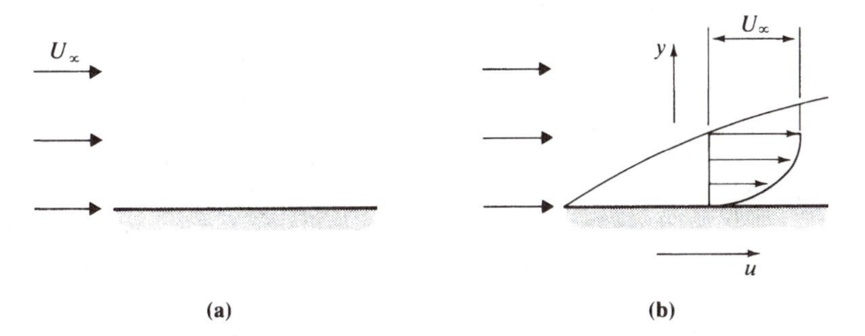

Figure 3.8 Development of a boundary layer.

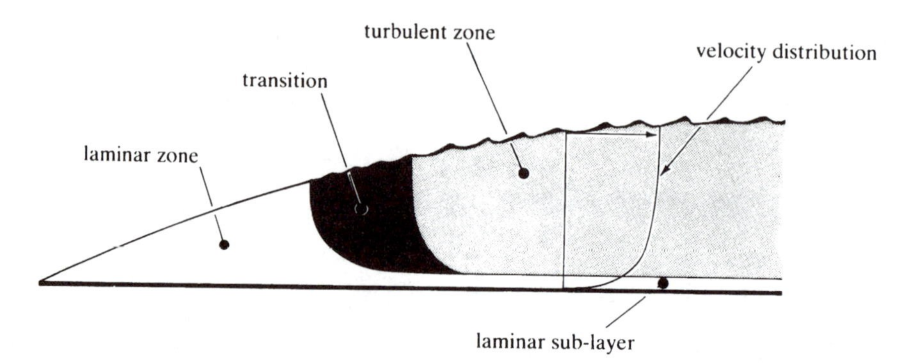

Figure 3.9 Structure of a boundary layer.

possibility of laminar flow here. The structure of the boundary layer is therefore as is shown in Figure 3.9. A graph (or 'distribution') of velocity variation with y may be drawn. This reveals that:

1. in the laminar zone there is a smooth velocity distribution to which a mathematical function can be fitted with good accuracy;
2. in the turbulent zone the mixing or eddying action produces a steeply sheared profile near the surface of the plate, but a flatter, more uniform profile further out towards the boundary layer edge.

In both cases, the velocity distributions are assumed to be asymptotic to the free stream velocity, U_∞.

Boundary layer equations

Although the basic structure of a boundary layer is clear, the engineer usually needs a precise numerical description for each particular problem. The basic parameters and equations required will now be developed. In the interests of simplicity, this treatment will be restricted to a two-dimensional incompressible flow with constant pressure.

(1) The boundary layer thickness, δ, is the distance in the y-direction from the solid surface to the outer edge of the boundary layer. Since the velocity distribution in the boundary layer is asymptotic to U_∞, it is difficult to measure an exact value for δ. The usual convention is to assume that the edge of the boundary layer occurs where $u/U_\infty = 0.99$.

(2) The displacement thickness, δ_*, is the distance by which a streamline is displaced due to the boundary layer. Consider the velocity distribution at a section in the boundary layer (Fig. 3.10). Inside the boundary layer, the velocity is everywhere less than in the free stream. The discharge through this cross-section is correspondingly less than the discharge through the

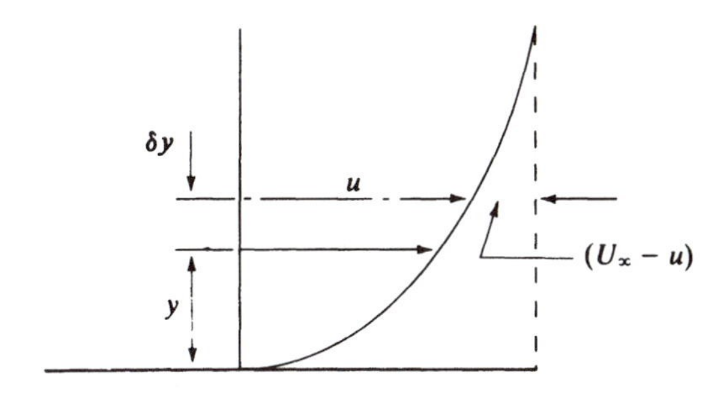

Figure 3.10 Velocity distribution in a boundary layer.

same cross-sectional area in the free stream. This deficit in discharge can be quantified for unit width, and an equation may then be developed for δ_*:

$$\text{deficit of discharge through an element} = (U_\infty - u)\,\delta y$$

$$\text{deficit through whole boundary layer section} = \int_0^\delta (U_\infty - u)\,\mathrm{d}y$$

In the free stream an equivalent discharge would pass through a layer of depth δ_*, so

$$U_\infty \delta_* = \int_0^\delta (U_\infty - u)\,\mathrm{d}y$$

Therefore

$$\delta_* = \int_0^\delta \left(1 - \frac{u}{U_\infty}\right) \mathrm{d}y \qquad (3.8)$$

(3) The momentum thickness, θ, is analogous to the displacement thickness. It may be defined as the depth of a layer in the free stream which would pass a momentum flux equivalent to the deficit due to the boundary layer:

$$\text{mass flow through element} = \rho u\,\delta y$$

$$\text{deficit of momentum flux} = \rho u\,\delta y\,(U_\infty - u)$$

$$\text{deficit through whole boundary layer section} = \int_0^\delta \rho u\,\mathrm{d}y\,(U_\infty - u)$$

In the free stream, an equivalent momentum flux would pass through a layer of depth θ and unit width, so that

$$\rho U_\infty^2 \theta = \int_0^\delta \rho u (U_\infty - u)\, \mathrm{d}y$$

$$\theta = \int_0 \frac{u}{U_\infty}\left(1 - \frac{u}{U_\infty}\right) \mathrm{d}y \tag{3.9}$$

(4) The definition of kinetic energy thickness δ_{**} follows the same pattern, leading to the equation

$$\delta_{**} = \int_0^\delta \frac{u}{U_\infty}\left(1 - \left(\frac{u}{U_\infty}\right)^2\right) \mathrm{d}y \tag{3.10}$$

(5) The momentum integral equation is used to relate certain boundary layer parameters so that numerical estimates may be made. Consider the longitudinal section through a boundary layer (Fig. 3.11), the section is bounded on its outer side by a streamline, BC, and is 1 m wide. The discharge across CD is

$$Q_{\mathrm{CD}} = \int_0^\delta u\, \mathrm{d}y$$

The momentum flux ($= \rho Q \times$ velocity) is therefore

$$\frac{\mathrm{d}M_{\mathrm{CD}}}{\mathrm{d}t} = \int_0^\delta \rho u^2\, \mathrm{d}y$$

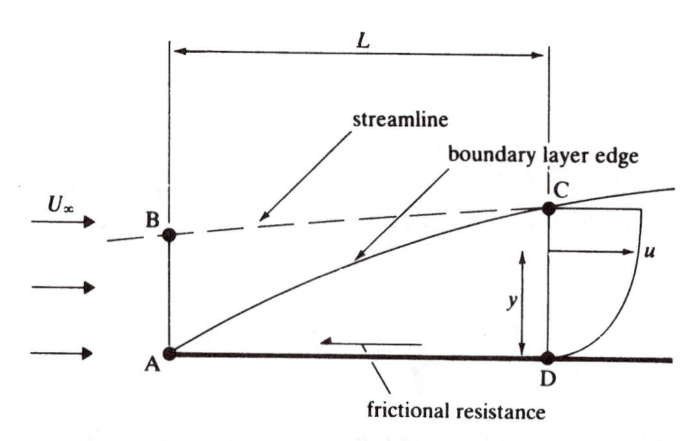

Figure 3.11 Longitudinal section through a boundary layer.

As BC is a streamline, the discharge across AB must be the same as that across CD:

$$Q_{AB} = \int_0^\delta u \, \mathrm{d}y$$

The incident velocity at AB is U_∞, so the momentum flux is

$$\frac{\mathrm{d}M_{AB}}{\mathrm{d}t} = \int_0^\delta \rho U_\infty u \, \mathrm{d}y$$

Boundary layers are actually very thin, so it is reasonable to assume the velocities are in the x-direction. The loss of momentum flux is due to the frictional shear force (F_S) at the solid surface. Therefore

$$-F_S = \int_0^\delta \rho u^2 \, \mathrm{d}y - \int_0^\delta \rho U_\infty u \, \mathrm{d}y$$

The negative sign follows from the fact that the frictional resistance acts in the opposite sense to the velocity. This equation may be rearranged to give

$$F_S = \int_0^\delta \rho(U_\infty u - u^2) \, \mathrm{d}y = \rho U_\infty^2 \int_0^\delta \frac{u}{U_\infty}\left(1 - \frac{u}{U_\infty}\right) \mathrm{d}y$$
$$= \rho U_\infty^2 \theta$$

The frictional shear at the solid surface is not a constant, but varies with x, due to the growth of the boundary layer. The shear force may therefore be expressed as

$$F_S = \int_0^L \tau_0 \, \mathrm{d}x$$

where τ_0 is the shear stress between the fluid and the solid surface. The momentum integral equation is therefore

$$\int_0^L \tau_0 \, \mathrm{d}x = \rho U_\infty^2 \theta \qquad (3.11)$$

Solution of the momentum integral equation

In order to solve the momentum equation, some further information regarding θ and τ_0 is required. Both of these quantities are related to the velocity distribution in the boundary layer. Since velocity distribution depends on the nature of the flow, i.e. whether it is laminar or turbulent, the two cases are now considered separately.

Laminar flow. The velocity distribution in a laminar boundary layer may be expressed in a number of forms. A typical equation is

$$\frac{u}{U_\infty} = \left[A\frac{y}{\delta} - B\left(\frac{y}{\delta}\right)^2\right]$$

Therefore,

$$\begin{aligned}\theta &= \int_0^\delta \frac{u}{U_\infty}\left(1 - \frac{u}{U_\infty}\right)\mathrm{d}y \\ &= \int_0^\delta \left[A\frac{y}{\delta} - B\left(\frac{y}{\delta}\right)^2\right]\left[1 - \left[A\frac{y}{\delta} - B\left(\frac{y}{\delta}\right)^2\right]\right]\mathrm{d}y \qquad (3.12)\end{aligned}$$

If A and B are known, the above integral can be evaluated. Furthermore:

$$\begin{aligned}\tau_0 &= \mu\frac{\mathrm{d}u}{\mathrm{d}y} = \mu U_\infty \frac{\mathrm{d}}{\mathrm{d}y}\left[A\frac{y}{\delta} - B\left(\frac{y}{\delta}\right)^2\right]_{y=0} \\ &= \mu U_\infty\left[\frac{A}{\delta} - \frac{2By}{\delta}\right]_{y=0} = \frac{\mu U_\infty A}{\delta} \qquad (3.13)\end{aligned}$$

If (3.12) and (3.13) are substituted into (3.11), a solution can be obtained.

Turbulent flow. Experimental investigations have shown that the velocity distribution outside the laminar sub-layer may be approximately represented by

$$\frac{u}{U_\infty} = \left(\frac{y}{\delta}\right)^{1/n}, \quad \text{where } 6 < n < 11 \qquad (3.14)$$

Typically (after Prandtl) a value $n = 7$ is used. This can be substituted into the equation for the momentum thickness (equation (3.9)) and a solution obtained. This equation cannot, however, be differentiated to obtain τ_0, since

$$\frac{\mathrm{d}}{\mathrm{d}y}\left(\frac{y}{\delta}\right)^{1/7}_{y=0} = 0$$

It is again necessary to use experimental data to fill the gap in the mathematical procedure. This provides us with the alternative relationship

$$\tau_0 = \frac{0.023\rho U_\infty^2}{(\mathrm{Re}_\delta)^{1/m}} \qquad (3.15)$$

where

$$\mathrm{Re}_\delta = \rho U_\infty \delta/\mu$$

Example 3.2 Turbulent boundary layer

Water flows down a smooth wide concrete apron into a river. Assuming that a turbulent boundary layer forms, estimate the shear stress and the boundary layer thickness 50 m downstream of the entrance to the apron. Use the following data:

$$U_\infty = 7\,\mathrm{m/s} \quad \mu = 1.14 \times 10^{-3}\,\mathrm{kg/m\,s} \quad m = 4$$

$$\frac{u}{U_\infty} = \left(\frac{y}{\delta}\right)^{1/7} \qquad \tau_0 = \frac{0.0225\rho U_\infty^2}{(\mathrm{Re}_\delta)^{1/4}}$$

Solution

From (3.11),

$$\int_0^L \tau_0\,\mathrm{d}x = \rho U_\infty^2 \theta$$

This may be rewritten as

$$\tau_0 = \frac{\mathrm{d}}{\mathrm{d}x}(\rho U_\infty^2 \theta)$$

From (3.9),

$$\theta = \int_0^\delta \frac{u}{U_\infty}\left(1 - \frac{u}{U_\infty}\right)\mathrm{d}y$$

Substituting $u/U_\infty = (y/\delta)^{1/7}$,

$$\theta = \int_0^\delta \left(\frac{y}{\delta}\right)^{1/7}\left[1 - \left(\frac{y}{\delta}\right)^{1/7}\right]\mathrm{d}y = \frac{7}{72}\,\delta$$

Hence

$$\tau_0 = \frac{\mathrm{d}}{\mathrm{d}x}\left[\rho U_\infty^2\left(\frac{7\delta}{72}\right)\right] = \frac{7}{72}\rho U_\infty^2 \frac{\mathrm{d}\delta}{\mathrm{d}x}$$

But it has been stated that $\tau_0 = 0.0225\rho U_\infty^2/(\mathrm{Re}_\delta)^{1/4}$. Therefore

$$\frac{0.0225\rho U_\infty^2}{(\mathrm{Re}_\delta)^{1/4}} = \frac{7}{72}\rho U_\infty^2 \frac{\mathrm{d}\delta}{\mathrm{d}x}$$

Substituting $\rho = 1000\,\mathrm{kg/m^3}$ and $U_\infty = 7\,\mathrm{m/s}$, and rearranging,

$$\frac{0.2314}{(\mathrm{Re}_\delta)^{1/4}} = \frac{\mathrm{d}\delta}{\mathrm{d}x}$$

$\mathrm{Re}_\delta = \rho U_\infty \delta/\mu$, therefore $(\mathrm{Re}_\delta)^{1/4} = 49.78\,\delta^{1/4}$, so, integrating,

$$x = \int \frac{49.78\,\delta^{1/4}}{0.2314}\,\mathrm{d}\delta$$

$$= \frac{39.82\,\delta^{5/4}}{0.2314}$$

or

$$\delta = \left(\frac{0.2314x}{39.82}\right)^{4/5}$$

so if $x = 50\,\mathrm{m}$, then $\delta = 0.372\,\mathrm{m}$ and

$$\tau_0 = \frac{0.0225\rho U_\infty^2}{\mathrm{Re}_\delta^{1/4}} = \frac{0.0225 \times 1000 \times 7^2}{\left(\dfrac{1000 \times 7 \times 0.372}{1.14 \times 10^{-3}}\right)^{1/4}}$$

$$= 28.36\,\mathrm{N/m^2}$$

3.6 Some implications of the boundary layer concept

'Flow separation'

Returning to the problem of flow round a 'bluff' shape, which was considered at the beginning of the chapter, it is now possible to investigate the pattern of events in the light of our knowledge of boundary layer formation (see Fig. 3.12).

Around the upstream half of the body, the fluid is deflected outwards, the streamlines converge as the flow accelerates and a boundary layer grows progressively. After the fluid passes the Y–Y axis, the flow is decelerating. The fluid in the boundary layer is travelling at a lower speed than the fluid in the free stream, and a point is reached at which negative velocities arise at the inner part of the boundary layer. The line traced by the points of zero velocity downstream of the body divides the zones of the positive and negative velocity, and indicates that flow separation has occurred. The development of the negative velocity zone further implies that the pressures within the zone are low compared with those in the free stream. Fluid from further out in the boundary layer is therefore drawn inwards to the low pressure zone. The effect of all this is that powerful eddies are generated,

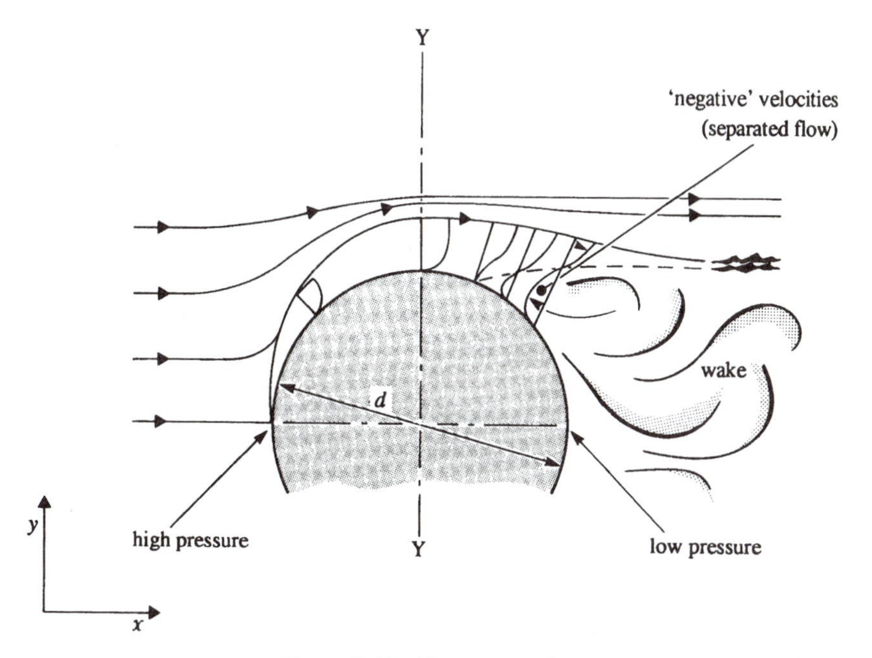

Figure 3.12 Flow separation.

which are then drawn downstream by the flow, thus forming the 'wake' zone.

Surface roughness and boundary layer development

A question may be posed: Different materials exhibit different degrees of roughness (e.g. plastic or concrete) – does this have any effect on the boundary layer? There are, broadly, three answers which could be given.

1. In laminar flow, the friction is transmitted by pure shearing action. Consequently, the roughness of the solid surface has no effect, except to trap small 'pools' of stationary fluid in the interstices, and thus slightly increase the thickness of the stationary layer of fluid.
2. In a turbulent flow, a laminar sub-layer forms close to the solid surface. If the average height of the surface roughness is smaller than the height of the laminar sub-layer, there will be little or no effect on the overall flow.
3. Turbulent flow embodies a process of momentum transfer from layer to layer. Consequently, if the surface roughness protrudes through the laminar region into the turbulent region, then it will cause additional eddy formation and therefore greater energy loss in the turbulent flow. This implies that the apparent frictional shear will be increased.

'Drag' forces on a body

From the outline given in the two preceding paragraphs, it will be evident that a body immersed in an external flow may be subjected to two distinct force patterns:

1. Due to the frictional shearing action between the body and the flow.
2. Due to flow separation, if the body is 'bluff'. Under such conditions, the deflection of the incident flow around the upstream face of the body creates a region of locally increased pressure, whilst the separation zone downstream creates a region of locally low pressure. This pressure difference exerts a force on the body. The component of this force in the direction of the incident flow is known as the 'form drag'. If there is a component perpendicular to this direction, it is often referred to as 'lift'. The terminology originated in the aeronautical industry, which sponsored much of the research subsequent to Prandtl's poineering work.

Measurements have been carried out on many body shapes, ranging from rectangular sections to aerofoil (wing) shapes. The measured forces are usually reduced to coefficient form, for example drag is expressed as 'drag coefficient' C_D, where

$$C_D = \frac{\text{drag force}}{\frac{1}{2}\rho A U_\infty^2}$$

A is a 'representative area', e.g. the cross-sectional area of a body perpendicular to the direction of the incident flow. Note that the drag force is usually the total force due to both friction and form components. C_D is therefore a function of the body shape and the Reynolds Number. Graphs of C_D versus the Reynolds Number for various shapes may be found in Schlichting *et al.* (1999). Civil engineers encounter 'drag' problems in connection with wind flows around buildings, bridge piers in rivers, etc.

It is worth drawing attention to one further characteristic of separated flows. Under certain conditions it is possible for the eddy-forming mechanism to generate a rhythmic pattern in the wake. Eddies are shed alternately, first from one side of the body and then from the other. This eddy-shedding process sets up a transverse (lift) force on the body, whereas in a symmetrically generated wake the forces generated in this way are also symmetrical, and therefore effectively cancel. If the rhythmical pattern is set up it generates an alternating transverse force which can cause vibration problems. The frequency of the eddy-shedding is related to the Strouhal Number:

$$\text{St} = \text{frequency} \times d/U_\infty$$

Experiments have shown that, for circular cylinders,

$$\mathrm{St} = 0.198(1 - 19.7/\mathrm{Re}), \quad \text{for } 250 < \mathrm{Re} < 2 \times 10^5$$

Vibration problems of this type tend to occur when the natural frequency of a flexible structure (e.g. a suspension bridge or metal-clad chimney) coincides with the eddy frequency.

Bounded flows

Most civil engineering flows are 'bounded', either completely (as in pipe flows) or partly (as in open channels). Boundary layer growth will commence at the entry to the pipe or channel system, and continue downstream. Providing that the system is of sufficient length, a point will be reached at which the boundary layer thickness, δ, extends as far as the centreline (Fig. 3.13). From this point onwards it is clear that no boundary layer growth is possible. The flow has therefore become virtually uniform, and the boundary shear stress, τ_0, must, within close limits, be constant. This is known as the 'fully developed flow' condition. In most practical cases the 'entry length' is short compared with the overall length of the pipe or channel (for example, in laminar pipe flow the entry length is approximately $110d$, while for turbulent pipe flow it is approximately $50d$). It is therefore conventional and simplest to treat most problems as if the flow were fully developed throughout. In fully developed flows it is usual for the boundary shear stress to be expressed in terms of a 'coefficient of friction' (symbol C_f), which is defined by the equation

$$C_f = \tau_0 \bigg/ \frac{1}{2}\rho V^2 (= \lambda/4, \text{see Chapter 4})$$

where V is the mean velocity of flow (i.e. Q/A).

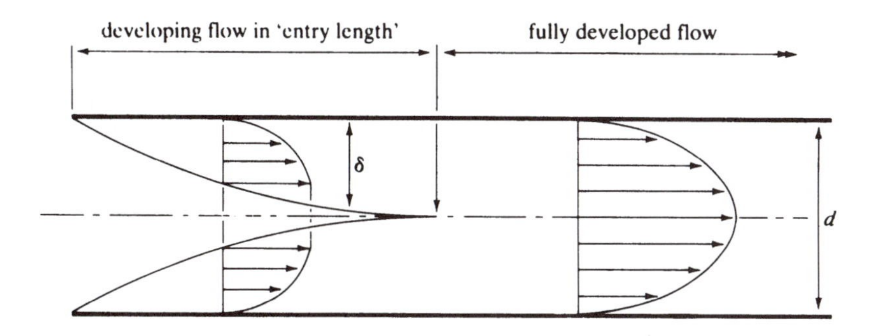

Figure 3.13 Development of flow pattern in the entrance to a pipe.

In view of what was said previously, it is not surprising that the friction factor is:

1. a function of the flow type (laminar or turbulent), and consequently a function of the Reynolds Number;
2. a function of the boundary roughness (in turbulent flows), providing the roughness elements are sufficiently high – the evaluation and applications of friction factors will be developed more fully in Chapter 4.

3.7 Cavitation

At the beginning of this chapter, reference was made to the characteristics of an ideal fluid. One of those characteristics specified that an ideal fluid constituted a homogeneous continuum under all conditions of flow. This is also frequently true of real fluids – frequently, but not always. The principal exception to homogeneous flow will now be examined.

If the pressure of a liquid falls, the temperature at which boiling occurs also falls. It follows that a fall in pressure of sufficient magnitude will induce boiling, even at normal atmospheric temperatures, and this phenomenon is called cavitation. Cavitation arises under certain conditions, some typical examples being:

1. In severely sheared separated flows of liquids. Such flows can occur on spillways or in control valves in pipelines.
2. In fluid machines such as pumps or turbines. Fluid machines involve an energy interchange process (e.g. pumps may convert electrical power into fluid power). The interchange involves a moving element that is immersed in the fluid. Under some circumstances, very low pressures may occur at the interface between the element and the fluid.

The reason for regarding cavitation as a problem is that the boiling process involves the formation and collapse of vapour bubbles in the liquid. When these bubbles collapse, they cause a considerable local shock (or 'hammer blow'), i.e. a sharp rise and fall in the local pressure. The peak pressure during such a shock may be up to $400 \times 10^6\,\mathrm{N/m^2}$. Although one such event lasts only a few milliseconds, under cavitating conditions the event may be repeated in rapid succession. The potential for producing damage to concrete or metal surfaces is thus considerable. There have been a number of cases of major (and therefore expensive!) damage in large hydroelectric schemes. Theoretically, the onset of cavitation coincides with the vapour pressure of the liquid. However, the presence of dissolved gases or small solid particles in suspension frequently means that cavitation can occur at higher pressures. It is usually assumed that with water at 20° C, for example, it is inadvisable to operate at pressure heads lower than 3 m absolute (i.e. 7 m vacuum).

3.8 Surface tension effects

Surface tension has been briefly discussed in the Introductory Notes. Surface tension, T_S, is usually very small compared with other forces in fluid flows (e.g. for a water surface exposed to air, $T_S \simeq 0.073\,\mathrm{N/m}$). In the vast majority of cases, engineers ignore surface tension. However, it may assume importance in low flows (for example, in hydraulic models or in weir flows under very low heads). Research into such situations led to the formation of a dimensionless parameter known as the Weber Number, We (named after the German naval architect M. Weber (1871–1951)):

$$\mathrm{We} = \sqrt{\frac{\text{inertia force}}{\text{surface tension}}} = \sqrt{\frac{\rho u^2 l}{T_S}}$$

For a given flow, the magnitude of We indicates whether T_S is likely to be significant.

3.9 Summary

The major topics covered in this chapter have related to turbulence and the boundary layer. Within this brief account, it is impossible to achieve more than an introductory coverage. More advanced studies may be pursued by consulting the work of Schlichting *et al.* (1999) or Cebeci and Bradshaw (1977) on boundary layers. Also, Hinze (1975), McComb (1991) and Tennekes and Lumley (1972) provide more comprehensive treatments of the physics and mathematical modelling of turbulence.

References and further reading

Bradshaw, P., Cebeci, T. and Whitelaw, J. H. (1981) *Engineering Calculation Methods for Turbulent Flow*, Academic Press, New York.

Cebeci, T. and Bradshaw, P. (1977) *Momentum Transfer in Boundary Layers*, McGraw-Hill, New York.

Hinze, J. O. (1975) *Turbulence*, 2nd edn, McGraw-Hill, New York.

McComb, W. D. (1991) *The Physics of Fluid Turbulence*, Oxford University Press, Oxford.

Rodi, W. (1984) *Turbulence Models and their Application in Hydraulics – A State of the Art Review*. International Association for Hydraulic Research, Delft.

Schlichting, H., Gersten, K., Krause, E. and Oertel, H. (Jr) (trans. Mayes, C.) (1999) *Boundary Layer Theory*, 8th edn, Springer Verlag, Berlin.

Tennekes, H. and Lumley, J. L. (1972) *A First Course in Turbulence*. MIT Press, Cambridge, Massachusetts.

Vardy, A. E. (1990) *Fluid Principles*, McGraw-Hill, London.

4

Flow in pipes and closed conduits

4.1 Introduction

The flow of water, oil and gas in pipes is of immense practical significance in civil engineering. Water is conveyed from its source, normally in pressure pipelines (Fig. 4.1), to water treatment plants where it enters the distribution system and finally arrives at the consumer. Surface water drainage and sewerage is conveyed by closed conduits, which do not usually operate under pressure, to sewage treatment plants, from where it is usually discharged to a river or the sea. Oil and gas are often transferred from their source by pressure pipelines to refineries (oil) or into a distribution network for supply (gas).

Surprising as it may seem, a comprehensive theory of the flow of fluids in pipes was not developed until the late 1930s, and practical design methods for the evaluation of discharges, pressures and head losses did not appear until 1958. Until these design tools were available, the efficient design of pipeline systems was not possible.

This chapter describes the theories of pipe flow, beginning with a review of the historical context and ending with the practical applications.

4.2 The historical context

Table 4.1 lists the names of the main contributors, and their contributions, to pipe flow theories in chronological order.

The Colebrook–White transition formula represents the culmination of all the previous work, and can be applied to any fluid in any pipe operating under turbulent flow conditions. The later contributions of Moody, Ackers and Barr are mainly concerned with the practical application of the Colebrook–White equation.

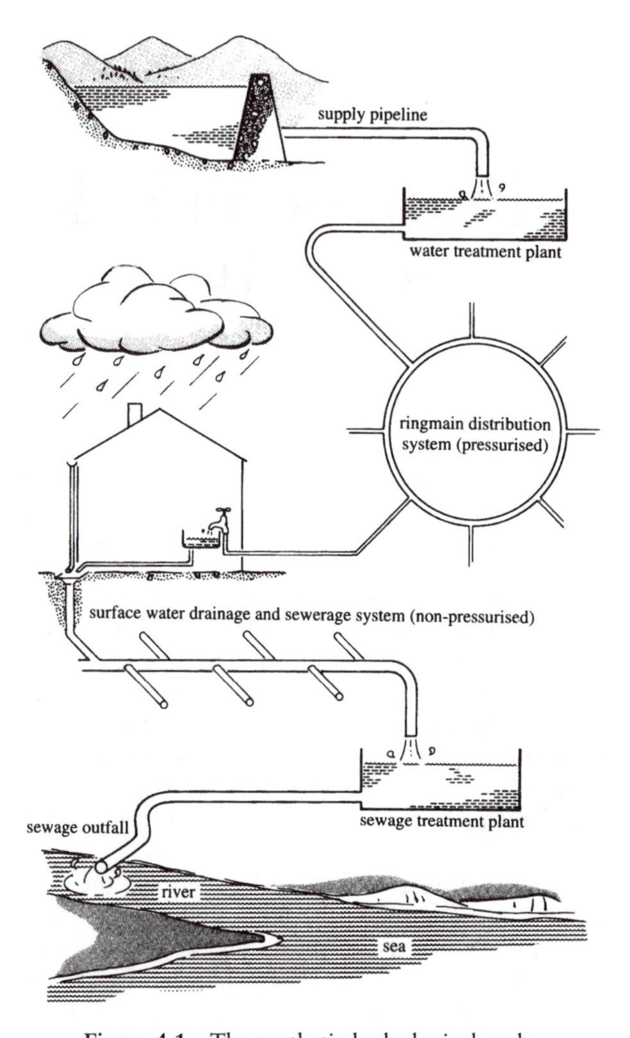

Figure 4.1 The synthetic hydrological cycle.

There are three major concepts described in the table. These are:

1. the distinction between laminar and turbulent flow;
2. the distinction between rough and smooth pipes;
3. the distinction between artificially roughened pipes and commercial pipes.

To understand these concepts, the best starting point is the contribution of Reynolds, followed by the laminar flow equations, before proceeding to the more complex turbulent flow equations.

Table 4.1 The chronological development of pipe flow theories.

Date	Name	Contribution
1839–41	Hagen and Poiseuille	laminar flow equation
1850	Darcy and Weisbach	turbulent flow equation
1884	Reynolds	distinction between laminar and turbulent flow – Reynolds' Number
1913	Blasius	friction factor equation for smooth pipes
1914	Stanton and Pannell	experimental values of the friction factor for smooth pipes
1930	Nikuradse	experimental values of the friction factor for artificially rough pipes
1930s	Prandtl and von Kármán	equations for rough and smooth friction factors
1937–39	Colebrook and White	experimental values of the friction factor for commercial pipes and the transition formula
1944	Moody	the Moody diagram for commercial pipes
1958	Ackers	the Hydraulics Research Station Charts and Tables for the design of pipes and channels
1975	Barr	direct solution of the Colebrook–White equation

Laminar and turbulent flow

Reynolds' experiments demonstrated that there were two kinds of flow – laminar and turbulent – as described in Chapter 3. He found that transition from laminar to turbulent flow occurred at a critical velocity for a given pipe and fluid. Expressing his results in terms of the dimensionless parameter $\mathrm{Re} = \rho DV/\mu$, he found that for Re less than about 2000 the flow was always laminar, and that for Re greater than about 4000 the flow was always turbulent. For Re between 2000 and 4000, he found that the flow could be either laminar or turbulent, and termed this the transition region.

In a further set of experiments, he found that for laminar flow the frictional head loss in a pipe was proportional to the velocity, and that for turbulent flow the head loss was proportional to the square of the velocity.

These two results had been previously determined by Hagen and Poiseuille ($h_f \infty V$) and Darcy and Weisbach ($h_f \infty V^2$), but it was Reynolds who put these equations in the context of laminar and turbulent flow.

4.3 Fundamental concepts of pipe flow

The momentum equation

Before proceeding to derive the laminar and turbulent flow equations, it is instructive to consider the momentum (or dynamic) equation of flow and the influence of the boundary layer.

Referring to Figure 4.2, showing an elemental annulus of fluid, thickness δr, length δl, in a pipe of radius R, the forces acting are the pressure forces, the shear forces and the weight of the fluid. The sum of the forces acting is equal to the change of momentum. In this case momentum change is zero, since the flow is steady and uniform. Hence

$$p2\pi r\,\delta r-\left(p+\frac{\mathrm{d}p}{\mathrm{d}l}\delta l\right)2\pi r\,\delta r+\tau 2\pi r\,\delta l-\left(\tau+\frac{\mathrm{d}\tau}{\mathrm{d}r}\delta r\right)2\pi(r+\delta r)\,\delta l$$
$$+\rho g 2\pi r\,\delta l\,\delta r\sin\theta=0$$

Setting $\sin\theta=-\mathrm{d}z/\mathrm{d}l$ and dividing by $2\pi r\,\delta r\,\delta l$ gives

$$-\frac{\mathrm{d}p}{\mathrm{d}l}-\frac{\mathrm{d}\tau}{\mathrm{d}r}-\frac{\tau}{r}-\rho g\frac{\mathrm{d}z}{\mathrm{d}l}=0$$

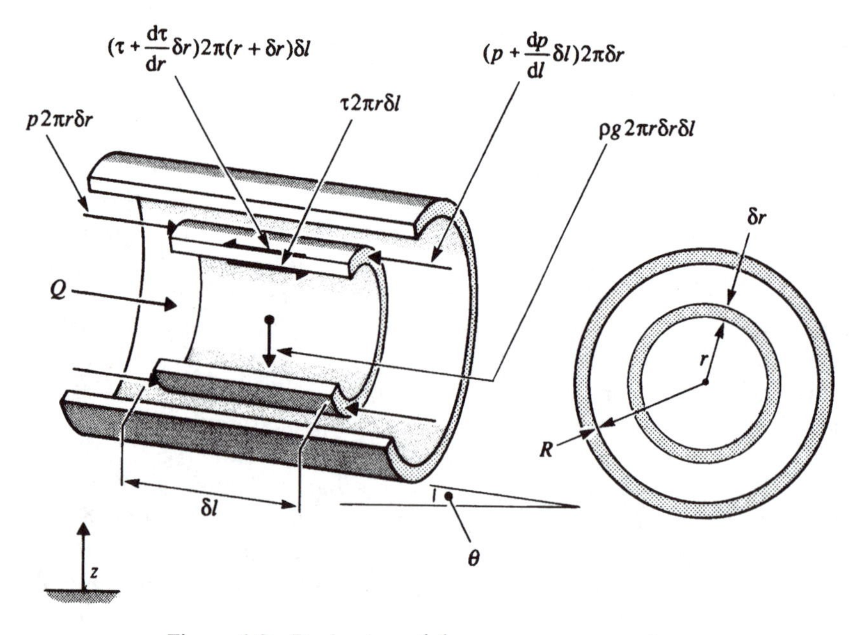

Figure 4.2 Derivation of the momentum equation.

(ignoring second-order terms), or

$$-\frac{dp^*}{dl} - \frac{\tau}{r} - \frac{d\tau}{dr} = 0$$

where $p^*(= p + \rho g z)$ is the piezometric pressure measured from the datum $z = 0$. As

$$\frac{1}{r}\frac{d}{dr}(\tau r) = \frac{1}{r}\left(r\frac{d\tau}{dr} + \tau\right) = \frac{d\tau}{dr} + \frac{\tau}{r}$$

then

$$-\frac{dp^*}{dl} - \frac{1}{r}\frac{d}{dr}(\tau r) = 0$$

Rearranging,

$$\frac{d}{dr}(\tau r) = -r\frac{dp^*}{dl}$$

Integrating both sides with respect to r,

$$\tau r = -\frac{dp^*}{dl}\frac{r^2}{2} + \text{constant}$$

At the centreline $r = 0$, and therefore constant $= 0$. Hence

$$\tau = -\frac{dp^*}{dl}\frac{r}{2} \tag{4.1}$$

Equation (4.1) is the momentum equation for steady uniform flow in a pipe. It is equally applicable to laminar or turbulent flow, and relates the shear stress τ at radius r to the rate of head loss with distance along the pipe. If an expression for the shear force can be found in terms of the velocity at radius r, then the momentum equation may be used to relate the velocity (and hence discharge) to head loss.

In the case of laminar flow, this is a simple matter. However, for the case of turbulent flow it is more complicated, as will be seen in the following sections.

The development of boundary layers

Figure 4.3(a) shows the development of laminar flow in a pipe. At entry to the pipe, a laminar boundary layer begins to grow. However, the growth

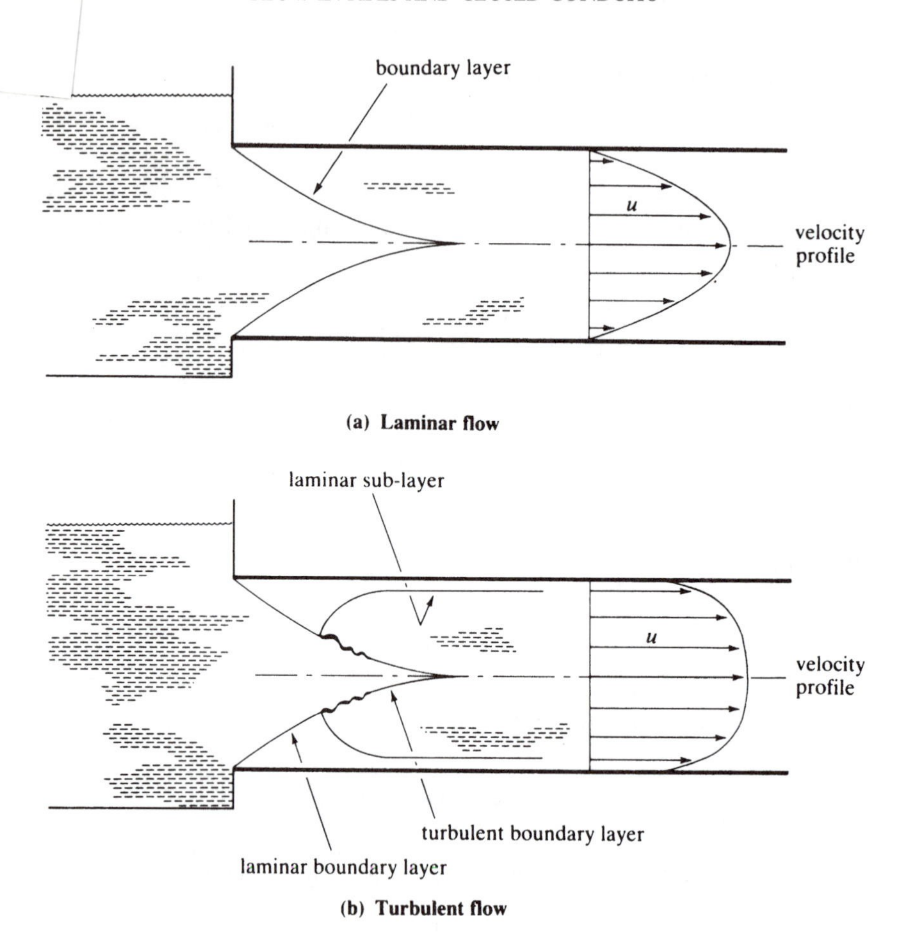

Figure 4.3 Boundary layers and velocity distributions.

of the boundary layer is halted when it reaches the pipe centreline, and thereafter the flow consists entirely of a boundary layer of thickness r. The resulting velocity distribution is as shown in Figure 4.3(a).

For the case of turbulent flow shown in Figure 4.3(b), the growth of the boundary layer is not suppressed until it becomes a turbulent boundary layer with the accompanying laminar sub-layer. The resulting velocity profile therefore differs considerably from the laminar case. The existence of the laminar sub-layer is of prime importance in explaining the difference between smooth and rough pipes.

Expressions relating shear stress to velocity have been developed in Chapter 3, and these will be used in explaining the pipe flow equations in the following sections.

4.4 Laminar flow

For the case of laminar flow, Newton's law of viscosity may be used to evaluate the shear stress (τ) in terms of velocity (u):

$$\tau = \mu \frac{\mathrm{d}u}{\mathrm{d}y} = -\mu \frac{\mathrm{d}u_r}{\mathrm{d}r}$$

Substituting into the momentum equation (4.1),

$$\tau = -\mu \frac{\mathrm{d}u_r}{\mathrm{d}r} = -\frac{\mathrm{d}p^*}{\mathrm{d}l}\frac{r}{2}$$

or

$$\frac{\mathrm{d}u_r}{\mathrm{d}r} = \frac{1}{2\mu}\frac{\mathrm{d}p^*}{\mathrm{d}l} r$$

Integrating,

$$u_r = \frac{1}{4\mu}\frac{\mathrm{d}p^*}{\mathrm{d}l} r^2 + \text{constant}$$

At the pipe boundary, $u_r = 0$ and $r = R$, hence

$$\text{constant} = -\frac{1}{4\mu}\frac{\mathrm{d}p^*}{\mathrm{d}l} R^2$$

and

$$u_r = -\frac{1}{4\mu}\frac{\mathrm{d}p^*}{\mathrm{d}l}(R^2 - r^2) \tag{4.2}$$

Equation (4.2) represents a parabolic velocity distribution, as shown in Figure 4.3(a). The discharge (Q) may be determined from (4.2). Returning to Figure 4.1 and considering the elemental discharge (δQ) through the annulus, then

$$\delta Q = 2\pi r\,\delta r u_r$$

Integrating

$$Q = 2\pi \int_0^R r\, u_r\, \mathrm{d}r$$

and substituting for u_r from (4.2) gives

$$Q = -\frac{2\pi}{4\mu}\frac{\mathrm{d}p^*}{\mathrm{d}l}\int_0^R r(R^2 - r^2)\,\mathrm{d}r$$

or

$$Q = -\frac{\pi}{8\mu}\frac{dp^*}{dl}R^4 \tag{4.3}$$

Also the mean velocity (V) may be obtained directly from Q:

$$V = \frac{Q}{A} = -\frac{\pi}{8\mu}\frac{dp^*}{dl}R^4\frac{1}{\pi R^2}$$

or

$$V = -\frac{1}{8\mu}\frac{dp^*}{dl}R^2 \tag{4.4}$$

In practice, it is usual to express (4.4) in terms of frictional head loss by making the substitution

$$h_f = -\frac{\Delta p^*}{\rho g}$$

Equation (4.4) then becomes

$$V = \frac{1}{8\mu}\frac{h_f}{L}\rho g\frac{D^2}{4}$$

or

$$h_f = \frac{32\mu LV}{\rho g D^2} \tag{4.5}$$

This is the Hagen–Poiseuille equation, named after the two people who first carried out (independently) the experimental work leading to it.

The wall shear stress (τ_0) may be related to the mean velocity (V) by eliminating dp^*/dl from (4.1) and (4.4) to give

$$\tau = 4\mu Vr/R^2 \tag{4.6}$$

As $\tau = \tau_0$ when $r = R$, then

$$\tau_0 = 4\mu V/R \tag{4.7}$$

Equation (4.6) shows that (for a given V) the shear stress is proportional to r, and is zero at the pipe centreline, with a maximum value (τ_0) at the pipe boundary.

Example 4.1 Laminar pipe flow

Oil flows through a 25 mm diameter pipe with a mean velocity of 0.3 m/s. Given that $\mu = 4.8 \times 10^{-2}$ kg/m s and $\rho = 800\,\text{kg/m}^3$, calculate (a) the pressure drop in a 45 m length and (b) the maximum velocity, and the velocity 5 mm from the pipe wall.

Solution

First check that flow is laminar, i.e. Re < 2000.

$$\begin{aligned}\text{Re} &= \rho DV/\mu = 800 \times 0.025 \times 0.3/4.8 \times 10^{-2}\\ &= 125\end{aligned}$$

(a) To find the pressure drop, apply (4.5):

$$\begin{aligned}h_f &= 32\mu LV/\rho g D^2\\ &= (32 \times 4.8 \times 10^{-2} \times 45 \times 0.3)/(800 \times 9.81 \times 0.025^2)\\ &= 4.228\,\text{m (of oil)}\end{aligned}$$

or $\Delta p = -\rho g h_f = -33.18\,\text{kN/m}^2$. (*Note*: the negative sign indicates that pressure reduces in the direction of flow.)

(b) To find the velocities, apply (4.2):

$$u_r = -\frac{1}{4\mu}\frac{\text{d}p^*}{\text{d}l}(R^2 - r^2)$$

The maximum velocity (U_{max}) occurs at the pipe centreline, i.e. when $r = 0$, hence

$$\begin{aligned}U_{\text{max}} &= -\frac{1}{4 \times 4.8 \times 10^{-2}} \times -\frac{33.18 \times 10^3}{45}(0.025/2)^2\\ &= 0.6\,\text{m/s}\end{aligned}$$

(*Note*: $U_{\text{max}} = 2\times$ mean velocity (compare (4.2) and (4.4).))

To find the velocity 5 mm from the pipe wall (U_5), use (4.2) with $r = (0.025/2) - 0.005$, i.e. $r = 0.0075$:

$$\begin{aligned}u_5 &= -\frac{1}{4 \times 4.8 \times 10^{-2}} \times -\frac{33.18 \times 10^3}{45}(0.0125^2 - 0.0075^2)\\ &= 0.384\,\text{m/s}\end{aligned}$$

4.5 Turbulent flow

For turbulent flow, Newton's viscosity law does not apply and, as described in Chapter 3, semi-empirical relationships for τ_0 were derived by Prandtl. Also, Reynolds' experiments, and the earlier ones of Darcy and Weisbach, indicated that head loss was proportional to mean velocity squared. Using the momentum equation (4.1), then

$$\tau_0 = -\frac{dp^*}{dl}\frac{R}{2}$$

and

$$-\frac{dp^*}{dl} = \frac{h_f \rho g}{L}$$

hence

$$\tau_0 = \frac{h_f}{L}\rho g \frac{R}{2}$$

Assuming $h_f = KV^2$, based on the experimental results cited above, then

$$\tau_0 = \frac{KV^2}{L}\rho g \frac{R}{2}$$

or

$$\tau_0 = K_1 V^2$$

(for $h_f = KV^2$).

Returning to the momentum equation and making the substitution $\tau_0 = K_1 V^2$, then

$$K_1 V^2 = -\frac{dp^*}{dl}\frac{R}{2}$$

hence

$$K_1 V^2 = \frac{h_f}{L}\rho g \frac{R}{2}$$

or

$$h_f = \frac{4K_1 L V^2}{\rho g D}$$

Making the substitution $\lambda = 8K_1/\rho$, then

$$h_f = \frac{\lambda L V^2}{2gD} \tag{4.8}$$

This is the Darcy–Weisbach equation, in which λ is called the pipe friction factor and is sometimes referred to as f (American practice) or $4f$ (early British practice). In current practice, λ is the normal usage and is found, for instance, in the Hydraulics Research Station charts and tables. It should be noted that λ is dimensionless, and may be used with any system of units.

The original investigators presumed that the friction factor was constant. This was subsequently found to be incorrect (as described in section 3.6). Equations relating λ to both the Reynolds Number and the pipe roughness were developed later.

Smooth pipes and the Blasius equation

Experimental investigations by Blasius and others early in the 20th century led to the equation

$$\lambda = 0.316/\mathrm{Re}^{0.25} \tag{4.9}$$

The later experiments of Stanton and Pannel, using drawn brass tubes, confirmed the validity of the Blasius equation for Reynolds' Numbers up to 10^5. However, at higher values of Re the Blasius equation underestimated λ for these pipes. Before further progress could be made, the distinction between 'smooth' and 'rough' pipes had to be established.

Artificially rough pipes and Nikuradse's experimental results

Nikuradse made a major contribution to the theory of pipe flow by objectively differentiating between smooth and rough turbulence in pipes. He carried out a painstaking series of experiments to determine both the friction factor and the velocity distributions at various Reynolds' Numbers up to 3×10^6. In these experiments, pipes were artificially roughened by sticking uniform sand grains on to smooth pipes. He defined the relative roughness (k_s/D) as the ratio of the sand grain size to the pipe diameter. By using pipes of different diameter and sand grains of different size, he produced a set of experimental results of λ and Re for a range of relative roughness of 1/30 to 1/1014.

He plotted his results as log λ against log Re for each value of K_s/D, as shown in Figure 4.4. This figure shows that there are five regions of flow, as follows:

(a) **Laminar flow.** The region in which the relative roughness has no influence on the friction factor. This was assumed in deriving the Hagen–Poiseuille equation (4.5). Equating this to the Darcy–Weisbach equation (4.8) gives

$$\frac{32\mu VL}{\rho g D^2} = \frac{\lambda L V^2}{2gD}$$

or

$$\lambda = \frac{64\mu}{\rho DV} = \frac{64}{\text{Re}} \tag{4.10}$$

Hence, the Darcy–Weisbach equation may also be used for laminar flow, provided that λ is evaluated by (4.10).

(b) **Transition from laminar to turbulent flow.** An unstable region between Re = 2000 and 4000. Fortunately, pipe flow normally lies outside this region.

(c) **Smooth turbulence.** The limiting line of turbulent flow, approached by all values of relative roughness as Re decreases.

(d) **Transitional turbulence.** The region in which λ varies with both Re and k_s/D. The limit of this region varies with k_s/D. In practice, most of pipe flow lies within this region.

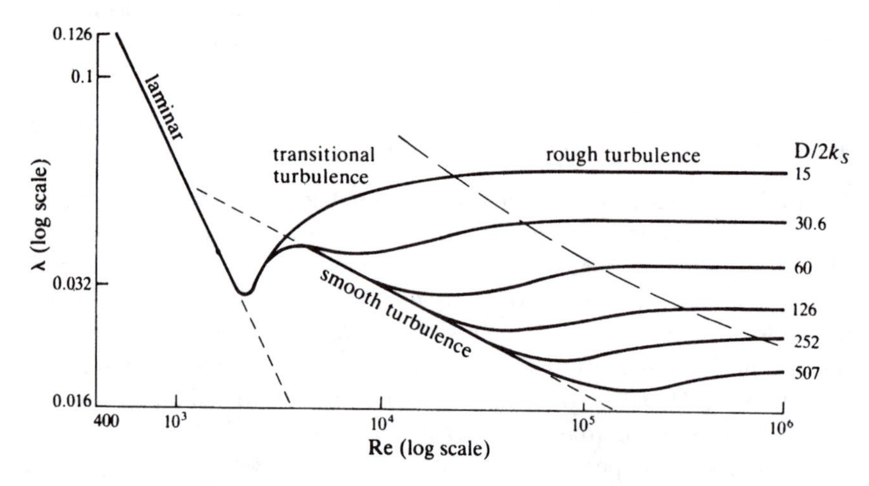

Figure 4.4 Nikuradse's experimental results.

(e) **Rough turbulence.** The region in which λ remains constant for a given k_s/D, and is independent of Re.

An explanation of why these five regions exist has already been given in section 3.6. It may be summarized as follows:

Laminar flow. Surface roughness has no influence on shear stress transmission.

Transitional turbulence. The presence of the laminar sub-layer 'smooths' the effect of surface roughness.

Rough turbulence. The surface roughness is large enough to break up the laminar sub-layer giving turbulence right across the pipe.

The rough and smooth laws of von Kármán and Prandtl

The publication of Nikuradse's experimental results (particularly his velocity distribution measurements) was used by von Kármán and Prandtl to supplement their own work on turbulent boundary layers. By combining their theories of turbulent boundary layer flows with the experimental results, they derived the semi-empirical rough and smooth laws. These were:

for smooth pipes

$$\frac{1}{\sqrt{\lambda}} = 2\log\frac{\mathrm{Re}\sqrt{\lambda}}{2.51} \tag{4.11}$$

for rough pipes

$$\frac{1}{\sqrt{\lambda}} = 2\log\frac{3.7D}{k_s} \tag{4.12}$$

The smooth law is a better fit to the experimental data than the Blasius equation.

The Colebrook–White transition formula

The experimental work of Nikuradse and the theoretical work of von Kármán and Prandtl provided the framework for a theory of pipe friction. However, these results were not of direct use to engineers because they applied only to artificially roughened pipes. Commercial pipes have roughness which is uneven both in size and spacing, and do not, therefore, necessarily correspond to the pipes used in Nikuradse's experiments.

Colebrook and White made two major contributions to the development and application of pipe friction theory to engineering design. Initially, they carried out experiments to determine the effect of non-uniform roughness as found in commercial pipes. They discovered that in the turbulent transition region the λ–Re curves exhibited a gradual change from smooth to rough turbulence in contrast to Nikuradse's 'S'-shaped curves for uniform roughness, size and spacing. Colebrook then went on to determine the 'effective roughness' size of many commercial pipes. He achieved this by studying published results of frictional head loss and discharge for commercial pipes, ranging in size from 4 inches (101.6 mm) to 61 inches (1549.4 mm), and for materials, including drawn brass, galvanized, cast and wrought iron, bitumen-lined pipes and concrete-lined pipes. By comparing the friction factor of these pipes with Nikuradse's results for uniform roughness size in the rough turbulent zone, he was able to determine an 'effective roughness' size for the commercial pipes equivalent to Nikuradse's results. He was thus able to publish a list of k_s values applicable to commercial pipes.

A second contribution of Colebrook and White stemmed from their experimental results on non-uniform roughness. They combined the von *Kármán*–Prandtl rough and smooth laws in the form

$$\frac{1}{\sqrt{\lambda}} = -2\log\left(\frac{k_s}{3.7D} + \frac{2.51}{\mathrm{Re}\sqrt{\lambda}}\right) \tag{4.13}$$

This gave predicted results very close to the observed transitional behaviour of commercial pipes, and is known as the Colebrook–White transition formula. It is applicable to the whole of the turbulent region for commercial pipes using an effective roughness value determined experimentally for each type of pipe.

The practical application of the Colebrook–White transition formula

Equation (4.13) was not at first used very widely by engineers, mainly because it was not expressed directly in terms of the standard engineering variables of diameter, discharge and hydraulic gradient. In addition, the equation is implicit and requires a trial-and-error solution. In the 1940s, slide-rules and logarithm tables were the main computational aids of the engineer, since pocket calculators and computers were not then available. So these objections to the use of the Colebrook–White equation were not unreasonable.

The first attempt to make engineering calculations easier was made by Moody. He produced a λ-Re plot based on (4.13) for commercial pipes, as shown in Figure 4.5 which is now known as the Moody diagram. He also presented an explicit formula for λ:

$$\lambda = 0.0055\left[1+\left(\frac{20\,000k_s}{D}+\frac{10^6}{\mathrm{Re}}\right)^{1/3}\right] \tag{4.14}$$

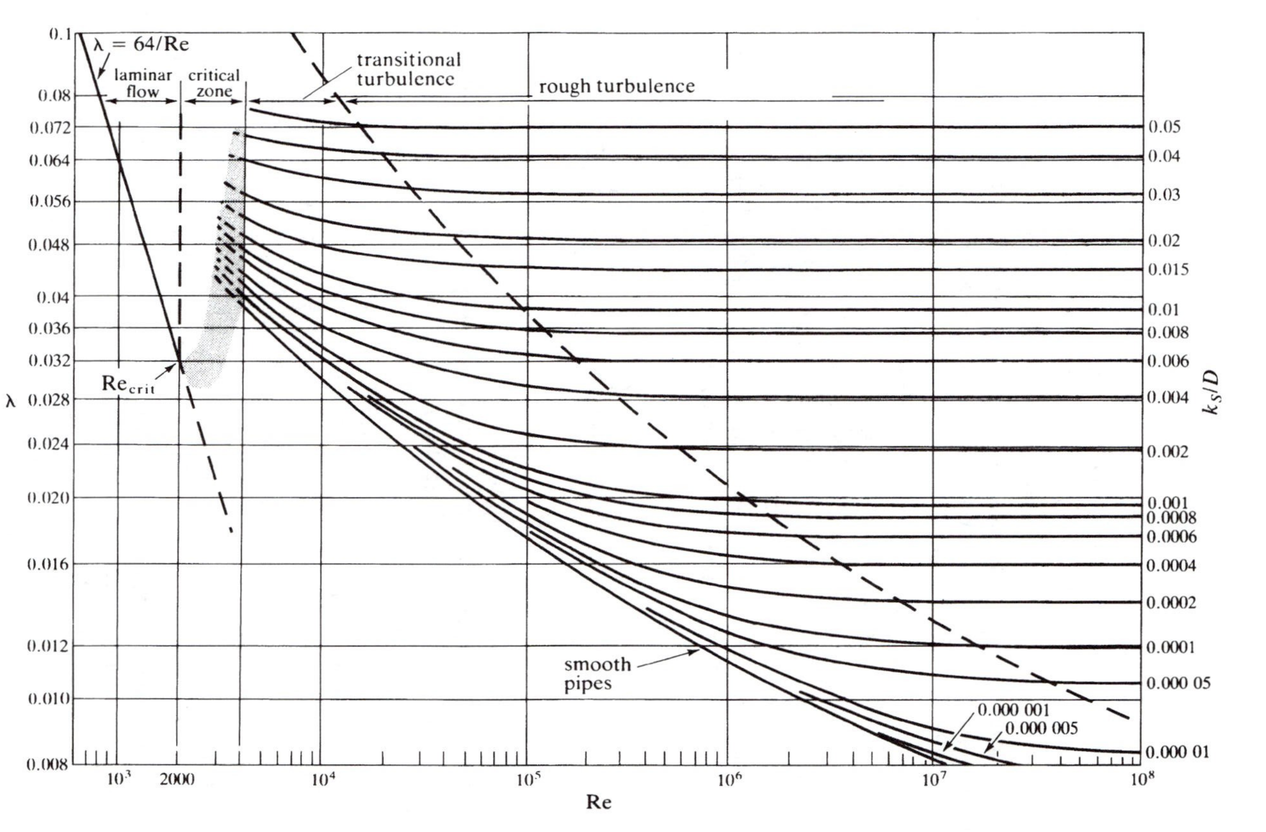

Figure 4.5 The Moody diagram.

which gives λ correct to $\pm 5\%$ for $4 \times 10^3 < \text{Re} < 1 \times 10^7$ and for $k_s/D < 0.01$.

In a more recent publication, Barr (1975) gives another explicit formulation for λ:

$$\frac{1}{\sqrt{\lambda}} = -2\log\left(\frac{k_s}{3.7D} + \frac{5.1286}{\text{Re}^{0.89}}\right) \tag{4.15}$$

In this formula the smooth law component ($2.51/\text{Re}\sqrt{\lambda}$) has been replaced by an approximation ($5.1286/\text{Re}^{0.89}$). For $\text{Re} > 10^5$ this provides a solution for $S_f(h_f/L)$ to an accuracy better than $\pm 1\%$.

However, the basic engineering objections to the use of the Colebrook–White equation were not overcome until the publication of *Charts for the Hydraulic Design of Channels and Pipes* in 1958 by the Hydraulics Research Station. In this publication, the three dependent engineering variables (Q, D and S_f) were presented in the form of a series of charts for various k_s values, as shown in Figure 4.6. Additional information regarding suitable design values for k_s and other matters was also included. Table 4.2 lists typical values for various materials.

These charts are based on the combination of the Colebrook–White equation (4.13) with the Darcy–Weisbach formula (4.8), to give

$$V = -2\sqrt{2gDS_f}\log\left(\frac{k_s}{3.7D} + \frac{2.51\upsilon}{D\sqrt{2gDS_f}}\right) \tag{4.16}$$

where $S_f = h_f/L$, the hydraulic gradient. (*Note*: for further details concerning the hydraulic gradient refer to Chapter 12.) In this equation the velocity (and hence discharge) can be computed directly for a known diameter and frictional head loss.

More recently, the Hydraulics Research Station have also produced *Tables for the Hydraulic Design of Pipes*.

In practice, any two of the three variables (Q, D and S_f) may be known, and therefore the most appropriate solution technique depends on circumstances. For instance, in the case of an existing pipeline, the diameter and available head are known and hence the discharge may be found directly from (4.16). For the case of a new installation, the available head and required discharge are known and the requisite diameter must be found. This will involve a trial-and-error procedure unless the HRS charts or tables are used. Finally, in the case of analysis of pipe networks, the required discharges and pipe diameters are known and the head loss must be computed. This problem may be most easily solved using an explicit formula for λ or the HRS charts.

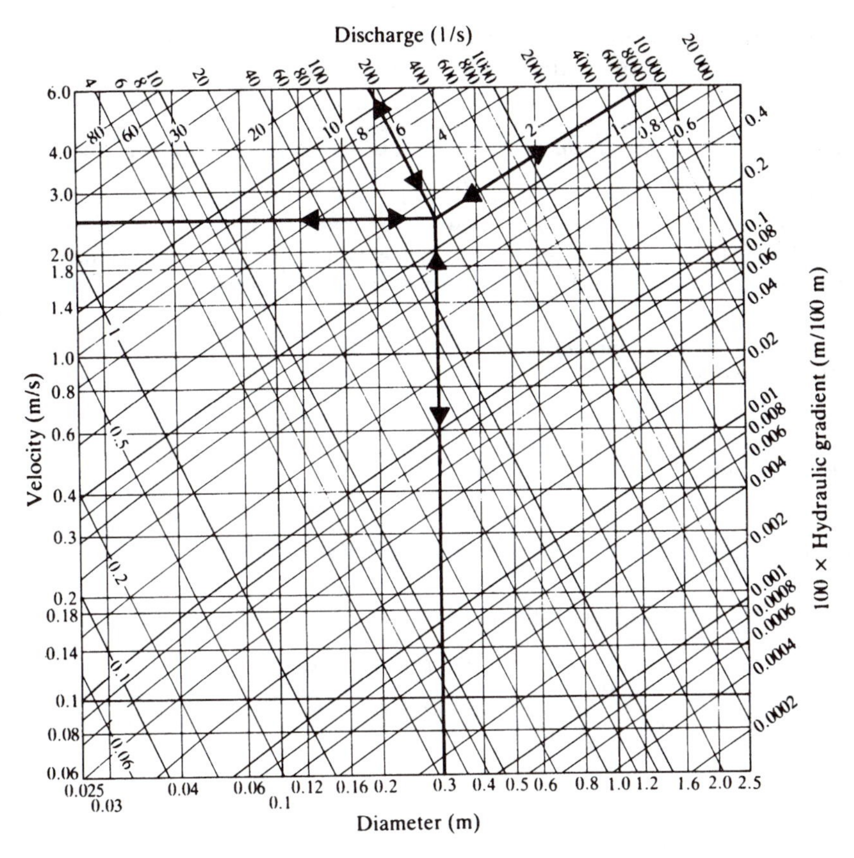

Figure 4.6 Hydraulics Research Station chart for $k_s = 0.03\,\text{mm}$.

Table 4.2 Typical k_s values.

Pipe material	k_s (mm)
brass, copper, glass, Perspex	0.003
asbestos cement	0.03
wrought iron	0.06
galvanized iron	0.15
plastic	0.03
bitumen-lined ductile iron	0.03
spun concrete lined ductile iron	0.03
slimed concrete sewer	6.0

Examples illustrating the application of the various methods to the solution of a simple pipe friction problem now follow.

Example 4.2 Estimation of discharge given diameter and head loss

A pipeline 10 km long, 300 mm in diameter and with roughness size 0.03 mm, conveys water from a reservoir (top water level 850 m above datum) to a water treatment plant (inlet water level 700 m above datum). Assuming that the reservoir remains full, estimate the discharge, using the following methods:

(a) the Colebrook–White formula;
(b) the Moody diagram;
(c) the HRS charts.

Note: Assume $\nu = 1.13 \times 10^{-6}\,\text{m}^2/\text{s}$.

Solution

(a) Using (4.16),

$$D = 0.3\ \text{m} \quad k_s = 0.03\,\text{mm}$$

$$S_f = (850 - 700)/10\ 000 = 0.015$$

hence

$$V = -2\sqrt{2g \times 0.3 \times 0.015}\log\left(\frac{0.03 \times 10^{-3}}{3.7 \times 0.3} + \frac{2.51 \times 1.13 \times 10^{-6}}{0.3\sqrt{2g \times 0.3 \times 0.015}}\right)$$

$$= 2.514\,\text{m/s}$$

and

$$Q = VA = \frac{2.514 \times \pi \times 0.3^2}{4} = 0.178\,\text{m}^3/\text{s}$$

(b) The same solution should be obtainable using the Moody diagram; however, it is less accurate since it involves interpolation from a graph. The solution method is as follows:

(1) calculate k_S/D
(2) guess a value for V
(3) calculate Re
(4) estimate λ using the Moody diagram
(5) calculate h_f
(6) compare h_f with the available head (H)
(7) if $H \neq h_f$, then repeat from step 2.

This is a tedious solution technique, but it shows why the HRS charts were produced!

(1) $k_S/D = 0.03 \times 10^{-3}/0.3 = 0.0001$.
(2) As the solution for V has already been found in part (a) take $V = 2.5\,\text{m/s}$.
(3)

$$\text{Re} = \frac{DV}{\nu} = \frac{0.3 \times 2.5}{1.13 \times 10^{-6}}$$
$$= 0.664 \times 10^6$$

(4) Referring to Figure 4.5, $\text{Re} = 0.664 \times 10^6$ and $k_s/D = 0.0001$ confirms that the flow is in the transitional turbulent region. Following the k_s/D curve until it intersects with Re yields

$$\lambda \simeq 0.014$$

(*Note*: Interpolation is difficult due to the logarithmic scale.)
(5) Using (4.8),

$$h_f = \frac{\lambda L V^2}{2gD} = \frac{0.014 \times 10^4 \times 2.5^2}{2g \times 0.3}$$
$$= 148.7\,\text{m}$$

(6) $H = (850 - 700) = 150 \neq 148.7$.
(7) A better guess for V is obtained by increasing V slightly. This will not significantly alter λ, but will increase h_f. In this instance, convergence to the solution is rapid because the correct solution for V was assumed initially!

(c) If the HRS chart shown in Figure 4.6 is used, then the solution of the equation lies at the intersection of the hydraulic gradient line (sloping downwards right to left) with the diameter (vertical), reading off the corresponding discharge (line sloping downwards left to right).

$S_f = 0.015 \qquad 100S_f = 1.5$
and $D = 300\,\text{mm}$
giving $Q = 180\ \text{l/s} = 0.18\,\text{m}^3/\text{s}$

Example 4.3 Estimation of pipe diameter given discharge and head

A discharge of 400 l/s is to be conveyed from a headworks at 1050 m above datum to a treatment plant at 1000 m above datum. The pipeline length is 5 km. Estimate the required diameter, assuming that $k_S = 0.03\,\text{mm}$.

Solution

This requires an iterative solution if methods (a) or (b) of the previous example are used. However, a direct solution can be obtained using the HRS charts.

$S_f = 50/5000 \qquad 100S_f = 1$
and $Q = 400$ l/s
giving $D = 440$ mm

In practice, the nearest (larger) available diameter would be used (450 mm in this case).

Example 4.4 Estimation of head loss given discharge and diameter

The known outflow from a branch of a distribution system is 30 l/s. The pipe diameter is 150 mm, length 500 m and roughness coefficient estimated at 0.06 mm. Find the head loss in the pipe, using the explicit formulae of Moody and Barr.

Solution

Again, the HRS charts could be used directly. However, if the analysis is being carried out by computer, solution is more efficient using an equation.

$$Q = 0.03\,\text{m}^3/\text{s}, \quad D = 0.15\,\text{m}$$

$$V = 1.7\,\text{m/s}$$

$$\text{Re} = 0.15 \times 1.7/1.13 \times 10^{-6}$$

$$\text{Re} = 0.226 \times 10^6$$

Using the Moody formula (4.14)

$$\lambda = 0.0055\left(1 + \left(\frac{20\,000 \times 0.06 \times 10^{-3}}{0.15} + \frac{1}{0.226}\right)^{1/3}\right)$$

$$\lambda = 0.0182$$

Using the Barr formula (4.15)

$$\frac{1}{\sqrt{\lambda}} = -2\log\left(\frac{0.06 \times 10^{-3}}{3.7 \times 0.15} + \frac{5.1286}{(0.226 \times 10^6)^{0.89}}\right)$$

$$\lambda = 0.0182$$

The accuracy of these formulae may be compared by substituting in the Colebrook–White equation (4.13) as follows:

λ	$1/\sqrt{\lambda}$	$-2\log\left(\dfrac{k_S}{3.7D}+\dfrac{2.51}{\mathrm{Re}\sqrt{\lambda}}\right)=\dfrac{1}{\sqrt{\lambda}}$
0.0182	7.415	7.441

This confirms that both formulae are accurate in this case.

The head loss may now be computed using the Darcy–Weisbach formula (4.8):

$$h_f = \frac{0.0182\times 500\times 1.7^2}{2g\times 0.15}$$

$$= 8.94\,\mathrm{m}$$

The Hazen–Williams formula

The emphasis here has been placed on the development and use of the Colebrook–White transition formula. Using the charts or tables it is simple to apply to single pipelines. However, for pipes in series or parallel or for the more general case of pipe networks it rapidly becomes impossible to use for hand calculations. For this reason, simpler empirical formulae are still in common use. Perhaps the most notable is the Hazen–Williams formula, which takes the form

$$V = 0.355CD^{0.63}(h_f/L)^{0.54}$$

or, alternatively,

$$h_f = \frac{6.78L}{D^{1.165}}\left(\frac{V}{C}\right)^{1.85}$$

where C is a coefficient. The value of C varies from about 70 to 150, depending on pipe diameter, material and age.

This formula gives reasonably accurate results over the range of Re commonly found in water distribution systems, and because the value of C is assumed to be constant, it can be easily used for hand calculation. In reality, C should change with Re, and caution should be exercised in its use. An interesting problem is to compare the predicted discharges as calculated by the Colebrook–White equation and by the Hazen–Williams formula over a large range of Re for a given pipe. The use of a microcomputer is recommended for this exercise.

4.6 Local head losses

Head losses, in addition to those due to friction, are always incurred at pipe bends, junctions and valves, etc. These additional losses are due to eddy formation generated in the fluid at the fitting, and, for completeness, they must be taken into account. In the case of long pipelines (e.g. several kilometres) the local losses may be negligible, but for short pipelines, they may be greater than the frictional losses.

A general theoretical treatment for local head losses is not available. It is usual to assume rough turbulence since this leads to the simple equation

$$h_L = k_L V^2/2g \tag{4.17}$$

where h_L is the local head loss and k_L is a constant for a particular fitting.

For the particular case of a sudden enlargement (for instance, exit from a pipe to a tank) an expression may be derived for k_L in terms of the area of the pipe. This result may be extended to the case of a sudden contraction (for instance, entry to a pipe from a tank). For all other cases (e.g. bends, valves, junctions, bellmouths, etc.) values for k_L must be derived experimentally

Figure 4.7(a) shows the case of a sudden enlargement. From position (1) to (2) the velocity decreases and therefore the pressure increases. At position

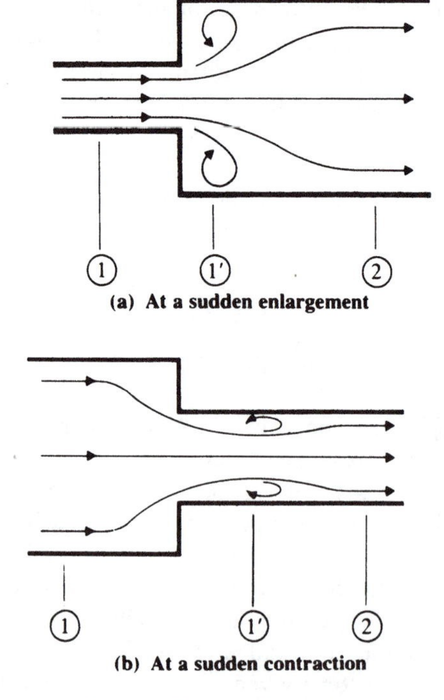

Figure 4.7 Local head loss.

(1′) turbulent eddies are formed, which gives rise to a local energy loss. As the pressure cannot change instantaneously at the sudden enlargement, it is usually assumed that at position (1′) the pressure is the same as at position (1). Applying the momentum equation between (1) and (2),

$$p_1A_2 - p_2A_2 = \rho Q(V_2 - V_1)$$

The continuity equation ($Q = A_2V_2$) is now used to eliminate Q, so, with some rearrangement,

$$\frac{p_2 - p_1}{\rho g} = \frac{V_2}{g}(V_1 - V_2) \qquad \text{(a)}$$

The local head loss may now be found by applying the energy equation from (1) to (2):

$$\frac{p_1}{\rho g} + \frac{V_1^2}{2g} = \frac{p_2}{\rho g} + \frac{V_2^2}{2g} + h_L$$

or

$$h_L = \frac{(V_1^2 - V_2^2)}{2g} - \frac{(p_2 - p_1)}{\rho g} \qquad \text{(b)}$$

If (a) and (b) are combined and rearranged,

$$h_L = (V_1 - V_2)^2/2g$$

The continuity equation may now be used again to express the result in terms of the two areas. Hence, substituting V_1A_1/A_2 for V_2

$$h_L = \left(V_1 - V_1\frac{A_1}{A_2}\right)^2 \Big/ 2g$$

or

$$h_L = \left(1 - \frac{A_1}{A_2}\right)^2 \frac{V_1^2}{2g} \qquad (4.18)$$

Equation (4.18) relates h_L to the areas and the upstream velocity. Comparing this equation with (4.17) yields

$$k_L = \left(1 - \frac{A_1}{A_2}\right)^2$$

For the case of a pipe discharging into a tank, A_2 is much greater than A_1, and hence $k_L = 1$. In other words, for a sudden large expansion, the head loss equals the velocity head before expansion.

Figure 4.7(b) shows the case of a sudden contraction. From position (1) to (1′) the flow contracts, forming a vena contracta. Experiments indicate that the contraction of the flow area is generally about 40%. If the energy loss from (1) to (1′) is assumed to be negligible, then the remaining head loss occurs in the expansion from (1′) to (2). Since an expansion loss gave rise to (4.18), that equation may now be applied here. As

$$A_1' \simeq 0.6A_2$$

then

$$h_L = \left(1 - \frac{0.6A_2}{A_2}\right)^2 \frac{(V_2/0.6)^2}{2g}$$

or

$$h_L = 0.44V_2^2/2g \tag{4.19}$$

i.e. $k_L = 0.44$.

Typical k_L values for other important local losses (bends, tees, bellmouths and valves) are given in Table 4.3.

Table 4.3 Local head loss coefficients.

Item	k_L value		Comments
	Theoretical	Design practice	
bellmouth entrance	0.05	0.10	V = velocity in pipe
exit	0.2	0.5	
90° bend	0.4	0.5	
90° tees			
in-line flow	0.35	0.4	(for equal diameters)
branch to line	1.20	1.5	(for equal diameters)
gate valve (open)	0.12	0.25	

Example 4.5 Discharge calculation for a simple pipe system including local losses

Solve Example 4.2 allowing for local head losses incurred by the following items:

20 90° bends
2 gate valves
1 bellmouth entry
1 bellmouth exit

Solution

The available static head (150 m) is dissipated both by friction and local head losses. Hence

$$H = h_f + h_L$$

Using Table 4.3,

$$h_L = [(20 \times 0.5) + (2 \times 0.25) + 0.1 + 0.5]V^2/2g$$
$$= 11.1V^2/2g$$

Using the Colebrook–White formula (as in Example 4.2) now requires an iterative solution, since h is initially unknown. A solution procedure is as follows:

(1) assume $h_f \simeq H$ (i.e. ignore h_L)
(2) calculate V
(3) calculate h_L using V
(4) calculate $h_f + h_L$
(5) if $h_f + h_L \neq H$, set $h_f = H - h_L$ and return to (2)

Using Example 4.2, an initial solution for V has already been found, i.e.

$$V = 2.514\,\text{m/s}$$

Hence,

$$h_L = 11.1 \times 2.514^2/2g = 3.58\,\text{m}$$

Adjust h,

$$h_f = 150 - 3.58 = 146.42\,\text{m}$$

Hence,

$$S_f = 146.42/10\,000 = 0.01464$$

Substitute in (4.16),

$$V = -2\sqrt{2g \times 0.3 \times 0.01464} \log \left(\frac{0.03 \times 10^{-3}}{3.7 \times 0.15} + \frac{2.51 \times 1.13 \times 10^{-6}}{0.3\sqrt{2g \times 0.3 \times 0.01464}} \right)$$

$$= 2.386\,\text{m/s}$$

Recalculate h_L,

$$h_L = 11.1 \times 2.386^2/2g = 3.22$$

Check

$$h_L + h_f = 146.42 + 3.22 = 149.64 \simeq 150$$

This is sufficiently accurate to be acceptable.

Hence,

$$Q = 2.386 \times \pi \times 0.3^2/4 = 0.17\,\text{m}^3/\text{s}$$

Note: Ignoring h_L gives $Q = 0.18\,\text{m}^3/\text{s}$.

4.7 Partially full pipes

Pipe systems for surface water drainage and sewerage are normally designed to flow full, but not under pressure. This contrasts with water mains, which are normally full and under pressure. The Colebrook–White equation may be used for drainage pipes by noting that, because the pipe flow is not pressurized, the water surface is parallel to the pipe invert, so the hydraulic gradient equals the pipe gradient:

$$h_f/L = S_0$$

where S_0 is the pipe gradient.

Additionally, an estimate of the discharge and velocity for the partially full condition is required. This enables the engineer to check if self-cleansing velocities are maintained at the minimum discharge. Self-cleansing velocities are of crucial importance in the design of surface water drainage and sewerage networks, where the flow may contain a considerable suspended solids load.

A free surface flow has one more variable than full pipe flow, namely the height of the free surface. This can introduce considerable complexity (refer to Chapter 5). However, for the case of circular conduits, the Colebrook–White equation may be modified to provide a solution.

Starting from the assumption that the friction factor for the partially full condition behaves similarly to that for the full condition, it remains to find a parameter for the partially full pipe which is equivalent to the diameter for the full pipe case. The hydraulic radius R is such a parameter:

$$R = A/P$$

where A is the water cross-sectional area and P is the wetted perimeter. For a pipe flowing full,

$$A/P = \pi D^2/4\pi D = D/4$$

or

$$4R = D$$

Hence the Colebrook–White transition law applied to partially full pipes becomes

$$\frac{1}{\sqrt{\lambda}} = -2\log\left(\frac{k_s}{3.7\times 4R} + \frac{2.51}{\mathrm{Re}\sqrt{\lambda}}\right) \tag{4.20}$$

where $\mathrm{Re} = 4RV/\nu$.

Figure 4.8 shows a pipe with partially full flow (at a depth d). Starting from the Darcy–Weisbach equation (4.8) and replacing h_f/L by S_0 gives

$$V^2 = 2gS_0D/\lambda$$

Hence, for a given pipe with partially full flow,

$$V = (2gS_0 4R/\lambda)^{1/2}$$

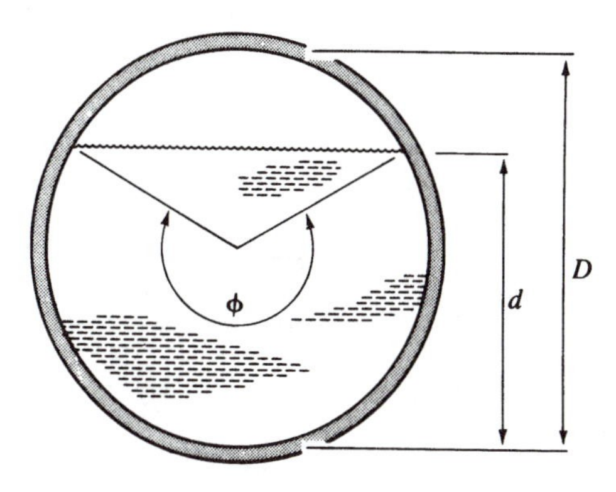

Figure 4.8 Pipe running partially full.

or

$$V = \text{constant} \cdot R^{1/2}/\lambda^{1/2}$$

Forming the ratio $V_d/V_D = V_p$ gives

$$V_p = \lambda_D^{1/2} R_p^{1/2} / \lambda_d^{1/2} \quad (4.21)$$

where the subscripts p, D and d refer, respectively, to the proportional value, the full depth (D) and the partially full depth (d). Similarly,

$$Q_p = \lambda_D^{1/2} A_p R_p^{1/2} / \lambda_d^{1/2} \quad (4.22)$$

For a circular pipe,

$$A_d = \left(\frac{\phi - \sin\phi}{8}\right) D^2$$

$$P_d = \phi D/2$$

$$R_d = \left(1 - \frac{\sin\phi}{\phi}\right)\frac{D}{4}$$

and hence

$$A_p = \left(\frac{\phi - \sin\phi}{2\pi}\right) \quad (4.23)$$

$$R_p = \left(1 - \frac{\sin\phi}{\phi}\right) \quad (4.24)$$

Substitution of (4.23) and (4.24) into (4.21) and (4.22) allows calculation of the proportional velocity and discharge for any proportional depth (d/D). The expression for λ (equation (4.20)) is, however, rather awkward to manipulate. Consider first the case of rough turbulence. Then,

$$\frac{1}{\sqrt{\lambda}} = 2\log\left(\frac{3.7D}{k_S}\right)$$

Hence,

$$\frac{\sqrt{\lambda_D}}{\sqrt{\lambda_d}} = \frac{2\log(3.7 \times 4R_d/k_S)}{2\log(3.7D/k_S)}$$

This may be expressed by its equivalent:

$$\frac{\sqrt{\lambda_D}}{\sqrt{\lambda_d}} = 1 + \frac{\log R_p}{\log(3.7D/k_S)} \quad (4.25)$$

as

$$1+\frac{\log R_p}{\log(3.7D/k_S)}=\frac{\log(3.7D/k_S)+\log R_p}{\log(3.7D/k_S)}$$

$$=\frac{\log[(3.7D/k_S)(R_d/R_D)]}{\log(3.7D/k_S)}$$

$$=\frac{\log[3.7\times 4R_d/k_S]}{\log(3.7D/k_S)}$$

Equation (4.25) may be substituted into (4.21) and (4.22) to yield

$$V_p=\left(1+\frac{\log R_p}{\log(3.7D/k_S)}\right)R_p^{1/2}$$

and

$$Q_p=\left(1+\frac{\log R_p}{\log(3.7D/k_S)}\right)A_pR_p^{1/2}$$

The equivalent expressions for the transition region (as derived in Hydraulics Research Paper No. 2, published in 1959) are

$$V_p=\left(1+\frac{\log R_p}{\log 3.7\theta}\right)R_p^{1/2} \tag{4.26}$$

and

$$Q_p=\left(1+\frac{\log R_p}{\log 3.7\theta}\right)A_pR_p^{1/2} \tag{4.27}$$

where

$$\theta\simeq\left(\frac{k_S}{D}+\frac{1}{3600DS_0^{1/3}}\right)^{-1} \tag{4.28}$$

These results for $\theta = 1000$ are plotted in Figure 4.9. Tabulated values for various θ may be found in Hydraulics Research Ltd (1983a). Neither V_p nor Q_p are very sensitive to θ.

Figure 4.9 shows that the discharge in a partially full pipe may be greater than the discharge for a full pipe. This is because the wetted perimeter reduces rapidly immediately the pipe ceases to be full whereas the area does not, with a consequent increase in velocity. However, this condition is usually ignored for design purposes because, if the pipe runs full at any

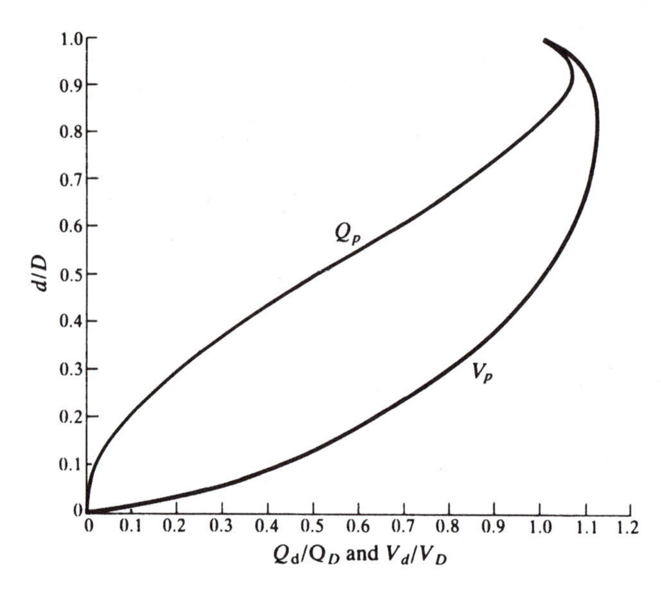

Figure 4.9 Proportional discharge and velocity for pipes flowing partially full (with $\theta = 1000$).

section (e.g. due to wave action or unsteady conditions), then the discharge will rapidly reduce to the full pipe condition and cause a 'backing up' of the flow upstream.

Example 4.6 Hydraulic design of a sewer

A sewerage pipe is to be laid at a gradient of 1 in 300. The design maximum discharge is 75 l/s and the design minimum flow is estimated to be 10 l/s. Determine the required pipe diameter to both carry the maximum discharge and maintain a self-cleansing velocity of 0.75 m/s at the minimum discharge.

Solution

The easiest way to solve this problem is to use the HRS design charts or tables. For a sewer, $k_S = 6.00\,\text{mm}$ (Table 4.2). However, to illustrate the solution, Figure 4.6 is used (for which $k_S = 0.03\,\text{mm}$):

$$Q = 75\,\text{l/s}$$

$$100h_f/L = 100/300 = 0.333$$

Using Figure 4.6

$$D = 300\,\text{mm} \quad \text{and} \quad V = 1.06\,\text{m/s}$$

Next check the velocity for $Q = 10$ l/s

$$Q_p = 10/75 = 0.133$$

Using Figure 4.9 (neglecting the effect of θ),

$$d/D = 0.25 \text{ for } Q_p = 0.133$$

Hence $V_p = 0.72$ and

$$V_d = 0.72 \times 1.06 = 0.76\,\text{m/s}$$

This value exceeds the self-cleansing velocity, and hence the solution is $D = 300$ mm. In cases where the self-cleansing velocity is not maintained, it is necessary to increase the diameter or the pipe gradient.

Note: The solution using $k_S = 6$ mm and accounting for θ gives the following values:

$D = 375$ mm for $Q = 81$ l/s and $V = 0.73$ m/s
$\theta = 45$
$Q_p = 10/81 = 0.123$
$d/D = 0.024$
$V_p = 0.67$ m/s
$V_d = 0.49$ m/s

Hence it would be necessary to increase D or S_0. In this case, increasing S_0 would be preferable.

References and further reading

Ackers, P. (1958) *Resistance of Fluids Flowing in Channels and Pipes*. Hydraulics Research Paper No. 1, HMSO, London.

Barr, D. I. H. (1975) Two additional methods of direct solution of the Colebrook–White function. *Proc. Instn Civ. Engrs*, **59**, 827.

Colebrook, C. F. (1939) Turbulent flows in pipes, with particular reference to the transition region between the smooth and rough pipe laws. *J. Instn Civ. Engrs*, **11**, 133.

Colebrook, C. F. and White, C. M. (1937) Experiments with fluid friction in roughened pipes. *Proc. Roy. Soc.*, **A161**, 367.

Hydraulics Research Limited (1983a) *Tables for the Hydraulic Design of Pipes*, 4th edn, Thomas Telford, London.

Hydraulics Research Limited (1983b) *Charts for the Hydraulic Design of Channels and Pipes*, 5th edn, Thomas Telford, London.

Moody, L. F. (1944) Friction factors for pipe flows. *Trans. Am. Soc. Mech. Engrs.*, **66**, 671.

Webber, N. B. (1971) *Fluid Mechanics for Civil Engineers*, Chapman and Hall, London.

5

Open channel flow

5.1 Flow with a free surface

Open channel flow is characterized by the existence of a free surface (the water surface). In contrast to pipe flow, this constitutes a boundary at which the pressure is atmospheric, and across which the shear forces are negligible. The longitudinal profile of the free surface defines the hydraulic gradient and determines the cross-sectional area of flow, as is shown in Figure 5.1. It also necessitates the introduction of an extra variable – the stage (see Fig. 5.1) – to define the position of the free surface at any point in the channel.

In consequence, problems in open channel flow are more complex than those of pipe flow, and the solutions are more varied, making the study of such problems both interesting and challenging. In this chapter the basic concepts are introduced and a variety of common engineering applications are discussed.

5.2 Flow classification

The fundamental types of flow have already been discussed in Chapter 2. However, it is appropriate here to expand those descriptions as applied to open channels. Recalling that flow may be steady or unsteady and uniform or non-uniform, the major classifications applied to open channels are as follows:

Steady uniform flow, in which the depth is constant, both with time and distance. This constitutes the fundamental type of flow in an open channel in which the gravity forces are in equilibrium with the resistance forces. It is considered in section 5.6.

Steady non-uniform flow, in which the depth varies with distance, but not with time. The flow may be either (a) gradually varied or (b) rapidly varied. Type (a) requires the joint application of energy and frictional resistance

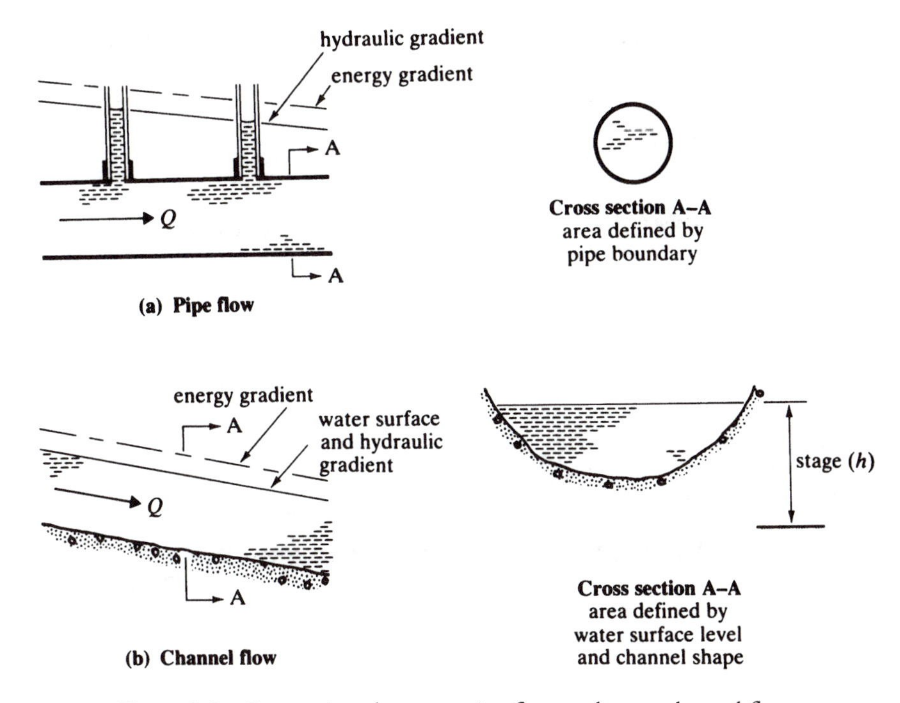

Figure 5.1 Comparison between pipe flow and open channel flow.

equations, and is considered in section 5.10. Type (b) requires the application of energy and momentum principles, and is considered in sections 5.7 and 5.8.

Unsteady flow, in which the depth varies with both time and distance (unsteady uniform flow is very rare). This is the most complex flow type, requiring the solution of energy, momentum and friction equations through time. It is considered in section 5.11.

The various flow types are all shown in Figure 5.2.

5.3 Natural and artificial channels and their properties

Artificial channels comprise all man-made channels, including irrigation and navigation canals, spillway channels, sewers, culverts and drainage ditches. They are normally of regular cross-sectional shape and bed slope, and as such are termed prismatic channels. Their construction materials are varied, but commonly used materials include concrete, steel and earth. The surface roughness characteristics of these materials are normally well defined within engineering tolerances. In consequence, the application of hydraulic theories to flow in artificial channels will normally yield reasonably accurate results.

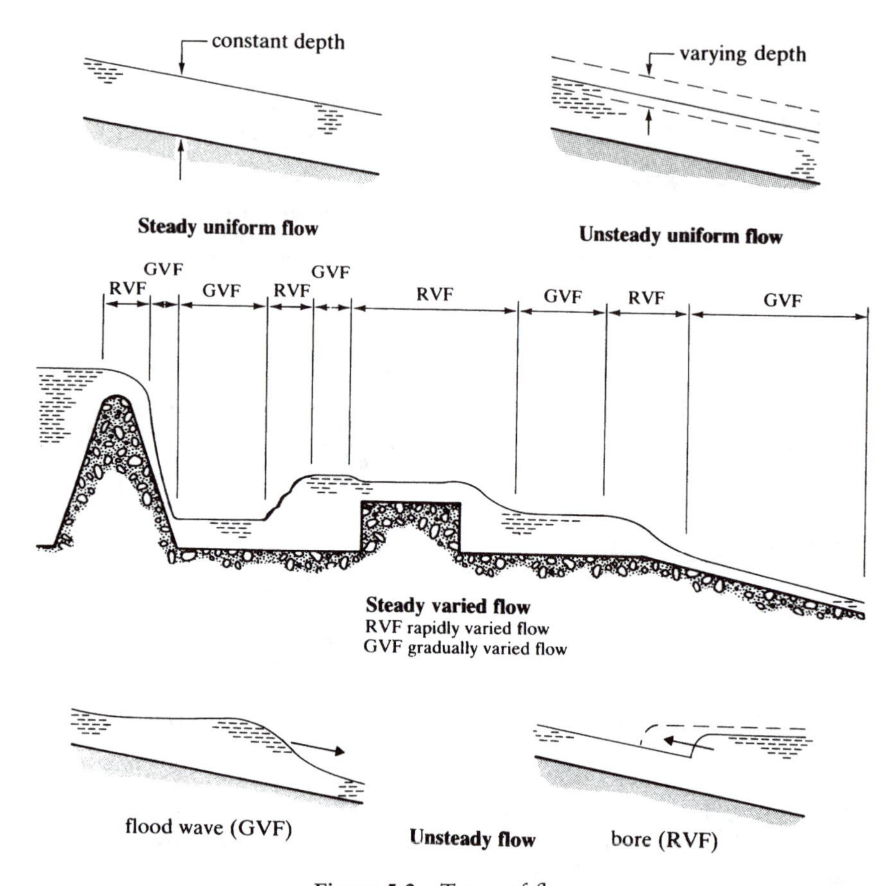

Figure 5.2 Types of flow.

In contrast, natural channels are normally very irregular in shape, and their materials are diverse. The surface roughness of natural channels changes with time, distance and water surface elevation. Therefore, it is more difficult to apply hydraulic theory to natural channels and obtain satisfactory results. Many applications involve man-made alterations to natural channels (e.g. river control structures and flood alleviation measures). Such applications require an understanding not only of hydraulic theory, but also of the associated disciplines of sediment transport, hydrology and river morphology (refer to Chapters 9, 10 and 15).

Various geometric properties of natural and artificial channels need to be determined for hydraulic purposes. In the case of artificial channels, these may all be expressed algebraically in terms of the depth (y), as is shown in Table 5.1. This is not possible for natural channels, so graphs or tables relating them to stage (h) must be used.

Table 5.1 Geometric properties of some common prismatic channels.

	Rectangle	Trapezoid	Circle
area, A	by	$(b+xy)y$	$\frac{1}{8}(\phi - \sin\phi)D^2$
wetted perimeter, P	$b+2y$	$b+2y\sqrt{1+x^2}$	$\frac{1}{2}\phi D$
top width, B	b	$b+2xy$	$\left(\sin\frac{\phi}{2}\right)D$
hydraulic radius, R	$\frac{by}{b+2y}$	$\frac{(b+xy)y}{b+2y\sqrt{1+x^2}}$	$\frac{1}{4}\left(1-\frac{\sin\phi}{\phi}\right)D$
hydraulic mean depth, D_m	y	$\frac{(b+xy)y}{b+2xy}$	$\frac{1}{8}\left(\frac{\phi-\sin\phi}{\sin(1/2\phi)}\right)D$

The commonly used geometric properties are shown in Figure 5.3 and defined as follows:

Depth (y) – the vertical distance of the lowest point of a channel section from the free surface;
Stage (h) – the vertical distance of the free surface from an arbitrary datum;
Area (A) – the cross-sectional area of flow normal to the direction of flow;
Wetted perimeter (P) – the length of the wetted surface measured normal to the direction of flow;
Surface width (B) – the width of the channel section at the free surface;
Hydraulic radius (R) – the ratio of area to wetted perimeter (A/P);
Hydraulic mean depth (D_m) – the ratio of area to surface width (A/B).

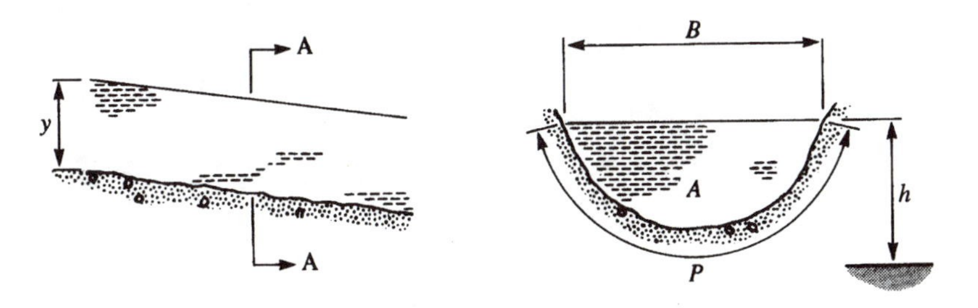

Figure 5.3 Definition sketch of geometric channel properties.

5.4 Velocity distributions, energy and momentum coefficients

The point velocity in an open channel varies continuously across the cross-section because of friction along the boundary. However, the velocity distribution is not axisymmetric (as in pipe flow) due to the presence of the free surface. One might expect to find the maximum velocity at the free surface, where the shear stress is negligible, but the maximum velocity normally occurs below the free surface. Typical velocity distributions are shown in Figure 5.4 for various channel shapes.

The depression of the point of maximum velocity below the free surface may be explained by the presence of secondary currents which circulate from the boundaries towards the channel centre. Detailed experiments on velocity distributions have demonstrated the existence of such secondary currents and recent theoretical studies concerning three-dimensional turbulence have illuminated the mechanisms for their existence (refer to section 15.7 for further details).

The energy and momentum coefficients (α and β) defined in Chapter 2 can only be evaluated for a channel if the velocity distribution has been measured. This contrasts with pipe flow, where theoretical velocity distributions for laminar and turbulent flow have been derived, which enable direct integration of the defining equations to be made.

For turbulent flow in regular channels, α rarely exceeds 1.15 and β rarely exceeds 1.05. In consequence, these coefficients are normally assumed to be unity. However, in irregular channels where the flow may divide into distinct regions, α may exceed 2 and should therefore be included in flow computations. Referring to Figure 5.5, which shows a natural channel with two flood banks, the flow may be divided into three regions. By making the

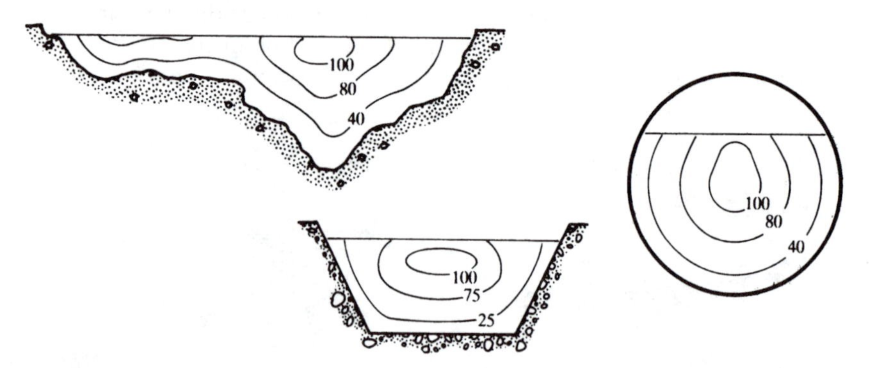

Figure 5.4 Velocity distributions in open channels. (*Note*: Contour numbers are expressed as a percentage of the maximum velocity.)

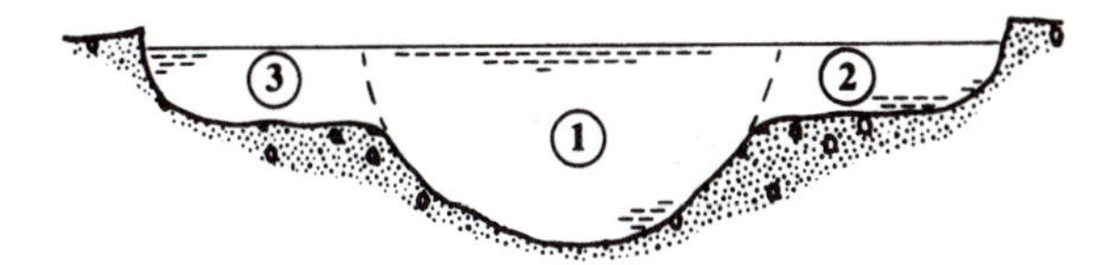

Figure 5.5 Division of a channel into a main channel and flood banks.

assumption that $\alpha = 1$ for each region, the value of α for the whole channel may be found as follows:

$$\alpha = \frac{\int u^3 \, \mathrm{d}A}{\overline{V}^3 A} = \frac{V_1^3 A_1 + V_2^3 A_2 + V_3^3 A_3}{\overline{V}^3 (A_1 + A_2 + A_3)} \tag{5.1}$$

where

$$\overline{V} = \frac{Q}{A} = \frac{V_1 A_1 + V_2 A_2 + V_3 A_3}{A_1 + A_2 + A_3}$$

5.5 Laminar and turbulent flow

In section 5.2, the state of flow was not discussed. It may be either laminar or turbulent, as in pipe flow. The criterion for determining whether the flow is laminar or turbulent is the Reynolds Number (Re), which was introduced in Chapters 3 and 4:

for pipe flow $\qquad \mathrm{Re} = \rho D V / \mu \qquad$ (3.2)

for laminar pipe flow $\qquad \mathrm{Re} < 2000$

for turbulent pipe flow $\qquad \mathrm{Re} > 4000$

These results can be applied to open channel flow if a suitable form of the Reynolds Number can be found. This requires that the characteristic length dimension, the diameter (for pipes), be replaced by an equivalent characteristic length for channels. The one adopted is termed the hydraulic radius (R) as defined in section 5.3.

Hence, the Reynolds Number for channels may be written as

$$\mathrm{Re}_{(\text{channel})} = \rho R V / \mu \tag{5.2}$$

For a pipe flowing full, $R = D/4$, so

$$\mathrm{Re}_{(\text{channel})} = \mathrm{Re}_{(\text{pipe})}/4$$

and for laminar channel flow

$$\mathrm{Re}_{(\mathrm{channel})} < 500$$

and for turbulent channel flow

$$\mathrm{Re}_{(\mathrm{channel})} > 1000$$

In practice, the upper limit of Re is not so well defined for channels as it is for pipes, and is normally taken to be 2000.

In Chapter 4, the Darcy–Weisbach formula for pipe friction was introduced, and the relationship between laminar, transitional and turbulent flow was depicted on the Moody diagram. A similar diagram for channels has been developed. Starting from the Darcy–Weisbach formula,

$$h_f = \lambda L V^2/2gD \tag{4.8}$$

and making the substitutions $R = D/4$ and $h_f/L = S_0$ (where S_0 = bed slope), then for uniform flow in an open channel

$$S_0 = \frac{\lambda V^2}{2g4R}$$

or

$$\lambda = 8gRS_0/V^2 \tag{5.3}$$

The λ–Re relationship for pipes is given by the Colebrook–White transition law, and by substituting $R = D/4$ the equivalent formula for channels is

$$\frac{1}{\sqrt{\lambda}} = -2\log\left(\frac{k_S}{14.8R} + \frac{0.6275}{\mathrm{Re}\sqrt{\lambda}}\right) \tag{5.4a}$$

Also combining (5.4a) with (5.3) yields

$$V = -2\sqrt{8gRS_0}\log\left(\frac{k_S}{14.8R} + \frac{0.6275\nu}{R\sqrt{8gRS_0}}\right) \tag{5.4b}$$

A λ–Re diagram for channels may be derived using (5.4a) and channel velocities may be found directly from (5.4b). However, the application of these equations to a particular channel is more complex than the pipe case, due to the extra variables involved (i.e. for channels, R changes with depth and channel shape). In addition, the validity of this approach is questionable because the presence of the free surface has a considerable effect on the velocity distributions, as previously discussed. Hence, the frictional resistance is non-uniformly distributed around the boundary, in contrast to pressurized pipe flow, where the frictional resistance is uniformly distributed around the pipe wall.

In practice, the flow in open channels is normally in the rough turbulent zone, and consequently it is possible to use simpler formulae to relate frictional losses to velocity and channel shape, as is discussed in the next section.

5.6 Uniform flow

The development of friction formulae

Historically, the development of uniform flow resistance equations preceded the detailed investigations of pipe flow resistance. These developments are outlined below, and comparisons are made with pipe flow theory.

For uniform flow to occur, the gravity forces must exactly balance the frictional resistance forces which constitute the boundary shear force. Figure 5.6 shows a small longitudinal section in which uniform flow exists.

The gravity force resolved in the direction of flow $= \rho g A L \sin\theta$ and the shear force resolved in the direction of flow $= \tau_0 P L$, where τ_0 is the mean boundary shear stress. Hence,

$$\tau_0 P L = \rho g A L \sin\theta$$

Considering channels of small slope only, then

$$\sin\theta \simeq \tan\theta = S_0$$

hence,

$$\tau_0 = \rho g A S_0 / P$$

or

$$\tau_0 = \rho g R S_0 \qquad (5.5)$$

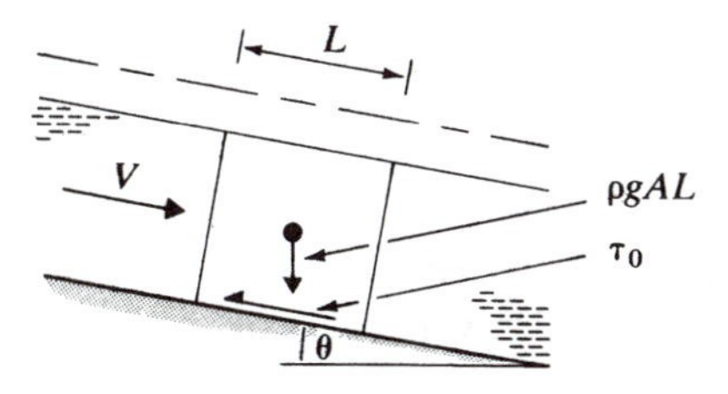

Figure 5.6 Derivation of uniform flow equations.

The Chézy equation

To interpret (5.5), an estimate of the magnitude of τ_0 is required. Assuming a state of rough turbulent flow, then

$$\tau_0 \propto V^2 \quad \text{or} \quad \tau_0 = KV^2$$

Substituting into (5.5) for τ_0,

$$V = \sqrt{\frac{\rho g}{K} R S_0}$$

which may be written as

$$V = C\sqrt{RS_0} \tag{5.6}$$

This is known as the Chézy equation. It is named after the French engineer who developed the formula when designing a canal for the Paris water supply in 1768. The Chézy coefficient C is not, in fact, constant but depends on the Reynolds Number and boundary roughness, as can be readily appreciated from the previous discussions of the λ–Re diagram. A direct comparison between C and λ can be found by substituting (5.6) into (5.3) to yield

$$C = \sqrt{8g/\lambda}$$

In 1869, an elaborate formula for Chézy's C was published by two Swiss engineers: Ganguillet and Kutter. It was based on actual discharge data from the River Mississippi and a wide range of natural and artificial channels in Europe. The formula (in metric units) is

$$C = 0.552\left(\frac{41.6 + 1.811/n + 0.00281/S_0}{1 + [41.65 + (0.000281/S_0)]n/\sqrt{R}}\right) \tag{5.7}$$

where n is a coefficient known as Kutter's n, and is dependent solely on the boundary roughness.

The Manning equation

In 1889, the Irish engineer Robert Manning presented another formula (at a meeting of the Institution of Civil Engineers of Ireland) for the evaluation of the Chézy coefficient, which was later simplified to

$$C = R^{1/6}/n \tag{5.8}$$

This formula was developed from seven different formulae, and was further verified by 170 observations. Other research workers in the field derived similar formulae independently of Manning, including Hagen in 1876, Gauckler in 1868 and Strickler in 1923. In consequence, there is some confusion as to whom the equation should be attributed to, but it is generally known as the Manning equation.

Substitution of (5.8) into (5.6) yields

$$V = (1/n)R^{2/3}S_0^{1/2}$$

and the equivalent formula for discharge is

$$Q = \frac{1}{n}\frac{A^{5/3}}{P^{2/3}}S_0^{1/2} \tag{5.9}$$

where n is a constant known as Manning's n (it is numerically equivalent to Kutter's n).

Manning's formula has the twin attributes of simplicity and accuracy. It provides reasonably accurate results for a large range of natural and artificial channels, given that the flow is in the rough turbulent zone and that an accurate assessment of Manning's n has been made. It has been widely adopted for use by engineers throughout the world.

Evaluation of Manning's n

The value of the roughness coefficient n determines the frictional resistance of a given channel. It can be evaluated directly by discharge and stage measurements for a known cross-section and slope. However, for design purposes, this information is rarely available, and it is necessary to rely on documented values obtained from similar channels.

For the case of artificially lined channels, n may be estimated with reasonable accuracy. For natural channels, the estimates are likely to be rather less accurate. In addition, the value of n may change with stage (particularly with flood flows over flood banks) and with time (due to changes in bed material as a result of sediment transport) or season (due to presence of vegetation). In such cases, a suitably conservative design value is normally adopted. Table 5.2 lists typical values of n for various materials and channel conditions. More detailed guidance is given by Chow (1959).

Uniform flow computations

Manning's formula may be used to determine steady uniform flow. There are two types of commonly occurring problems to solve. The first is to determine the discharge given the depth, and the second is to determine the depth given the discharge. The depth is referred to as the normal depth, which is

Table 5.2 Typical values of Manning's n.

Channel type	Surface material and alignment	
river	earth, straight	0.02–0.025
	earth, meandering	0.03–0.05
	gravel (75–150 mm), straight	0.03–0.04
	gravel (75–150 mm), winding or braided	0.04–0.08
unlined canals	earth, straight	0.018–0.025
	rock, straight	0.025–0.045
lined canals	concrete	0.012–0.017
models	mortar	0.011–0.013
	Perspex	0.009

Note: See Chow (1959) for a discussion regarding the units of Manning's n.

synonymous with steady uniform flow. As uniform flow can only occur in a channel of constant cross-section, natural channels should be excluded. However, in solving the equations of gradually varied flow applicable to natural channels, it is still necessary to solve Manning's equation. Therefore it is useful to consider the application of Manning's equation to irregular channels in this section. The following examples illustrate the application of the relevant principles.

Example 5.1 Discharge from depth for a trapezoidal channel

The normal depth of flow in a trapezoidal concrete lined channel is 2 m. The channel base width is 5 m and has side slopes of 1:2. Manning's n is 0.015 and the bed slope, S_0, is 0.001. Determine the discharge (Q), mean velocity (V) and the Reynolds Number (Re).

Solution

Using Table 5.1,

$$A = (5 + 2y)y \quad P = 5 + 2y\sqrt{1 + 2^2}$$

Hence, applying (5.9) for $y = 2\,\text{m}$

$$\begin{aligned} Q &= \frac{1}{0.015} \times \frac{[(5+4)2]^{5/3}}{[5 + (2 \times 2\sqrt{5})]^{2/3}} \times 0.001^{1/2} \\ &= 45\,\text{m}^3/\text{s} \end{aligned}$$

To find the mean velocity, simply apply the continuity equation:

$$V = \frac{Q}{A} = \frac{45}{(5+4)2} = 2.5\,\text{m/s}$$

The Reynolds Number is given by

$$\text{Re} = \rho RV/\mu$$

where $R = A/P$. In this case,

$$R = \frac{(5+4)2}{[5+(2\times 2\sqrt{5})]} = 1.29\,\text{m}$$

and

$$\text{Re} = \frac{10^3 \times 1.29 \times 2.5}{1.14\times 10^{-3}} = 2.83\times 10^6$$

Note: Re is very high, and corresponds to the rough turbulent zone. Therefore, Manning's equation is applicable. The interested reader may care to check the validity of this statement by applying the Colebrook–White equation (5.4b). Firstly calculate a k_s value equivalent to $n = 0.015$ for $y = 2\,\text{m}[k_s = 2.225\,\text{mm}]$. Then using these values of k_s and n compare the discharges as calculated using the Manning and Colebrook–White equations for a range of depths. Provided the channel is operating in the rough turbulent zone, the results are very similar.

Example 5.2 Depth from discharge for a trapezoidal channel

If the discharge in the channel given in Example 5.1 were $30\,\text{m}^3/\text{s}$, find the normal depth of flow.

Solution

From Example 5.1

$$Q = \frac{1}{0.015} \times \frac{[(5+2y)y]^{5/3}}{[5+2\sqrt{5}y]^{2/3}} \times 0.001^{1/2}$$

or

$$Q = 2.108 \times \frac{[(5+2y)y]^{5/3}}{[5+2\sqrt{5}y]^{2/3}}$$

At first sight this may appear to be an intractable equation, and it will also be different for different channel shapes. The simplest method of solution is to adopt a trial-and-error procedure. Various values of y are tried, and the resultant Q is compared with that required. Iteration ceases when reasonable agreement is found. In this case $y < 2$ as $Q < 45$, so an initial value of 1.7 is tried.

Trial depths (m)	Resultant discharge (m^3/s)
1.7	32.7
1.6	29.1
1.63	30.1

Hence the solution is $y = 1.63\,m$ for $Q = 30\,m^3/s$.

Channel conveyance

Channel conveyance (K) is a measure of the discharge carrying capacity of a channel, defined by the equation.

$$Q = KS_0^{1/2} \tag{5.10}$$

For any given water depth (or stage) its value may be found by equating (5.10) with Manning's equation (5.9) to give

$$K = \frac{A^{5/3}}{nP^{2/3}} \tag{5.11}$$

Its principal use is in determining the discharge and the energy and momentum coefficients in compound channels. It is also a convenient parameter to use in the computational procedures for evaluating gradually varying (steady and unsteady) flow problems in compound channels.

The equation for the energy coefficient α in a compound channel (5.1) may be expressed in general terms as

$$\alpha = \frac{\sum_{i=1}^{N} V_i^3 A_i}{\overline{V}^3 \sum_{i=1}^{N} A_i}$$

where N is the number of subsections.

This may be rewritten as

$$\alpha = \frac{\left(\sum_{i=1}^{N} A_i\right)^2}{\left(\sum_{i=1}^{N} Q_i\right)^3} \cdot \sum\left(\frac{Q_i^3}{A_i^2}\right) \tag{5.12}$$

where

$$\frac{1}{\overline{V}^3 \sum_{i=1}^{N} A_i} = \frac{\left(\sum_{i=1}^{N} A_i\right)^2}{\left(\sum_{i=1}^{N} Q_i\right)^3} \quad \text{and} \quad \sum_{i=1}^{N} V_i^3 A_i = \sum_{i=1}^{N} \left(\frac{Q_i^3}{A_i^2}\right)$$

As S_0 is constant and recalling (5.10) then

$$\frac{Q_1}{K_1} = \frac{Q_2}{K_2} = \cdots \frac{Q_N}{K_N} = \text{const} = S_0^{1/2} = \frac{\sum_{i=1}^{N} Q_i}{\sum_{i=1}^{N} K_i} \tag{5.13}$$

Hence as

$$\sum_{i=1}^{N} Q_i = \text{const} \sum_{i=1}^{N} K_i$$

Then (5.12) may be rewritten as

$$\alpha = \frac{\left(\sum_{i=1}^{N} A_i\right)^2}{\left(\sum_{i=1}^{N} K_i\right)^3} \sum_{i=1}^{N} \left(\frac{K_i^3}{A_i^2}\right) \tag{5.14}$$

Similarly it may be shown that

$$\beta = \frac{\sum_{i=1}^{N} A_i}{\left(\sum_{i=1}^{N} K_i\right)^2} \sum_{i=1}^{N} \left(\frac{K_i^2}{A_i}\right) \tag{5.15}$$

Also (5.13) may be rewritten as

$$S_0 = \frac{Q^2}{\left(\sum_{i=1}^{N} K_i\right)^2} \tag{5.16}$$

Thus both α and β can be evaluated for any given stage without explicitly determining Q_i. In addition (5.16) may be used to find the friction slope S_f, a quantity defined in section 5.10 and used in determining gradually varied flow profiles.

Compound channels

Example 5.3 illustrates the calculation of the discharge and the energy coefficient in a compound channel. It may be noted that although trapezoidal sections are used, the same principles apply to natural sections except that the areas and wetted perimeters must be evaluated from tables of stage versus area and perimeter rather than from depths and side slopes.

Example 5.3 Compound channels

During large floods, the water level in the channel given in Example 5.1 exceeds the bank-full level of 2.5 m. The flood banks are 10 m wide and are grassed with side slopes of 1:3. The estimated Manning's n for these flood banks is 0.035. Estimate the discharge for a maximum flood level of 4 m and the value of the velocity coefficient α.

Solution

In this case, it is necessary to split the section into subsections (1), (2), (3) as shown in Figure 5.7. Manning's formula may be applied to each one in turn, and the discharges can be summed. The division of the section into subsections is a little arbitrary. If the shear stress across the arbitrary divisions is small compared with the bed shear stresses, it may be ignored to obtain an approximate solution.

For section (1)

$$A_1 = \left(\frac{5+15}{2}\right)2.5 + (15 \times 1.5) = 47.5\,\text{m}^2$$

and

$$P_1 = 5 + (2\sqrt{5} \times 2.5) = 16.18\,\text{m}$$

hence

$$K_1 = \frac{47.5^{5/3}}{(16.18^{2/3} \times 0.015)} = 6492.5$$

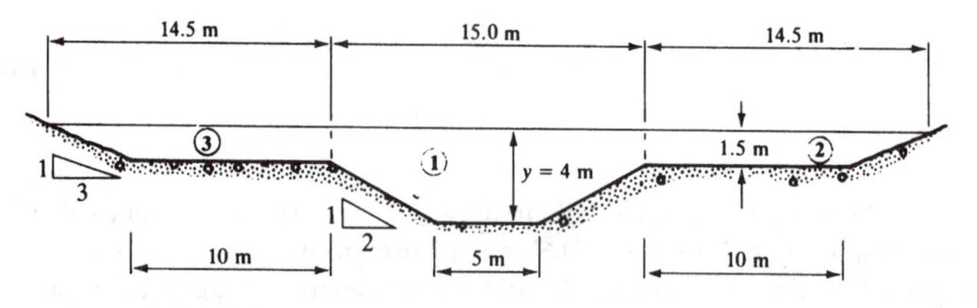

Figure 5.7 Compound channel section for Example 5.3.

Sections (2) and (3) have the same dimensions, hence

$$A_2 = A_3 = \left(\frac{10+14.5}{2}\right) \times 1.5 = 18.38\,\text{m}^2$$

$$P_2 = P_3 = 10 + (\sqrt{10} \times 1.5) = 14.74\,\text{m}$$

and

$$K_2 = K_3 = \frac{18.38^{5/3}}{14.74^{2/3} \times 0.035} = 608.4$$

hence

$$Q_1 = \frac{1}{0.015} \times \frac{47.5^{5/3}}{16.18^{2/3}} \times 0.001^{1/2} \quad \text{(from (5.9))}$$

or

$$\begin{aligned} Q_1 &= K_1 \times 0.001^{1/2} \quad \text{(from (5.10))} \\ &= 205.3\,\text{m}^3/\text{s} \end{aligned}$$

$$Q_2 = Q_3 = \frac{1}{0.035} \times \frac{18.38^{5/3}}{14.74^{2/3}} \times 0.001^{1/2}$$

or

$$\begin{aligned} Q_2 = Q_3 &= K_2 \times 0.001^{1/2} \\ &= 19.2\,\text{m}^3/\text{s} \end{aligned}$$

hence

$$\begin{aligned} Q = Q_1 + Q_2 + Q_3 &= (K_1 + K_2 + K_3) \times 0.001^{1/2} \\ &= 243.7\,\text{m}^3/\text{s} \end{aligned}$$

The velocity coefficient may be found directly from (5.1) or equivalently from (5.14).
From (5.1)

$$\alpha = \frac{V_1^3 A_1 + V_2^3 A_2 + V_3^3 A_3}{\overline{V}^3 A}$$

$$V_1 = \frac{Q_1}{A_1} = \frac{205.3}{47.5} = 4.32\,\text{m/s}$$

$$V_2 = V_3 = \frac{Q_2}{A_2} = \frac{19.2}{18.38} = 1.04\,\text{m/s}$$

$$\overline{V} = \frac{Q}{A} = \frac{243.7}{47.5 + 18.38 + 18.38} = 2.89\,\text{m/s}$$

Hence

$$\alpha = \frac{4.32^3 \times 47.5 + [2(1.04^3 \times 18.38)]}{2.89^3 \times 84.26}$$
$$= 1.9$$

from (5.14)

$$\alpha = \frac{(47.5 + 2 \times 18.38)^2}{(6492.4 + 2 \times 608.4)^3} \times \left(\frac{6492.4^3}{47.5^2} + 2\left(\frac{608.4^3}{18.38^2}\right)\right) = 1.9$$

Note: This example illustrates a case in which the velocity coefficient should not be ignored.

Example 5.3 demonstrates how a **first estimate** of the relationship between stage and discharge may be obtained for a compound channel. At the time of writing, flow in compound channels is the subject of an intensive and collaborative research programme (using the SERC flood channel facility at Hydraulics Research Ltd, Wallingford). One preliminary finding of this research is that the method outlined in Example 5.3 may lead to errors of up to ±20% (or even more) in the predicted discharge at a given stage. The first interim report concerning this research is that of Ramsbottom (1989) in which currently used methods for flow estimation in compound channels, estimates of their accuracy and the variation of Manning's n with stage are described. Additionally Knight (1989) contains an excellent description of the problems of (and some of the solutions to) the hydraulics of flood channels. The interested reader should also refer to Chapter 15 in which some of the complexities of flow in compound channels are further described.

5.7 Rapidly varied flow: the use of energy principles

Applications and methods of solution

Rapidly varied flow occurs whenever there is a sudden change in the geometry of the channel or in the regime of the flow. Typical examples of the first type include flow over sharp-crested weirs and flow through regions of rapidly varied cross-section (e.g. venturi flumes and broad-crested weirs). The second type is normally associated with the hydraulic jump phenomenon in which flow with high velocity and small depth is rapidly changed to flow with low velocity and large depth. The regime of flow is defined by the Froude Number, a concept which is explained later in this section.

In regions of rapidly varied flow, the water surface profile changes suddenly and therefore has pronounced curvature. The pressure distribution under these circumstances departs considerably from the hydrostatic distribution. The assumptions of parallel streamlines and hydrostatic pressure distribution which are used for uniform and gradually varied flow do not apply. Solutions to rapidly varied flow problems have been found using the concepts of ideal fluid flow coupled with the use of finite-element techniques. However, such solutions are complex and do not include the boundary layer effects in real fluids.

Many rapidly varied flow problems may be solved approximately using energy and momentum concepts, and for engineering purposes this is often sufficiently accurate. This section describes and explains the use of these concepts.

The energy equation in open channels

Bernoulli's equation, derived in Chapter 2, may be applied to any streamline. If the streamlines are parallel, then the pressure distribution is hydrostatic.

Referring to Figure 5.8, which shows uniform flow in a steep channel, consider point A on a streamline. The pressure force at point A balances the component of weight normal to the bed, i.e.

$$p_A \Delta S = \rho g d \Delta S \cos\theta$$

$$p_A = \rho g d \cos\theta = \rho g y_2$$

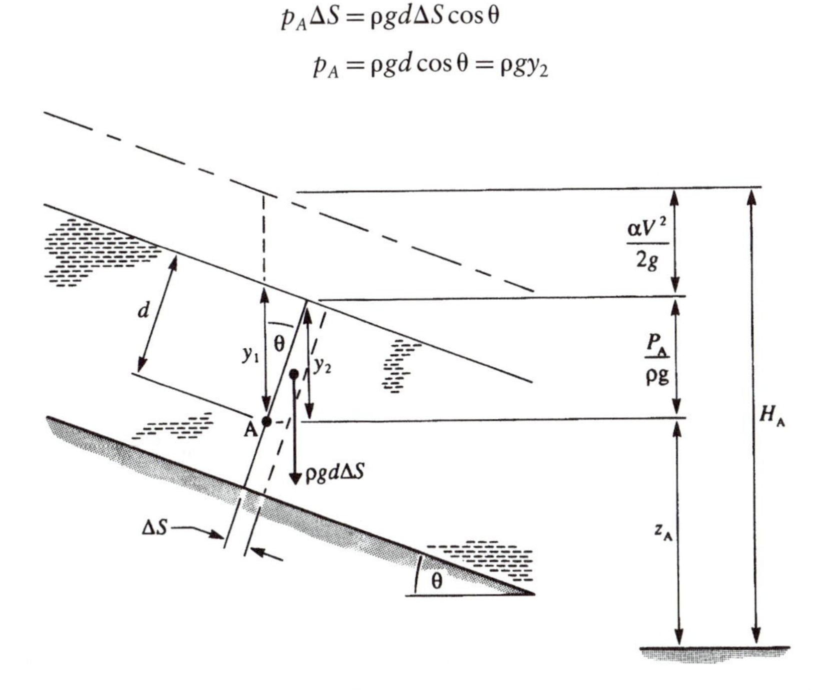

Figure 5.8 Application of Bernoulli's equation to uniform flow in open channels.

It is more convenient to express this pressure force in terms of y_1 (the vertical distance from the streamline to the free surface). Since

$$d = \frac{y_2}{\cos\theta} \quad \text{and} \quad d = y_1 \cos\theta$$

then

$$y_2 = y_1 \cos^2 \theta$$

and hence

$$p_A = \rho g y_1 \cos^2 \theta$$

or

$$p_A/\rho g = y_1 \cos^2 \theta$$

Most channels have very small bed slopes (e.g. less than 1:100, corresponding to θ less than 0.57° or $\cos^2\theta$ greater than 0.9999) and therefore for such channels

$$\cos^2 \theta \simeq 1$$

and

$$p_A/\rho g \simeq y_1 \tag{5.17}$$

Hence, Bernoulli's equation becomes

$$H = y + \frac{\alpha V^2}{2g} + z \tag{5.18}$$

Application of the energy equation

Consider the problem shown in Figure 5.9. Steady uniform flow is interrupted by the presence of a hump in the streambed. The upstream depth and the discharge are known, and it simply remains to find the depth of flow at position (2).

Applying the energy equation (5.18) and assuming that frictional energy losses between (1) and (2) are negligible, then

$$y_1 + \frac{V_1^2}{2g} = y_2 + \frac{V_2^2}{2g} + \Delta z \tag{5.19}$$

(taking $\alpha = 1$).

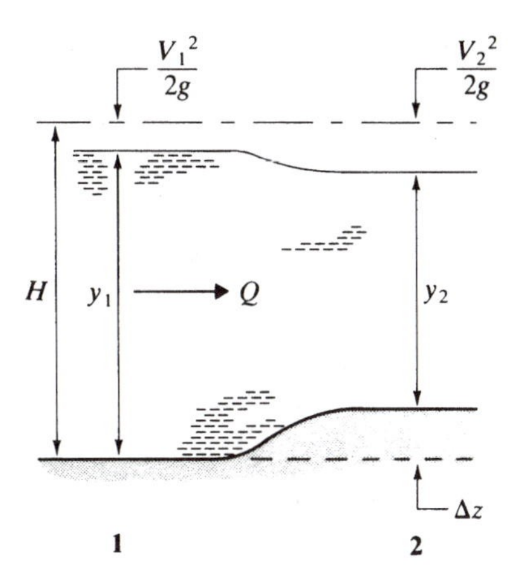

Figure 5.9 Flow transition problem.

There are two principal unknown quantities – V_2 and y_2 – and hence to solve this a second equation is required: the continuity equation:

$$V_1 y_1 = V_2 y_2 = q$$

where q is the discharge per unit width. Combining these equations,

$$y_1 + \frac{q^2}{2gy_1^2} = y_2 + \frac{q^2}{2gy_2^2} + \Delta z$$

and rearranging,

$$2gy_2^3 + y_2^2\left(2g\Delta z - 2gy_1 - \frac{q^2}{y_1^2}\right) + q^2 = 0$$

This is a cubic equation, in which y_2 is the only unknown, and to which there are three possible mathematical solutions. As far as the fluid flow is concerned, however, only one solution is possible. To determine which of the solutions for y_2 is correct requires a more detailed knowledge of the flow.

Specific energy

To solve the above problem, the concept of specific energy was introduced by Bakhmeteff in 1912. Specific energy (E_s) is defined as the energy of the flow referred to the channel bed as datum:

$$E_s = y + \alpha V^2/2g \tag{5.20}$$

Application of the specific energy equation provides the solution to many rapidly varied flow problems.

For steady flow, (5.20) may be rewritten as

$$E_s = y + \alpha \frac{(Q/A)^2}{2g}$$

Now consider a rectangular channel:

$$\frac{Q}{A} = \frac{bq}{by} = \frac{q}{y}$$

where b is the width of the channel. Hence

$$E_s = y + \alpha q^2/(2gy^2)$$

As q is constant,

$$(E_s - y)y^2 = \alpha q^2/2g = \text{constant}$$

or

$$(E_s - y) = \text{constant}/y^2$$

Again, this is a cubic equation for the depth y for a given E_s. Considering only positive solutions, then the equation is a curve with two asymptotes:

$$\text{as } y \to 0, \qquad E_s \to \infty$$
$$\text{as } y \to \infty, \qquad E_s \to y$$

This curve is now used to solve the problem given in Figure 5.9. In Figure 5.10, the problem has been redrawn alongside a graph of depth versus specific energy.

Equation (5.19) may be written as

$$E_{s1} = E_{s2} + \Delta z \tag{5.21}$$

This result is plotted on the specific energy diagram in Figure 5.10. Point A on the curve corresponds to conditions at point (1) in the channel. Point (2) in the channel must be at either point B or B′ on the specific energy curve (from (5.21)). All points between (1) and (2) must lie on the specific energy curve between A and B or B′. To arrive at point B′ would imply that at some intermediate point $E_{s1} - E_{s2} > \Delta z$, which is physically impossible. Hence, the flow depth at (2) must correspond to point B on the specific energy curve. This is a very significant result, so an example of such a flow transition follows.

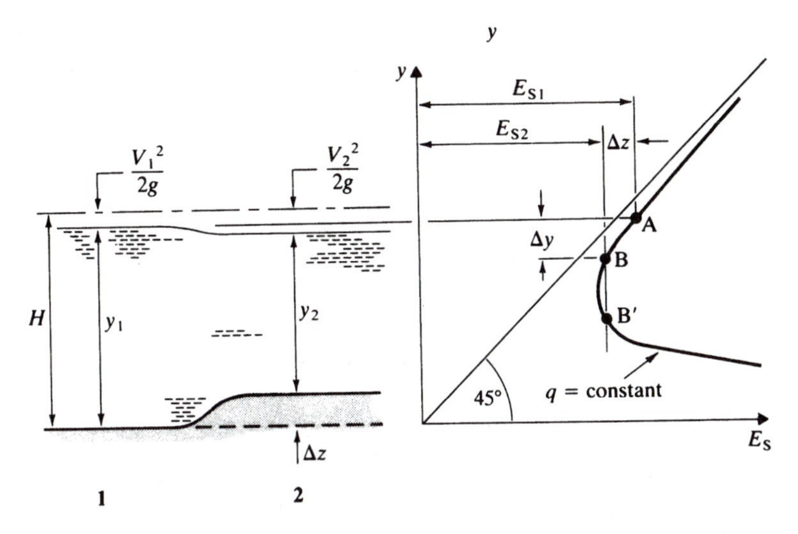

Figure 5.10 The use of specific energy to solve flow transition problems.

Example 5.4 Flow transition

The discharge in a rectangular channel of width 5 m and maximum depth 2 m is 10 m^3/s. The normal depth of flow is 1.25 m. Determine the depth of flow downstream of a section in which the bed rises by 0.2 m over a distance of 1 m.

Solution

The solution is shown graphically in Figure 5.10. Assuming frictional losses are negligible, then (5.21) applies, i.e.

$$E_{s1} = E_{s2} + \Delta z$$

In this case,

$$E_{s1} = y_1 + V_1^2/2g = 1.25 + [10/(5 \times 1.25)]^2/2g = 1.38\,\text{m}$$
$$E_{s2} = y_2 + [10/(5y_2)]^2/2g = y_2 + 2^2/2gy_2^2$$
$$\Delta z = 0.2$$

hence

$$1.38 = y_2 + (2^2/2gy_2^2) + 0.2$$

or

$$1.18 = y_2 + 2/gy_2^2$$

This is a cubic equation for y_2, but the correct solution is that given by point B in Figure 5.10, which in this case is about 0.9 m. This is used as the initial estimate in a trial-and-error solution, as follows:

y_2 (m)	$E_{s2} = y_2 + 2/gy_2^2$ (m)
0.9	1.15
1.0	1.2
0.96	1.18

Hence the solution is $y_2 = 0.96$ m.

This result is often a source of puzzlement, and a simple physical explanation is called for. The answer lies in realizing that the fluid is accelerating. Consider, initially, that the water surface remains at the same level over the bed rise. As the water depth is less (over the bed rise) and the discharge constant, then the velocity must have increased; i.e. the fluid must have accelerated. However, to accelerate the fluid, a force is required. This is provided by a fall in the water surface elevation, which implies that the hydrostatic force acting in the downstream direction is greater than that acting in the upstream direction.

Subcritical, critical and supercritical flow

In Figure 5.11, the specific energy curve (for constant discharge) has been redrawn alongside a second curve of depth against discharge for constant E_s. The figure is now used to illustrate several important principles of rapidly varied flow.

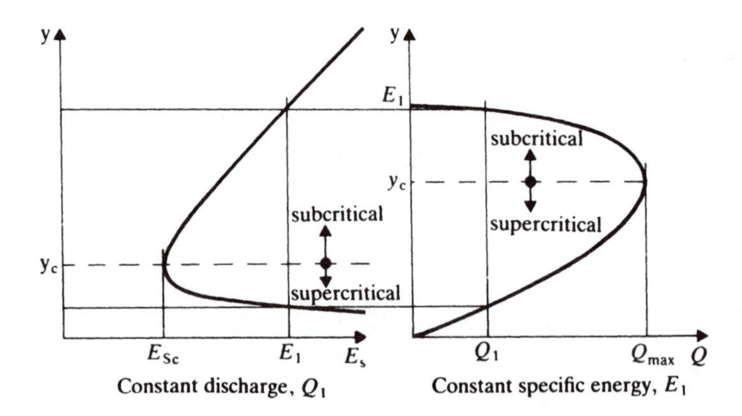

Figure 5.11 Variation of specific energy and discharge with depth.

(a) For a given constant discharge:

1. The specific energy curve has a minimum value E_{Sc} at point c with a corresponding depth y_c – known as the critical depth;
2. For any other value of E_s there are two possible depths of flow known at alternate depths, one of which is termed subcritical and the other supercritical;

$$\text{for supercritical flow, } y < y_c$$

$$\text{for subcritical flow, } y > y_c$$

(b) For a given constant specific energy:

1. The depth–discharge curve shows that discharge is a maximum at the critical depth.
2. For all other discharges there are two possible depths of flow (sub- and supercritical) for any particular value of E_s.

The general equation of critical flow

Referring to Figure 5.11, the general equation for critical flow may be derived by determining E_{Sc} and Q_{max}, independently, from the specific energy equation. It will be seen that the two methods both result in the same solution.

For $Q =$ constant

$$E_s = y + \alpha V^2/2g$$

or

$$E_s = y + \frac{\alpha}{2g}\frac{Q^2}{A^2} \qquad \text{(a)}$$

For a minimum value (i.e. E_{Sc}),

$$\frac{dE_s}{dy} = 0$$

$$= 1 + \frac{\alpha Q^2}{2g}\frac{d}{dA}\left(\frac{1}{A^2}\right)\frac{dA}{dy} \qquad \text{(b)}$$

For $E_s =$ constant

$$E_s = y + \alpha V^2/2g$$

or

$$Q = \sqrt{\frac{2g}{\alpha}}A(E_s - y)^{1/2} \qquad \text{(a)}$$

For Q_{max} with $E_s = E_0 =$ constant,

$$\frac{dQ}{dy} = \sqrt{\frac{2g}{\alpha}}\left(\frac{A(E_0 - y)^{-1/2}}{-2} + \frac{dA}{dy}(E_0 - y)^{1/2}\right) = 0 \qquad \text{(b)}$$

Since $\delta A = B\,\delta y$,
then in the limit

$$\frac{dA}{dy} = B$$

substituting, into (b)

$$0 = 1 - \frac{\alpha Q^2}{2g} B_c\, 2A_c^{-3}$$

or

$$\frac{\alpha Q^2 B_c}{gA_c^3} = 1$$

Since $\delta A = B\,\delta y$,
then in the limit

$$\frac{dA}{dy} = B$$

substituting into (b)

$$E_0 = \frac{A}{2B} + y$$

and substituting into (a)

$$\frac{\alpha Q_{max}^2 B}{gA^3} = 1$$

Comparing,

$$\frac{\alpha Q_{max}^2 B_c}{gA_c^3} = 1 \tag{5.22}$$

In other words, at the critical depth the discharge is a maximum and the specific energy is a minimum.

Critical depth and critical velocity (for a rectangular channel)

Determination of the critical depth (y_c) in a channel is necessary for both rapidly and gradually varied flow problems. The associated critical velocity (V_c) will be used in the explanation of the significance of the Froude Number. Both of these parameters may be derived directly from (5.22) for any shape of channel. For illustrative purposes, the simplest artificial channel shape is used here, i.e. rectangular. For critical flow,

$$\frac{\alpha Q^2 B}{gA^3} = 1$$

For a rectangular channel, $Q = qb$, $B = b$, and $A = by$. Substituting in (5.22) and taking $\alpha = 1$, then

$$y_c = (q^2/g)^{1/3} \tag{5.23}$$

and as

$$V_c y_c = q$$

then

$$V_c = \sqrt{gy_c} \tag{5.24}$$

Also, as

$$E_{Sc} = y_c + V_c^2/2g$$

then

$$E_{Sc} = y_c + y_c/2$$

or

$$y_c = \frac{2}{3}E_{Sc} \tag{5.25}$$

An application of the critical depth line

The application of critical depth is illustrated in Figure 5.12, which shows the effect of a local bed rise on the free surface elevation for various initial depths of flow. It should be noted from (5.23) that the critical depth is dependent only on the discharge. Hence, for a given discharge the critical depth line may be drawn on the longitudinal channel section as shown in Figure 5.12. If the upstream flow is initially subcritical, then a depression of the water surface will occur above the rise in bed level. Conversely, if the upstream water level is initially supercritical, then an increase in the

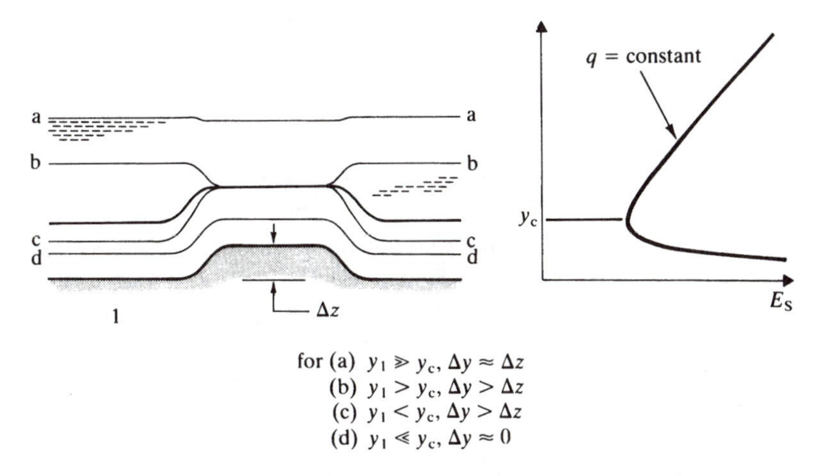

Figure 5.12 Effect of a local bed rise on the water surface elevation.

water surface level will occur. As the upstream water level approaches the critical depth, then the change in elevation of the free surface becomes more marked. All of these results may easily be verified by reference to the specific energy diagram drawn alongside in Figure 5.12. A practical example of this phenomenon often occurs in rivers under flood conditions. The normal depth of flow may be nearly critical, and any undulations in the bed result in large standing wave formations.

A further result may be gleaned from this example. If the local bed rise is sufficiently large, then critical flow will occur. It has already been shown that discharge is a maximum for critical flow and hence, under these circumstances, the local bed rise is acting as a 'choke' on the flow, so called because it limits the discharge. It is left to the reader to decide what happens if Δz is so large that E_{s2} is less than any point on the E_s curve. The question will be discussed again later in section 5.9 and in Example 5.10.

The Froude Number

The Froude Number is defined as

$$\mathrm{Fr} = V/\sqrt{gL}$$

where L is a characteristic dimension. It is attributed to William Froude (1810–97), who used such a relationship in model studies for ships. If L is replaced by D_m, the hydraulic mean depth, then the resulting dimensionless parameter

$$\mathrm{Fr} = V/\sqrt{gD_m} \tag{5.26}$$

is applicable to open channel flow. This is extremely useful, as it defines the regime of flow, and as many of the energy and momentum equations may be written in terms of the Froude Number.

The physical significance of Fr may be understood in two different ways. First, from dimensional analysis,

$$\mathrm{Fr}^2 = \frac{\text{inertial force}}{\text{gravitational force}} \equiv \frac{\rho L^2 V^2}{\rho g L^3} = \frac{V^2}{gL}$$

Secondly, by consideration of the speed of propagation, c, of a wave of low amplitude and long wavelength. Such waves may be generated in a channel as oscillatory waves or surge waves. Oscillatory waves (e.g. ocean waves) are considered in Chapter 8, and surge waves in section 5.11. Both types of wave lead to the result that $c = \sqrt{gy}$. For a rectangular channel $D_m = y$, and hence

$$\mathrm{Fr} = \frac{V}{\sqrt{gy}} = \frac{\text{water velocity}}{\text{wave velocity}}$$

Also, for a rectangular channel $V_c = \sqrt{gy_c}$, and hence, for critical flow,

$$\mathrm{Fr} = \frac{V}{\sqrt{gy}} = \frac{\sqrt{gy_c}}{\sqrt{gy_c}} = 1$$

This is a general result for all channels (as it can be shown that for non-rectangular channels $V_c = \sqrt{gD_m}$ and $c = \sqrt{gD_m}$).

For subcritical flow,

$$V < V_c \quad \text{and} \quad \mathrm{Fr} < 1$$

For supercritical flow

$$V > V_c \quad \text{and} \quad \mathrm{Fr} > 1$$

The Froude Number therefore defines the regime of flow. There is a second consequence of major significance. Flow disturbances are propagated at a velocity of $c = \sqrt{gy}$. Hence, if the flow is supercritical, any flow disturbance can only travel downstream as the water velocity exceeds the wave velocity. The converse is true for subcritical flow. Such flow disturbances are introduced by all channel controls and local features. For example, if a pebble is dropped into a still lake, then small waves are generated in all directions at equal speeds, resulting in concentric wave fronts. Now imagine dropping the pebble into a river. The wave fronts will move upstream slower than they do downstream, due to the current. If the water velocity exceeds the wave velocity, the wave will not move upstream at all.

An example of how channel controls transmit disturbances upstream and downstream is shown in Figure 5.13. To summarize:

for Fr > 1
- supercritical flow
- water velocity > wave velocity
- disturbances travel downstream
- upstream water levels are unaffected by downstream control

for Fr < 1
- subcritical flow
- water velocity < wave velocity
- disturbances travel upstream and downstream
- upstream water levels are affected by downstream control

When the Froude Number is close to one in a channel reach, flow conditions tend to become unstable, resulting in wave formations. If the channel is a compound channel, for example flood flows in a main channel and its flood plains, then some very interesting and little investigated phenomena

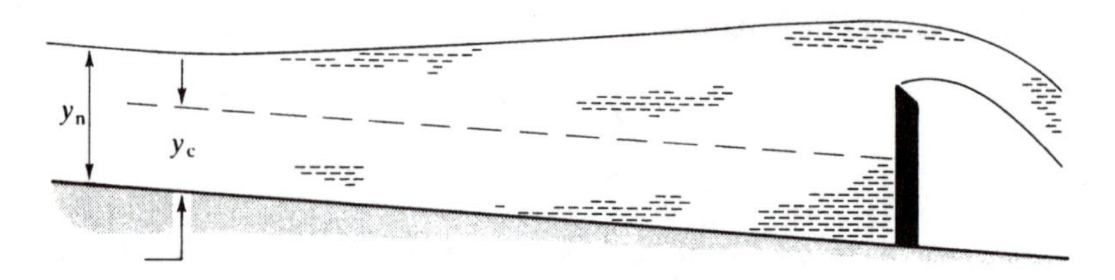

(a) Caused by the weir is transmitted upstream as Fr $<$ 1 **and** $y_n > y_c$

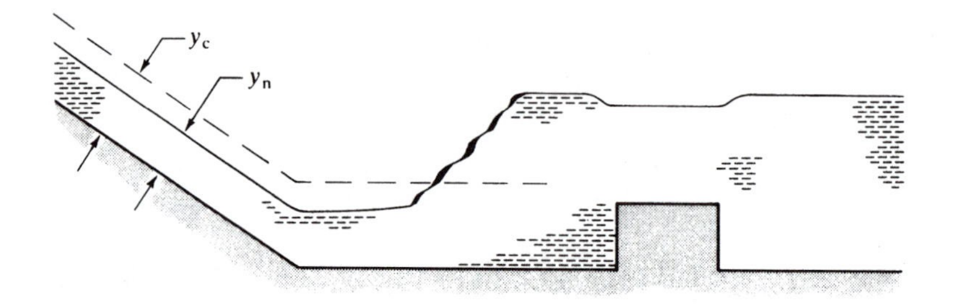

(b) Caused by a change of slope is not transmitted upstream as Fr $>$ 1 **and** $y_n < y_c$

Figure 5.13 Flow disturbance.

can occur. For example, it is possible to imagine that flow in the main channel is very close to critical flow and that flow on the flood plains is still subcritical (due to much lower velocities). In fact, what happens is that at the interface between flood plain and main channel, flow on the **flood plain** side of the interface becomes critical first because the velocity is close to that of the main channel but the depth is much less. Studies by Knight and Yuen (1990) have shown that both sub- and supercritical flow can exist simultaneously in a compound channel.

5.8 Rapidly varied flow: the use of momentum principles

The hydraulic jump

The hydraulic jump phenomenon is an important type of rapidly varied flow. It is also an example of a stationary surge wave. A hydraulic jump occurs when a supercritical flow meets a subcritical flow. The resulting flow transition is rapid, and involves a large energy loss due to turbulence. Under these circumstances, a solution to the hydraulic jump problem cannot be found using a specific energy diagram. Instead, the momentum equation is used. Figures 5.14(a) and (b) depict a hydraulic jump and the associated specific energy and force–momentum diagrams. Initially, ΔE_s is unknown,

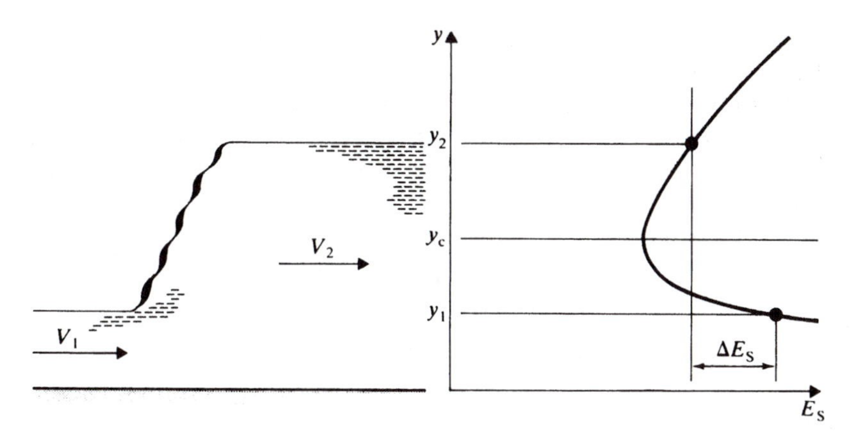

(a) With associated specific energy diagram

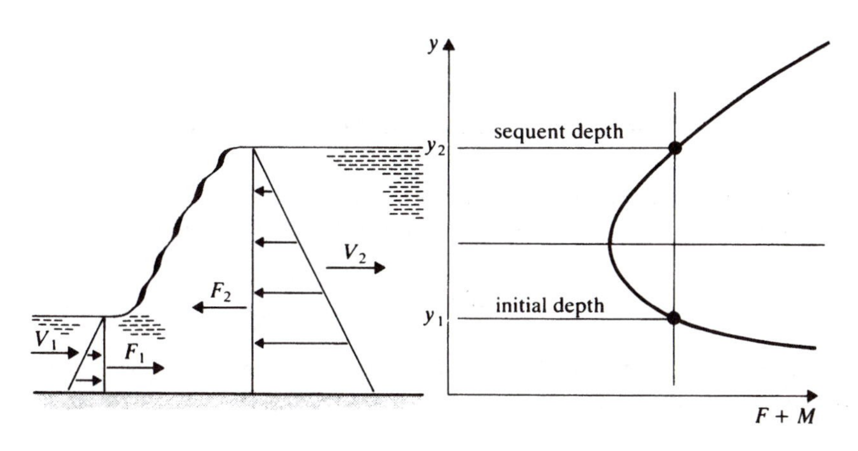

(b) With associated force–momentum diagram

Figure 5.14 The hydraulic jump.

as only the discharge and upstream depth are given. By using the momentum principle, the sequent depth y_2 may be found in terms of the initial depth y_1 and the upstream Froude Number (Fr_1).

The momentum equation

The momentum equation derived in Chapter 2 may be written as

$$F_m = \Delta M$$

In this case, the forces acting are the hydrostatic pressure forces upstream (F_1) and downstream (F_2) of the jump. Their directions of action are as shown in Figure 5.14(b). This is a point which often causes difficulty. It may be understood most easily by considering the jump to be inside a control volume, and to consider the external forces acting on the control volume (as described in Chapter 2).

Ignoring boundary friction, and for small channel slopes:

$$\text{net force in } x\text{-direction} = F_1 - F_2$$

$$\text{momentum change} = M_2 - M_1$$

hence

$$F_1 - F_2 = M_2 - M_1$$

or

$$F_1 + M_1 = F_2 + M_2 = \text{constant (for constant discharge)}$$

If depth (y) is plotted against force + momentum ($F + M$) for a constant discharge, as in Figure 5.14(b), then for a stable hydraulic jump

$$F + M = \text{constant} \tag{5.27}$$

Therefore, for any given initial depth, the sequent depth is the corresponding depth on the force–momentum diagram.

Solution of the momentum equation for a rectangular channel

For a rectangular channel, (5.27) may be evaluated as follows:

$$F_1 = \rho g(y_1/2)y_1 b \qquad F_2 = \rho g(y_2/2)y_2 b$$

$$M_1 = \rho Q V_1 \qquad M_2 = \rho Q V_2$$

$$= \rho Q \frac{Q}{y_1 b} \qquad = \rho Q \frac{Q}{y_2 b}$$

Substituting in (5.27) and rearranging,

$$\frac{\rho g b}{2}(y_1^2 - y_2^2) = \frac{\rho Q^2}{b}\left(\frac{1}{y_2} - \frac{1}{y_1}\right)$$

Substituting $q = Q/b$ and simplifying,

$$\frac{1}{2}(y_1^2 - y_2^2) = \frac{q^2}{g}\left(\frac{1}{y_2} - \frac{1}{y_1}\right)$$

$$\frac{1}{2}(y_1 + y_2)(y_1 - y_2) = \frac{q^2}{g}\frac{(y_1 - y_2)}{y_2 y_1}$$

$$\frac{1}{2}(y_1 + y_2) = \frac{q^2}{g}\frac{1}{y_2 y_1}$$

Substituting $q = V_1 y_1$ and dividing by y_1^2,

$$\frac{1}{2}\frac{y_2}{y_1}\left(1 + \frac{y_2}{y_1}\right) = \frac{V_1^2}{g y_1} = \mathrm{Fr}_1^2$$

This is a quadratic equation in y_2/y_1, whose solution is

$$y_2 = (y_1/2)(\sqrt{1 + 8\mathrm{Fr}_1^2} - 1) \tag{5.28a}$$

Alternatively it may be shown that

$$y_1 = (y_2/2)(\sqrt{1 + 8\mathrm{Fr}_2^2} - 1) \tag{5.28b}$$

Energy dissipation in a hydraulic jump

Using (5.28a), y_2 may be evaluated in terms of y_1 and hence the energy loss through the jump determined:

$$\Delta E = E_1 - E_2 = \left(y_1 + \frac{V_1^2}{2g}\right) - \left(y_2 + \frac{V_2^2}{2g}\right)$$

Substituting $q = Vy$,

$$\Delta E = y_1 - y_2 + \frac{q^2}{2g}\left(\frac{1}{y_1^2} - \frac{1}{y_2^2}\right) \tag{5.29}$$

Equation (5.28a) may be rewritten as

$$\mathrm{Fr}_1^2 = \frac{1}{2}\frac{y_2}{y_1}\left(\frac{y_2}{y_1} + 1\right) \tag{5.30}$$

The upstream Froude Number is related to q as follows:

$$\mathrm{Fr}_1 = V_1/\sqrt{gy_1} = (q/y_1)/\sqrt{gy_1}$$

$$\mathrm{Fr}_1^2 = q^2/gy_1^3 \tag{5.31}$$

Substituting (5.30) and (5.31) into (5.29), the solution is

$$\Delta E = (y_2 - y_1)^3/4y_1y_2 \tag{5.32}$$

It should be noted that the energy loss increases very sharply with the relative height of the jump.

Significance of the hydraulic jump equations

In the preceding section, the hydraulic jump equation was deliberately formulated in terms of the upstream Froude Number. The usefulness of this form of the equation is now demonstrated. For critical flow $\mathrm{Fr} = 1$, and substitution into (5.28a) gives $y_2 = y_1$. This is an unstable condition resulting in an undular standing wave formation. For supercritical flow upstream, $\mathrm{Fr} > 1$ and hence $y_2 > y_1$. This is the required condition for the formation of a hydraulic jump. For subcritical flow upstream, $\mathrm{Fr}_1 < 1$ and hence $y_2 < y_1$. This is not physically possible without the presence of some obstruction in the flow path. Such an obstruction would introduce an associated resistance force (F_R) and the momentum equation would then become $F_1 - F_2 - F_R = M_2 - M_1$. Under these circumstances a hydraulic jump would not form, but the momentum equation could be used to evaluate the resistance force (F_R). To summarize, for a hydraulic jump to form, the upstream flow must be supercritical, and the higher the upstream Froude Number, the greater the height of the jump and the greater the loss of energy through the jump.

Stability of the hydraulic jump

The hydraulic jump equation contains three independent variables: y_1, y_2 and Fr_1. A stable (i.e. stationary) jump will form only if these three independent variables conform to the relationship given in (5.28a). To illustrate this, Figure 5.15 shows a hydraulic jump formed between two sluice gates. The upstream depth (y_1) and Froude Number (Fr_1) are controlled by the upstream sluice gate for a given discharge. The downstream depth (y_2) is controlled by the downstream sluice gate, and not by the hydraulic jump. Denoting the sequent depth by y_2', then the following observations may be made:

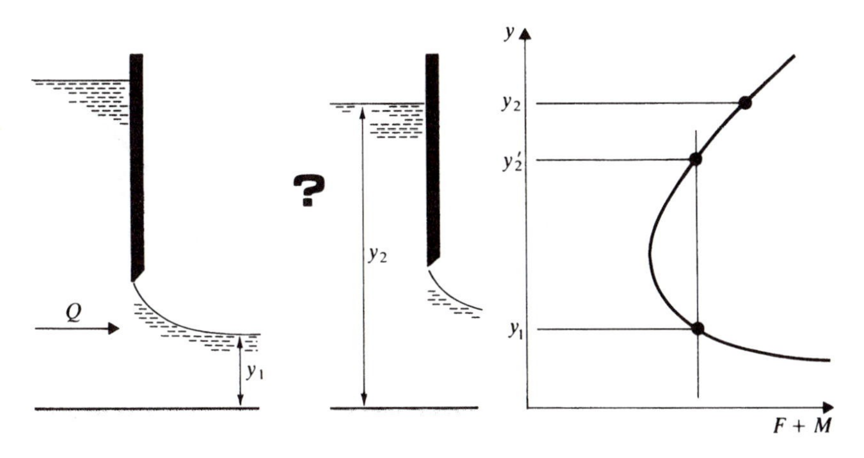

Figure 5.15 Stability of the hydraulic jump.

if $y_2 = y_2'$, then a stable jump forms;
if $y_2 > y_2'$, then the downstream force + momentum is greater than the upstream force + momentum, and the jump moves upstream;
if $y_2 < y_2'$, then the downstream force + momentum is less than the upstream force + momentum and the jump moves downstream.

Even if a stable jump forms, it may occur anywhere between the two sluice gates.

The occurrence and uses of a hydraulic jump

The hydraulic jump phenomenon has important applications in hydraulic engineering. Jumps may form downstream of hydraulic structures such as spillways, sluice gates and venturi flumes. They also sometimes occur downstream of bridge piers (normally under flood flow conditions).

Jumps may be deliberately induced to act as an energy dissipation device (e.g. in stilling bays) and to localize bed scour. These uses are discussed in Chapter 13.

5.9 Critical depth meters

The effect of a local bed rise on the flow regime has previously been discussed. It was concluded that critical flow over the bed rise would occur if the change in specific energy was such that it was reduced to the minimum

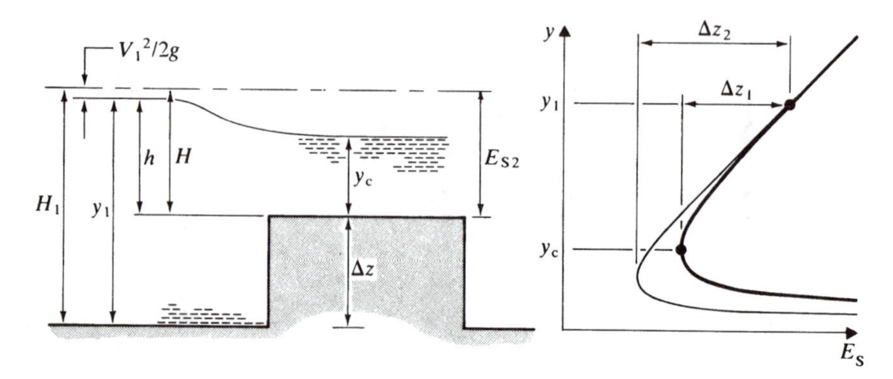

Figure 5.16 The broad-crested weir.

specific energy. This situation is shown in Figure 5.16. This condition of critical depth over the bed rise, at first sight, may be thought only to occur for one particular step height (Δz_1) and upstream depth (y_1).

Consider what would happen if Δz_1 was further increased to Δz_2. The change in energy from (1) to (2) would be such that no position on the specific energy curve corresponding to (2) could be found. This apparent impossibility may be explained by reconsidering the specific energy curve. It was drawn for a particular fixed value of discharge (Q_1). In Figure 5.16, a second (thinner) curve is drawn for a smaller discharge (Q_2) for which the specific energy at position (2) is a minimum. Hence, if the step height was Δz_2, then the discharge would initially be limited to Q_2. However, if the discharge in the channel was Q_1, then the difference in the discharges ($Q_1 - Q_2$) would have to be stored in the channel upstream, resulting in an increase in the depth y_1. After a short time, the new upstream depth (y_1') would be sufficient to pass the required discharge Q_1 over the bed rise. In other words, the upstream energy will adjust itself such that the given discharge (Q_1) can pass over the bed rise with the minimum specific energy obtained at the critical depth. Of course, this will only apply provided that Δz is sufficiently large. Under these conditions, the local bed rise is acting as a control on the discharge by providing a choke. The upstream water depth is controlled by the bed rise, not by the channel. Such a bed rise is known as a broad-crested weir. Its usefulness lies in its ability to control the discharge and hence it may be used as a discharge measuring device.

Derivation of the discharge equation for broad-crested weirs

Referring to Figure 5.16, and assuming that the depth is critical at position (2), then from (5.24)

$$V_2 = V_c = \sqrt{g y_c}$$

and

$$Q = VA = \sqrt{gy_c}by_c$$

where b is the width of the channel at position (2). Also, from (5.25)

$$y_c = \frac{2}{3}E_{S2}$$

Assuming no energy losses between (1) and (2), then

$$E_{S2} = h + V_1^2/2g = H$$

$$y_c = \frac{2}{3}H$$

Substituting for y_c

$$Q = g^{1/2}b\left(\frac{2}{3}\right)^{3/2}H^{3/2}$$

or

$$Q = \frac{2}{3}\sqrt{\frac{2g}{3}}bH^{3/2} \qquad (5.33)$$

In practice, there are energy losses and it is the upstream depth rather than energy that is measured. Equation (5.33) is modified by the inclusion of the coefficient of discharge (C_d) to account for energy losses and the coefficient of velocity (C_v) to account for the upstream velocity head, giving

$$Q = C_d C_v \frac{2}{3}\sqrt{\frac{2g}{3}}bh^{3/2} \qquad (5.34)$$

Broad-crested weirs are discussed in more detail in Chapter 13.

Venturi flumes

This is a second example of a critical depth meter. In this case, the channel width is contracted to choke the flow, as shown in Figure 5.17. To demonstrate that critical flow is produced, it is necessary to draw two specific energy curves. Although the discharge (Q) is constant, the discharge per unit width (q) is greater in the throat of the flume than it is upstream. To force critical depth to occur in the throat, the specific energy in the throat must be a minimum. Neglecting energy losses, then the upstream specific energy (E_{S1}) will have the same value. Providing that the upstream energy (E_n) for

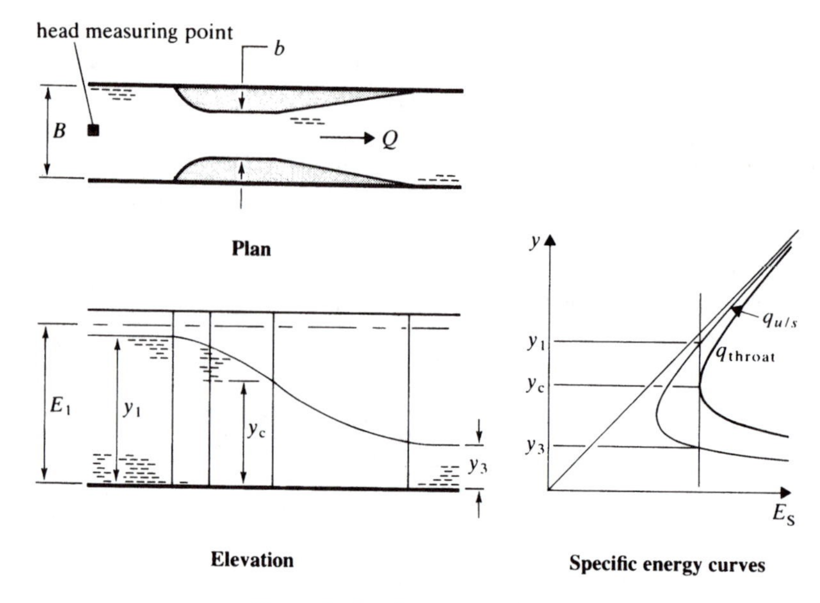

Figure 5.17 The venturi flume.

uniform flow is less than E_{S1}, then the venturi flume, not the channel, will control the discharge. The equation for the discharge is the same as for the broad-crested weir (i.e. equation (5.34)), remembering that b is the width at the section at which flow is critical, i.e. the throat.

Downstream of the throat the width expands and the flow, rather than returning to subcritical flow, continues to accelerate and becomes supercritical. As channel flow is normally subcritical, then at some point downstream a hydraulic jump will form. If the hydraulic jump moves upstream into the throat, then critical flow will no longer be induced and the flume will operate in the submerged condition. Under these circumstances, the discharge equation (5.34) no longer applies, and venturi flumes are therefore designed to prevent this happening for as wide a range of conditions as possible.

Venturi flumes are discussed in more detail in Chapter 13.

5.10 Gradually varied flow

The significance of bed slope and channel friction

In the discussions of rapidly varied flow, the influence of bed slope and channel friction was not mentioned. It was assumed that frictional effects may be ignored in a region of rapidly varied flow. This is a reasonable assumption, since the changes take place over a very short distance.

However, bed slope and channel friction are very important because they determine the flow regime under gradually varied flow conditions.

The discussion of the specific energy curve and the criteria for maximum discharge indicated that for a given specific energy or discharge there are two possible flow depths at any point in a channel. The solution of Manning's equation results in only one possible flow depth (the normal depth). This apparent paradox is resolved by noting the influence of the bed slope and channel friction. These parameters determine which of the two possible flow depths will occur at any given point for uniform flow. In other words, for uniform flow, the bed slope and channel friction determine whether the flow regime is sub- or supercritical.

For a given channel and discharge, the normal depth of flow may be found using Manning's equation and the critical depth using (5.23). The normal depth of flow may be less than, equal to, or greater than, the critical depth. For a given channel shape and roughness, only one value of slope will produce the critical depth, and this is known as the critical slope (S_c). If the slope is steeper than S_c, then the flow will be supercritical and the slope is termed a steep slope. Conversely, if the slope is less steep than S_c, then the flow will be subcritical and the slope is termed a mild slope.

Critical bed slope in a wide rectangular channel

To illustrate this concept, the equation for the critical bed slope is now derived for the case of a wide rectangular channel.

For uniform flow,

$$Q = \frac{1}{n}\frac{A^{5/3}}{P^{2/3}} S_0^{1/2} \tag{5.9}$$

and for critical flow

$$\frac{Q^2 B}{gA^3} = 1$$

(taking $\alpha = 1$ in (5.22)).

Hence, combining (5.9) and (5.22) to eliminate Q,

$$\frac{1}{n}\frac{A^{5/3}}{P^{2/3}} S_c^{1/2} = \sqrt{\frac{gA^3}{B}}$$

where S_c is the critical bed slope.

For a wide rectangular channel of width b, $B = b$, $A = by$ and $P \simeq b$.

Making these substitutions for a wide rectangular channel,

$$\frac{1}{n}\frac{(by_c)^{5/3}}{b^{2/3}} S_c^{1/2} = \frac{g^{1/2}(by_c)^{3/2}}{b^{1/2}}$$

where y_c is the critical depth. Hence

$$S_c = gn^2/y_c^{1/3} \tag{5.35}$$

The implications of this equation are best seen from a numerical example.

Example 5.5 Determination of critical bed slope

Given a wide rectangular channel of width 20 m, determine the critical bed slope and discharge for critical depths of 0.2, 0.5 and 1.0 m. Assume that $n = 0.035$.

Solution

Using

$$S_c = gn^2/y_c^{1/3} \tag{5.35}$$

and

$$Q = \frac{1}{n}\frac{A^{5/3}}{P^{2/3}} S_0^{1/2} \tag{5.9}$$

For a particular y_c, substitute into (5.35) to find S_c, and then substitute into (5.9) to find Q, hence

y (m)	S_c (m/m)	Q (m^3/s)
0.2	0.02	5.5
0.5	0.015	21.3
1.0	0.012	58.7

The results of Example 5.5 demonstrate that the critical bed slope is dependent on discharge. In other words, for a given channel with a given slope, it is the discharge which determines whether that slope is mild or steep.

Example 5.6 Critical depth and slope in a natural channel

The data given below were derived from the measured cross-section of a natural stream channel. Using the data, determine the critical stage and associated critical bed slope for a discharge of 60 m^3/s assuming $n = 0.04$.

Stage (m)	Area (m^2)	Perimeter (m)	Surface width (m)
0.5	3.5	9.5	9.0
1.0	9.0	13.9	13.0
1.5	16.0	16.7	15.0
2.0	24.0	19.5	17.0

Solution

In this case, (5.23) and (5.35) for critical depth and slope cannot be used, as these were both derived assuming a rectangular channel. Instead, (5.22) – the general equation for critical flow – must be used, i.e.

$$\frac{\alpha Q^2 B_c}{g A_c^3} = 1 \tag{5.22}$$

Both A and B are functions of stage (h), as given in the above table. In this case, the best method of solution is a graphical one. Rearranging (5.22),

$$\frac{\alpha Q^2}{g} = \frac{A_c^3}{B_c}$$

Hence, if h is plotted against A^3/B, then for $h = h_c$, $A_c^3/B_c = \alpha Q^2/g$, and h_c may be read directly from the graph for any value of $\alpha Q^2/g$ on the A^3/B axis. From the given data, values of A^3/B for various h may be calculated, i.e.

h (m)	A^3/B (m^5)
0.5	4.8
1.0	56.1
1.5	273.1
2.0	813.2

and for critical flow $\alpha Q^2/g = 60^2/g = 367$ (assuming that $\alpha = 1$).

By inspection, the critical stage must be between 1.5 and 2.0. Using linear interpolation,

$$\frac{(h_c - 1.5)}{(2.0 - 1.5)} = \frac{(367 - 273.1)}{(813.2 - 273.1)}$$

$$h_c = 1.59\,\text{m}$$

To find the critical bed slope, apply Manning's equation with $h = h_c = 1.59$, $Q = 60$:

$$Q = \frac{1}{n}\frac{A_c^{5/3}}{P_c^{2/3}} S_c^{1/2}$$

or, in this case,

$$S_c = \left(\frac{QnP_c^{2/3}}{A_c^{5/3}}\right)^2$$

For $h_c = 1.59$, again using linear interpolation,

$$P_c = 16.7 + \frac{0.09}{0.5}(19.5 - 16.7) = 17.2$$

$$A_c = 16 + \frac{0.09}{0.5}(24 - 16) = 17.44$$

hence

$$S_c = \left(\frac{60 \times 0.04 \times 17.2^{2/3}}{17.44^{5/3}}\right)^2$$

$$= 0.0186$$

Flow transitions

Having developed the idea of mild and steep slopes, the critical depth line and subcritical and supercritical flow, the concept of flow transitions is now introduced. Figure 5.18 shows two types of transition, due to changes of bed slope. In Figure 5.18(a) it is presupposed that the channel is of mild slope upstream and steep slope downstream. The critical depth (for a given discharge) is constant. Upstream, the flow is subcritical and the depth is greater than the critical depth. Downstream, the converse is true. In the vicinity of the intersection of the mild and steep slopes, gradually varied flow is taking place and the flow regime is in transition from sub- to supercritical. At the intersection the flow is critical.

In Figure 5.18(b), the slopes have been reversed and the resulting flow transition is both more spectacular and more complex. Upstream, the flow is

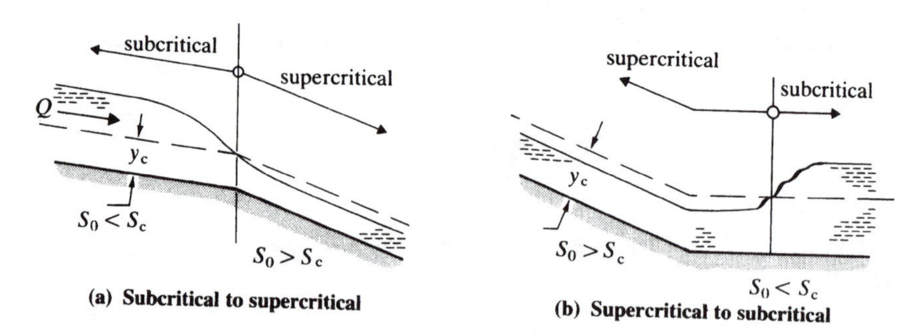

Figure 5.18 Flow transitions.

supercritical, and downstream the flow is subcritical. This type of transition is only possible through the mechanism of the hydraulic jump. Gradually varied flow takes place between the intersection of the slopes and the upstream end of the jump.

An explanation as to why these two types of transition exist may be found in terms of the Froude Number. Consider what would happen if a flow disturbance was introduced in the transition region shown in Figure 5.18(a). On the upstream (mild) slope $\mathrm{Fr} < 1$, and the disturbance would propagate both upstream and downstream. On the downstream (steep) slope the disturbance would propagate downstream. The net result is that all flow disturbances are swept away from the transition region, resulting in the smooth flow transition shown.

Conversely, for the transition shown in Figure 5.18(b), flow disturbances introduced upstream (on the steep slope) propagate downstream only. Those introduced downstream propagate both upstream and downstream. The net result in this case is that the disturbances are concentrated into a small region, which is the hydraulic jump.

The general equation of gradually varied flow

To determine the flow profile through a region of gradually varied flow, due to changes of slope or cross-section, the general equation of gradually varied flow must first be derived.

The equation is derived by assuming that for gradually varied flow the change in energy with distance is equal to the frictional losses. Hence

$$\frac{\mathrm{d}H}{\mathrm{d}x} = \frac{\mathrm{d}}{\mathrm{d}x}\left(y + \frac{\alpha V^2}{2g} + z\right) = -S_f \tag{5.36}$$

where S_f is the friction slope. Rewriting,

$$\frac{\mathrm{d}}{\mathrm{d}x}\left(y + \frac{\alpha V^2}{2g}\right) = -\frac{\mathrm{d}z}{\mathrm{d}x} - S_f$$

or

$$\frac{\mathrm{d}E_S}{\mathrm{d}x} = S_0 - S_f \tag{5.37}$$

where S_0 is the bed slope. From section (5.7),

$$\frac{\mathrm{d}E_S}{\mathrm{d}y} = 1 - \frac{Q^2 B}{g A^3} \quad (\text{taking } \alpha = 1),$$

and as

$$\mathrm{Fr}^2 = \frac{V^2}{gA/B} = \frac{Q^2B}{gA^3}$$

then

$$\frac{\mathrm{d}E_S}{\mathrm{d}y} = 1 - \mathrm{Fr}^2$$

Combining this with (5.37) gives

$$\frac{\mathrm{d}y}{\mathrm{d}x} = \frac{S_0 - S_f}{1 - \mathrm{Fr}^2} \tag{5.38}$$

which is the general equation of gradually varied flow.

S_f represents the slope of the total energy line ($\mathrm{d}H/\mathrm{d}x$). Since the bed slope (S_0) and the friction slope (S_f) are coincident for uniform flow, the friction slope (S_f) may be evaluated using Manning's equation or the Colebrook–White equation (5.4(b)).

Equations (5.37) and (5.38) are differential equations relating depth to distance. There is no general explicit solution (although particular solutions are available for prismatic channels). Numerical methods of solution are normally used in practice. These methods are considered in a later section.

Classification of flow profiles

Before examining methods of solution of (5.37) and (5.38), a deeper understanding of this type of flow may be gained by taking a general overview of (5.38).

For a given discharge, S_f and Fr^2 are functions of depth (y), e.g.

$$\left.\begin{aligned} S_f &= n^2Q^2P^{4/3}/A^{10/3} \\ \mathrm{Fr}^2 &= Q^2B/gA^3 \end{aligned}\right\} \text{ both decrease with increasing A and hence increasing } y$$

Hence it follows that as

$$S_f = S_0 \text{ when } y = y_n \quad \text{(uniform flow)}$$

then

$$S_f > S_0 \text{ when } y < y_n$$

$$S_f < S_0 \text{ when } y > y_n$$

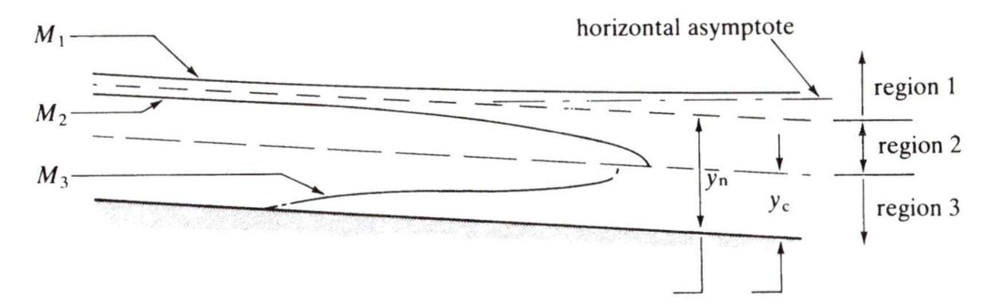

Figure 5.19 Profile types for a mild slope.

and

$$\mathrm{Fr}^2 > 1 \text{ when } y < y_c$$
$$\mathrm{Fr}^2 < 1 \text{ when } y > y_c$$

These inequalities may now be used to find the sign of dy/dx in (5.38) for any condition.

Figure 5.19 shows a channel of mild slope with the critical and normal depths of flow marked. For gradually varied flow, the surface profile may occupy the three regions shown, and the sign of dy/dx can be found for each region:

Region 1
$y > y_n > y_c$, $S_f < S_0$ and $\mathrm{Fr}^2 < 1$, hence dy/dx is positive.

Region 2
$y_n > y > y_c$, $S_f > S_0$ and $\mathrm{Fr}^2 < 1$, hence dy/dx is negative.

Region 3
$y_n > y_c > y$, $S_f > S_0$ and $\mathrm{Fr}^2 > 1$, hence dy/dx is positive.
The boundary conditions for each region may be determined similarly.

Region 1. As $y \to \infty$, S_f and $\mathrm{Fr} \to 0$ and $dy/dx \to S_0$, hence the water surface is asymptotic to a horizontal line (as y is referred to the channel bed).

For $y \to y_n$, $S_f \to S_0$ and $dy/dx \to 0$, hence the water surface is asymptotic to the line $y = y_n$.

This water surface profile is termed an M1 profile. It is the type of profile which would form upstream of a weir or reservoir, and is known as a backwater curve.

Regions 2 and 3. The profiles may be derived in a similar manner. They are shown in Figure 5.19. However, there are two anomalous results for

Regions 2 and 3. First, for the M2 profile, $dy/dx \rightarrow \infty$ as $y \rightarrow y_c$. This is physically impossible, and may be explained by the fact that as $y \rightarrow y_c$ the fluid enters a region of rapidly varied flow, and hence (5.37) and (5.38) are no longer valid. The M2 profile is known as a drawdown curve, and would occur at a free overfall.

Secondly, for the M3 profile, $dy/dx \rightarrow \infty$ as $y \rightarrow y_c$. Again, this is impossible, and in practice a hydraulic jump will form before $y = y_c$.

So far, the discussion of surface profiles has been restricted to channels of mild slope. For completeness, channels of critical, steep, horizontal and adverse slopes must be considered. The resulting profiles can all be derived by similar reasoning, and are shown in Figure 5.20.

Outlining surface profiles and determining control points

An understanding of these flow profiles and how to apply them is an essential prerequisite for numerical solution of the equations. Before particular types of flow profiles can be determined for any given situation, two things must be ascertained:

(a) Whether the channel slope is mild, critical or steep. To determine its category, the critical and normal depth of flow must be found for the particular design discharge.
(b) The position of the control point or points must be established. A control point is defined as any point where there is a known relationship between head and discharge. Typical examples are weirs, flumes and gates or, alternatively, any point in a channel where critical depth occurs (e.g. at the brink of a free overfall), or the normal depth of flow at a suitably remote distance from the point of interest.

Having established the slope category and the position of any control points, the flow profile(s) may then be sketched. For subcritical flow, the profiles are controlled from a point downstream. For supercritical flow, the profiles are controlled from upstream.

Figure 5.21 shows three typical flow profiles. Figure 5.21 (a) shows the influence of a broad-crested weir on upstream water levels. Numerical solution of this problem proceeds upstream from the weir (refer to Example 5.7).

Figure 5.21(b) shows the influence of bridge piers under flood conditions. Many old masonry bridges (with several bridge piers) act in a similar manner to venturi flumes in choking the flow (particularly at high discharges). Flow through the bridge is rapidly varied and, on exit from the bridge piers, the flow is supercritical. However, supercritical flow cannot exist for long, as the downstream slope is mild and the downstream flow uniform (assuming that it is unaffected by downstream control). A hydraulic jump must form to return the flow to the subcritical condition. The position and height of

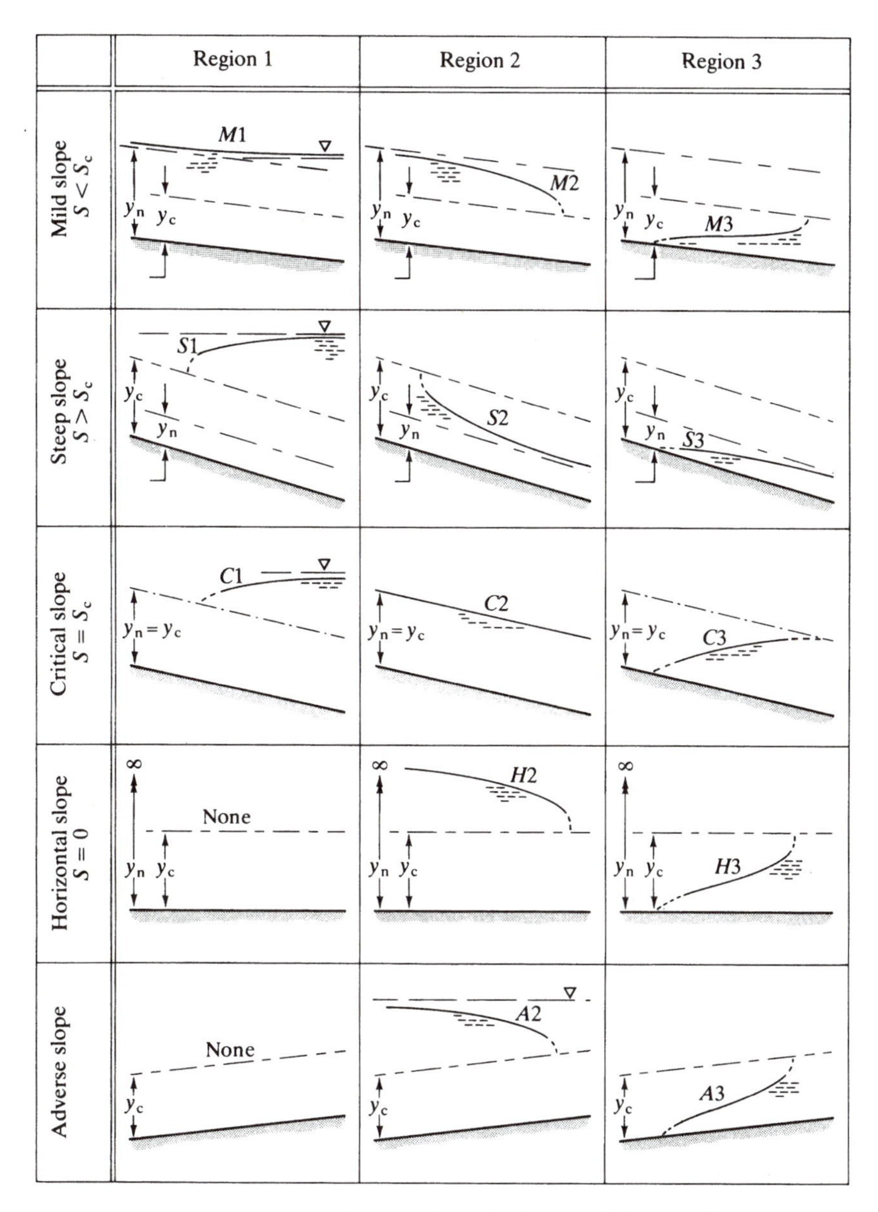

Figure 5.20 Classification of gradually varied flow profiles.

the jump are determined as shown in Figure 5.21(b). An example of this type of problem is given in Example 5.8.

Figure 5.21(c) shows the flow profile for a side channel spillway and stilling basin. In this example, it is assumed that the side channel is not

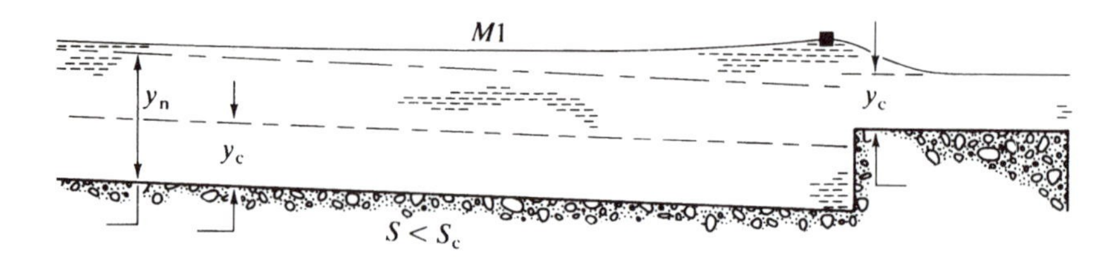

(a) Influence of broadcrested weir on river flow

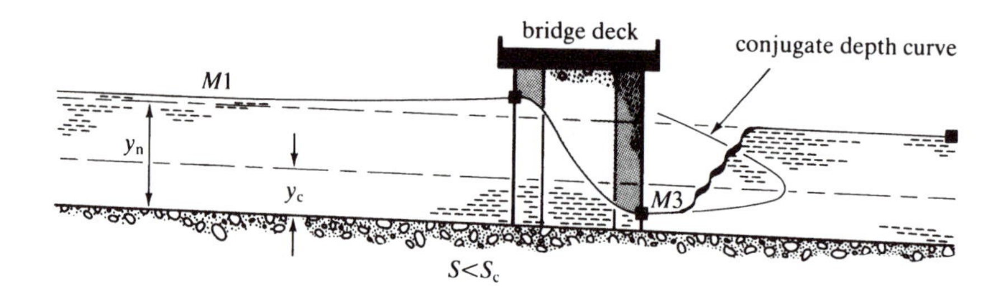

(b) Influence of bridge piers on river flow under flood conditions

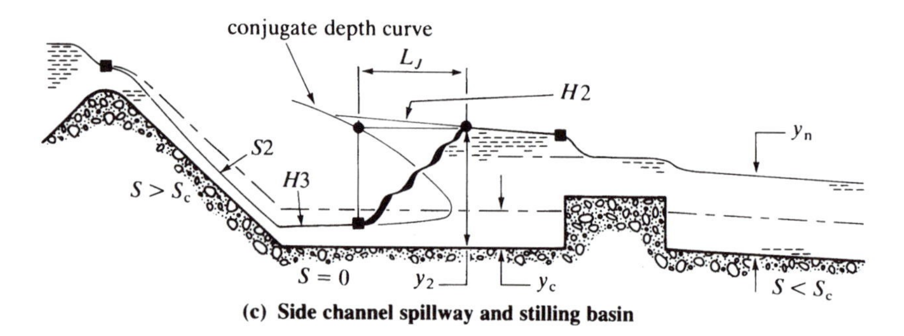

(c) Side channel spillway and stilling basin

Figure 5.21 Examples of typical surface profiles. (*Note*: ■ denotes a control point.)

so steep as to invalidate the equations of gradually varied flow, and that aeration does not occur (refer to Chapter 13 for more details).

Methods of solution of the gradually varied flow equation

The three forms of the gradually varied flow equation are:

$$\frac{\mathrm{d}H}{\mathrm{d}x} = -S_f \tag{5.36}$$

$$\frac{\mathrm{d}E_S}{\mathrm{d}x} = S_0 - S_f \tag{5.37}$$

$$\frac{dy}{dx} = \frac{S_0 - S_f}{1 - Fr^2} \tag{5.38}$$

There are three types of solution to the above equations:

1. direct integration;
2. graphical integration;
3. numerical integration.

These three methods of solution are discussed, and examples are given where appropriate.

Direct integration. Equation (5.38) may be solved directly only for regular channels. Integration methods have been developed by various workers, starting with Dupuit in 1848, who found a solution for wide rectangular channels. He used the Chézy resistance equation and ignored changes of kinetic energy. In 1860, Bresse found a solution including changes of kinetic energy (again for wide rectangular channels using the Chézy equation). Bakhmeteff, starting in 1912, extended the work to trapezoidal channels. Using Manning's resistance equation, various other workers have extended Bakhmeteff's method.

Full details of these methods may be found in Henderson (1966) or Chow (1959). They are not discussed here as they have been superseded by numerical integration methods which may be used for both regular and irregular channels.

Graphical integration. Equation (5.38) may be rewritten as

$$\frac{dx}{dy} = \frac{1 - Fr^2}{S_0 - S_f}$$

Hence,

$$\int_0^X dx = \int_{y_1}^{y_2} \frac{1 - Fr^2}{S_0 - S_f} dy$$

or

$$X = \int_{y_1}^{y_2} f(y) dy$$

where

$$f(y) = \frac{1 - Fr^2}{S_0 - S_f}$$

If a graph of y against $f(y)$ is plotted, then the area under the curve is equivalent to X. The value of the function $f(y)$ may be found by substitution of A, P, S_0 and S_f for various y for a given Q. Hence, the distance X between given depths (y_1 and y_2) may be found graphically.

This method was quite popular until the widespread use of computers facilitated the use of the more versatile numerical methods.

Numerical integration. Using simple numerical techniques, all types of gradually varied flow problems may be quickly and easily solved using only a microcomputer. A single program may be written which will solve most problems.

However, as an aid to understanding, the numerical methods are discussed under three headings:

(a) the direct step method (distance from depth for regular channels);
(b) the standard step method, regular channels (depth from distance for regular channels); and
(c) the standard step method, natural channels (depth from distance for natural channels).

The direct step method. Equation (5.38) may be rewritten in finite difference form as

$$\Delta x = \Delta y \left(\frac{1 - \mathrm{Fr}^2}{S_0 - S_f} \right)_{\text{mean}} \tag{5.39}$$

where 'mean' refers to the mean value for the interval (Δx). This form of the equation may be used to determine, directly, the distance between given differences of depth for any trapezoidal channel. The method is best illustrated by an example.

Example 5.7 Determining a backwater profile by the direct step method

Using Figure 5.21(a), showing a backwater curve, determine the profile for the following (flood) conditions:

$$Q = 600\ \mathrm{m^3/s} \quad S_0 = 2\ \mathrm{m/km} \quad n = 0.04$$

channel: rectangular, width 50 m

weir: $C_d = 0.88$

sill height: $P_s = 2.5\,\mathrm{m}$

Solution

First, establish a control point as follows.
(a) Find normal depth (y_n) from Manning's equation:

$$Q = \frac{1}{n}\frac{A^{5/3}}{P^{2/3}} S_0^{1/2}$$

In this case,

$$600 = \frac{1}{0.04} \times \frac{(50y_n)^{5/3}}{(50+2y_n)^{2/3}} \times 0.002^{1/2}$$

Solution by trial and error yields

$$y_n = 4.443\,\text{m}$$

(b) Find the critical depth (y_c):

$$y_c = \left(\frac{q^2}{g}\right)^{1/3} \quad \text{(for a rectangular channel)}$$

In this case,

$$y_c = \left(\frac{(600/50)^2}{g}\right)^{1/3} = 2.448\,\text{m}$$

(c) Find the depth over the weir (y_w):

$$Q = 1.705 C_d B h^{3/2}$$

In this case,

$$600 = 1.705 \times 0.88 \times 50 \times h^{3/2}$$

so

$$h = 4\,\text{m}$$

As

$$y_w = h + P_S$$
$$y_w = 6.5\,\text{m}$$

Hence $y_w > y_n > y_c$ (i.e. Region 1), and $S < S_c$ (i.e. a mild slope).

This confirms (a) that the control is at the weir and (b) that there is an M1 type profile.

The profile is found by taking $y_w = 6.5$ m as the initial depth and $y = 4.5$ m (slightly greater than y_n) as the final depth, and proceeding upstream at small intervals of

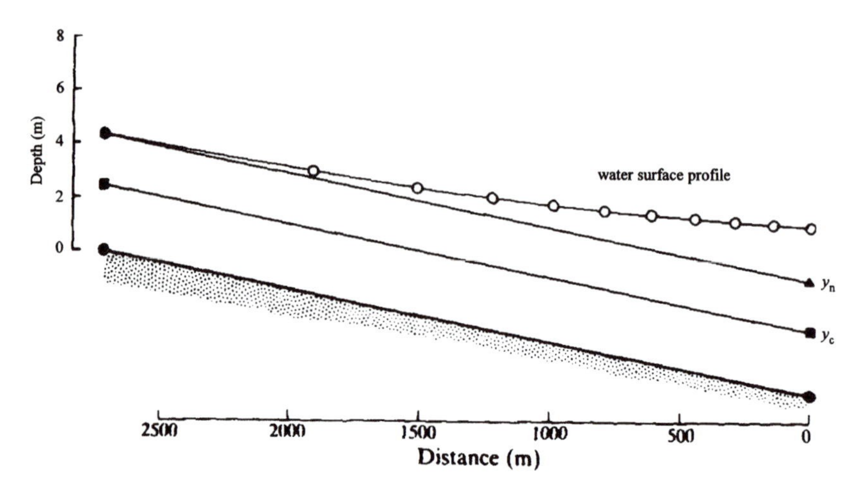

Figure 5.22 Backwater profile for Example 5.7.

depth Δy. Fr, S_0 and S_f are evaluated at each intermediate depth, and a solution is found for Δx using the finite difference equation. A tabular solution is shown in Table 5.3, and the resulting profile is shown in Figure 5.22.

Table 5.3 is self-explanatory, but the following points should be kept in mind:

(a) *Signs.*

x	is positive in the direction of flow
S_0	is positive (except for adverse slopes)
S_f	is positive by definition
Δy	is positive if final depth > initial depth
Δy	is negative if initial depth > final depth

In Example 5.7, Δy is negative and S_0 is positive, which makes Δx negative.

(b) *Accuracy.* Normal depth is always approached asymptotically, so for depths approaching normal depth $S_0 - S_f$ is very small and must be calculated accurately. Of course, it is not possible to calculate the distance of normal depth from the control point as this is theoretically infinite. In Example 5.7, a depth slightly larger than normal depth was chosen for this reason.

In regions of large curvature (i.e. approaching critical depth) equation (5.38) is no longer valid as the pressure distribution departs from hydrostatic pressure. Thus the accuracy of the solution is impaired as critical depth is approached and the solution should be terminated before critical depth is reached.

Table 5.3 Computer solution of Example 5.7.

Discharge (cumecs)?600
Channel width (m)?50
Mannings's n?0.04
Number of intervals (max 99)?10
Slope (in decimals)?0.002
SLOPE + OR – ?1
Normal depth 4.4333
Critical depth 2.4483

Initial depth?6.5
Final depth?4.5

y (m)	A (m^2)	P (m)	Fr	$(1 - Fr^2)$ mean	S_f	$(S_0 - S_f)$ mean	x (m)
6.5000	325.0000	63.0000	0.2312		0.0006		0.0000
				0.9439		0.0014	
6.3000	315.0000	62.6000	0.2423		0.0007		−139.1034
				0.9383		0.0013	
6.1000	305.0000	62.2000	0.2543		0.0007		−284.4225
				0.9319		0.0012	
5.9000	295.0000	61.8000	0.2673		0.0008		−437.6747
				0.9246		0.0011	
5.7000	285.0000	61.4000	0.2815		0.0009		−601.3459
				0.9163		0.0010	
5.5000	275.0000	61.0000	0.2970		0.0010		−779.2213
				0.9066		0.0009	
5.3000	265.0000	60.6000	0.3140		0.0011		−977.4688
				0.8954		0.0008	
5.1000	255.0000	60.2000	0.3327		0.0013		−1207.1478
				0.8823		0.0006	
4.9000	245.0000	59.8000	0.3532		0.0015		−1491.1912
				0.8669		0.0004	
4.7000	235.0000	59.4000	0.3760		0.0017		−1890.7515
				0.8488		0.0002	
4.5000	225.0000	59.0000	0.4014		0.0019		−2695.5699

(c) *Choice of step interval.* Numerical solutions always involve approximations, and here the choice of step interval affects the solution. The smaller the step interval, the greater the accuracy. If the calculations are carried out by computer, then successively smaller steps may be chosen until the solution converges to the desired degree of accuracy. In Example 5.7, such an exercise yielded the following results:

No. of steps	Δy	Length of reach	Percentage change
5	−0.4	2587.0	
			4.0
10	−0.2	2696.0	
			2.2
20	−0.1	2757.0	
			0.6
30	−0.067	2775.0	
			0.5
60	−0.033	2788.4	
			0.14
125	−0.016	2792.2	
			0.03
250	−0.008	2793.1	
			0.007
500	−0.004	2793.3	
			0.004
1000	−0.002	2793.4	
			0.000
2000	−0.001	2793.4	

The results of this exercise suggest that 1000 steps are necessary to achieve convergence. However the 10 steps originally selected give a solution to within 3.5% of the converged solution.

(d) *Validity of solution.* It has been suggested that the identification of a control point is of paramount importance in determining gradually varied flow profiles. This is reasonable if a broad understanding of the flow pattern is the aim. However, the equations may be solved between any two depths provided they are within the same region of flow. Thus, in the case of Example 5.7, the solution may proceed upstream or downstream, provided that both the initial and final depths are greater than the normal depth. If this condition is not met, then the sign of Δx will change at some intermediate point, demonstrating that the solution has passed through an asymptote and the solution is no longer valid.

(e) *Composite profiles*. Channels may have more than one control point, as already shown in Figure 5.21(b) and (c). In the case of Figure 5.21(b), a hydraulic jump effects the transition from supercritical to subcritical flow, but its height and position are determined by the upstream and downstream control points. An example of how to solve such problems follows.

Example 5.8 Composite profiles

For the situation shown in Figure 5.21(b), determine the distance of the jump from the bridge for the following conditions:

$$Q = 600\,\mathrm{m^3/s} \quad S_0 = 3\,\mathrm{m/km} \quad n = 0.04$$

channel: rectangular, width 50 m

depth at exit from the bridge = 1.2 m

Solution

First, normal and critical depth must be found using the same methods as in Example 5.7. These are:

$$y_n = 3.897\,\mathrm{m} \quad y_c = 2.448\,\mathrm{m}$$

In this case, the sequent depth of the jump equals the normal depth of flow. Utilizing the hydraulic jump equation (5.28b)

$$y_1 = (y_2/2)(\sqrt{1 + 8\mathrm{Fr}_2^2} - 1)$$

gives $y_1 = 1.41\,\mathrm{m}$ for $y_2 = y_n = 3.897\,\mathrm{m}$.

Hence, to find the distance of the jump from the bridge an M3 profile must be calculated starting from the bridge (at $y = 1.2$) and ending at the initial depth of the jump ($y = 1.41$). This is most easily achieved by using the direct step method. Table 5.4 shows the calculations using a step length of 0.05 m. The position of $y = 1.41\,\mathrm{m}$ is founded by interpolation:

$$y = 1.4 \quad x = 11.368$$

$$y = 1.45 \quad x = 14.151$$

$$\frac{(1.41 - 1.4)}{(X - 11.368)} = \frac{(1.45 - 1.4)}{(14.151 - 11.368)}$$

$$X = 11.925\,\mathrm{m}$$

Table 5.4 Computer solution of Example 5.8.

Discharge (cumecs)?600
Channel width (m)?50
Mannings's n?0.04
Number of intervals (max 99)?5
Slope (in decimals)?0.003
SLOPE+OR−?1
Normal depth 3.8967
Critical depth 2.4483

Initial depth?1.2
Final depth?1.45

y (m)	A (m^2)	P (m)	Fr	$(1-\mathrm{Fr}^2)$ mean	S_f	(S_0-S_f) mean	x (m)
1.2000	60.0000	52.4000	2.9146		0.1336		0.0000
				−7.0052		−0.1222	
1.2500	62.5000	52.5000	2.7415		0.1169		2.8657
				−6.0985		−0.1068	
1.3000	65.0000	52.6000	2.5848		0.1028		5.7196
				−5.3237		−0.0939	
1.3500	67.5000	52.7000	2.4426		0.0909		8.5558
				−4.6578		−0.0828	
1.4000	70.0000	52.8000	2.3129		0.0807		11.3682
				−4.0822		−0.0734	
1.4500	72.5000	52.9000	2.1943		0.0720		14.1507

An alternative approach, necessary in cases where the jump exists between two profiles, is to plot the conjugate depth curve for the jump on the profile to find the points of intersection, as shown in Figure 5.23.

The standard step method (for regular channels). Equation (5.37) may be rewritten in finite difference form as

$$\Delta E_S = \Delta x(S_0 - S_f)_{\text{mean}} \tag{5.40}$$

where 'mean' refers to the mean values for the interval Δx.

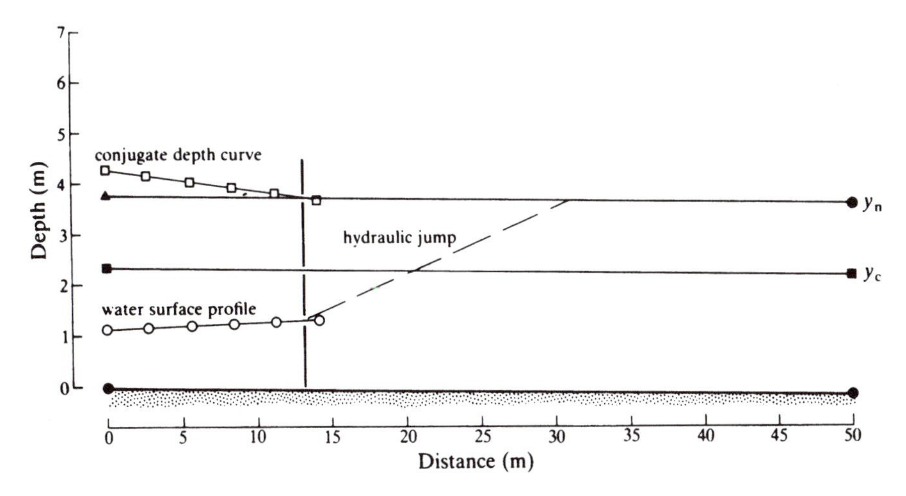

Figure 5.23 Composite profiles for Example 5.8.

This form of the equation may be used to determine the depth at given distance intervals. The solution method is an iterative procedure as follows:

1. assume a value for depth (y);
2. calculate the corresponding specific energy ($E_{S_{xG}}$);
3. calculate the corresponding friction slope (S_f);
4. calculate ΔE_S over the interval Δx using (5.40);
5. calculate $E_{S_{x+\Delta x}} = E_{S_x} + \Delta E_S$;
6. compare $E_{S_{x+\Delta x}}$ and $E_{S_{xG}}$;
7. if $E_{S_{x+\Delta x}} \neq E_{S_{xG}}$, then return to 1.

Hence, to determine the depth at a given distance may require several iterations. However, this method has the advantage over the direct step method that depth is calculated from distance, which is the more usual problem. Example 5.9 illustrates the method by solving again the problem of Example 5.7.

Example 5.9 Determining a backwater profile by the standard step method

Solve Example 5.7 again by using a standard step length of 150 m.

The establishment of a control point and the normal and critical depths are as in Example 5.7. A tabular solution is shown in Table 5.5. Only the correct values of depth (y) are shown in the table (to save space). The solution is most easily carried out by computer.

Solution

Table 5.5 Computer solution of Example 5.9.

INPUT Q,N,SO,B,Y1,DX,L 600,.04,.002,50,6.5,–150,1950								
Y (m)	A (m^2)	P (m)	EG (m)	S_f	$(S_0 - S_f)$ mean	ΔE (m)	EC (m)	X (m)
6.5000	325.0000	63.0000	6.6737	0.0006			6.6737	0
					0.0014	−0.2032		
6.2848	314.2386	62.5695	6.4706	0.0007			6.4705	−150
					0.0013	−0.1928		
6.0795	303.9756	62.1590	6.2781	0.0008			6.2778	−300
					0.0012	−0.1814		
5.8852	294.2583	61.7703	6.0971	0.0008			6.0967	−450
					0.0011	−0.1692		
5.7027	285.1330	61.4053	5.9283	0.0009			5.9279	−600
					0.0010	−0.1561		
5.5328	276.6424	61.0657	5.7726	0.0010			5.7722	−750
					0.0009	−0.1424		
5.3765	268.8232	60.7529	5.6304	0.0011			5.6302	−900
					0.0009	−0.1283		
5.2351	261.7527	60.4701	5.5029	0.0012			5.5020	−1050
					0.0008	−0.1141		
5.1078	255.3910	60.2156	5.3891	0.0013			5.3887	−1200
					0.0007	−0.1002		
4.9958	249.7889	59.9916	5.2899	0.0014			5.2890	−1350
					0.0006	−0.0867		
4.8975	244.8755	59.7950	5.2035	0.0015			5.2032	−1500
					0.0005	−0.0740		
4.8134	240.6717	59.6269	5.1302	0.0015			5.1295	−1650
					0.0004	−0.0623		
4.7416	237.0775	59.4831	5.0680	0.0016			5.0679	−1800
					0.0003	−0.0518		
4.6818	234.0911	59.3636	5.0167	0.0017			5.0162	−1950

The standard step method (for natural channels). A common application of flow profiles is in determining the effects of channel controls in natural channels. For several reasons this application is more complex than the preceding cases. The discharge is generally more variable and difficult to quantify, and the assessment of Manning's n is less accurate. These aspects are dealt with elsewhere (see section 5.6 and Chapter 10).

Another main difficulty lies in relating areas and perimeters to depth. This can only be accomplished by a detailed cross-sectional survey at known locations. In this way, tables of area and perimeter for a given stage may be prepared. Notice that depth must be replaced by stage, as depth is not a meaningful quantity for natural sections. Alternatively, the cross-sectional data may be stored on a computer, values of area and perimeter being computed as required for a given stage.

The solution technique is to use (5.36) in finite difference form as

$$\Delta H = \Delta x(-S_f)_{\text{mean}} \tag{5.41}$$

where 'mean' refers to the mean value over the interval Δx.

This may be solved iteratively to find the stage at a given distance in a similar manner to the standard step method for regular channels, if depth is replaced by stage and specific energy by total energy.

Two further complications arise in applying equation (5.41) to natural channels. The first one is in determining the mean friction slope ($S_{f\,\text{mean}}$). At any particular cross-section S_f may be determined either directly from Manning's equation (see page 155) or by using the conveyance function (equation (5.16) with S_f replacing S_0).

However, in a natural channel each cross-section is likely to be different and hence it is necessary to find a **representative** friction slope for each **reach** from the cross-sectional data at each end of the reach. It is possible to conceive the representative friction slope as being some kind of weighted average value derived either directly from the friction slopes at each cross-section or directly from a weighted average value of the channel conveyances.

Cunge *et al.* (1980, pages 129–30) compare four methods of finding a representative friction slope. Differences between results from the methods can be very large if the upstream and downstream cross-sections are very different. He concludes that if this is the case then a smaller distance step should be introduced. In equation (5.41) the mean value of S_f has been chosen as it is the simplest representation conceptually.

The second complication in solving equation (5.41) in natural channels arises from the presence of flood banks. When flood discharges are being considered, the flow will overtop the main channel and flow on the flood plain will occur. Under these circumstances the energy coefficient (α) must be computed. Where natural channels have well defined flood plains (as shown in Figure 5.5), then α is conveniently estimated by equation (5.14) which uses the channel conveyance function.

An iterative solution method for two-stage natural sections can, therefore, be set up as follows:

1. assume a value for stage (h_G) at $x+\Delta x$;
2. calculate the corresponding values of A_i, P_i and K_i at $x+\Delta x$ from the tabulated cross-sectional data and hence find α (using (5.14)) at $x+\Delta x$ for the assumed stage (h_G);
3. calculate the mean cross-sectional velocity ($\overline{V}=Q/A$) at $x+\Delta x$ and hence find the total energy at $x+\Delta x$ ($H_G = h_G + \alpha\overline{V}^2/2g$);
4. calculate the friction slope $S_{fG\ (x+\Delta x)}$ at $x+\Delta x$ using equation (5.16) and hence find the mean friction slope $S_{f\text{mean}} = (S_{fG\ (x+\Delta x)} + S_{f(x)})/2$;
5. calculate ΔH over the interval Δx using (5.41);
6. calculate $H_{x+\Delta x} = H_x + \Delta H$;
7. compare $H_{x+\Delta x}$ and H_G;
8. if $H_{x+\Delta x} \neq H_G$ then repeat from 1 until suitable convergence is obtained.

Several commercially available mathematical models have been developed for application to natural channels. One such model (FLUCOMP1 & 2), developed by the Hydraulics Research Station (England), allows for calibration of Manning's n values from recorded discharge and water levels, incorporates the effects of weirs and bridge piers, and includes lateral inflows. This model also computes one-dimensional gradually varied unsteady flow (see section 5.11 for further details). Complete details of the model are given in Samuels and Gray (1982).

5.11 Unsteady flow

Types of unsteady flow

Unsteady flow is the normal state of affairs in nature, but for many engineering applications the flow may be considered to be steady. However, under some circumstances, it is necessary to consider unsteady flow. Such circumstances may include the following:

Translatory waves. The movement of flood waves down rivers.

Surges and bores. Produced by sudden changes in depth and/or discharge (e.g. tidal effects or control gates).

Oscillatory waves. Waves produced by vertical movement rather than horizontal movement (e.g. ocean waves).

Translatory waves are considered in a simplified way in Chapter 10 (hydrology, flood routing) and they are considered in more detail later in this section as an example of gradually varied unsteady flow. Surge waves are an example of rapidly varied unsteady flow and a simple treatment is presented in the following section. Oscillatory waves are considered in Chapter 8.

Surge waves

A typical surge wave is shown in Figure 5.24(a), in which a sudden increase in the downstream depth has produced a steep-fronted wave moving upstream at velocity V. The wave is, in fact, a moving hydraulic jump, and the solution for the speed of the wave is most easily obtained by transposing the problem to that of the hydraulic jump. This may be achieved by the use of the technique of the travelling observer, as shown in Figure 5.24(b). To an observer travelling on the wave at velocity V, the wave is stationary but the upstream and downstream velocities are increased to $V_1 + V$ and $V_2 + V$, respectively (and the river bed is moving at velocity V). The travelling observer sees a (stationary) hydraulic jump. Hence, the hydraulic jump equation may be applied as follows:

$$\frac{y_2}{y_1} = \frac{1}{2}(\sqrt{1 + 8\mathrm{Fr}_1^2} - 1) \tag{5.28a}$$

or

$$\mathrm{Fr}_1^2 = \frac{1}{2}\frac{y_2}{y_1}\left(\frac{y_2}{y_1} + 1\right) \text{ (see derivation of (5.28a))}$$

i.e.

$$\frac{(V_1 + V)^2}{gy_1} = \frac{1}{2}\frac{y_2}{y_1}\left(\frac{y_2}{y_1} + 1\right) \tag{5.42}$$

This equation may be solved for known upstream conditions and the downstream depth. If y_2 is unknown, then the continuity equation may be used:

$$(V_1 + V)y_1 = (V_2 + V)y_2$$

In this case, Q_2 must be known for solution of V.

Considering (5.42) in more detail, three useful results may be obtained. First, as y_2/y_1 tends to unity, then $(V_1 + V)$ tends to $\sqrt{gy}$. As $(V_1 + V) = c$ (the

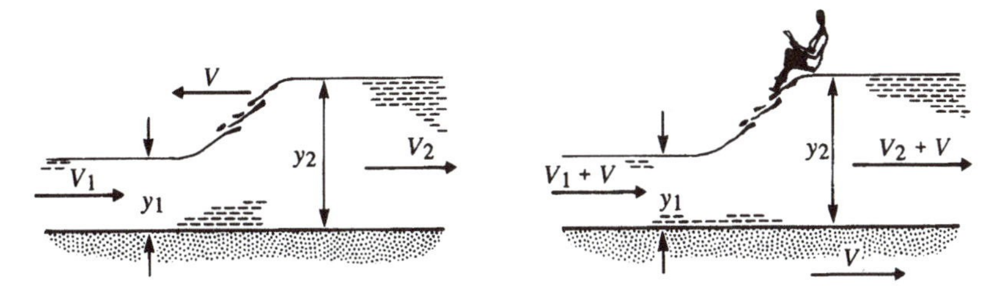

Figure 5.24 Surge waves.

wave speed relative to the water) this confirms the result given in section 5.7 that small disturbances are propagated at a speed $c = \sqrt{gy}$ (for rectangular channels).

Secondly, for $y_2/y_1 > 1$, then $(V_1 + V) > \sqrt{gy_1}$. This implies that surge waves can travel upstream even for supercritical flow. This is confirmed by Example 5.10. However, after the passage of such a surge wave, the resulting flow is always subcritical.

Thirdly, for $(V_1 + V) > \sqrt{gy_1}$, then the surge wave will overtake any upstream disturbances, and conversely any downstream disturbances will overtake the surge. Hence, this type of surge wave remains stable and steep fronted.

There are, in fact, four possible types of surge waves; upstream and downstream, each of which may be positive or negative. Positive surges result in an increase in depth and are stable and steep fronted (as discussed above). Negative surges result in a decrease in depth. These are unstable and tend to die out as disturbances travel faster than the surge wave.

Example 5.10 Speed of propagation of an upstream positive surge wave

Uniform flow in a steep rectangular channel is interrupted by the presence of a hump in the channel bed which produces critical flow for the initial discharge of $4\,m^3/s$. If the flow is suddenly reduced to $3\,m^3/s$ some distance upstream, determine the new depth and speed of propagation of the surge wave. (*Channel data*: $b = 3\,m$, $n = 0.015$, $S_0 = 0.01$, $\Delta z = 0.083\,m$).

Solution

Referring to Figure 5.12, this problem is an example of what happens when $\Delta z > E_{S1} - E_{S2}$ and $y_1 < y_c$. In this case, for the initial discharge flow over the hump is critical, but for the reduced discharge E_{S1} (for $y_1 = y_n$) is insufficient. The result is that choked flow is produced by y_1 increasing to a new value above critical depth, which produces a positive surge wave travelling upstream.

For $Q = 4\,m^3/s$:

$y_n = 0.421\,m$ (using Manning's equation);

$y_c = 0.566\,m$ (using (5.23));

$E_{Sc} = 0.849\,m$ (using (5.25));

$E_{Sn} = 0.932\,m$ (using (5.20));

$E_{Sn} - E_{Sc} = 0.083 = \Delta z$ (i.e. critical flow is produced).

For $Q = 3\,m^3/s$:

$y_n = 0.348\,m$;

$y_c = 0.467\,m$;

$E_{Sc} = 0.701\,\text{m};$

$E_{Sn} = 0.769\,\text{m};$

$E_{Sn} - E_{Sc} = 0.068 < \Delta z.$

Hence, normal depth cannot be maintained. The new depth upstream of the hump (y_2) is found from

$$E_{S2} = E_{Sc} + \Delta z$$

i.e.

$$E_{S2} = 0.784\,\text{m}$$

giving

$$y_2 = 0.67\,\text{m (by iteration)}$$

The equation of the surge wave may now be applied:

$$\frac{(V_1 + V)^2}{gy_1} = \frac{1}{2}\left[\frac{y_2}{y_1}\left(\frac{y_2}{y_1} + 1\right)\right]$$

where

$$y_1 = y_n = 0.348\,\text{m} \quad y_2 = 0.67\,\text{m}$$
$$V_1 = Q/A_1 = 3/(3 \times y_n) = 2.874\,\text{m/s}$$

Substituting into (5.42) yields

$$V = 0.23\,\text{m/s}$$

This example shows that a surge wave may travel upstream even for supercritical flow, but leaves behind subcritical flow. The surge wave will proceed upstream until either it submerges the upstream control or until a stationary hydraulic jump forms (by y_2 decreasing to the sequent depth).

Gradually varied unsteady flow: the equations of one-dimensional motion

This type of flow implies translatory wave motion of long wavelength and low amplitude. In this case the assumption of parallel streamlines and hydrostatic pressure distributions is reasonable.

Strictly speaking, the equations of motion apply only to truly one-dimensional flow in which the water surface is horizontal at any cross-section and the velocity is uniform over the cross-section. This is only an approximation to natural river flow. Supplementary coefficients (e.g. the momentum coefficient β) are often introduced into the equations to simulate quasi two-dimensional flow. For flood flows in natural channels with flood plains further approximations have to be made.

The derivation and solution of the gradually varied unsteady flow equations is a complicated matter, even for the simplest case of a rectangular channel. General solutions are made practicable only by the use of a computer. An introductory discussion of some of the techniques used may be found in Chapter 14, and the Reference list there includes some of the more advanced texts, which give details of the range of numerical techniques in current use. Practical aspects of computational modelling of river flows are discussed in Chapter 15.

References and further reading

Ackers, P. (1992) Hydraulic Design of Two-stage Channels, *Proc. Instn. Civil Engrs., Water, Maritime and Energy*, **96**, Dec., 247–57.

Chow Ven te (1959) *Open Channel Hydraulics*, McGraw-Hill, Tokyo.

Cunge, J. A., Holly, F. H. (Jr) and Verwey, A. (1980) *Practical Aspects of Computational River Hydraulics*, Pitman, London.

French, R. H. (1986) *Open Channel Hydraulics*, McGraw-Hill, Singapore.

Henderson, F. M. (1966) *Open Channel Flow*, Macmillan, New York.

Jansen, P.Ph. *et al.* (eds) (1979) *Principles of River Engineering*, Pitman, London.

Knight, D. (1989) Hydraulics of flood channels, Chapter 6 in *Floods: Hydrological, Sedimentological and Geomorphological Implications* (eds K. Beven and P. Carling), Wiley, Chichester.

Knight, D. W. and Yuen, K. W. H. (1990) Critical flow in a two stage channel, in *International Conference on River Flood Hydraulics* (ed. W. R. White), Wiley, Chichester.

Ramsbottom, D. M. (1989) *Flood Discharge Assessment – Interim Report*, Report SR195, Hydraulics Research, Wallingford.

Townson, J. M. (1991) *Free Surface Hydraulics*, Unwin Hyman, London.

6

Pressure surge in pipelines

6.1 Introduction

All of the pipe flows considered in Chapter 4 were steady flows, i.e. discharge was assumed to remain constant with time. This corresponds to the situation when the control valves in the system are held at a fixed setting. From time to time it will be necessary to alter the valve setting. Immediately following such an alteration the flow will be accelerating (or decelerating), and will therefore be unsteady. Unsteady (or 'surge') conditions often continue for only a very short period. Nevertheless, the effects on the system may, under some circumstances, be dramatic (including pipe bursts).

For the sake of simplicity, consider first a case involving flow of an incompressible liquid. Civil engineers commonly deal with pipelines of considerable length and diameter. Such a pipeline will contain a large mass of flowing fluid, and the momentum of the fluid will therefore be correspondingly large, as in the case of a long pipeline from a reservoir (Fig. 6.1(a)). Note that the velocity of flow is here denoted by 'u':

$$\text{mass of liquid in pipeline} = \rho AL$$

$$\text{momentum of liquid} = M = \rho ALu \quad (u = Q/A)$$

If the flow is retarded by a control valve at the downstream end, then the fluid undergoes a rate of change of momentum of

$$\frac{\mathrm{d}M}{\mathrm{d}t} = \rho AL\frac{\mathrm{d}u}{\mathrm{d}t} \tag{6.1}$$

A force, F, is required to produce this change of momentum. This force is applied to the fluid by the control valve. This force exists only during the deceleration of the fluid, and it is therefore only a short-lived phenomenon. According to the laws of fluid pressure a force (or pressure) applied in one plane must be transmitted in all directions (Fig. 6.1(b)), therefore

$$\delta pA = F$$

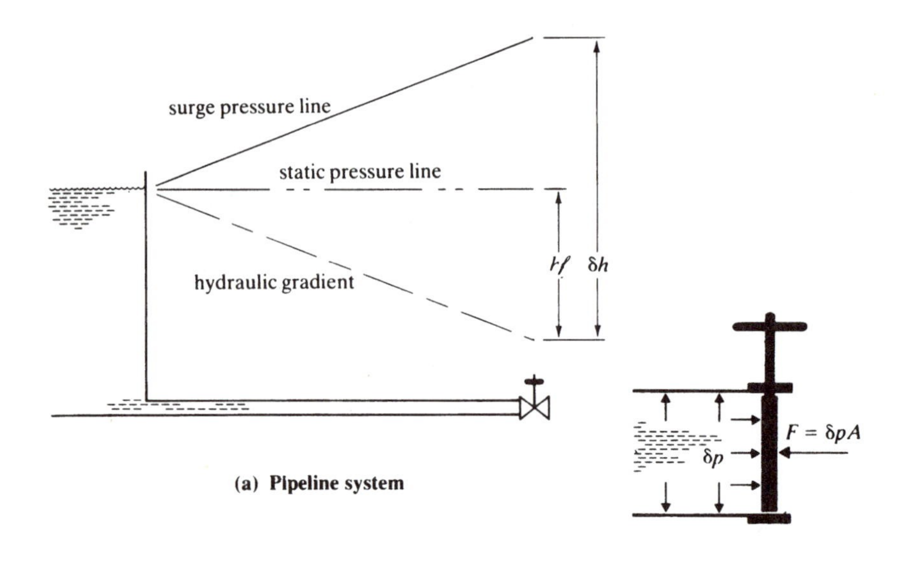

Figure 6.1 Incompressible surge.

where δp is the instantaneous rise in pressure corresponding to F (this is known as a 'surge pressure' or 'transient pressure'). Hence

$$F = \delta p A = \rho A L \frac{\mathrm{d}u}{\mathrm{d}t}$$

and so

$$\delta p = \rho L \frac{\mathrm{d}u}{\mathrm{d}t}$$

or

$$\frac{\delta p}{\rho g} = \delta h = \frac{L}{g}\frac{\mathrm{d}u}{\mathrm{d}t} \qquad (6.2)$$

As L is a constant for a given system, $\delta p = f(\mathrm{d}u/\mathrm{d}t)$, i.e. δp is a function of the rate of closure of the valve. Upstream of the valve, δp (or δh) varies linearly with distance. An example will now be developed to illustrate the use of this approach.

Example 6.1 Incompressible unsteady flow

Water flows from a reservoir along a rigid horizontal pipeline. The pipe intake is 20 m below the water surface elevation in the reservoir. The pipe is 0.15 m in diameter and 1500 m long, and $\lambda = 0.02$. The pipe discharges to atmosphere through a valve at its downstream end. The rate of valve adjustment is such that it can be completely closed in 4 s and gives a uniform deceleration of the water in the pipe. Calculate the pressure just upstream of the valve and 500 m upstream of the valve if the valve aperture is adjusted from the fully open to the half-open position (in 2 s).

Solution

Before valve closure commences the velocity of flow is u_0, so

$$h_f = \frac{\lambda L u_0^2}{2gD} = \frac{0.02 \times 1500 \times u_0^2}{2 \times 9.81 \times 0.15} = 20\,\text{m}$$

Therefore

$$u_0 = 1.4\,\text{m/s}$$

During valve closure

$$\frac{\mathrm{d}u}{\mathrm{d}t} = \frac{u_0}{t} = \frac{1.4}{4}$$

For half closure, $u = \frac{1}{2}u_0 = 0.7\,\text{m/s}$, therefore

$$h_f = \frac{0.02 \times 1500 \times 0.7^2}{2 \times 9.81 \times 0.15} = 5\,\text{m}$$

The surge pressure (δh) is given by

$$\delta h = \frac{L}{g}\frac{\delta u}{\delta t} = \frac{1500}{9.81} \times \frac{1.4}{4} = 53.5\,\text{m}$$

Total head at valve = (static head + $\delta h - h_f$) is

$$20 + 53.5 - 5 = 68.5\,\text{m}$$

At a point 500 m upstream of the valve

$$\delta h = \frac{1000}{9.81} \times \frac{1.4}{4} = 35.7\,\text{m}$$

$$h_f = 5 \times \frac{1000}{1500} = 3.3\,\text{m}$$

Therefore, total head = $20 + 35.7 - 3.3 = 52.4$ m. This example presupposes that the liquid is incompressible, and that the pipe is rigid.

6.2 Effect of 'rapid' valve closure

Taking the pipe system used in the preceding example, consider now the effect of increasing the speed of valve closure:

Time of closure (s)	H (m)
4	53.5
3	71.3
2	107.0
1	214.0
0.5	428.0
0	∞

The above figures do not give a completely accurate picture. If pressure measurements were to be taken, it would be found that the increase in head was, in fact, self-limiting, and that the maximum head was 214 m, not infinity. Therefore, it is necessary to ask why the theory outlined above is so much in error for the fast valve closure times ($t < 1$ s) and yet is reasonably accurate at the slower closure times. The answer lies in the assumptions made about the liquid and the pipeline (i.e. incompressible liquid and rigid pipe material). When rapid changes occur in the flow, the system no longer behaves in this 'inelastic' manner. To elucidate this point, the case of a compressible liquid contained in a rigid pipe is now considered (the effect of pipe material elasticity is considered subsequently).

6.3 Unsteady compressible flow

General description

As the rate of valve closure (or opening) increases, so do the inertia forces. A point is reached at which the liquid is being subjected to pressures which are sufficient to cause it to compress. Once compressibility has been brought into play, a radical change takes place in the pressure surge process in the body of fluid. The rapid alteration of the valve setting does not cause a uniform deceleration along the pipeline. Instead, the alteration generates a shock wave in the fluid. A shock wave is a zone in which the fluid is rapidly compressed (i.e. p and ρ increase) (see Fig. 6.2). It travels through a fluid at the speed (or celerity) of sound (symbol c).

Through the mechanism of compressibility, the kinetic energy of the fluid before acceleration is transformed into elastic energy (thus upholding the principle of conservation of energy). Note that the fluid upstream of the

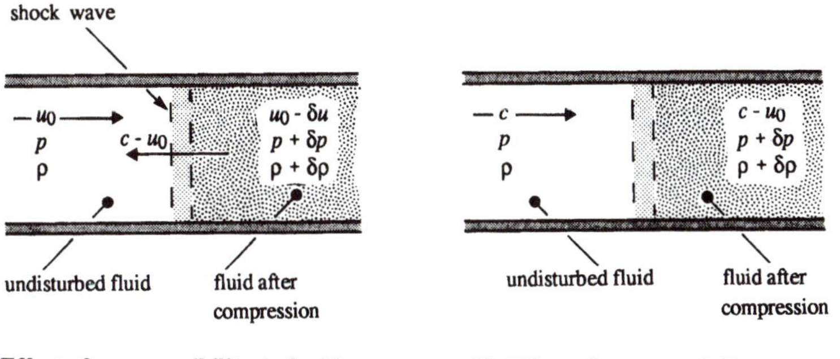

Figure 6.2 Surge conditions – compressible fluid and rigid pipe.

wave is still flowing at its original velocity, pressure and density. In fact, one way of viewing the process is to regard the shock wave as a 'messenger' which is carrying the 'news' of the change in valve setting. This information is 'received' by the fluid at a given point only when the shock wave arrives at that point. The progress of the shock wave is now described, assuming a rigid pipe and complete valve closure, in a pipeline whose upstream end is connected to a reservoir.

(a) The valve closes, generating the shock wave (Fig. 6.3(a)).
(b) The shock wave travels back along the pipeline (Fig. 6.3(b)). Fluid between the wave and the valve is now compressed, with the pressure therefore raised to $p + \delta p$ (i.e. the head raised by δh).
(c) The wave arrives at the reservoir (Fig. 6.3(c)). The pipe is therefore now full of compressed and pressurized fluid.
(d) There is an inequilibrium at the pipe–reservoir interface since the reservoir head remains unchanged. The fluid in the pipe begins to discharge in reverse into the reservoir (Fig. 6.3(d)). A decompression wave is generated, which travels back towards the valve. The fluid between the reservoir and the decompression wave is therefore at the original p and ρ, but u is reversed. Once again, the fluid between the wave and the valve has not yet received the 'message' and is still at $(p + \delta p)$ and $(\rho + \delta p)$.
(e) The control valve (if closed) constitutes a dead end. When the decompression wave arrives, the reversed flow can proceed no further, so the fluid here cannot flow back to the reservoir. A negative pressure is generated at the valve (Fig. 6.3(e)), which produces a 'negative' shock wave. This, in turn, is transmitted towards the reservoir. Fluid between the wave and the valve is at rest, but under a negative pressure (which is theoretically equal in magnitude to δp under (b), above). Fluid between

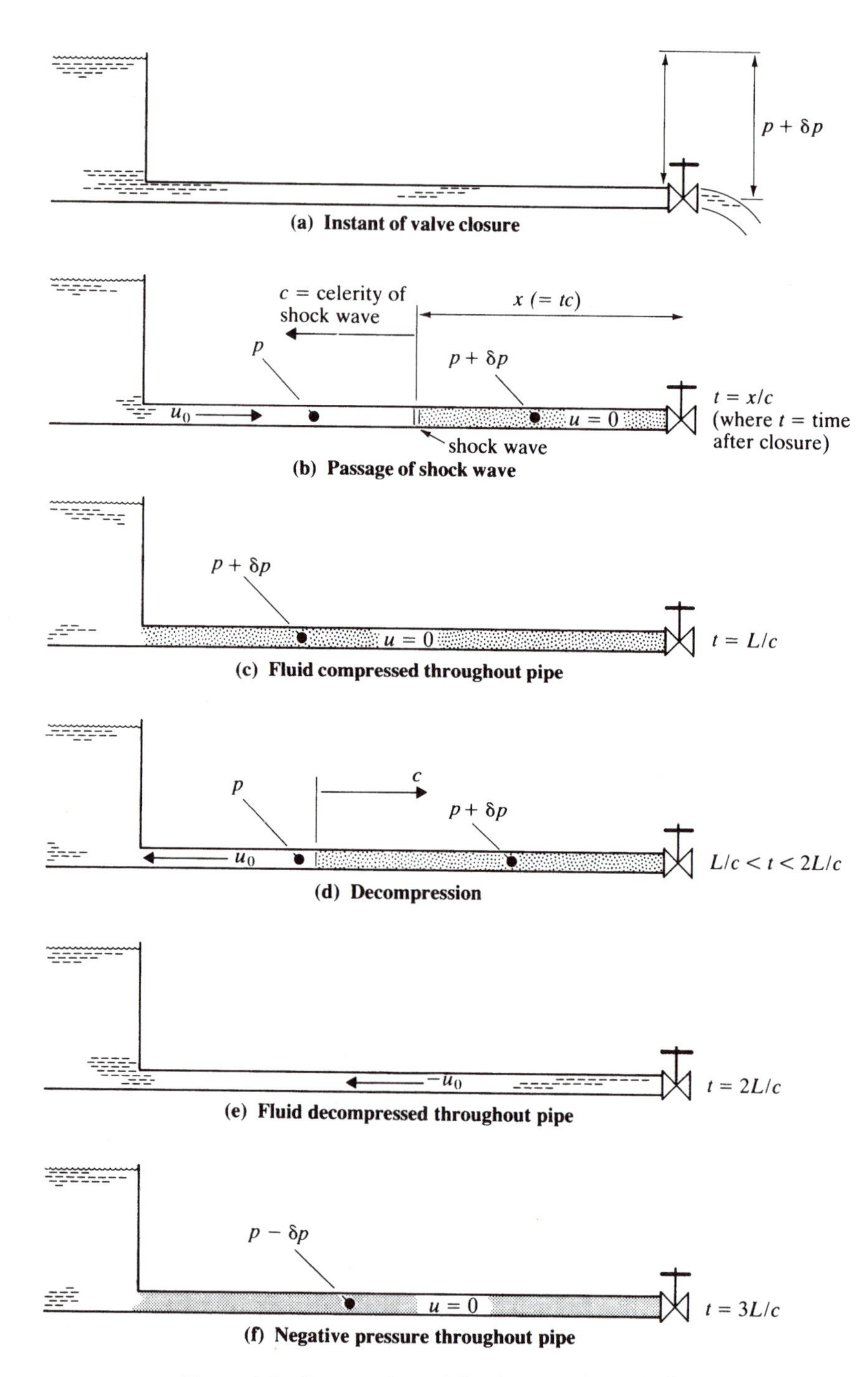

Figure 6.3 Propagation of shock waves in a pipeline.

the wave and the reservoir is still at p, ρ, and the flow is still reversed. (For simplicity, the possibility of cavitation has been ignored here.)

(f) When the negative wave arrives at the reservoir (Fig. 6.3(f)), the pipeline pressure is lower than the reservoir pressure. Fluid therefore flows from the reservoir into the pipe. Theoretically, the velocity is now restored to its original magnitude and direction, since we have so far ignored all hydraulic losses.

The whole cycle (a) to (f) then repeats.

Strictly, the celerity of the shock wave is measured relative to the fluid, and therefore the shock wave travels past a stationary point at $(c - u)$. However, since $c \gg u$, $(c - u) \rightarrow c$. The time taken for the shock wave to traverse the length of the pipeline is therefore

$$t = L/c$$

At any point in the pipeline the timing of pressure and velocity changes may be conveniently measured in multiples of L/c.

Simple equations for 'instantaneous' alteration of valve setting in a rigid pipeline

This is the simplest case to treat. To develop the equations, consider the conditions on either side of a shock wave. In order to view these conditions, it is convenient to use co-ordinates which move with the shock wave. From this viewpoint, velocities are as shown in Figure 6.2(b). Because of the compression, it is imperative that the equation of mass continuity (i.e. mass conservation) be used:

$$\text{mass entering control volume} = \text{mass leaving control volume}$$

$$\rho A c = (\rho + \delta\rho)(c - u_0)A$$

which may be rewritten as

$$\rho u_0 = \delta\rho(c - u_0) \tag{6.3}$$

Similarly, applying the momentum equation

$$\text{force} = \text{mass flow} \times \text{change in velocity}$$

$$(p + \delta p)A - pA = \rho A c\,[c - (c - u_0)]$$

Hence

$$\delta p = \rho c u_0 \tag{6.4}$$

The measure of the elasticity of a liquid is its bulk modulus (K):

$$K = -\frac{\delta p}{\delta V/V}$$

V being the volume of liquid. The negative sign arises because increasing pressure causes a reduction (negative increase) in volume.

By definition,

$$\rho = \frac{\text{mass}}{\text{volume}} = \frac{m}{V}$$

therefore

$$\frac{d\rho}{dV} = -\frac{m}{V^2} = -\frac{\rho}{V}$$

so

$$\frac{\delta\rho}{\rho} = -\frac{\delta V}{V}$$

therefore

$$K = \frac{\delta p}{\delta\rho/\rho} \tag{6.5}$$

From (6.3),

$$\frac{\delta\rho}{\rho} = \frac{u_0}{(c - u_0)} \tag{6.6}$$

From (6.4),

$$u_0 = \frac{\delta p}{\rho c}$$

From (6.5),

$$\frac{\delta\rho}{\rho} = \frac{\delta p}{K}$$

Substituting for u_0 and $\delta\rho/\rho$ in (6.6),

$$\frac{\delta p}{K} = \frac{(\delta p/\rho c)}{[c - (\delta p/\rho c)]}$$

Rearranging,

$$\frac{\rho c^2 - \delta p}{K} = 1$$

If $\delta p \ll K$, then

$$\rho c^2/K = 1$$

therefore

$$K = \rho c^2 \tag{6.7}$$

Equations (6.4) and (6.7) may then be solved simultaneously to obtain a value for δp. Note that no allowance has been made for straining of the pipe material, since the pipe is assumed to be rigid (i.e. $E \to \infty$). It should be further noted that the passage of events is extremely rapid, as will become apparent in the following example.

Example 6.2 Surge in a simple pipeline

A valve is placed at the downstream end of a 3 km long pipeline. Water is initially flowing along the pipe at a mean velocity of 2.5 m/s. What is the magnitude of the surge pressure generated by a sudden and complete valve closure? Sketch the variation in pressure at the valve and at the mid-point of the pipeline after valve closure. Take celerity of sound as 1500 m/s.

Solution

Increase in pressure is estimated from (6.4):

$$\delta p = \rho c u_0 = 1000 \times 1500 \times 2.5 = 3.75 \times 10^6\ \mathrm{N/m^2}$$

To sketch the pressure variation, it is necessary to know the transmission time for the shock wave:

$$t = L/c = 3000/1500 = 2\ \mathrm{s}$$

Following the cycle of events (a) to (f), above, at the valve the increase in pressure will be maintained while the shock wave travels to the reservoir and while the decompression wave returns to the valve, this takes $2L/c = 4$ s (Fig. 6.4(a)). Similarly, the pressure will be negative for another 4 s, etc.

At the mid-point, it will take $t = L/2c\,(= 1\ \mathrm{s})$ for the shock wave 'messenger' to arrive and 'deliver' the pressure increase. The initial increase is then maintained while the shock wave travels to the reservoir and the reflected decompression wave

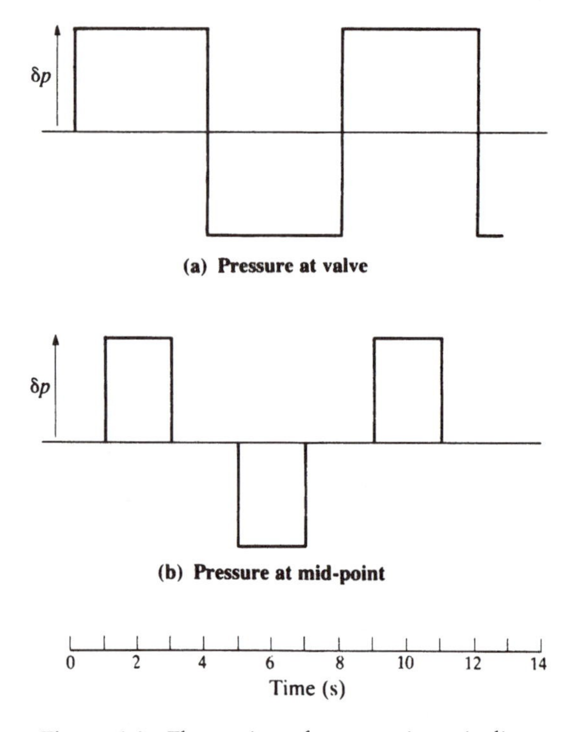

Figure 6.4 Fluctuation of pressure in a pipeline.

returns (this takes $2 \times (L/2c) = L/c$), whereupon the pressure reverts to its original value. The negative pressure from the valve arrives $L/2c\,(= 1\,\text{s})$ after it has been generated at the valve, and so on (Fig. 6.4(b)).

Equations for 'instantaneous' valve closure in an elastic pipeline

We now proceed to a slightly more realistic setting for the pressure surge phenomenon, viz. a compressible fluid flowing in an elastic pipeline. The effect of the pressure increase behind the shock wave is:

1. to compress the fluid;
2. to strain the pipe walls.

Due to the strain effect, the pipe cross-section behind the shock wave is greater than the unstrained cross-section ahead of the wave (Fig. 6.5(a)). The mass continuity equation must therefore be written as

$$\rho A c = (\rho + \delta\rho)\,(A + \delta A)\,(c - u_0)$$

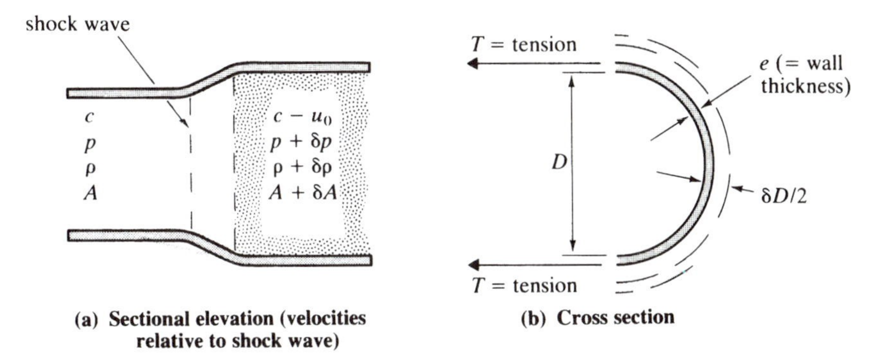

Figure 6.5 Surge conditions – compressible fluid and elastic pipe material.

which may be rearranged to give

$$\rho u_0(A+\delta A)-\rho c\,\delta A=\delta\rho(c-u_0)\,(A+\delta A) \tag{6.8}$$

The momentum equation is

$$(p+\delta p)\,(A+\delta A)-pA=\rho Acu_0$$

Ignoring small quantities, this may be written as

$$\delta p=\rho cu_0$$

which is equation (6.4). The equation relating to the elasticity of the fluid (equation (6.5)) remains unchanged:

$$K=\frac{\delta p}{\delta\rho/\rho}$$

From (6.8),

$$\frac{\delta\rho}{\rho}=\frac{u_0(A+\delta A)-c\,\delta A}{(c-u_0)(A+\delta A)} \tag{6.9}$$

From (6.4),

$$u_0=\frac{\delta p}{\rho c}$$

From (6.5),

$$\frac{\delta\rho}{\rho}=\frac{\delta p}{K}$$

Substituting the above two equations in (6.9),

$$\frac{\delta p}{K} = \frac{(\delta p/\rho c)(A+\delta A) - c\,\delta A}{[c-(\delta p/\rho c)](A+\delta A)} = \frac{\delta p(A+\delta A) - \rho c^2\,\delta A}{(\rho c^2 - \delta p)(A+\delta A)}$$

Therefore

$$\frac{\delta p}{K}\rho c^2(A+\delta A) + \rho c^2\,\delta A = \delta p(A+\delta A) + \frac{\delta p^2}{K}(A+\delta A)$$

and rearranging,

$$\rho c^2 = \frac{\delta p\left[1+\dfrac{\delta A}{A}\right]\left[1+\dfrac{\delta p}{K}\right]}{\dfrac{\delta p}{K}\left[1+\dfrac{\delta A}{A}\right]+\dfrac{\delta A}{A}}$$

If $\delta p \ll K$ and $\delta A/A \ll 1$, then the above equation simplifies to

$$\rho c^2 = \frac{\delta p}{(\delta p/K)+(\delta A/A)} \tag{6.10}$$

Note that if the $\delta A/A$ term is ignored, this becomes identical to (6.7). However, as the elasticity of the pipe is to be taken into account, it is necessary to find some relationship between δA and A in terms of the pipe characteristics. In developing the relationship it is assumed that the pipe is axially rigid (due, say, to pipe supports) but free to move radially.

Referring to Figure 6.5(b), an increase in pressure δp inside a pipe produces a bursting (or 'hoop') tension T. Thus, for a pipe of length δx and diameter D,

$$\delta p\,D\,\delta x = 2T = 2\sigma e\,\delta x$$

where σ is the hoop stress in the pipe wall. Since

$$\frac{\text{stress}}{\text{strain}}\left(=\frac{\sigma}{s}\right) = E$$

then

$$s = \sigma/E$$

Therefore

$$\frac{\delta p\,D\,\delta x}{E} = \frac{2\sigma e\,\delta x}{E}$$

so

$$\frac{\delta p\, D}{2Ee} = s \tag{6.11}$$

But

$$s = \frac{\text{change in length}}{\text{original length}} = \frac{\pi\, \delta D}{\pi D} = \frac{\delta D}{D}$$

and

$$\frac{\delta A}{A} = \frac{\pi(D+\delta D)^2 - \pi D^2}{\pi D^2} = \frac{2\,\delta D}{D} = 2\,s$$

Therefore

$$\frac{\delta A}{A} = \frac{\delta p\, D}{Ee}$$

$$\rho c^2 = \frac{\delta p}{(\delta p/K) + (\delta p D/Ee)} = \frac{1}{(1/K) + (D/Ee)} \tag{6.12}$$

Hence, if K, ρ, E and e are known, c can be calculated.

The only difference between the rigid pipe case and the elastic pipe case is therefore the speed at which the shock wave travels down the pipe. In every other respect, the solutions follow the pattern of Example 6.2.

Some installations incorporate supports which allow some axial movement. When this is so, a hoop strain will produce a corresponding strain in the axial direction. This will slightly influence the magnitude of c. An appropriate allowance for this effect may be made by incorporating a term with the Poisson ratio in (6.11) (and hence in (6.12)).

6.4 Analysis of more complex problems

Description

For a realistic system under surge conditions the analysis must take account of:

1. elasticity (of fluid and pipe material);
2. effect of hydraulic losses;
3. non-instantaneous valve movement.

The method given in the preceding section can be suitably extended, but it becomes rather cumbersome. It is therefore worth turning to a method which is readily adapted for computer use. Such a method is now outlined, and although the actual process of developing the equations can seem rather long, it should be borne in mind that we are really only interested in the final equations. These will introduce the 'method of characteristics' which may, in principle, be applied to any unsteady flow. To clarify matters as far as possible, the equations are developed step by step from first principles.

It is worth reviewing the events which take place as a pressure wave passes along a short length of pipe (Fig. 6.6(a)). The valve closure (or change in setting) is now not instantaneous, though it is fast enough to generate a shock wave. However, since the change in setting takes a finite time, it will also take a finite time for the pressure rise δp to be generated. Consequently, the shock wave now occupies a significant length of pipeline, δx. This length will clearly be directly related to the pressure rise time ($\delta x/c = \delta t$). Due to the increase in pressure, two changes will occur:

1. the fluid will compress;
2. the (elastic) pipe will experience a strain.

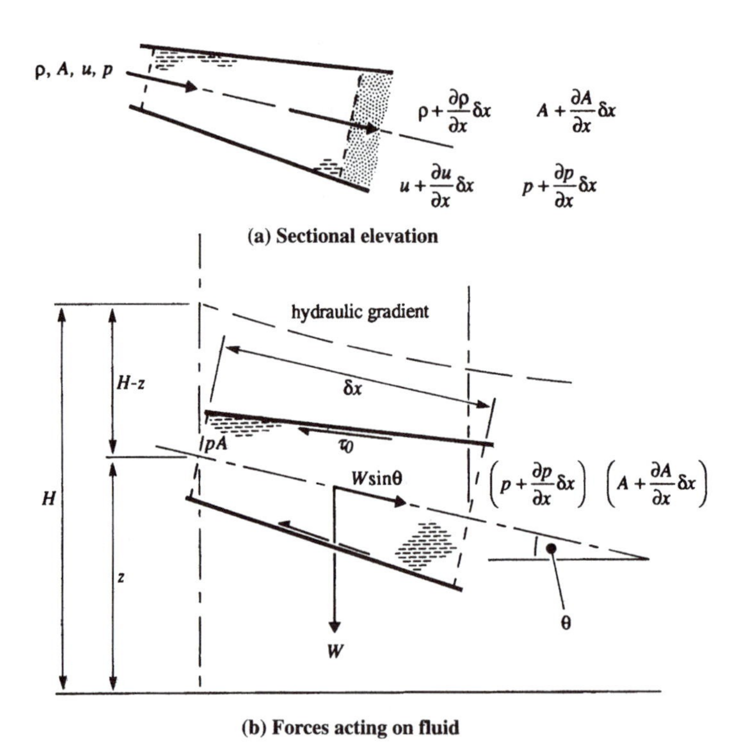

Figure 6.6 Surge conditions – compressible fluid, elastic pipe and 'rapid' (but not instantaneous) valve closure.

Differential equations may therefore be developed based on the conservation of mass and momentum equations.

Conservation of mass. For the control volume shown in Figure 6.6(a), (mass flow in) – (mass flow out) = (rate of change of mass in control volume)

$$\rho A u - \left[\rho A u + \frac{\partial}{\partial x}(\rho A u)\,\delta x\right] = \frac{\partial}{\partial t}(\rho A\,\delta x)$$

Hence

$$\frac{\partial(\rho A)}{\partial t} + \rho A\frac{\partial u}{\partial x} + u\frac{\partial(\rho A)}{\partial x} = 0$$

or

$$\frac{\partial p}{\partial t}\frac{\mathrm{d}(\rho A)}{\mathrm{d}p} + \rho A\frac{\partial u}{\partial x} + u\frac{\partial p}{\partial x}\frac{\mathrm{d}(\rho A)}{\mathrm{d}p} = 0$$

Taking

$$c^2 = \frac{A}{\mathrm{d}(\rho A)/\mathrm{d}p} \quad \text{(cf. equation (6.10))}$$

then the conservation of mass equation is

$$\frac{\partial p}{\partial t} + \rho c^2\frac{\partial u}{\partial x} + u\frac{\partial p}{\partial x} = 0 \tag{6.13}$$

The momentum equation. For the control volume shown in Figure 6.6(b),

$$\text{Sum of forces} = \text{rate of change of momentum}$$

$$pA - \left(p + \frac{\partial p}{\partial x}\delta x\right)\left(A + \frac{\partial A}{\partial x}\delta x\right) + \left(p + \frac{1}{2}\frac{\partial p}{\partial x}\delta x\right)\frac{\partial A}{\partial x}\delta x - \tau_0 P\,\delta x + W\sin\theta$$
$$= \frac{W}{g}\left(u\frac{\partial u}{\partial x} + \frac{\partial u}{\partial t}\right) \tag{6.14}$$

This can be simplified by ignoring higher order terms and dividing by $W/g(W = \rho g \delta A)$. Also, recalling that $\tau_0 = \lambda\rho\, u|u|/8$ (where λ is the friction factor and $|u|$ is the modulus or positive value of u), and that, for a circular pipe, the wetted perimeter $P = \pi D$, and the cross-sectional area is $\pi D^2/4$:

$$\frac{1}{\rho}\frac{\partial p}{\partial x} + \frac{\lambda u|u|}{2D} - g\sin\theta + u\frac{\partial u}{\partial x} + \frac{\partial u}{\partial t} = 0 \tag{6.15}$$

The u^2 term (in the expression for shear stress) is replaced by $u|u|$ to allow for the fact that the velocity may be positive or negative, and the direction of the shear stress must be in the opposite direction to motion.

Equations (6.13) and (6.15) may be expressed in terms of H instead of pressure, p. Taking $p = \rho g(H - z)$,
Then

$$\frac{\partial p}{\partial x} = \rho g \frac{\partial (H - z)}{\partial x} = \rho g \left(\frac{\partial H}{\partial x} + \sin\theta \right)$$

and

$$\frac{\partial p}{\partial t} = \rho g \frac{\partial (H - z)}{\partial t} = \rho g \frac{\partial H}{\partial t},$$

since $\partial z / \partial t = 0$
Hence the conservation of mass equation becomes:

$$g\frac{\partial H}{\partial t} + c^2 \frac{\partial u}{\partial x} + ug\frac{\partial H}{\partial x} + gu\sin\theta = 0 \qquad (6.16)$$

and the momentum equation becomes:

$$g\frac{\partial H}{\partial x} + \frac{\lambda u|u|}{2D} + u\frac{\partial u}{\partial x} + \frac{\partial u}{\partial t} = 0 \qquad (6.17)$$

Equations (6.16) and (6.17) form the basis for a numerical model representing the velocities and pressures in a pipeline under conditions of unsteady compressible flow. A number of graphical and manual solution techniques have been developed and some are still useful under some circumstances (see, for example, Fox (1989), Chapter 2). However, a computer-based method is frequently the most convenient. The above equations are transformed into a computational model in Chapter 14.

6.5 Concluding remarks

(1) The intensity of surge pressures depends on the rate of change at a (variable) controlling boundary. If the rate of change is sufficiently fast, then the liquid will behave in the elastic mode. Therefore, it is important to have some general rules which indicate whether shock waves are likely to be generated. To illustrate this point, consider the effect of the time T required for complete closure of a valve at the downstream end of a pipeline of length L:

(a) If $T < 2L/c$, then the maximum surge pressure is incurred, since a returning decompression wave cannot reach the valve before complete closure. Closure is therefore effectively instantaneous;

(b) at the other extreme, if $T > 20L/c$, then there will be little or no shock wave activity, and the liquid will behave in the incompressible mode – this is 'slow' closure;
(c) for $2L/c < T < 20L/c$, conditions are intermediate between (a) and (b), above – this is known as a 'rapid' valve closure.

(2) In this chapter it has been assumed that the surge conditions have been generated by the closing of a valve. However, any variable boundary can produce a surge. An example of a boundary which is variable (but might not be thought of as such) is a pump under power failure conditions.
(3) In designing a major pipeline system, it is necessary to evaluate the surge pressures under all foreseeable changes of boundary conditions. If excessively high values are predicted, then it is necessary to incorporate a system for reducing the surge to an acceptable level. This aspect of unsteady flow is discussed in section 12.6.
(4) The treatment of unsteady flows as presented here is only introductory. For a more advanced approach, reference may be made to Fox (1989) and to Wylie and Streeter (1995). Parmakian (1963) gives some alternative methods, together with a number of worked examples. Kranenburg (1974) may be consulted regarding the effects of cavitation and gas release during surge.

References and further reading

Fox, J. A. (1989) *Transient Flow in Pipes, Open Channels and Sewers*, Ellis Horwood, Hemel Hempstead.

Kranenburg, C. (1974) Gas release during transient cavitation in pipes. *Am. Soc. Civ. Engrs, J. Hydraulics Divn.*, **100** (HY 10), 1383–98.

McInnis, D. and Karney, B. W. (1996) Transients in distribution networks: field tests and demand models. *Am. Soc. Civ. Engrs, J. Hydraulic Engng.*, **121** (3), 218–31.

Parmakian, J. (1963) *Waterhammer Analysis*, Dover, New York.

Streeter, V. L. and Wylie, E. B. (1978) *Hydraulic Transients*, McGraw-Hill, New York.

Vardy, A. E., and Hwang, K-L. (1991) A characteristics model of transient friction in pipes. *J. Hydraulic Res.*, **29** (5), 669–84.

Wylie, E. B. and Streeter, V. L. (1995) *Fluid Transients in Systems*, Prentice Hall, New Jersey.

7

Hydraulic machines

7.1 Classification of machines

Civil engineers are not usually involved in the design of hydraulic machines. However, they are frequently called upon to design systems of which these machines form an integral part. Examples of such systems are pipelines for water supply, for surface and foul water drainage, for hydroelectric schemes, and so on. It is therefore important that the civil engineer should have a sound basic understanding of hydraulic machines and their applications.

It is convenient to divide these machines into two categories:

1. pumps, i.e. machines which transform a power input (e.g. from an electric motor) into a hydraulic power output;
2. turbines, i.e. machines which transform a hydraulic power input into a mechanical power output which is mainly utilised for generation of electrical power.

Machines can be further subdivided into 'positive displacement' units and 'continuous flow' units. Typical positive displacement units are piston and diaphragm pumps, which are often used for pumping out groundworks. This chapter concentrates on the continuous flow units, which are by far the most widely used for permanent installations.

7.2 Continuous flow pumps

Continuous flow or 'rotodynamic' pumps comprise two elements:

1. a rotating element (impeller) which transfers energy to the fluid;
2. some form of casing which encloses the impeller and to which the pipeline is connected.

In many cases the impeller is driven by an electric motor.

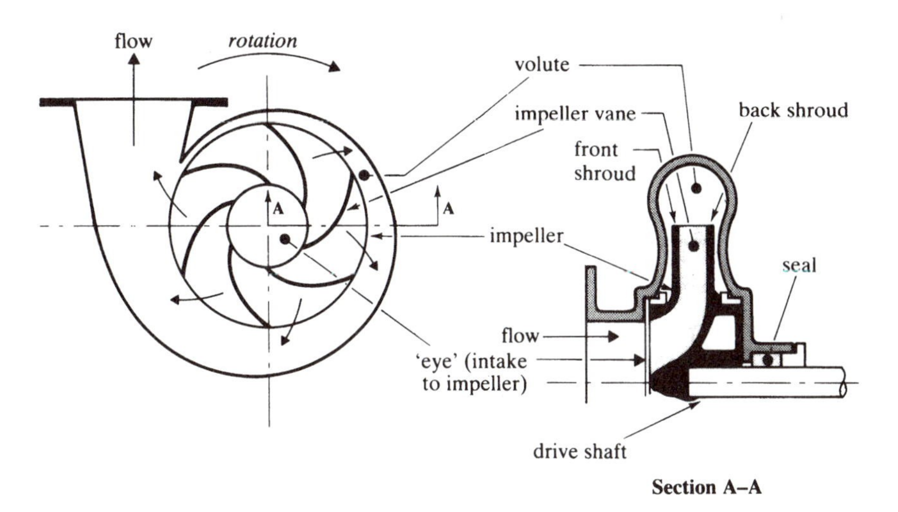

Figure 7.1 Centrifugal pump.

The radial flow (centrifugal) pump

This is one of the most widely utilized hydraulic machines (Fig. 7.1). Like most continuous flow machines, it has the advantage of being able to pass either liquids or liquids with suspended solids (there are, of course, limits to the size and nature of the solids). These machines can thus be applied in such environments as sewage plants, drainage schemes and irrigation schemes. The casing must initially be full of liquid (i.e. the pump must be 'primed') in order to function. When the impeller rotates at its design speed, it imparts a radial component of motion to the volume of fluid trapped in the passages between the vanes. That fluid therefore passes outwards into the outer part of the casing (the volute) and then out into the high pressure (delivery) pipeline. As fluid continually evacuates the impeller, replenishment fluid is continually drawn from the low pressure (suction) pipeline into the casing.

Energy transfer in radial flow pumps. An insight into the energy transfer process may be obtained by producing vector diagrams representing the flow as it enters and as it leaves the impeller. The flow through a radial flow pump impeller provides a relatively straightforward example. It is usually assumed that the incoming liquid enters the casing axially and then turns outwards, so that the flow is in the radial plane as it approaches the impeller. The curvature of the impeller vanes is designed to produce a change in the direction of the flow as the fluid passes outwards through the impeller. Vectors may be constructed which represent:

1. the absolute velocity of flow at a point, i.e. the velocity relative to a stationary point;

2. the relative velocity of flow, i.e. the velocity as 'seen' from a point on the impeller.

The principal velocity vectors are

1. the tangential velocity of the impeller itself at radius r

$$V_{\mathrm{I}} = 2\pi rn$$

where n is the rotational speed of impeller (rev/s);
2. the absolute velocity V_{A} of the fluid at radius r;
3. the relative velocity, V_{R} (i.e. the velocity of the fluid relative to the impeller);
4. the tangential component of V_{A} is V_{W}, often known as the 'whirl' velocity;
5. the radial component of V_{A} is V_{F}.

In order to produce the vector diagrams, it is simplest to assume that, as the flow passes through the impeller, the relative velocity traces a path which is everywhere congruent to the vane. Vector diagrams (Fig. 7.2) can then be drawn.

Based on these diagrams, equations describing the energy transfer are now developed. At any radius r in the impeller, the flow will possess a tangential component of velocity V_{W}. It will therefore have a corresponding tangential momentum flux, ρQV_{W}. The moment of this flux about the impeller axis is $\rho QV_{\mathrm{W}}r$. Just as force equals change of momentum flux, so torque equals

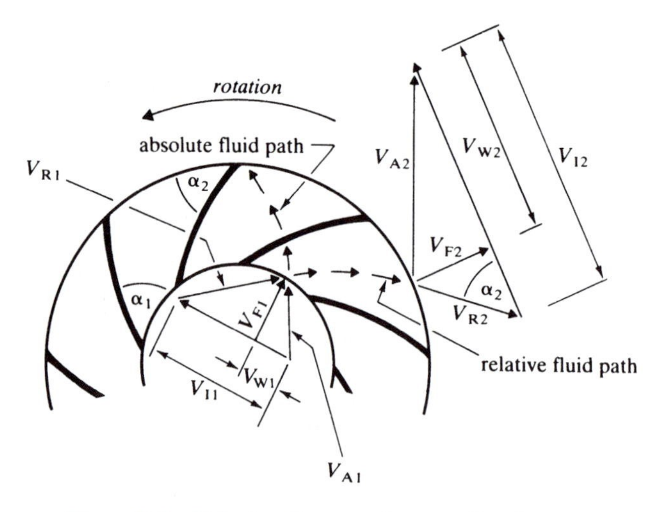

Figure 7.2 Velocity diagrams for centrifugal pump.

change of moment of momentum flux. In this case, the change takes place between radii r_1 and r_2, so

$$\text{torque}(T) = \rho Q V_{W2} r_2 - \rho Q V_{W1} r_1$$
$$\text{power} = 2\pi n T = 2\pi n(\rho Q V_{W2} r_2 - \rho Q V_{W1} r_1)$$

But $2\pi rn = V_I$, therefore

$$\text{power} = \rho Q(V_{W2} V_{I2} - V_{W1} V_{I1}) \tag{7.1}$$

Therefore ideal energy head, H_{ideal}, imparted to the fluid is

$$H_{\text{ideal}} = \frac{\text{power}}{\rho g Q} = \frac{1}{g}(V_{W2} V_{I2} - V_{W1} V_{I1}) \tag{7.2}$$

If V_{A1} is radial in direction, then $V_{W1} = 0$.

Energy losses in radial flow pumps. The overall rate of energy transfer actually attained is always less than the ideal values predicted by (7.2). There are a number of contributory factors, of which the major ones are:

1. During its passage through the impeller and volute, the fluid will suffer both frictional and local energy losses.
2. The liquid in the volute is at a much higher pressure than the liquid at the entry so, unless the internal sealing arrangements are extremely good, some of the pressurized fluid leaks back to the impeller eye and then recirculates through the impeller. This clearly wastes a certain amount of energy.
3. In the vector analysis, it had to be assumed that the flow through the impeller was axisymmetric. However, in order to produce a tangential acceleration, the force (and therefore pressure) must be higher on the front of each vane than on the back (Fig. 7.3). At any given radius in the impeller, the pressure is therefore non-uniform. This results in a nonuniform relative velocity distribution, with the highest velocities at the back of the vanes and the lowest velocities at the front of the vanes. The effect of this is to distort the pattern of the vectors at the exit from the impeller and to reduce V_{W2}.
4. If the pressure in the suction line is too low, cavitation problems may arise. The low pressure zone near the backs of the vanes is particularly prone to cavitation, and if the problem is severe, then the pump performance will fall off dramatically.

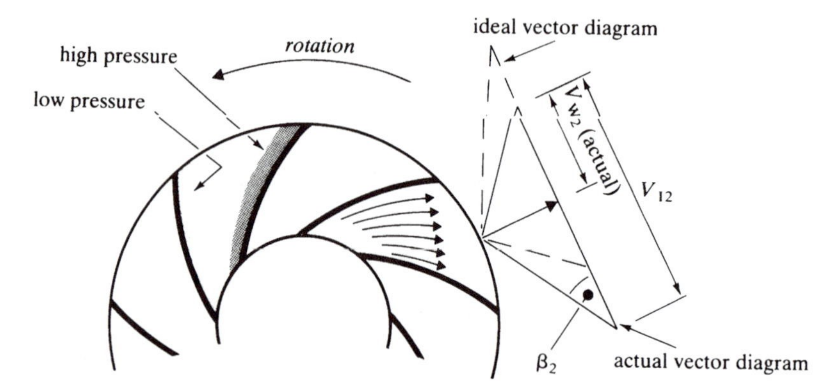

Figure 7.3 Effect of pressure distribution on velocity diagram for centrifugal pump.

Example 7.1 Centrifugal pump

A centrifugal pump is required to deliver 350 l/s against a 15 m head. The impeller is of 400 mm outer diameter and 200 mm inner diameter, and rotates at 1400 rev/min. The intake and delivery pipelines are of the same diameter. The water flows in a radial direction as it approaches the impeller, and the radial velocity component remains constant at 5 m/s as the flow passes through the impeller. Assume that the hydraulic losses account for a 10% head loss. Calculate the required vane angles at the outlet of the impeller. If the mechanical efficiency is 88%, estimate the power requirement for an electric motor.

Solution

(a) The ideal head $= 15/0.9 = 16.67$ m, since the 15 m head was a net head (i.e. head after deduction of losses), and the hydraulic efficiency $= (100 - 10)\% = 90\%$.

(b) From (7.2)

$$H_{\text{ideal}} = (1/g)(V_{W2}V_{I2} - V_{W1}V_{I1})$$

Since the incoming flow is in a radial direction, $V_{W1} = 0$. Therefore,

$$H_{\text{ideal}} = 16.67\,\text{m} = (1/g)(V_{W2}V_{I2})$$

Now,

$$V_{I2} = 2\pi r_2 n = 2\pi \times \frac{0.4}{2} \times \frac{1400}{60} = 29.32\,\text{m/s}$$

Hence

$$V_{W2} = 5.58\,\text{m/s}$$

(c) From Figure 7.2, the outlet vane angle may be derived by taking

$$\cot \alpha_2 = \frac{V_{I2} - V_{W2}}{V_{F2}} = \frac{29.32 - 5.58}{5} = 4.748$$

therefore,

$$\alpha_2 = 11.9°$$

(d) Power absorbed by impeller is

$$\rho g Q H_{\text{ideal}} = 1000 \times 9.81 \times 0.35 \times 16.67$$
$$= 57\,236\,\text{W}$$

Since mechanical efficiency is 88%, the power absorbed by the pump will be

$$\frac{57\,236}{0.88} = 65\,041\,\text{W}$$
$$\text{say, } 65\,\text{kW}.$$

Axial flow machines

The axial flow pump comprises an impeller in the form of a propeller sited on the pipeline axis (Fig. 7.4). The pipeline itself therefore forms a 'casing'. A cross-section through one of the propeller blades has a shape which resembles a section through an aerofoil (aircraft wing). The section is arranged at some angle to the plane of the pipe cross-section. This angle and the blade section shape combine to impart the required tangential acceleration to the fluid as it passes through the rotating impeller. Vector diagrams can be constructed to represent the flow at impeller entry and exit,

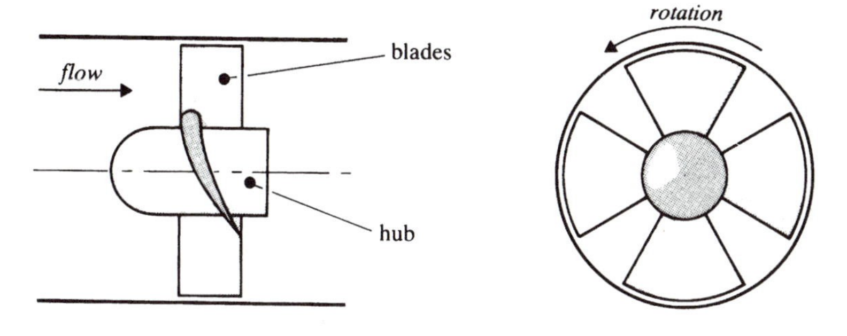

Figure 7.4 Axial flow pump.

using exactly the same principles as for the radial flow machine. However, for almost all practical cases, the exit vector diagram varies with radius, so the procedure for estimating rate of transfer of energy is more complicated. In any case, modern design techniques are based on an adaptation of aerofoil theory known as 'cascade theory', rather than on vector diagrams. A study of cascade theory is beyond the scope of this chapter, so the discussion will be curtailed here.

For axial flow machines with fixed blades, high efficiency is attained only over a fairly narrow range of discharge. For applications where a wider operational range is required, pumps with variable blade angle are available. Such machines are clearly more expensive, since a mechanical control system is required, together with more sophisticated instrumentation.

'Mixed flow' machines

'Mixed flow' machines are, in effect, a compromise between axial and radial flow machines. The impeller induces a three-dimensional flow pattern such that the path of a fluid particle would lie approximately along the surface of a diverging cone. The performance characteristics of these units lie between those of radial and axial flow machines. Some fairly large installations have been designed to incorporate mixed flow pumps.

7.3 Performance data for continuous flow pumps

The equations developed in section 7.2 cannot readily be extended to enable realistic estimates of pump performance to be made. Actual performance data are invariably derived from a comprehensive laboratory testing programme. Such tests provide values of pump head, H_p, and power input, P, for the range of discharge, Q, at the design speed of the impeller. The energy imparted to the fluid is $\rho g Q H_p$, so that a pump efficiency, η, may be derived:

$$\eta = \rho g Q H_p / P \tag{7.3}$$

The performance data may be presented in tabular or graphical form. Figures 7.5 and 7.6 show typical performance curves for a centrifugal pump and an axial flow pump, respectively. In this connection, some general observations may be made:

1. The performance of a radial pump (or a fixed blade axial pump) depends substantially on the exit vane angle α_2. For many radial pumps, $\alpha_2 < 90°$, and such designs have a peak power requirement corresponding to a particular value of Q. The electrical (or other) drive motor must be rated accordingly.

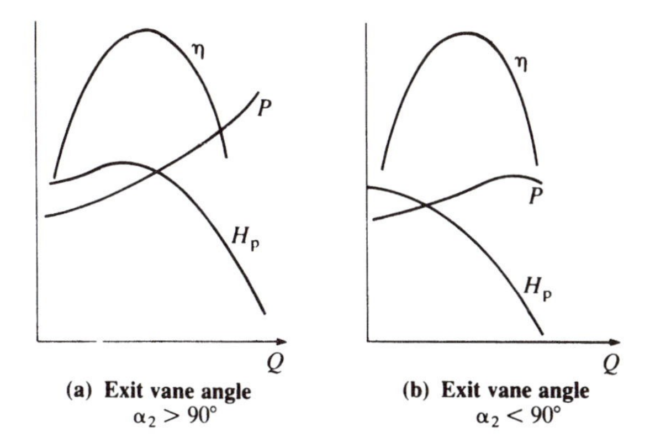

Figure 7.5 Performance data for centrifugal pump.

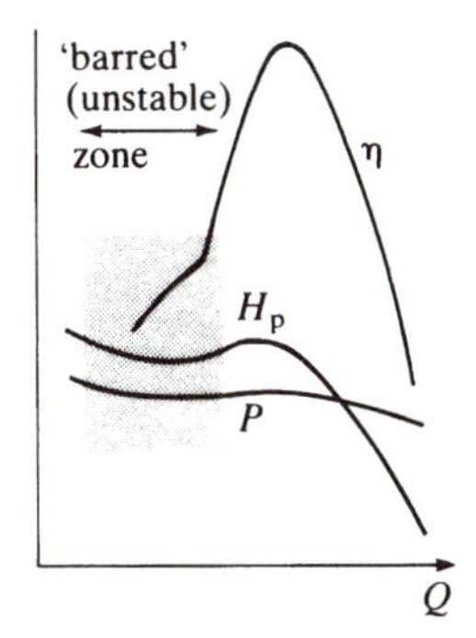

Figure 7.6 Performance data for axial pump.

2. Where machines are designed with $\alpha_2 > 90°$, then the power requirement increases with Q. This can pose difficult motor selection problems.
3. Pressure at the intake to a centrifugal pump should not be allowed to fall below about 3 m absolute head of water, or cavitation problems will arise. Axial machines have a much more limited suction capacity, since the flow around the blade section is very sensitive to low pressure. Too low an intake pressure therefore provokes flow separation and consequent loss of energy.
4. The relationship between H_p and Q for some axial designs is such that there may be parts of the range which are 'barred'. The reason for this is that the pump would be prone to unsteady flow in the barred range.
5. In order to obtain the pump performance required by the system, it is sometimes necessary to use a multiple array of pumps. The pumps may be arranged either in series or in parallel. This is discussed further in section 12.5.

6. In order to function satisfactorily, the pump must be matched to the characteristics of the pipeline system in which it is installed. Again, this is covered in section 12.5.

7.4 Pump selection

An indication of the type of pump which is appropriate to a particular duty may be obtained by reference to a performance parameter known as the 'specific speed' N_s. For pumps,

$$N_s = NQ^{1/2}/H_p^{3/4}$$

The derivation and significance of N_s is discussed in section 11.6. The 'specific speed' is not really representative of any meaningful or measurable speed in a machine, so some practitioners refer to it as 'type number', since it is used in selection of pump type. The value of N_s for a particular machine is calculated for the conditions obtaining at its point of optimum efficiency, since ideally this should coincide with the installed operating point of the pump.

Unfortunately, a wide variety of units are used in calculating values of N_s:

N may be in rad/s, rev/s or rev/min;
Q may be in m^3/s, m^3/min, l/s or l/min, UK gallons/min, US gallons/min, etc.;
H_p may be in m or ft.

The engineer needs to check this point carefully with the pump manufacturer.

Taking N in rev/min, Q in m^3/s and H in m, a very rough guide to the range of duties covered by the different machines is as follows:

$10 < N_s < 70$ centrifugal (high head, low to moderate discharge);
$70 < N_s < 165$ mixed flow (moderate head, moderate discharge);
$110 < N_s$ axial flow (low head, high discharge).

In making the selection, the points raised in the preceding section should also be borne in mind.

7.5 Hydro-power turbines

There are many hydroelectric power installations throughout the world. The various modern water turbine designs are capable of covering a very wide range of operating conditions. Thus, hydroelectric schemes may now be found in mountain, river or estuarine environments. Rudimentary water power machines have been in use for centuries, but it was not until the later part of the 19th century that efficient designs emerged. Subsequently, the range of machines has widened, but for large scale power generation, three basic types are dominant. These are the radial flow (Francis) turbine, the axial flow turbine and the Pelton turbine.

Of the three machines mentioned above, the Francis turbine (Fig. 7.7) may be regarded as a centrifugal pump operating 'in reverse'. The water enters the outer part of the casing and flows radially inwards through the rotating element (this time called a 'runner'), transferring energy to the runner. The water then drains out of the casing through the central outlet passage. There are two major differences between the turbine and the pump:

1. The rotating element is surrounded by guide vanes, so that the direction of the absolute velocity vector of the water can be adjusted. This adjustable angle provides for optimum flow conditions over a wide range of discharge and power output (N.B. electrical requirements vary over a 24 h period).
2. The water leaving the turbine passes through a 'draft tube', which is tapered so as to minimise loss of kinetic energy. This ensures that the maximum energy is available at the turbine.

The axial flow turbine (Fig. 7.8) may similarly be regarded as the reversal of the axial flow pump. Once again, guide vanes are installed at the turbine inlet for the same reasons as outlined above. Virtually all modern units incorporate mechanical means of varying the propeller blade angle of the

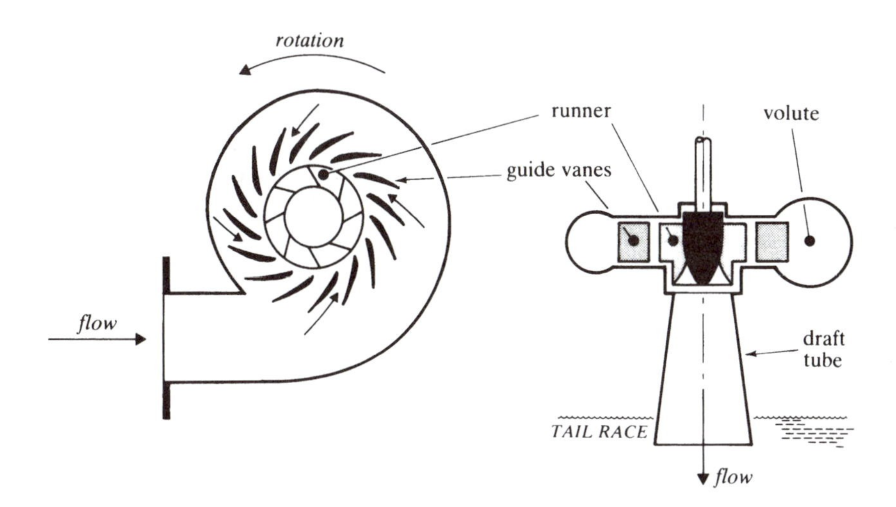

Figure 7.7 Radial flow (Francis) turbine.

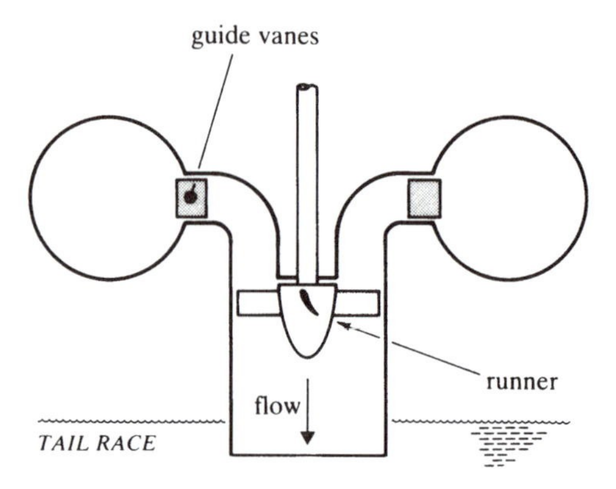

Figure 7.8 Axial flow turbine.

runner, so that high runner efficiency may be maintained over the specified range of discharge and power output. Variable angle designs are called 'Kaplan' turbines. In recent years, a number of 'pumped storage' schemes have utilized a special type of Kaplan unit. This unit operates as a turbine during periods when the demand for electricity is high. However, demand is low during the night. The blade angle is then reversed and the unit is motor-driven so as to act as a pump. Water is then pumped from low level back up to the high level reservoir in readiness for the next period of high demand.

Both the Kaplan and Francis units run full of water. There is, therefore, a continuous flow between the reservoir and the tailwater. The gross head, H, is the difference between the reservoir and the tail water levels. Ideally, all of this energy would be converted by the turbine into mechanical energy. In practice, hydraulic losses are incurred in the pipeline system, so the head available at the turbine is $H_t(= H - \text{losses})$. Losses also occur in the machine itself, and there are, additionally, mechanical losses (bearing friction, etc.). Nevertheless, turbines are very efficient, commonly attaining figures in excess of 90%. Kaplan units are generally applied where there is a large discharge at low pressure, whereas the Francis unit is appropriate for moderate discharge at moderate head.

The Pelton Wheel (Fig. 7.9) is applied where a high pressure water supply is available. By contrast with the other turbines, the casing does not run full of water. Essentially, the runner comprises a disc, with a series of flow deflectors or 'buckets' around its periphery. The water enters the casing through one or more nozzles as a high velocity jet. As the runner rotates, the jet impinges upon each bucket in turn. The water is deflected by the buckets, and the change in its momentum transfers a force to the bucket, and hence a torque to the drive shaft. The jet incorporates a streamlined

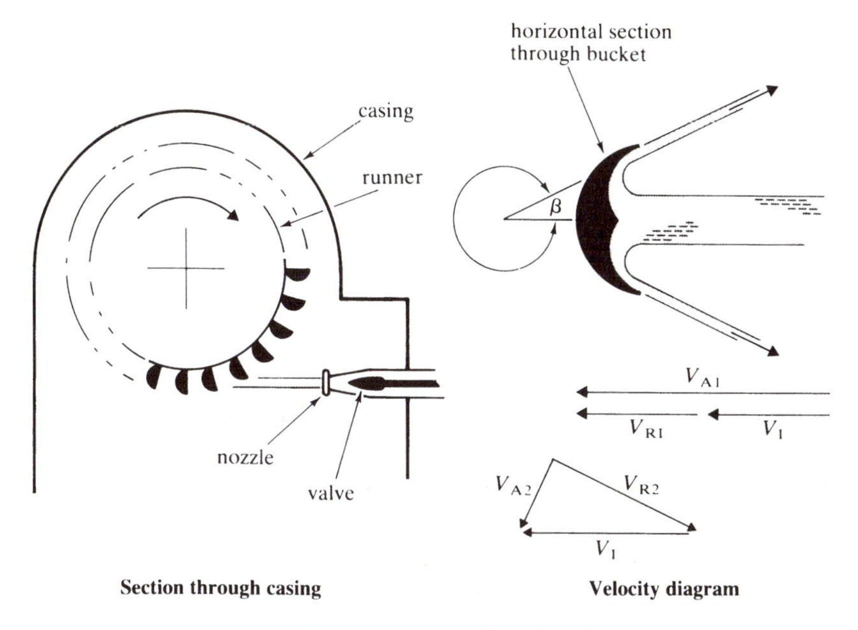

Figure 7.9 The 'Pelton Wheel' turbine.

control valve which is used to regulate the discharge, and hence the power output. Peak efficiency coincides with a particular runner speed.

The gross head for a Pelton Wheel is due only to the height of the reservoir above the nozzle(s). The head, H_t, is again defined as gross head minus losses.

Readers who require a slightly more detailed coverage of pump and turbine analysis (together with worked examples) are referred to Dugdale and Morfett (1983). For a more advanced treatment of pumps, see Stepanoff (1957), and for a broad practical treatment of both pumps and turbines see Davis and Sorensen (1969).

7.6 Turbine selection

Turbines, like pumps, may be selected on the basis of the specific speed appropriate to their duty. For turbines, the definition of specific speed is

$$N_S = NP^{1/2}/H^{5/4}$$

(see section 11.6) for which, again, there are a variety of units.

Taking N in rev/min, power (P) in kW and H in m, the range of duties appropriate to each of the different types of turbine is:

$8 < N_S < 80$ Pelton Wheel (high head, low discharge);
$55 < N_S < 320$ radial flow (Francis) (moderate head, moderate discharge);
$250 < N_S < 750$ axial flow (low head, high discharge).

7.7 Cavitation in hydraulic machines

Before closing the discussion, it is worth taking a slightly more detailed glance at the subject of cavitation. It is possible for cavitation to occur in both pumps and turbines if the liquid pressure falls sufficiently far below atmospheric pressure at any point in the machine. The vacuum head, H_C, at which cavitation will commence is defined by the equation

$$H_C = (p_{atm} - p_{vap})/\rho g$$

where p_{vap} is absolute vapour pressure and p_{atm} is the absolute pressure of the atmosphere. For the case of a pump sited above a sump (see Fig. 12.6), the suction head at the pump equals the sum of the height of pump above sump, the friction loss in suction pipe and the kinetic head in suction pipe or, expressed algebraically,

$$H_{ps} = H_s + h_{f_s} + V_s^2/2g$$

The 'net positive suction head' (NPSH) is defined as

$$\text{NPSH} = H_C - H_{ps}$$

A cavitation parameter σ_{Th} which was suggested by Thoma (a German engineer) relates NPSH to pump head

$$\sigma_{Th} = \text{NPSH}/H_p$$

For a turbine, the definition of NPSH is usually related to the height of the turbine (z) above tailrace, so that $\text{NPSH} = (H_C - z)$ and

$$\sigma_{Th} = \text{NPSH}/H_t$$

where H_t is the head available at the turbine (= gross head minus losses).

A considerable amount of data has been amassed on cavitation, and critical values of $\sigma_{Th}(= \sigma_{crit})$ can be estimated for pumps or turbines. For freedom from cavitation, the operational σ_{Th} should exceed σ_{crit}. Some typical equations for σ_{crit} are as follows:

for a single intake radial flow pump,

$$\sigma_{crit} = \left(\frac{NQ^{1/2}}{H_p \times 191}\right)^{4/3} = \left(\frac{N_S}{191}\right)^{4/3}$$

for a radial flow turbine,

$$\sigma_{crit} = 0.006 + 0.55(N_S/381)^{1.8}$$

For an axial flow turbine,

$$\sigma_{crit} = 0.10 + 0.3(N_S/381)^{2.5}$$

These values give only an approximate guide, and for more detailed information refer to Davis and Sorensen (1969).

References and further reading

Davis, C. V. and Sorensen, K. E. (1969) *Handbook of Applied Hydraulics*, McGraw-Hill, New York.

Dugdale, R. H. and Morfett, J. C. (1983) *Mechanics of Fluids*, George Godwin, London.

Stepanoff, A. J. (1957) *Centrifugal and Axial Flow Pumps*, 2nd edn, Wiley, New York.

8

Wave theory

8.1 Wave motion

Ocean waves are examples of periodic progressive waves, and as they approach the shores they are of particular interest to civil engineers involved in the design of marine structures or coastal defence works. Consequently, this chapter is concerned with the theories of periodic progressive waves and the interaction of waves with shorelines and coastal structures.

Ocean waves are generated mainly by the action of wind on water. The waves are formed initially by a complex process of resonance and shearing action, in which waves of differing wave height, length, period and direction are produced. Once formed, ocean waves can travel for vast distances, spreading in area and reducing in height, but maintaining wavelength and period. This process is called **dispersion**, and is shown in Figure 8.1. For example, waves produced in the gales of the 'roaring forties' have been monitored all the way north across the Pacific Ocean to the shores of Alaska (a distance of 10 000 km).

Wave theory normally employs the term **wave frequency** in preference to wave period, where frequency is the inverse of period. As will be shown later, waves of differing frequencies travel at different speeds, and therefore outside the storm generation area the sea state is modified as the various frequency components separate. The low-frequency waves travel more quickly than the high-frequency waves, resulting in a swell sea condition as opposed to a storm sea condition. Thus wind waves may be characterized as irregular, short crested and steep, containing a large range of frequencies and directions. By contrast, swell waves may be characterized as fairly regular, long crested and not very steep, containing a small range of low frequencies and directions.

As waves approach a shoreline, their speed, wavelength, direction and height are altered by the processes of refraction and shoaling before breaking on the shore. Once waves have broken, they enter what is termed the surf zone. Here some of the most complex transformation and attenuation processes occur, including generation of cross and longshore currents,

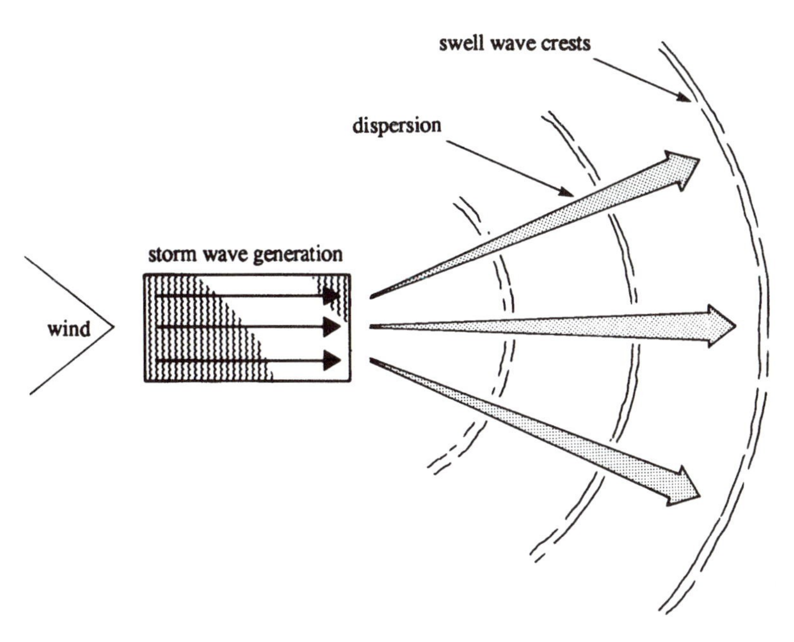

Figure 8.1 Wave generation and dispersion.

a set-up of the mean water level, and vigorous sediment transport of beach material.

Where coastal structures are present, either on the shoreline or in the near-shore zone, waves may also be diffracted and reflected, resulting in additional complexities in the wave motion. Figure 8.2 shows a simplified concept of the main wave transformation and attenuation processes that must be considered by coastal engineers in designing coastal defence schemes.

The traditional approach to determining a design wave condition at a specific site of interest has been first to estimate a representative wave height, period and direction offshore in deep water, and then to transform this representative wave by simulating the processes of refraction and shoaling to obtain the characteristics of the modified design wave at the point of interest. The more modern approach has been to replace the representative wave with what is termed a **directional energy density curve**. This is in recognition of the fact that, at any point in the sea, the sea state is composed of a (largely) random periodic motion consisting of a range of wave heights, periods and directions.

Additionally, the existence of wave groups is of considerable significance, as they have been shown to be responsible for the structural failure of some maritime structures designed using the traditional approach. The existence of wave groups also generates secondary wave forms of much lower

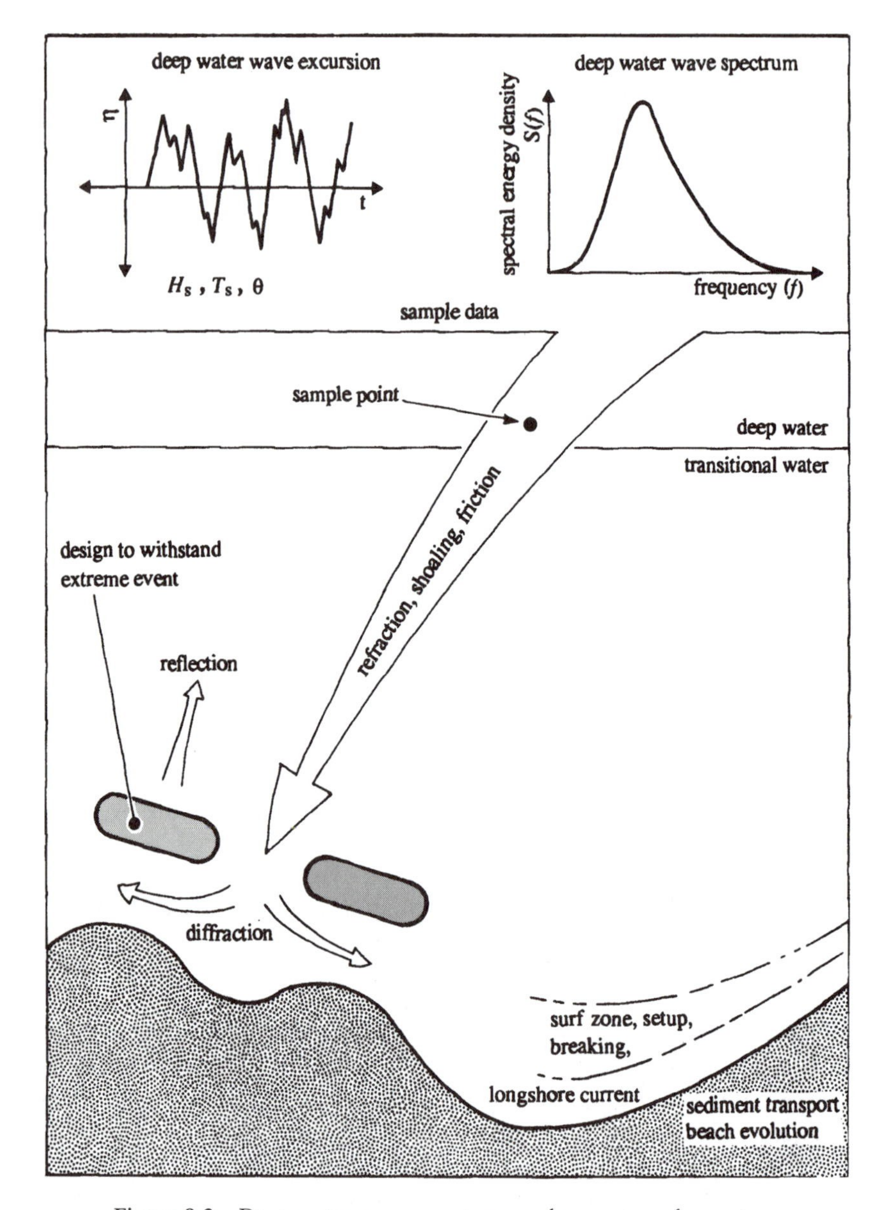

Figure 8.2 Deep water wave spectrum and wave transformation.

frequency and amplitude, called bound longwaves. Inside the surf zone these waves become separated from the 'short' waves, and have been shown to have a major influence on sediment transport and beach morphology, producing long-and cross-shore variations in the surf zone wave field.

To gain insight into and understanding of these complex processes it is perhaps best to begin by considering the simplest approaches and then to consider how these need to be modified and extended so that a more accurate, realistic and reliable methodology can be determined for the specification of a sea state for coastal defence.

The following sections describe some aspects of wave theory of particular application in coastal engineering. Many results are quoted without derivation, as the derivations are often long and complex. The interested reader should consult the references provided for further details.

8.2 Linear wave theory

The mathematical description of periodic progressive waves is complicated, and is still the subject of research. Fortunately, the earliest (and simplest) description, attributed to Airy in 1845, is sufficiently accurate for many engineering purposes. Airy wave theory is commonly referred to as linear or first-order wave theory, because of the simplifying assumptions made in its derivation. Other examples of wave theories are Stokes' (second-, third-, and fifth-order) and Cnoidal, both of which are approximations to the wave form of steep waves.

Airy waves

The Airy wave was derived using the concepts of two-dimensional ideal fluid flow. This is a reasonable starting point for ocean waves, which are not greatly influenced by viscosity, surface tension or turbulence.

Figure 8.3 shows a sinusoidal wave of wavelength L, height H and period T. The variation of surface elevation with time, from the still water level, is denoted by η (referred to as excursion) and given by

$$\eta = \frac{H}{2} \cos 2\pi \left(\frac{x}{L} - \frac{t}{T} \right) \tag{8.1}$$

The wave celerity, c, is given by

$$c = L/T \tag{8.2}$$

Equation (8.1) represents the surface solution to the Airy wave equations. The derivation of the Airy wave equations starts from the Laplace equation for irrotational flow of an ideal fluid. The Laplace equation is simply an expression of the continuity equation applied to a flow net, as described in Chapter 2, and is given by

$$\underbrace{\frac{\partial u}{\partial x} - \frac{\partial w}{\partial z}}_{\text{continuity}} = 0 = \underbrace{\frac{\partial^2 \phi}{\partial x^2} + \frac{\partial^2 \phi}{\partial z^2}}_{\text{Laplace}}$$

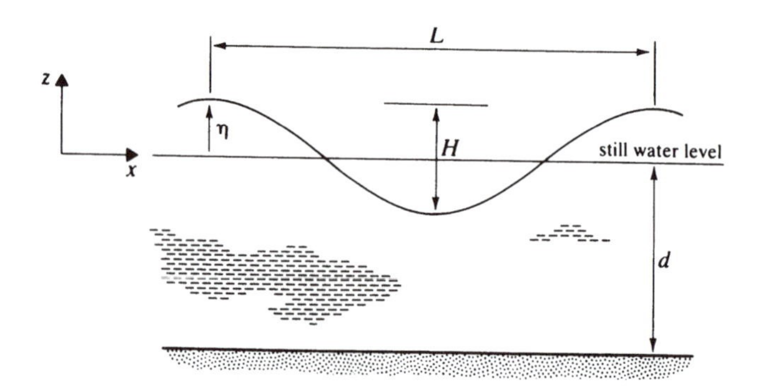

Figure 8.3 Definition sketch for a sinusoidal wave.

where u is the velocity in the x direction, w is the velocity in the z direction, and ϕ is the velocity potential.

A solution for ϕ is sought that satisfies the Laplace equation throughout the body of the flow. Additionally this solution must satisfy the 'boundary conditions': that is, the conditions at the bed and on the surface. At the bed, assumed horizontal, the vertical velocity w must be zero. At the surface, any particle on the surface must remain on the surface, and the (unsteady) Bernoulli's energy equation must be satisfied. Making the assumptions that $H \ll L$ and $H \ll d$ results in the linearized boundary conditions (in which the smaller, higher-order terms are neglected). The resulting solution for ϕ is given by

$$\phi = -gH\left(\frac{T}{4\pi}\right)\frac{\cosh\left(\frac{2\pi}{L}\right)(d+z)}{\cosh\left(\frac{2\pi}{L}\right)d}\sin\left(\frac{2\pi x}{L} - \frac{2\pi t}{T}\right)$$

Substituting this solution for ϕ into the two linearized surface boundary conditions yields the surface profile given in equation (8.1) and the wave celerity c given by

$$c = (gT/2\pi)\tanh(2\pi d/L) \tag{8.3a}$$

Most modern texts concerning wave theory use the terms **wave number** $(k)(k = 2\pi/L)$ and **wave frequency** $(\omega)(\omega = 2\pi/T)$. Thus equation (8.3a) may be more compactly stated as

$$c = (g/\omega)\tanh(kd) \tag{8.3b}$$

Substituting for c from equation (8.2) gives

$$c = \frac{L}{T} = \frac{\omega}{k} = \left(\frac{g}{\omega}\right)\tanh(kd)$$

or

$$\omega^2 = gk\tanh(kd) \tag{8.3c}$$

Equation (8.3c) is known as the **wave dispersion equation.** It may be solved, iteratively, for the wave number (k) and hence wavelength and celerity given the wave period and depth. Further details of the solution and its implications are given in section 8.3.

The corresponding equations for the horizontal (ζ) and vertical (ξ) displacements and velocities of a particle at a mean depth $-z$ below the still water level are

$$\zeta = -\frac{H}{2}\left[\frac{\cosh k(z+d)}{\sinh kd}\right]\sin 2\pi\left(\frac{x}{L}-\frac{t}{T}\right) \tag{8.4a}$$

$$u = \frac{\pi H}{T}\left[\frac{\cosh k(z+d)}{\sinh kd}\right]\cos 2\pi\left(\frac{x}{L}-\frac{t}{T}\right) \tag{8.4b}$$

and

$$\xi = \frac{H}{2}\left[\frac{\sinh k(z+d)}{\sinh kd}\right]\cos 2\pi\left(\frac{x}{L}-\frac{t}{T}\right) \tag{8.5a}$$

$$w = \frac{\pi H}{T}\left[\frac{\sinh k(z+d)}{\sinh kd}\right]\sin 2\pi\left(\frac{x}{L}-\frac{t}{T}\right) \tag{8.5b}$$

Equations (8.4a) and (8.5a) describe an ellipse, which is the path line of a particle according to linear theory. Equations (8.4b) and (8.5b) give the corresponding velocity components of the particle as it travels along its path. These equations are illustrated graphically in Figure 8.4.

Readers who wish to see a full derivation of the Airy wave equations are referred to Sorensen (1993) and Dean and Dalrymple (1991), in the first instance, for their clarity and engineering approach.

Strictly, the Airy wave equations apply only to waves of relatively small height in comparison with their wavelength and water depth. For steep waves and shallow water waves the profile becomes asymmetric, with high crests and shallow troughs. For such waves, celerity and wavelength are affected by wave height. For details concerning finite amplitude wave theories the reader is referred to Sorensen (1993) for the reasons already cited.

Deep water approximations. The particle displacement equations (8.4) and (8.5) describe approximately circular patterns of motion in deep water. At a depth ($-z$) of $L/2$, the diameter is only 4% of the surface value. Hence deep water waves are unaffected by depth, and have little influence on the sea bed (see Fig. 8.4).

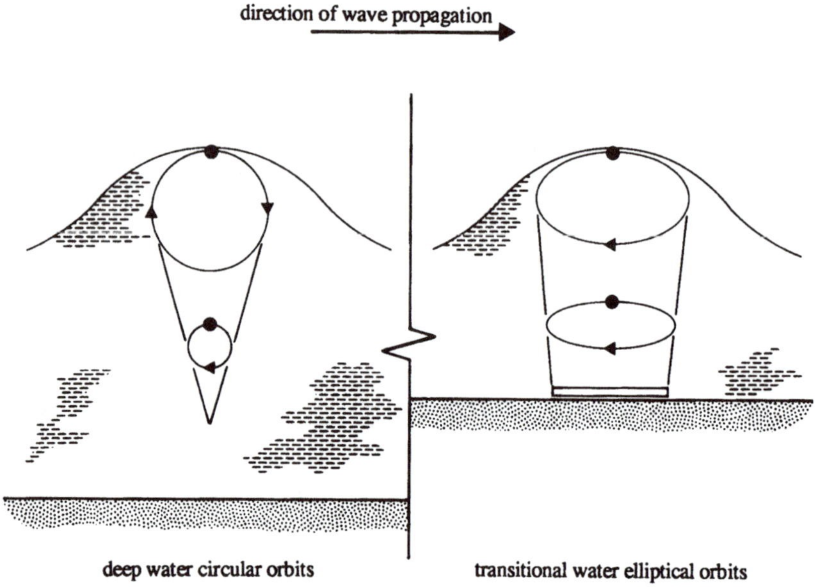

direction of wave propagation

deep water circular orbits

transitional water elliptical orbits

Figure 8.4 Particle displacements for deep and transitional waves.

For $d/L > 0.5$, $\tanh(kd) \simeq 1$. Hence (8.3a) reduces to

$$c_0 = gT/2\pi \tag{8.6}$$

where the subscript 0 refers to deep water. Alternatively, using (8.2),

$$c_0 = (gL_0/2\pi)^{1/2}$$

Thus the deep water wave celerity and wavelength are determined solely by the wave period.

Shallow water approximations. For $d/L < 0.04$, $\tanh\ (kd) \simeq 2\pi d/L$. Hence (8.3a) reduces to

$$c = gTd/L$$

and substituting this into (8.2) gives $c = \sqrt{gd}$.

Thus the shallow water wave celerity is determined by depth, and not by wave period. Hence shallow water waves are not frequency dispersive whereas deep water waves are.

Transitional water depth. This is the zone between deep water and shallow water: that is, $0.5 > d/L > 0.04$. In this zone $\tanh(kd) < 1$: hence

$$c = \frac{gT}{2\pi}\tanh(kd) = c_0 \tanh(kd) < c_0$$

This has important consequences, exhibited in the phenomena of refraction and shoaling, which are discussed in section 8.3.

In addition, the particle displacement equations show that at the seabed vertical components are suppressed, so only horizontal displacements now take place (see Fig. 8.4). This has important implications regarding sediment transport.

Wave energy and wave power

The energy contained within a wave is the sum of the potential, kinetic and surface tension energies of all the particles within a wavelength, and it is quoted as the total energy per unit area of the sea surface. For Airy waves, the potential (E_P) and kinetic (E_K) energies are equal, and $E_P = E_K = \rho g H^2 L/16$. Hence the energy E per unit area of ocean is

$$E = \rho g H^2/8 \tag{8.7}$$

(ignoring surface tension energy, which is negligible for ocean waves). This is a considerable amount of energy. For example, a (Beaufort) force 8 gale blowing for 24 h will produce a wave height in excess of 5 m, giving a wave energy exceeding $30\,\text{kJ/m}^2$.

One might expect that wave power (or the rate of transmission of wave energy) would be equal to wave energy times the wave celerity. This is incorrect, and the derivation of the equation for wave power leads to an interesting result, which is of considerable importance. Wave energy is transmitted by individual particles, which possess potential, kinetic and pressure energy. Summing these energies and multiplying by the particle velocity in the x direction for all particles in the wave gives the rate of transmission of wave energy or wave power, P, and leads to the result (for an Airy wave)

$$P = \frac{\rho g H^2}{8}\frac{c}{2}\left(1 + \frac{2kd}{\sinh 2kd}\right) \tag{8.8}$$

or

$$P = EC_G$$

where C_G is the group wave celerity, given by

$$C_G = \frac{c}{2}\left(1 + \frac{2kd}{\sinh 2kd}\right) \tag{8.9}$$

In deep water ($d/L > 0.5$) the group wave velocity $C_G = c/2$, and in shallow water $C_G = c$.

Hence in deep water wave energy is transmitted forward at only half the wave celerity. This is a difficult concept to grasp, and therefore it is useful to examine it in more detail.

Consider a wave generator in a model bay supplying a constant energy input of 128 units, and assume deep water conditions. In the time corresponding to the first wave period all of the energy supplied by the generator must be contained within one wavelength from the generator. After two wave periods, half of the energy contained within the first wavelength from the generator (64 units) will have been transmitted a further wavelength (that is, two wavelengths in total). Also, the energy within the first wavelength will have gained another 128 units of energy from the generator and will have lost half of its previous energy in transmission (64 units). Hence the energy level within the first wavelength after two wave periods will be $128 + 128 - 64 = 192$ units. The process may be repeated indefinitely. Table 8.1 shows the result after eight wave periods. This demonstrates that although energy has been radiated to a distance of eight wavelengths, the energy level of 128 units is propagating only one wavelength in every two wave periods. Also, the eventual steady wave energy at the generator corresponds to 256 units of energy in which 128 units is continuously being supplied and half of the 256 units is continuously being transmitted.

Table 8.1 Wave generation: to show group wave speed.

	Wave energy within various wavelengths from generator								Total wave energy
Number of wave periods	1	2	3	4	5	6	7	8	Generated energy
1	128	0	0	0	0	0	0	0	1
2	192	64	0	0	0	0	0	0	2
3	224	**128**	32	0	0	0	0	0	3
4	240	176	80	16	0	0	0	0	4
5	248	208	**128**	48	8	0	0	0	5
6	252	228	168	88	28	4	0	0	6
7	254	240	198	**128**	58	16	2	0	7
8	255	247	219	163	93	37	9	1	8

The appearance of the waveform to an observer is therefore one in which the leading wave front moves forward but continuously disappears. If the wave generator were stopped after eight wave periods, the wave group (of eight waves) would continue to move forward, but in addition wave energy would remain at the trailing edge in the same way as it appears at the leading edge. Thus the wave group would appear to move forward at half the wave celerity, with individual waves appearing at the rear of the group and moving through the group to disappear again at the leading edge.

Returning to our example of a force 8 gale, a typical wave celerity is 14 m/s (for a wave period of 9 s): the group wave celerity is thus 7 m/s, giving a wave power of 210 kW/m.

8.3 Wave transformation and attenuation processes

As waves approach a shoreline, they enter the transitional depth region in which the wave motions are affected by the seabed. These effects include reduction of the wave celerity and wavelength, and thus alteration of the direction of the wave crests (refraction) and wave height (shoaling), with wave energy dissipated by sea bed friction, and finally breaking.

Refraction

Wave celerity and wavelength are related through two equations ((8.2) and (8.3a)) to wave period (which is the only parameter that remains constant). This can be appreciated by postulating a change in wave period (from T_1 to T_2) over an area of sea. The number of waves entering the area in a fixed time t would be t/T_1, and the number leaving would be t/T_2. Unless T_1 equals T_2 the number of waves within the region could increase or decrease indefinitely. Thus

$$c/C_0 = \tanh(kd) \quad \text{(from (8.3a))}$$

and

$$c/C_0 = L/L_0 \quad \text{(from (8.2a))}$$

To find the wave celerity and wavelength at any depth d, these two equations must be solved simultaneously. The solution is always such that $c < C_0$ and $L < L_0$ for $d < d_0$ (where the subscript 0 refers to deep water conditions).

Consider a deep water wave approaching the transitional depth limit ($d/L = 0.5$), as shown in Figure 8.5. A wave travelling from A to B (in deep water) traverses a distance L_0 in one wave period T. However, the wave travelling from C to D traverses a smaller distance L in the same time, as

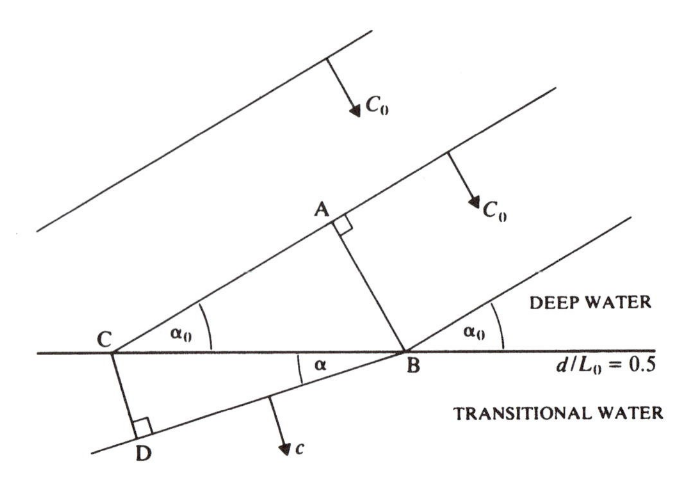

Figure 8.5 Wave refraction.

it is in the transitional depth region. Hence the new wave front is now BD, which has rotated with respect to AC. Letting the angle α represent the angle of the wave front to the depth contour, then

$$\sin\alpha = L/BC \quad \text{and} \quad \sin\alpha_0 = L_0/BC$$

Combining,

$$\frac{\sin\alpha}{\sin\alpha_0} = \frac{L}{L_0}$$

Hence

$$\frac{\sin\alpha}{\sin\alpha_0} = \frac{L}{L_0} = \frac{c}{C_0} = \tanh(kd) \tag{8.10}$$

As $c < C_0$ then $\alpha < \alpha_0$, which implies that as a wave approaches a shoreline from an oblique angle the wave fronts tend to align themselves with the underwater contours. Figure 8.6 shows the variation of c/C_0 with d/L_0 and of α/α_0 with d/L_0 (the latter specifically for the case of parallel contours). Note that L_0 is used in preference to L as the former is a fixed quantity.

For non-parallel contours, individual wave rays (that is, the orthogonals to the wave fronts) must be traced. Figure 8.6 can still be used to find α at each contour if α_0 is taken as the angle (say α_1) at one contour and α is taken as the new angle (say α_2) to the next contour. The wave ray is usually taken to change direction midway between contours. This procedure may be carried out by hand using tables or figures (see Silvester, 1974) or by computer as described later in this section.

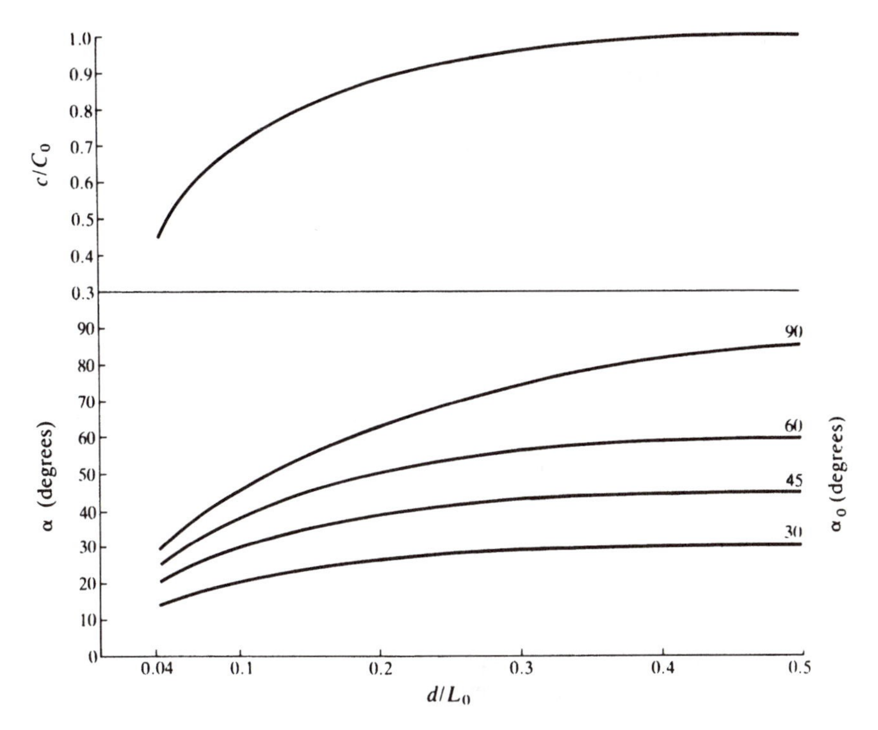

Figure 8.6 Variations of wave celerity and angle with depth.

Wave breaking. As the refracted waves enter the shallow water region, they break before reaching the shoreline. The foregoing analysis is not strictly applicable to this region, because the wave fronts steepen and are no longer described by the Airy waveform. However, as a general guideline, waves will break when

$$d_B \approx 1.28 H_B \tag{8.11}$$

where the subscript B refers to breaking.

The subject of wave breaking is of considerable interest both theoretically and practically. Further details are described in section 8.4.

It is common practice to apply refraction analysis up to the breaker line. This is justified on the grounds that the inherent inaccuracies are small compared with the initial predictions for deep water waves, and are within acceptable engineering tolerances. To find the breaker line, it is necessary to estimate the wave height as the wave progresses inshore. This can be estimated from the refraction diagram in conjunction with shoaling calculations, as is now described.

Shoaling. Consider first a wave front travelling parallel to the seabed contours. Making the assumption that wave energy is transmitted shorewards without loss due to bed friction or turbulence, then

$$\frac{P}{P_0} = 1 = \frac{EC_G}{E_0 C_{G_0}} \quad \text{(from (8.8))}$$

Substituting:

$$E = \rho g H^2/8 \quad \text{(equation (8.7))}$$

Then

$$\frac{P}{P_0} = 1 = \left(\frac{H}{H_0}\right)^2 \frac{C_G}{C_{G_0}}$$

or

$$\frac{H}{H_0} = \left(\frac{C_{G_0}}{C_G}\right)^{1/2} = K_S$$

where K_S is the shoaling coefficient.

The shoaling coefficient can be evaluated from the equation for the group wave celerity (equation (8.9)).

$$K_S = \left(\frac{C_{G_0}}{C_G}\right)^{1/2} = \left(\frac{C_0/2}{\frac{c}{2}[1 + (2kd/\sinh 2kd)]}\right)^{1/2} \qquad (8.12)$$

The variation of K_S with d/L_0 is shown in Figure 8.7.

Refraction and shoaling. Consider next a wave front travelling obliquely to the seabed contours, as shown in Figure 8.8. In this case, as the wave rays bend, they may converge or diverge as they travel shorewards. At the contour $d/L_0 = 0.5$,

$$\mathrm{BC} = \frac{b_0}{\cos\alpha_0} = \frac{b}{\cos\alpha}$$

or

$$\frac{b}{b_0} = \frac{\cos\alpha}{\cos\alpha_0}$$

Again, assuming that the power transmitted between any two wave rays is constant (that is, conservation of wave energy flux), then

$$\frac{P}{P_0} = 1 = \frac{Eb}{E_0 b_0} \frac{C_G}{C_{G_0}}$$

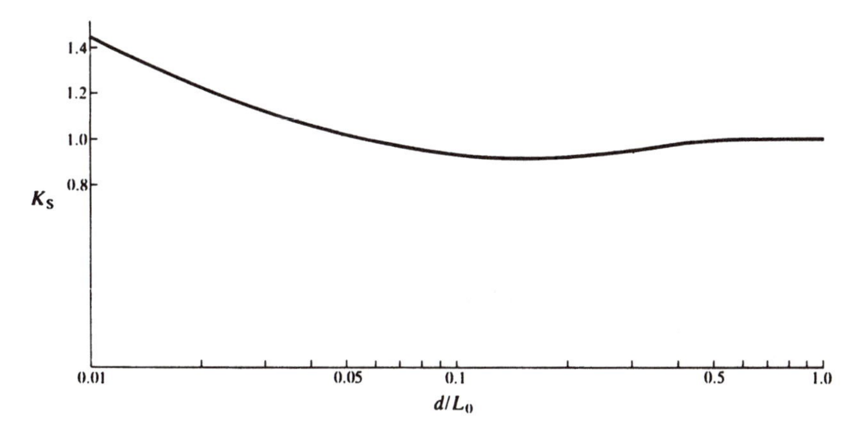

Figure 8.7 Variation of the shoaling coefficient with depth.

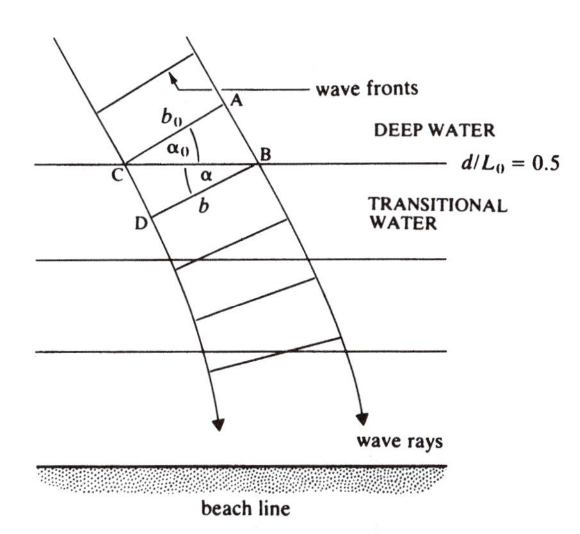

Figure 8.8 Divergence of wave rays over parallel contours.

Substituting for E and b,

$$\left(\frac{H}{H_0}\right)^2 \frac{\cos\alpha}{\cos\alpha_0}\frac{C_G}{C_{G_0}} = 1$$

or

$$\frac{H}{H_0} = \left(\frac{\cos\alpha_0}{\cos\alpha}\right)^{1/2}\left(\frac{C_{G_0}}{C_G}\right)^{1/2}$$

Hence

$$H/H_0 = K_R K_S \tag{8.13}$$

where

$$K_R = \left(\frac{\cos\alpha_0}{\cos\alpha}\right)^{1/2}$$

and is called the **refraction coefficient.**

For parallel contours, K_R can be found using Figure 8.6. In the more general case, K_R can be found from the refraction diagram directly by measuring b and b_0.

Example 8.1 Wave refraction and shoaling

A deep water wave has a period of 8.5 s, a height of 5 m and is travelling at 45° to the shoreline. Assuming that the seabed contours are parallel, find the height, depth, celerity and angle of the wave when it breaks.

Solution

(a) Find the deep water wavelength and celerity. From (8.6),

$$C_0 = gT/2\pi = 13.27\,\text{m/s}$$

From (8.2),

$$L_0 = C_0 T = 112.8\,\text{m}$$

(b) At the breaking point, the following conditions (from (8.11) and (8.13)) must be satisfied:

$$d_B = 1.28 H_B \quad \text{and} \quad H_B/H_0 = K_R K_S$$

For various trial values of d/L_0, H_B/H_0 can be found using Figures 8.6 and 8.7. The correct solution is when (8.11) and (8.12) are satisfied simultaneously. This is most easily seen by preparing a table, as shown in Table 8.2. For

Table 8.2 Tabular solution for breaking waves

d/L_0	d (m)	c/C_0	c (m/s)	K_S	α (degrees)	K_R	H/H_0	H (m)	d_B (m)
0.1	11.3	0.7	9.3	0.93	30	0.9	0.84	4.2	5.4
0.05	5.6	0.52	6.9	1.02	22	0.87	0.89	4.45	5.7

$d/L_0 = 0.05$, $d = 5.6$ m and $H = 4.45$, requiring a depth of breaking of 5.7 m. This is sufficiently accurate for an acceptable solution, so

$$H_B = 4.45\,\text{m} \quad c = 6.9\,\text{m/s} \quad d_B = 5.7\,\text{m} \quad \alpha_B = 22^\circ$$

Numerical solution of the wave dispersion equation. In order to solve this problem from first principles it is first necessary to solve the wave dispersion equation for L in any depth d. This may be done by a variety of numerical methods. Starting from equation (8.10),

$$\frac{L}{L_0} = \frac{c}{C_0} = \tanh(kd)$$

Hence

$$L = \frac{gT^2}{2\pi} \tanh\left(\frac{2\pi d}{L}\right)$$

Given T and d, an initial estimate of L (say L_1) can be found by substituting L_0 into the tanh term. Thereafter, successive estimates (say L_2) can be taken as the average of the current and previous estimates (for example, $L_2 = (L_0 + L_1)/2$) until sufficiently accurate convergence is obtained. A much more efficient technique is described by Goda (2000), based on Newton's method, given by

$$x_2 = x_1 - \frac{(x_1 - D\coth x_1)}{1 + D(\coth^2 x_1 - 1)}$$

where $x = 2\pi d/L$, $D = 2\pi d/L_0$ and the best estimate for the initial value is

$$x_1 = \left[\begin{array}{l} D \text{ for } D \geqslant 1 \\ D^{1/2} \text{ for } D < 1 \end{array}\right.$$

This provides an absolute error of less than 0.05% after three iterations.

A direct solution was derived by Hunt (1979), given by

$$\frac{c^2}{gd} = \left[y + (1 + 0.6522y + 0.4622y^2 + 0.0864y^4 + 0.0675y^5)^{-1}\right]^{-1}$$

where $y = k_0 d$, which is accurate to 0.1% for $0 < y \propto$.

Seabed friction. In the analysis of refraction and shoaling given above it was assumed that there was no loss of energy as the waves were transmitted inshore. In reality, waves in transitional and shallow water depths will be attenuated by wave energy dissipation through seabed friction. Such energy losses can be estimated, using linear wave theory, in an analogous way to pipe and open channel flow frictional relationships. First, the mean seabed shear stress τ_b may be found using

$$\tau_b = \frac{1}{2} f_w \rho u_m^2$$

where f_w is the wave friction factor and u_m is the maximum near-bed orbital velocity; f_w is a function of a local Reynolds number (Re_w) defined in terms of u_m (for velocity) and either a_b (wave amplitude at the bed) or the seabed grain size k_s (for the characteristic length). A diagram relating f_w to Re_w for various ratios of a_b/k_s, due to Jonsson, is given in Dyer (1986). This diagram is analogous to the Moody diagram for pipe friction factor (λ). Values of f_w range from about 0.5×10^{-3} to 5. Hardisty (1990) summarizes field measurements of f_w (from Sleath), and notes that a typical field value is about 0.1. Using linear wave theory, u_m is given by

$$u_m = \frac{\pi H}{T \sinh kd}$$

Second, the rate of energy dissipation may be found by combining the expression for τ_b with linear wave theory to obtain

$$\frac{dH}{dx} = -\frac{4 f_w k^2 H^2}{3\pi \sinh(kd)(\sinh(2kd) + 2kd)} \tag{8.14}$$

The wave height attenuation due to seabed friction is of course a function of the **distance travelled** by the wave as well as the depth, wavelength and wave height. Thus the total loss of wave height (ΔH_f) due to friction may be found by integrating over the path of the wave ray.

BS 6349 presents a chart from which a wave height reduction factor may be obtained. Except for large waves in shallow water, seabed friction is of relatively little significance. Hence, for the design of maritime structures in depths of 10 m or more, seabed friction is often ignored. However, in determining the wave climate along the shore, seabed friction is now normally included in numerical models, although an appropriate value for the wave friction factor remains uncertain, and is subject to change with wave-induced bed forms.

Wave–current interaction. So far, consideration of wave properties has been limited to the case of waves generated and travelling on quiescent water. In general, however, ocean waves are normally travelling on currents generated by tides and other means. These currents will also, in general, vary in both space and time. Hence two distinct cases need to be considered here. The first is that of waves travelling *on* a current and the second when waves generated in quiescent water encounter a current (or travel over a varying current field).

For waves travelling on a current, two frames of reference need to be considered. The first is a moving or *relative* frame of reference, travelling at the current speed. In this frame of reference, all the wave equations derived so far still apply. The second frame of reference is the stationary or *absolute* frame. The concept which provides the key to understanding this situation is that the wavelength is the *same* in both frames of reference. This is because the wavelength in the relative frame is determined by the dispersion equation and this wave is simply moved at a different speed in the absolute frame. In consequence, the absolute and relative wave periods are different.

Consider the case of a current with magnitude (u) following a wave with wave celerity (c), the wave speed with respect to the sea bed (c_a) becomes $c+u$. As the wavelength is the same in both reference frames, the absolute wave period will be less than the relative wave period. Consequently, if waves on a current are measured at a fixed location (e.g. in the absolute frame), then it is the absolute period (T_a) which is measured. The current magnitude must, therefore, also be known in order to determine the wavelength. This can be shown as follows:

Starting from the dispersion equation (8.3a) and noting that $c = L/T_r$ leads to

$$c = \left(\frac{gL}{2\pi} \tanh \frac{2\pi d}{L} \right)^{1/2}$$

As $c_a = c + u$ and $c_a = L/T_a$, then

$$L = \left[\left(\frac{gL}{2\pi} \tanh \frac{2\pi d}{L} \right)^{1/2} + u \right] T_a \qquad (8.15)$$

This equation thus provides an implicit solution for the wavelength in the presence of a current when the absolute wave period has been measured.

Conversely, when waves travelling in quiescent water encounter a current, changes in wave height and wavelength will occur. This is because as waves travel from one region to the other requires that the absolute wave period remains constant for waves to be conserved. Consider the case of an opposing current, the wave speed relative to the sea bed is reduced and therefore the wavelength will also decrease. Thus wave height and steepness will increase. In the limit the waves will break when they reach limiting steepness. In addition, as wave energy is transmitted at the group wave speed, waves cannot penetrate a current whose magnitude equals or exceeds the group wave speed and thus wave breaking and diffraction will occur under these circumstances. Such conditions can occur in the entrance channels to estuaries when strong ebb tides are running, creating a region of high, steep and breaking waves.

Current refraction. Another example of wave-current interaction is that of current refraction. This occurs when a wave obliquely crosses from a region of still water to a region in which a current exists or in a changing current field. The simplest case is illustrated in Figure 8.9 showing deep water wave refraction by a current. In an analogous manner to refraction caused by depth changes, Jonsson showed that in the case of current refraction

$$\sin \alpha_c = \frac{\sin \alpha}{\left(1 - \frac{u}{c} \sin \alpha\right)^2}$$

The wave height is also affected and will decrease if the wave orthogonals diverge (as shown) or increase if the wave orthogonals converge.

The generalized refraction equations for numerical solution techniques. The foregoing equations for refraction and shoaling may be generalized

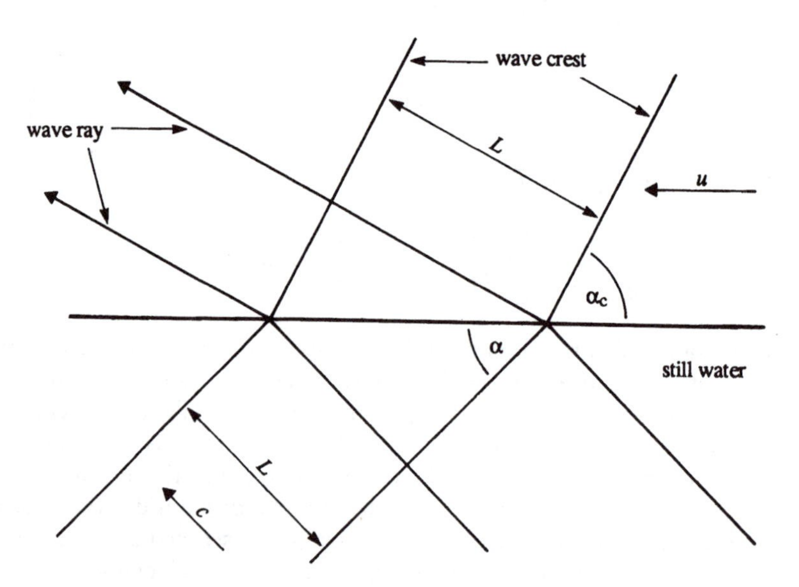

Figure 8.9 Deep water wave refraction by a current.

for application to irregular bathymetry, and then solved using a suitable numerical scheme. Two approaches have been developed. The first is the numerical equivalent of the ray (that is, wave orthogonal) tracing technique, and allows determination of individual ray paths, giving a clear picture of wave refraction patterns for any bathymetry. The wave height at any location, however, has to be calculated separately using the local ray spacing b to find the refraction coefficient K_R. The second method computes the local wave height and direction at each point on a regular grid using the wave and energy conservation equation in cartesian coordinates. This is much more useful as input to other models (for example, for wave-induced currents).

The wave conservation equation in wave ray form. Figure 8.10 shows a pair of wave crests and a corresponding pair of wave rays. The wave rays are everywhere at right angles to the wave crests, resulting in an orthogonal grid. This implies that only wave refraction can occur. The wave ray at point A is at an angle θ with the x axis and is travelling at speed c. The wave ray at **B** is a small distance δb from **A**, and is travelling at a speed $c+\delta c$, as it is in slightly deeper water than at point **A**.

In a small time δt, the wave ray at **A** moves to **E** at a speed c and the wave ray at **B** moves to **D** at speed $c+\delta c$. Thus the wave orthogonal rotates

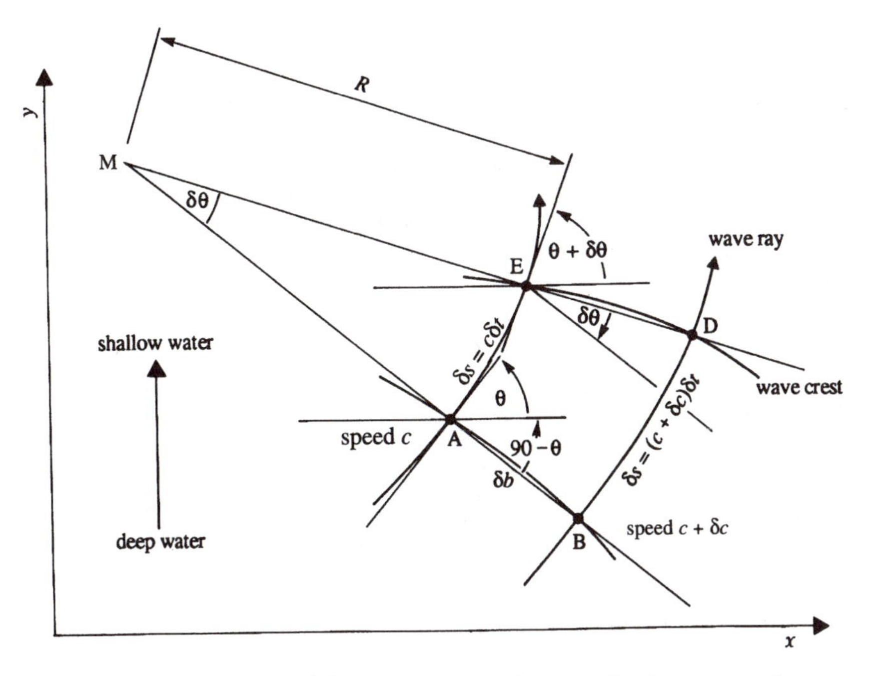

Figure 8.10 Derivation of the wave conservation equation in wave ray form.

through $\delta\theta$. Let point **M** be the centre of rotation at distance R from A and E. Using similar triangles,

$$\frac{c\delta t}{R} = \frac{(c+\delta c)\delta t - c\delta t}{\delta b}$$

Simplifying and rearranging,

$$\frac{\delta b}{R} = \frac{\delta c}{c} \tag{8.16}$$

Also

$$\frac{\delta s}{R} = -\delta\theta$$

The negative sign is introduced to ensure that the orthogonal bends in the direction of reducing c or

$$\frac{\delta\theta}{\delta s} = \frac{1}{R} \tag{8.17}$$

Combining (8.16) and (8.17) and in the limit

$$\frac{\mathrm{d}\theta}{\mathrm{d}s} = -\frac{1}{c}\frac{\mathrm{d}c}{\mathrm{d}b} \tag{8.18}$$

Considering a ray path, by trigonometry

$$\delta x = \delta s \cos\theta$$

$$\delta y = \delta s \sin\theta$$

and as

$$\delta s = c\delta t$$

then in the limit

$$\frac{\mathrm{d}x}{\mathrm{d}t} = c\cos\theta \tag{8.19}$$

$$\frac{\mathrm{d}y}{\mathrm{d}t} = c\sin\theta \tag{8.20}$$

Returning to (8.18) and given that

$$c = f(x, y) \quad \text{and} \quad x, y = f(b)$$

then applying the chain rule,

$$\frac{\mathrm{d}c}{\mathrm{d}b} = \frac{\partial c}{\partial x}\frac{\partial x}{\partial b} + \frac{\partial c}{\partial y}\frac{\partial y}{\partial c} \tag{8.21}$$

Along a wave crest,

$$\frac{\partial x}{\partial b} = \cos(90 - \theta) = -\sin\theta \tag{8.22}$$

$$\frac{\partial y}{\partial b} = \sin(90 - \theta) = \cos\theta \tag{8.23}$$

Substituting (8.21), (8.22) and (8.23) into (8.18) yields

$$\frac{\mathrm{d}\theta}{\mathrm{d}s} = \frac{1}{c}\left(\frac{\partial c}{\partial x}\sin\theta - \frac{\partial c}{\partial y}\cos\theta\right) \tag{8.24}$$

Finally, as $\partial s = c\partial t$ and substituting into (8.24),

$$\frac{\mathrm{d}\theta}{\mathrm{d}t} = \frac{\partial c}{\mathrm{d}x}\sin\theta - \frac{\partial c}{\partial y}\cos\theta \tag{8.25}$$

Equations (8.19), (8.20) and (8.25) may be solved numerically along a ray path sequentially through time. Koutitas (1988) gives a worked example of such a scheme. If two closely spaced ray paths are calculated, the local refraction coefficient may then be found and hence the wave heights along the ray path determined. However, a more convenient method to achieve this was developed by Munk and Arthur. They derived an expression for the orthogonal separation factor $\beta = b/b_c = K_r^{-1/2}$ given by

$$\frac{\mathrm{d}^2\beta}{\mathrm{d}s^2} + p\frac{\mathrm{d}\beta}{\mathrm{d}s} + q\beta = 0$$

where

$$p = \frac{\cos\theta}{c}\frac{\partial c}{\partial x} - \frac{\sin\theta}{c}\frac{\partial c}{\partial y}$$

$$q = \frac{\sin^2\theta}{c}\frac{\partial^2 c}{\partial x^2} - 2\frac{\sin\theta\cos\theta}{c}\frac{\partial^2 c}{\partial s\partial y} + \frac{\cos^2\theta}{c}\frac{\partial^2 c}{\partial y^2}$$

The derivation of these equations may be found in Dean and Dalrymple (1991) together with some references to the numerical solution techniques.

Wave conservation equation and wave energy conservation equation in Cartesian coordinates. The wave conservation equation (8.18) may be reformulated in Cartesian coordinates by transformation of the axes. The result, in terms of the wave number ($k = 2\pi/L = \omega/c$) is given by

$$\frac{\partial(k \sin\theta)}{\partial x} - \frac{\partial(k \cos\theta)}{\partial y} = 0 \tag{8.26}$$

The proof that equation (8.26) is equivalent to (8.18) is given in Dean and Dalrymple (1991).

The wave energy conservation equation is given by

$$\frac{\partial(EC_g \cos\theta)}{\partial x} + \frac{\partial(EC_g \sin\theta)}{\partial y} = -\varepsilon_d \tag{8.27}$$

where ε_d represents energy losses (due to seabed friction, cf. equation (8.14)). Again, Koutitas (1988) gives a worked example of a numerical solution to equations (8.26) and (8.27).

Wave reflection

Waves normally incident on solid vertical boundaries (such as harbour walls and sea walls) are reflected such that the reflected wave has the same phase (but opposite direction) and substantially the same amplitude as the incident wave. This fulfils the necessary condition that the horizontal velocity at the boundary is always zero. The resulting wave pattern set up is called a **standing wave,** as shown in Figure 8.11.

The equation of the surface excursion for the standing wave (subscript s) may be found by adding the two waveforms of the incident (subscript i) and reflected (subscript r) waves. Thus

$$\eta_i = \frac{H_i}{2}\cos 2\pi\left(\frac{x}{L} - \frac{t}{T}\right)$$

$$\eta_r = \frac{H_r}{2}\cos 2\pi\left(\frac{x}{L} + \frac{t}{T}\right)$$

$$\eta_s = \eta_i + \eta_r$$

Taking

$$H_r = H_i = H_s/2$$

then

$$\eta_s = H_s \cos(2\pi x/L)\cos(2\pi t/T) \tag{8.28}$$

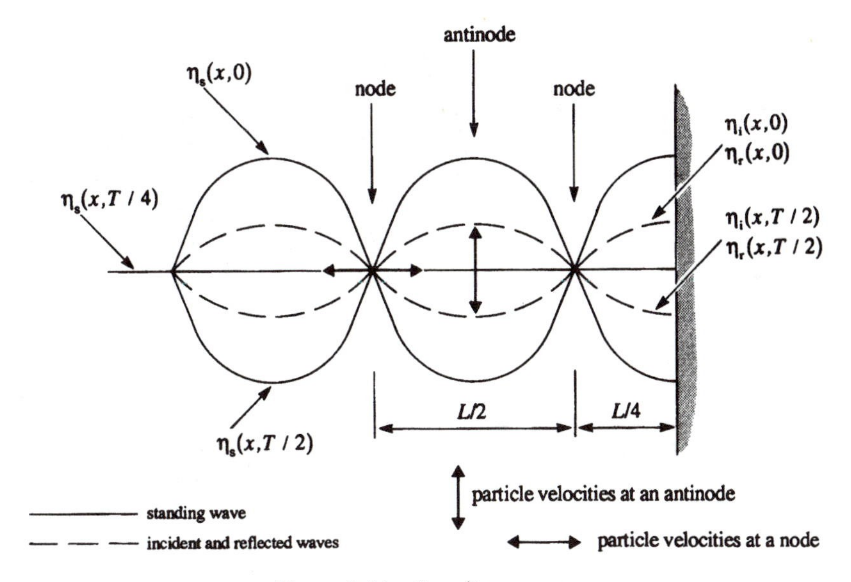

Figure 8.11 Standing waves.

At the nodal points there is no vertical movement with time, but the horizontal velocities are a maximum. By contrast, at the antinodes, crests and troughs appear alternately, but the horizontal velocities are zero. For the case of large waves in shallow water and if the reflected wave has a similar amplitude to the incident wave, then the advancing and receding crests collide in a spectacular manner, forming a plume known as a **clapotis**. This is commonly observed at sea walls. Standing waves can cause considerable damage to maritime structures, and bring about substantial erosion.

Clapotis gaufre. When the incident wave is at an angle α to the normal from a vertical boundary, then the reflected wave will be in a direction α on the opposite side of the normal. This is illustrated in Figures 8.12 and 8.13.

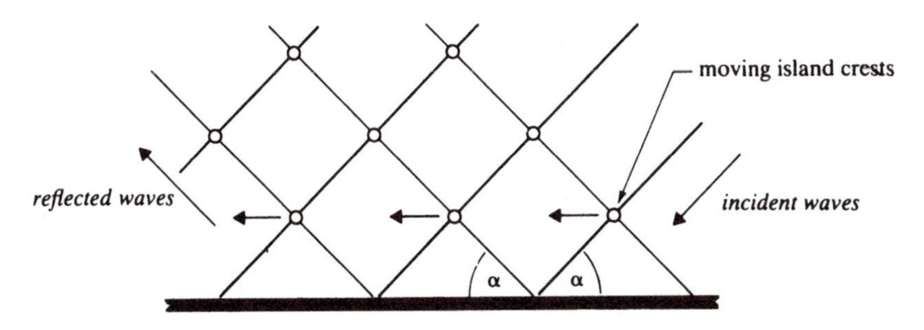

Figure 8.12 Plan view of oblique wave reflection.

Figure 8.13 Wave impact and reflection during a storm.

The resulting wave motion (the clapotis gaufre) is complex, but essentially consists of a diamond pattern of island crests, which move parallel to the boundary. It is sometimes referred to as a short-crested system. The crests form at the intersection of the incident and reflected wave fronts. The resulting particle displacements are also complex, but include the generation of a pattern of moving vortices. A detailed description of these motions may be found in Silvester (1974). The consequences of this in terms of sediment transport may be severe. Very substantial erosion and longshore transport may take place. Considering that oblique wave attack to sea walls is the norm rather than the exception, the existence of the clapotis gaufre has a profound influence on the long-term stability and effectiveness of coastal defence works. This does not seem to have been fully understood in traditional designs of sea walls, with the result that collapsed sea walls and eroded coastlines have occurred.

Wave reflection coefficients. Defining a reflection coefficient $K_r = H_r/H_i$, then typical values are as follows:

Reflection barrier	K_r
Concrete sea walls	0.7–1.0
Rock breakwaters	0.2–0.7
Beaches	0.05–0.2

Note that the reflected wave energy is equal to K_r^2, as energy is proportional to H^2.

Predictive equations for wave reflection from rock slopes. The CIRIA/CUR manual (1991) gives an excellent summary of the development of wave reflection equations based on laboratory data of reflection from rock breakwaters. This work clearly demonstrates that rock slopes considerably reduce reflection compared with smooth impermeable slopes. Based on this data, the best-fit equation was found to be

$$K_r = 0.125\xi_p^{0.7}$$

where ξ_p is the Iribarren no. $= \tan\beta/\sqrt{H/L_p}$ and p refers to peak frequency. Davidson *et al.* (1996) subsequently carried out an extensive field measurement programme of wave reflection at prototype scale at the Elmer break-waters (Sussex, UK), and after subsequent analysis proposed a new predictive as follows.

A new dimensionless reflection parameter was proposed, given by

$$R = \frac{d_t \lambda_0^2}{H_i D^2 \cot\beta} \tag{8.29}$$

where d_t(m) is the water depth at the toe of the structure, λ_0 is the deep water wavelength at peak frequency, H_i is the significant incident wave height, D is the characteristic diameter of rock armour ($= W_{50}/\rho$ median mass/density), and $\tan\beta$ is structure gradient. R was found to be a better parameter than ξ in predicting wave reflection.

The reflection coefficient is then given by

$$K_r = 0.151R^{0.11} \tag{8.30}$$

or

$$K_r = \frac{0.635R^{0.5}}{41.2 + R^{0.5}} \tag{8.31}$$

Wave diffraction

This is the process whereby waves bend round obstructions by radiation of the wave energy. Figure 8.14 shows an oblique wave train incident on the tip of a breakwater. There are three distinct regions:

1. the shadow region in which diffraction takes place;

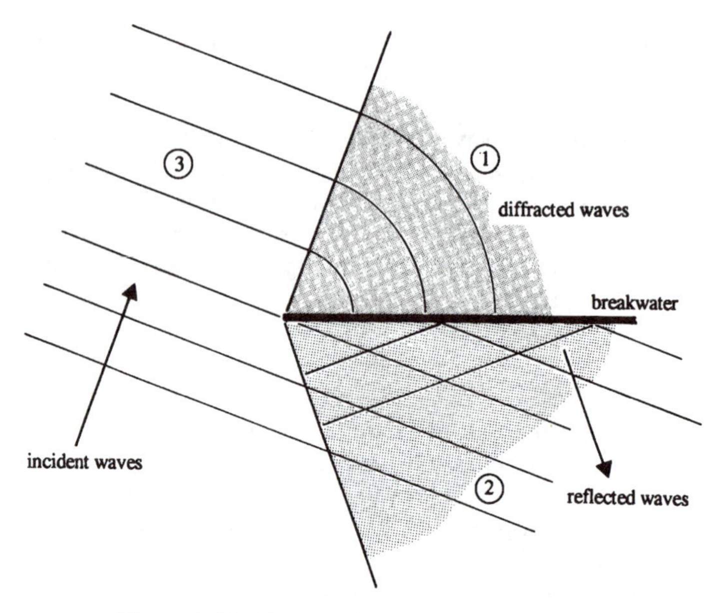

Figure 8.14 Wave diffraction at a breakwater.

2. the short-crested region in which incident and reflected waves form a clapotis gaufre;
3. an undisturbed region of incident waves.

In region (1), the waves diffract with the wave fronts, forming circular arcs centred on the point of the breakwater. When the waves diffract, the wave heights diminish as the energy of the incident wave spreads over the region. The real situation is, however, more complicated than that presented in Figure 8.14. The reflected waves in region (2) will diffract into region (3) and hence extend the short-crested system into region (3).

Mathematical formulation of wave diffraction. Mathematical solutions for wave diffraction have been developed for the case of constant water depth using linear wave theory. The basic differential equation for wave diffraction is known as the **Helmholtz equation.** This can be derived from the Laplace equation (refer to section 8.2) by making the appropriate substitutions as follows:

$$\nabla^2\phi = 0 \quad \text{(Laplace equation)}$$

Expanding in three dimensions:

$$\frac{\partial^2\phi}{\partial x^2} + \frac{\partial^2\phi}{\partial y^2} + \frac{\partial^2\phi}{\partial z^2} = 0$$

Now, let

$$\phi(x, y, z) = Z(z)F(x, y)\, e^{i\omega t}$$

(i.e. ϕ is a function of depth and horizontal coordinates, and is periodic, and i is the imaginary number $= \sqrt{-1}$).

For uniform depth an expression for $Z(z)$ satisfying the no-flow bottom boundary condition is

$$Z(z) = \cosh k(h + z)$$

Substituting for ϕ and Z in the Laplace equation leads (after further manipulation) to the Helmholtz equation

$$\frac{\partial^2 F}{\partial x^2} + \frac{\partial^2 F}{\partial y^2} + K^2 F(x, y) = 0 \tag{8.32}$$

Solutions to the Helmholtz equation. A solution to the Helmholtz equation was first found by Sommerfeld in 1896, who applied it to the diffraction of light (details may be found in Dean and Dalrymple, 1991). Somewhat later, Penney and Price (1952) showed that the same solution applied to water waves, and presented solutions for incident waves from different directions passing a semi-infinite barrier and for normally incident waves passing through a barrier gap. For the case of normal incidence on a semi-infinite barrier, it may be noted that, for a monochromatic wave, the diffraction coefficient K_d is approximately 0.5 at the edge of the shadow region, and that K_d exceeds 1.0 in the 'undisturbed' region because of diffraction of the reflected waves caused by the (perfectly) reflecting barrier. Their solution for the case of a barrier gap is essentially the superposition of the results from two mirror image semi-infinite barriers.

Their diagrams apply for a range of gap width to wavelength (b/L) from 1 to 5. When b/L exceeds 5 the diffraction patterns from each barrier do not overlap, and hence the semi-infinite barrier solution applies. For b/L less than 1 the gap acts as a point source, and wave energy is radiated as if it were coming from a single point at the centre of the gap.

It is important to note here that these diagrams should not be used for design. This is because of the importance of considering directional wave spectra, which are discussed in section 8.6.

Combined refraction and diffraction

Refraction and diffraction often occur together. For example, the use of a wave ray model over irregular bathymetry may produce a caustic (that is, a region where wave rays cross). Here diffraction will occur, spreading wave energy away from regions of large wave heights. Another example is around offshore breakwaters; here diffraction is often predominant close to the structure, with refraction becoming more important further away from the structure. In this latter case, an approximate solution can be obtained by applying diffraction diagrams over a distance of, say, up to 3–4 wavelengths from the breakwater gap and then refracting the waves thereafter.

The alternative is to find a solution to the Laplace equation over irregular bathymetry (mildly sloping bed) satisfying the various boundary equations. Such a solution was first derived in 1972 by Berkhoff, and is known as the **mild slope equation.**

It may be written as

$$\frac{\partial}{\partial x}\left(cCg\frac{\partial\phi}{\partial x}\right)+\frac{\partial}{\partial y}\left(cCg\frac{\partial\phi}{\partial y}\right)+\omega^2\frac{Cg}{c}\phi=0 \qquad (8.33)$$

where $\phi(x, y)$ is a complex wave potential function. The solution of this equation is highly complex, and beyond the scope of this text. However, the interested reader is directed to McDowell (1988) and Dodd and Brampton (1995) for reviews of the subject. One of the most recent developments in solving the mild slope equation is that due to Li (1994). This version of the mild slope equation allows the simultaneous solution of refraction, diffraction and reflection. It has also been the subject of a field validation study. Initial results may be found in Ilic and Chadwick (1995). They tested this model at the site of an offshore breakwater scheme where refraction and reflection are the main processes seaward of the breakwaters, with diffraction and refraction taking place shoreward of the breakwaters.

8.4 Surf zone processes

A general description of the surf zone

In section 8.3 reference was made to the fact that waves may break in the shallow waters near to the shoreline. It is worth considering some aspects of breaking waves a little more closely. Waves may break: (a) because the relationship between the water depth and the wave height produces an inherently unstable wave form; (b) because of the interaction between the waves and a current; (c) because of the presence of a natural or artificial barrier (such as a breakwater).

Consider the case of a coast with the seabed and beach consisting of sand. The bed slope will usually be fairly shallow (say $0.01 < \beta < 0.03$). Waves will therefore tend to start to break at some distance offshore of the beach or shoreline (that is, the beach contour line that corresponds to the still water level; see Fig. 8.15). At this initial break point the wave will be of height H_b and at an angle α_b to the beach line. The region between

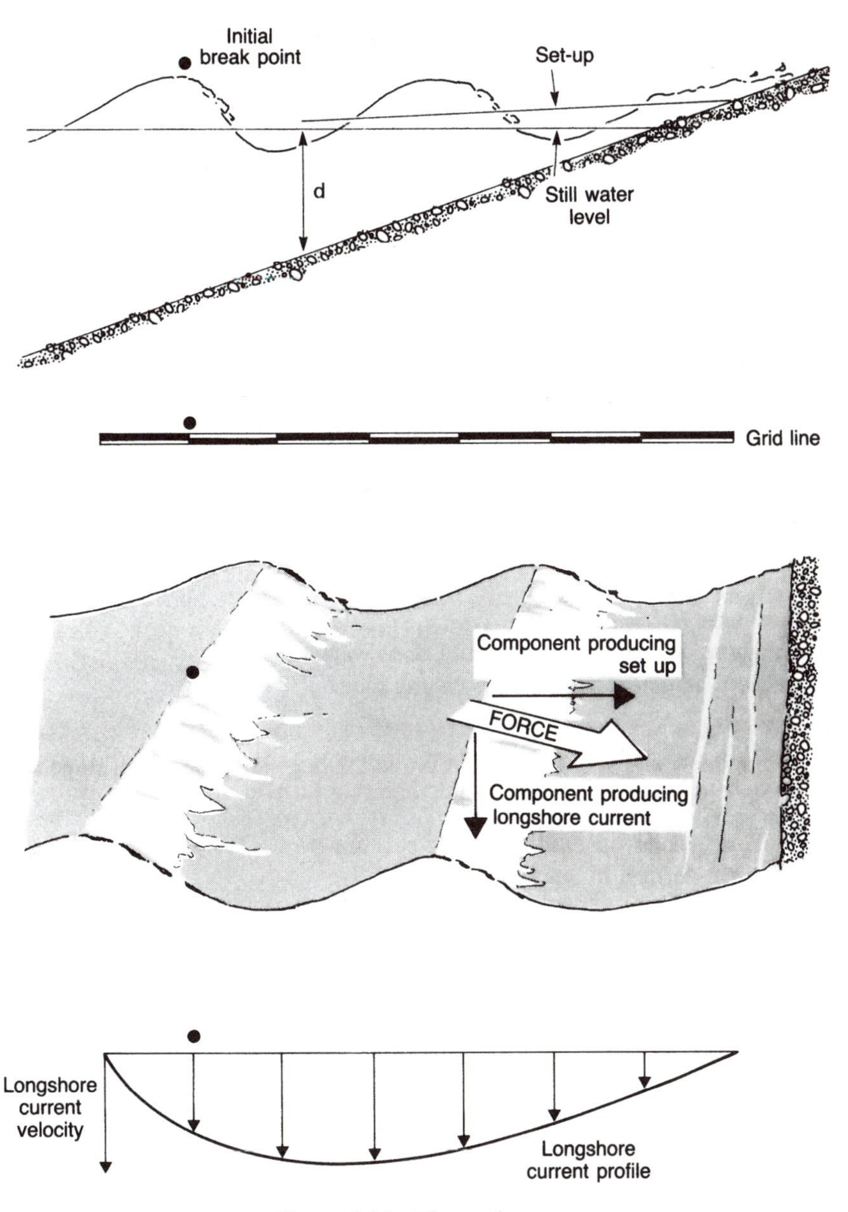

Figure 8.15 The surf zone.

this initial point and the beach is known as the **surf zone.** In this region the height of an individual wave is largely controlled by the water depth. The wave height will progressively attenuate as it advances towards the beach, and the characteristic foam or surf formation will be visible on the wave front. The mechanics of this progressive breaking are very complex. A brief summary is as follows:

1. Turbulence and aeration are produced.
2. Significant rates of change are induced in the momentum of the elements of fluid that constitute the wave. This produces a momentum force, which may be resolved into two components (Fig. 8.15). The component that lies parallel to the shoreline is the cause of a corresponding 'longshore current'. The component that is perpendicular to the shoreline produces an increase in the depth of water above the still water level, and this is usually called the **set-up**.
3. Energy is lost because of bed friction and because of the production of turbulence. The frictional losses are produced both by the oscillatory motion at the seabed due to the wave and by the unidirectional motion of the longshore current. The two motions are not completely independent, and their interaction has significant effects on the bed friction.

Wave breaking. There are two criteria that determine when a wave will break. The first is a limit to wave steepness, and the second is a limit on the ratio of wave height to water depth. Theoretical limits can be derived from 'solitary wave' theory. A 'solitary wave' is a single wave with a crest and no trough.

Such a wave was first observed by Russell in 1840, being produced by a barge on the Forth and Clyde canal. The two criteria are given by:

1. Steepness $H/L < 1/7$
   This normally limits the height of deep water waves.
2. Ratio of height to depth: the breaking index

$$\gamma = H/d = 0.78 \tag{8.34}$$

   In practice γ can vary from about 0.4 to 1.2 depending on beach slope and breaker type.

For random waves on mildly sloping beaches $\gamma \approx 0.6$ for H_s (the significant wave height defined in section 8.5).

Breaker types. Breaking waves may be classified as one of three types, as shown in Figure 8.16. The type can be approximately determined by the value of the surf similarity parameter (or Iribarren no.)

$$\xi_b = \tan\beta/\sqrt{H_b/L_b} \tag{8.35}$$

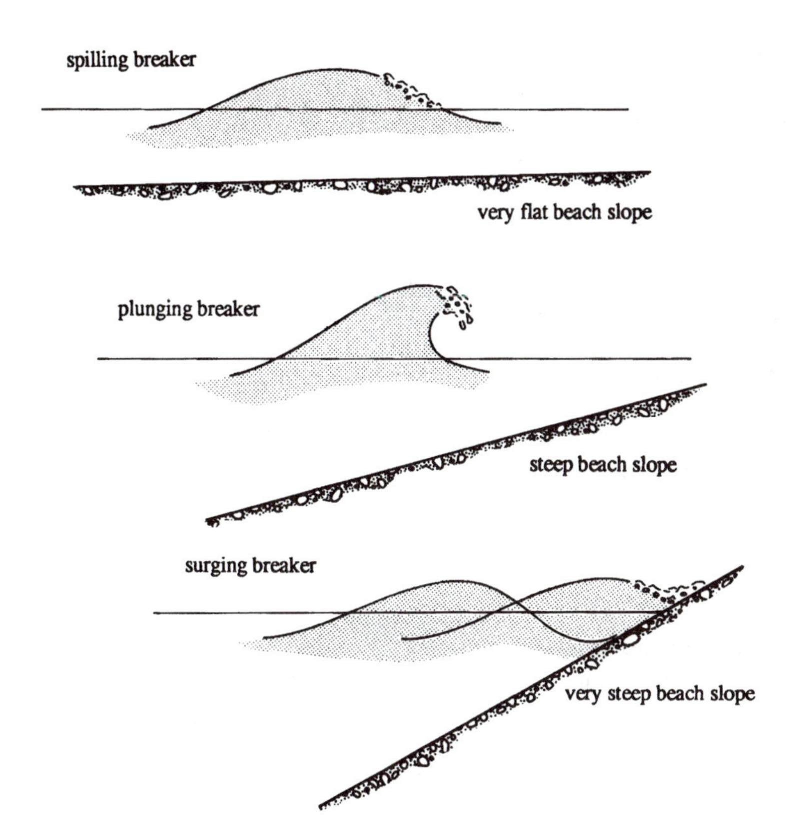

Figure 8.16 Principal types of breaking waves.

where $\tan \beta$ = beach slope, and for

spilling breaker	$\xi_b < 0.4$
plunging breaker	$0.4 < \xi_b < 2.0$
surging breaker	$\xi_b > 2.0$

Battjes found from real data that

$$\gamma \approx \xi_0^{0.17} + 0.08 \quad \text{for} \quad 0.05 < \xi < 2 \tag{8.36}$$

Further details may be found in Horikawa (1988) and Fredsoe and Deigaard (1992).

Radiation stress (momentum flux) theory

Wave set-up (and set-down) and longshore currents are all outcomes of the theory of radiation stress. This is defined as the **excess flow of momentum**

due to the presence of waves (with units of force/unit length). It arises from the orbital motion of individual water particles in the waves. These particle motions produce a net force in the direction of propagation (S_{XX}) and a net force at right angles to the direction of propagation (S_{YY}). The original theory was developed by Longuet-Higgins and Stewart (1964). Its application to longshore currents was subsequently developed by Longuet-Higgins (1970). The interested reader is strongly recommended to refer to these papers, which are both scientifically elegant and presented in a readable style. Further details may also be found in Horikawa (1978) and Komar (1976). Here only a summary of the main results is presented.

The radiation stresses were derived from the linear wave theory equations by integrating the dynamic pressure over the total depth under a wave and over a wave period, and subtracting from this the integral static pressure below the still water depth. Thus, using the notation of Figure 8.3,

$$S_{XX} = \overline{\int_{-d}^{\eta} (p + \rho u^2)\, \mathrm{d}z} - \int_{-d}^{0} p\, \mathrm{d}z \tag{8.37}$$

The first integral is the mean value of the integrand over a wave period, where u is the horizontal component of orbital velocity in the x direction. After considerable manipulation it may be shown that

$$S_{XX} = E\left(\frac{2kd}{\sinh 2kd} + \frac{1}{2}\right) \tag{8.38}$$

Similarly

$$S_{YY} = \overline{\int_{-d}^{\eta} (p + \rho v^2)\, \mathrm{d}z} - \int_{-d}^{0} p\, \mathrm{d}z$$

where v is the horizontal component of orbital velocity in the y direction.

For waves travelling in the x direction $v = 0$, and

$$S_{YY} = E\left(\frac{kd}{\sinh 2kd}\right) \tag{8.39}$$

In deep water

$$S_{XX} = \frac{1}{2}E \quad S_{YY} = 0$$

In shallow water

$$S_{XX} = \frac{3}{2}E \quad S_{YY} = \frac{1}{2}E$$

Thus both S_{XX} and S_{YY} increase in reducing water depths.

Wave set-down and set-up. The onshore momentum flux (that is, force) S_{XX} must be balanced by an equal and opposite force for equilibrium. This manifests itself as a **slope** in the mean still water level (given by $d\eta/dx$).

Consider the control volume shown in Figure 8.17, in which a set-up ($\overline{\eta}$) on the still water level exists, induced by wave action. The forces acting are the pressure forces, the reaction force on the bottom and the radiation stresses (all forces are wave period averaged). For equilibrium the net force in the x direction is zero. Hence

$$(F_{p_1} - F_{p_2}) + (S_{XX_1} - S_{XX_2}) - R_x = 0 \tag{8.40}$$

As

$$F_{p_2} = F_{p_1} + \frac{dF_p}{dx}\delta x$$

$$S_{XX_2} = S_{XX_1} + \frac{dS_{XX}}{dx}\delta x$$

then by substitution into (8.40)

$$\frac{dF_p}{dx}\delta x + \frac{dS_{XX}}{dx}\delta x = R_x \tag{8.41}$$

As

$$F_p = \frac{1}{2}\rho g(d+\overline{\eta})^2 \quad \text{(i.e. the hydrostatic pressure force)}$$

then

$$\frac{dF_p}{dx} = \frac{1}{2}\rho g\frac{d}{dx}(d+\overline{\eta})^2 = \rho g(d+\overline{\eta})\left(\frac{dd}{dx} + \frac{d\overline{\eta}}{dx}\right)$$

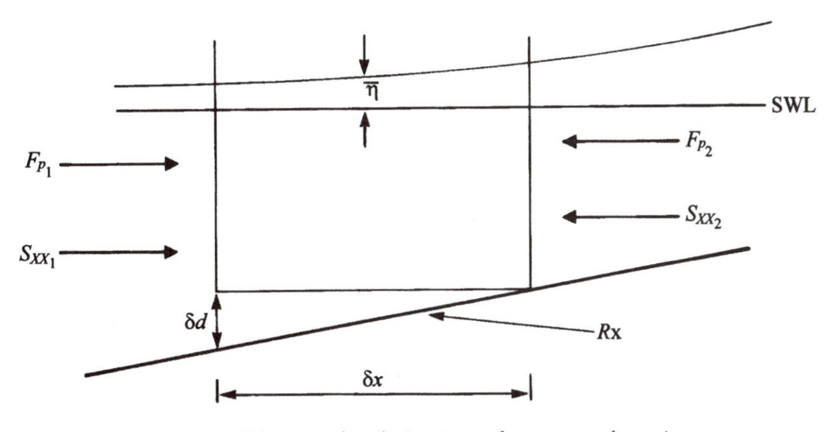

Figure 8.17 Diagram for derivation of wave set-down/set-up.

and as R_x for a mildly sloping bottom is due to bottom pressure,

$$
\begin{aligned}
R_x &= \bar{p}\delta d \\
&= \bar{p}\frac{\mathrm{d}d}{\mathrm{d}x}\delta x \\
&= \rho g(\mathrm{d}+\bar{\eta})\frac{\mathrm{d}d}{\mathrm{d}x}\delta x
\end{aligned}
$$

then substituting for F_p and R_x in (8.41)

$$\rho g(d+\bar{\eta})\left(\frac{\mathrm{d}d}{\mathrm{d}x}+\frac{\mathrm{d}\bar{\eta}}{\mathrm{d}x}\right)\delta x+\frac{\mathrm{d}S_{XX}}{\mathrm{d}x}\delta x=\rho g(d+\bar{\eta})\frac{\mathrm{d}d}{\mathrm{d}x}\delta x$$

and finally, simplifying we obtain

$$\frac{\mathrm{d}S_{XX}}{\mathrm{d}x}+\rho g(d+\bar{\eta})\frac{\mathrm{d}\bar{\eta}}{\mathrm{d}x}=0 \tag{8.42}$$

where $\bar{\eta}$ is the difference between the still water level and the mean water level in the presence of waves.

Outside the breaker zone, equation (8.42) (in which equation (8.38) is substituted for S_{XX} may be integrated to obtain

$$\bar{\eta}_\mathrm{d}=-\frac{1}{8}\frac{kH_\mathrm{b}^2}{\sinh(2kd)} \tag{8.43}$$

This is referred to as the **set-down** ($\bar{\eta}_\mathrm{d}$), and demonstrates that the mean water level decreases in shallower water.

Inside the breaker zone the momentum flux rapidly reduces as the wave height decreases. This manifests itself as a **set-up** ($\bar{\eta}_\mathrm{u}$) of the mean still water level. Making the assumption that inside the surf zone the broken wave height is controlled by depth such that

$$H=\gamma(\bar{\eta}+d) \tag{8.44}$$

where $\gamma \approx 0.8(\mathrm{cf.}H_\mathrm{b}/d_\mathrm{b}=1/1.28=0.78)$, then combining equations (8.38), (8.42) and (8.44) leads to the result

$$\frac{\mathrm{d}\bar{\eta}}{\mathrm{d}x}=\left(\frac{1}{1+\frac{8}{3\gamma^2}}\right)\tan\beta$$

where β is the beach slope angle. Thus for a uniform beach slope it may be shown that

$$\bar{\eta}_u=\left(\frac{1}{1+\frac{8}{3\gamma^2}}\right)(d_\mathrm{b}-d)+\bar{\eta}_{\mathrm{d}_\mathrm{b}} \tag{8.45}$$

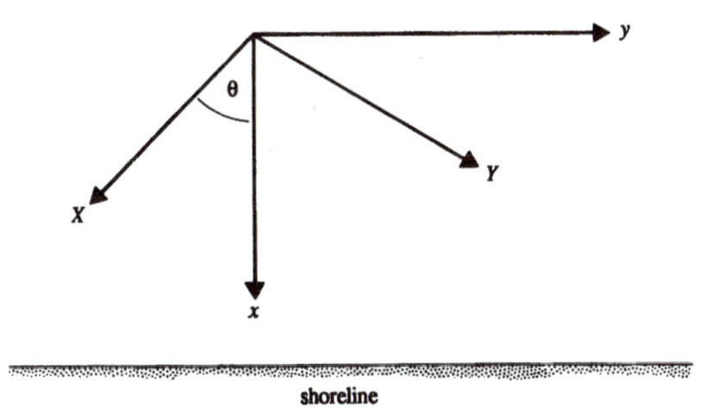

Figure 8.18 Relationships between principal axes and shoreline axes.

demonstrating that inside the surf zone there is a rapid increase in the mean water level.

Radiation stress components for oblique waves. The radiation stresses S_{XX}, S_{YY} are, in fact, principal stresses. Utilizing the theory of principal stresses, shear stresses will also act on any plane at an angle to the principal axes. This is illustrated in Figure 8.18 for the case of oblique wave incidence to a coastline.

The relationships between the principal radiation stresses and the direct and shear components in the x, y directions are

$$S_{xx} = S_{XX}\cos^2\theta + S_{YY}\sin^2\theta = \frac{1}{2}E\left[(1+G)\cos^2\theta + G\right]$$

$$S_{yy} = S_{XX}\sin^2\theta + S_{YY}\cos^2\theta = \frac{1}{2}E\left[(1+G)\sin^2\theta + G\right]$$

$$S_{xy} = S_{XX}\sin\theta\cos\theta - S_{YY}\sin\theta\cos\theta = \frac{1}{2}E\left[(1+G)\sin\theta\cos\theta\right]$$

Where

$$G = 2kd/\sinh(2kd)$$

Longshore currents

Radiation stress theory has also been successfully used to explain the presence of longshore currents. The original theory is eloquently explained by Longuet-Higgins (1970). Subsequently Komar (1976), as a result of his own theoretical and field investigations, developed the theory further and presented revised equations. All of the foregoing is succinctly summarized in Hardisty (1990). Here a summary of the main principles is given together with a statement of the main equations.

An expression for the mean wave **period-averaged** longshore velocity ($\overline{v}_l$) was derived from the following considerations. First, outside the surf zone the energy flux towards the coast (P_x) of a wave travelling at an oblique angle (α) is constant and given by

$$P_x = EC_G \cos\alpha \quad \text{(cf. equation (8.8))} \tag{8.46}$$

Second, the radiation stress (S_{xy}) that constitutes the flux of y momentum parallel to the shoreline across a plane $x =$ constant is given by

$$\begin{aligned} S_{xy} &= S_{XX}\sin\alpha\cos\alpha - S_{YY}\sin\alpha\cos\alpha \\ &= E\left(\frac{1}{2} + \frac{kd}{\sinh 2kd}\right)\cos\alpha\sin\alpha \\ &= E\left(\frac{C_G}{c}\right)\cos\alpha\sin\alpha \end{aligned} \tag{8.47}$$

Hence combining (8.46) and (8.47),

$$S_{xy} = P_x\left(\frac{\sin\alpha}{c}\right) \quad \text{(outside the surf zone)}$$

Inside the surf zone, the net thrust F_y per unit area exerted by the waves in the water is given by

$$F_y = \frac{-\partial S_{xy}}{\partial x} \tag{8.48}$$

Substituting for S_{xy} from (8.48) and taking conditions at the wave break point (at which $C_G = c = \sqrt{gd_b}$, $H_b/d_b = \gamma$ and $u_m = \gamma/2\sqrt{gd_b}$), Longuet-Higgins derived an expression for F_y:

$$F_y = \frac{5}{4}\rho u_{mb}^2 \tan\beta\sin\alpha \tag{8.49}$$

Finally, by assuming that this thrust was balanced by frictional resistance in the longshore (y) direction he derived an expression for the mean longshore velocity $\overline{v}_l$:

$$\overline{v}_l = \frac{5\pi}{8C}u_{mb}\tan\beta\,\sin\alpha_b \tag{8.50}$$

where C was a friction coefficient.

Subsequently Komar found from an analysis of field data that $\tan\beta/C$ was effectively constant, and he therefore proposed a modified formula:

$$\overline{v}_l = 2.7u_b\sin\alpha_b\cos\alpha_b \tag{8.51}$$

in which the $\cos\alpha_b$ term has been added to cater for larger angles of incidence (Longuet-Higgins assumed α small and therefore $\cos\alpha \to 1$).

The distribution of longshore currents within the surf zone was also studied by both Longuet-Higgins and Komar. The distribution depends upon the assumptions made concerning the horizontal eddy coefficient, which has the effect of transferring horizontal momentum across the surf zone. Komar (1976) presents a set of equations to predict the distribution.

Example 8.2 Wave set-down, set-up and longshore velocity

(a) A deep water wave of period 8.5 s and height 5 m is approaching the shoreline normally. Assuming the seabed contours are parallel, estimate the wave set-down at the breakpoint and the wave set-up at the shoreline.

(b) If the same wave has a deep water approach angle of 45°, estimate the mean longshore current in the surf zone.

Solution

(a) The first stage of the solution is analogous to Example 8.1, except that no refraction occurs; thus at the break point we obtain

d/L_0	d (m)	c/C_0	c (m/s)	K_S	α (degrees)	K_R	H/H_0	H (m)	d_B (S)
0.06	6.4	0.56	7.5	1.00	0	1.00	1.0	5	6.4

The set-down may now be calculated from equation (8.43), that is:

$$\bar{\eta}_d = -\frac{1}{8}\frac{H_b^2 k}{\sinh(2kd)}$$

as

$$k = 2\pi/L = 2\pi/(112.8 \times 0.56) = 0.099$$

and

$$2kd = 2 \times 0.099 \times 6.4 = 1.267$$

$$\bar{\eta}_d = -\frac{1}{8}\frac{5^2 \times 0.099}{\sinh(1.267)}$$

then

$$\bar{\eta}_d = -0.19\,\text{m}$$

The set-up may be calculated from equation (8.45):

$$\overline{\eta}_u = \left(\frac{1}{1+\frac{8}{3\gamma^2}} \right) (d_b - d) + \overline{\eta}_d$$

At the shoreline $d = 0$ and taking $\gamma = 0.78$ then

$$\overline{\eta}_u = \frac{1}{1+8/(3+0.78^2)} \times 6.4 - 0.19$$

$$\overline{\eta}_u = 1.0\,\text{m}$$

Thus it may be appreciated that set-down is quite small and the set-up much larger. In general, wave set-down is less than 5% of the breaking depth, and wave set-up is about 20 of the breaking depth. For a real sea, composed of varying wave heights and periods, the wave set-up will vary along a shoreline at any moment. This can produce the phenomenon referred to as **surf beats**. Wave set-up also contributes to the overtopping of sea defence structures during storm conditions, and may thus cause coastal flooding.

(b) Here the same wave as in Example 8.1 has been used. Recalling that at the wave breakpoint $\alpha_b = 22°$ and $d_b = 5.7\,\text{m}$, then equation (8.51) may be used to estimate $\overline{v}_l$:

$$\overline{v}_l = 2.7 u_{mb} \sin\alpha_b \cos\alpha_b$$

Recalling that

$$u_{mb} = \frac{\gamma}{2}\sqrt{g d_b}$$

then

$$u_{mb} = \frac{0.78}{2}\sqrt{9.81 \times 5.7}$$
$$= 2.92\,\text{m/s}$$

Hence

$$\overline{v}_l = 2.7 \times 2.92 \sin(22)\cos(22)$$
$$= 2.74\,\text{m/s}$$

Infragravity waves

Waves often travel in groups, as shown in Figure 8.19: hence under large waves the set-down is larger than under small waves. This results in a second order wave – the **bound long wave**. The bound long waves travel with the wave groups with a celerity corresponding to the group celerity

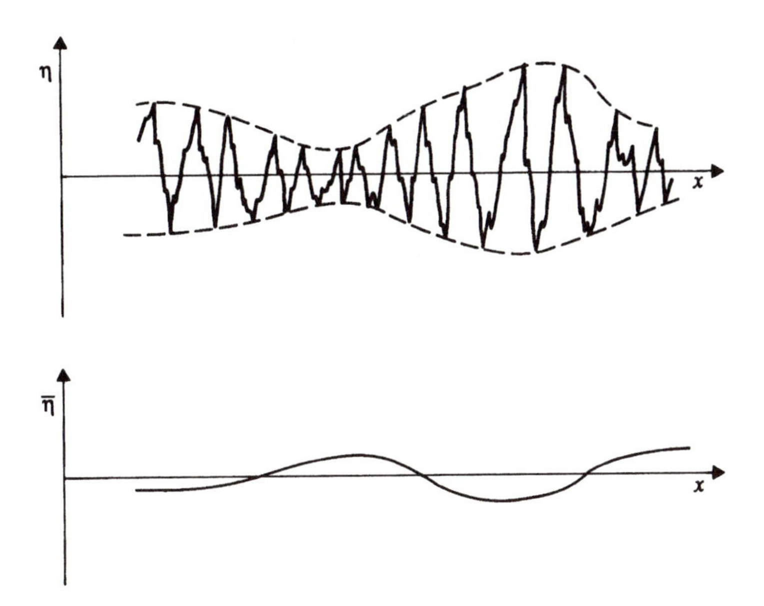

Figure 8.19 The wave groups and the associated mean water level.

of the short waves, and thus are refracted with the short waves. In shallow water the height of the bound long waves will increase quite dramatically because of shoaling.

In the surf zone the short waves lose height and energy, and can no longer balance the bound long waves, which are therefore released as free long waves. The free long waves are substantially reflected from the beach, and either progress back out to sea (for normally incident short waves), termed the **leaky mode**, or refract and turn back to the shore to be re-reflected, termed the **trapped mode**. The trapped free long waves then form 3D edge waves with a wave height that decreases with distance from the shore.

Another mechanism for generating long waves in the surf zone is variation in set-up caused by breaking wave groups. Surf beat is the variation of set-up on the shoreline, and may be caused by a combination of free long waves in the surf zone, generated at sea as bound long waves, and free long waves generated in the surf zone because of variations in set-up.

Cell circulation is the term used to describe currents within the surf zone that are not parallel to the shore. The existence of cell circulations is evidenced by rip currents (a common hazard for swimmers). Rip currents are a seaward return flow of water concentrated at points along the beach. They are caused by a longshore variation of wave height, and hence set-up, which provides the necessary hydraulic head to drive them. The longshore variation of wave height can be caused either by refraction effects or by the presence of edge waves. Under the latter circumstances a regular pattern of cell circulations and rip currents will exist, and beach cusps may be formed.

The interested reader is referred to Komar (1976) and Huntley *et al.* (1993) for further details.

8.5 Analysis of wave records: short-term wave statistics

Wind-generated waves are complex, incorporating many superimposed components of wave periods, heights and directions. If the sea state is recorded in a storm zone, then the resulting wave trace appears to consist of random periodic fluctuations. To find order in this apparent chaos, considerable research and measurement has been, and is being, undertaken.

Wave records are available for certain locations. These are normally gathered either by shipborne wave recorders (for fixed locations) or by wave rider buoys (which may be placed at specific sites of interest). These records generally consist of a wave trace for a short period (typically 20 min), recorded at fixed intervals (normally 3 h) and sampled at two readings per second (2 Hz). In this way, the typical sea state may be inferred without the necessity for continuous monitoring. An example wave trace is shown in Figure 8.20(a). (*Note*: this was recorded in shallow water.)

Short-term wave statistics

Using such wave trace records, two types of analysis may be performed. The first type is referred to as **time domain analysis** and the second as **frequency domain analysis**. Both methods assume a state of stationarity (that is, the sea state does not vary with time).

Time domain analysis. For a given wave record (for example, a 20 min record representing a 3 h period), the following parameters may be directly derived in the time domain (refer to Fig. 8.21) using either upcrossing or downcrossing analysis:

1. H_z (mean height between zero upward (or downward) crossings);
2. T_z (mean period between zero upward (or downward) crossings);
3. H_c (mean height between wave crests);
4. T_c (mean period between wave crests);
5. H_{max} (maximum difference between adjacent crest and trough);
6. H_{rms} (root-mean-square wave height);
7. $H_{1/3}$ (commonly referred to as H_s mean height of the highest one-third of the waves);
8. $H_{1/10}$ (mean height of the highest one-tenth of the waves).

Note that T_{max}, $T_{1/10}$ are the periods for the corresponding wave heights.

Based on previous experience, wave record analysis is greatly simplified if some assumptions are made regarding the probability distribution of wave

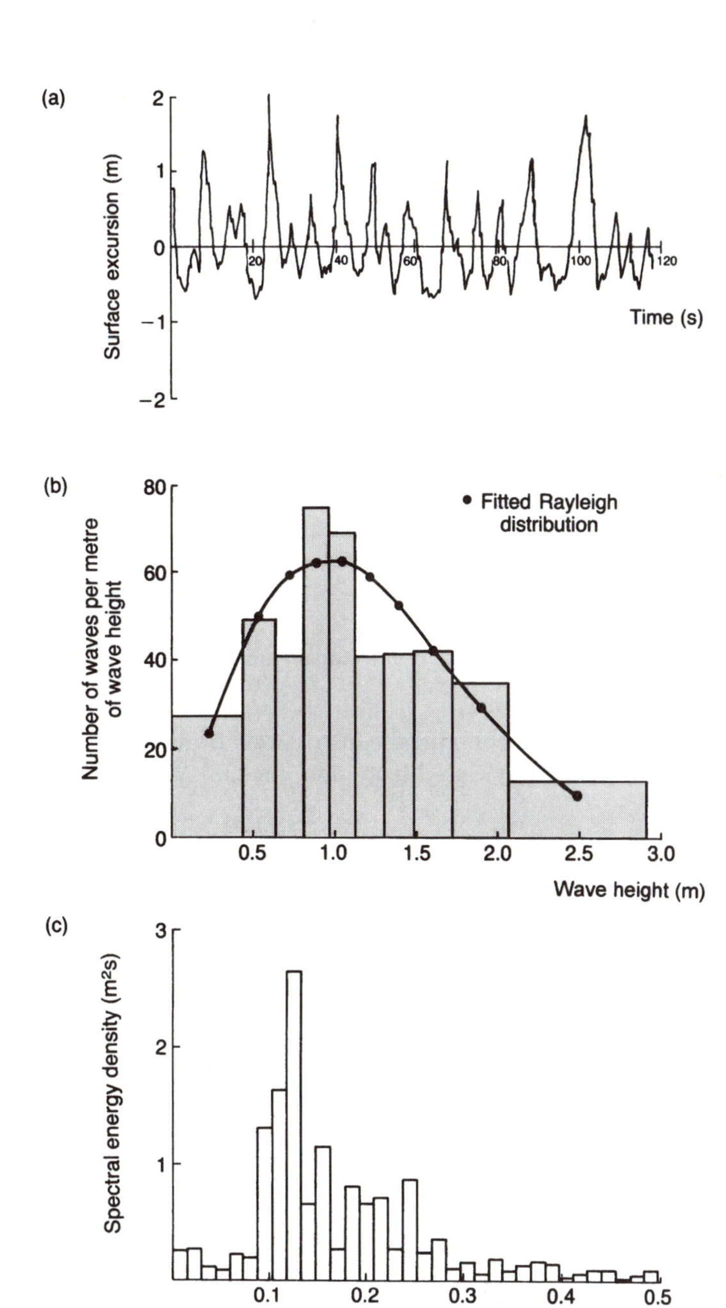

Figure 8.20 Analysis of a wave record.

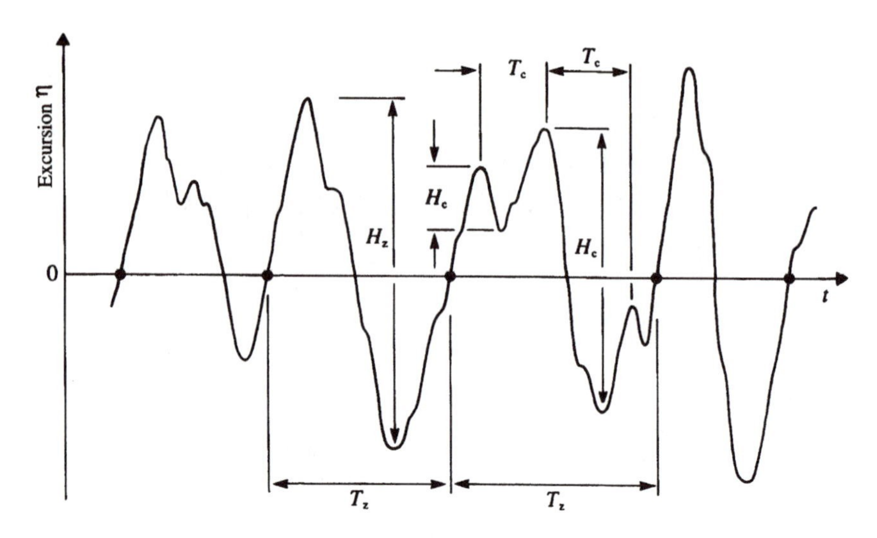

Figure 8.21 Time domain analysis.

heights and periods. For example, the distribution of wave heights is often assumed to follow the Rayleigh distribution: thus

$$P(h \geqslant H) = \exp[-2(H/H_s)^2] \tag{8.52}$$

where $P(h \geqslant H)$ is the probability that the wave height h will equal or exceed the given value H.

The corresponding probability density function $f(h)$ is given by

$$f(h) = (2h/H_{rms}^2)\exp[-(h/H_{rms})^2] \tag{8.53}$$

(*Note*: for a full discussion of probability density functions refer to Chapter 10.)

For a wave record of N waves, taking

$$P(h \geqslant H) = i/N$$

where i is the rank number and

$$i = \begin{cases} 1 & \text{for the largest wave} \\ N & \text{for the smallest wave} \end{cases}$$

then rearranging (8.52) and substituting for P gives

$$H = H_s\left[\frac{1}{2}\ln(N/i)\right]^{1/2}$$

For the case of $i = 1$, corresponding to $H = H_{max}$,

$$H_{max} = H_s \left(\frac{1}{2} \ln N \right)^{1/2}$$

However, H_{max} itself varies quite dramatically from one wave record to the next, and needs to be treated as a statistical quantity. Hence it may be shown that the most probable value of H_m is

$$H_{max} = H_s \left(\frac{1}{2} \ln N \right)^{1/2} \qquad \text{as above} \tag{8.54a}$$

and the mean value of H_{max} is

$$H_{max} = \frac{H_s}{\sqrt{2}} \left((\ln N^{1/2}) + 0.2886 (\ln N)^{1/2} \right) \tag{8.54b}$$

Other useful results that have been derived include:

$$H_{rms} \approx 1.13 H_z$$
$$H_s \simeq 2^{1/2} H_{rms} \simeq 1.414 H_{rms}$$
$$H_{1/10} \simeq 1.27 H_s \simeq 1.8 H_{rms}$$
$$H_{1/100} \simeq 1.67 H_s \simeq 2.36 H_{rms}$$
$$H_{max} \simeq 1.6 H_s \quad \text{(for a typical 20 min wave trace)}$$

Thus, if the value of H_{rms} is calculated from the record, the values of H_s etc. may easily be estimated.

The Rayleigh distribution was originally derived by Lord Rayleigh in the late 19th century for sound waves. It is commonly assumed to apply to wind waves and swell mixtures, and gives a good approximation to most sea states. However, the Rayleigh distribution is theoretically only a good fit to sine waves with a small range of periods with varying amplitudes and phases. This is more characteristic of swell waves than of storm waves. To determine what type of distribution is applicable, the parameter ε, known as the spectral width, may be calculated:

$$\varepsilon^2 = 1 - \left(\frac{T_c}{T_z} \right)^2$$

For the Rayleigh distribution (that is, a small range of periods) $T_c \simeq T_z$: hence $\varepsilon \to 0$. For a typical storm sea (containing many frequencies) the period between adjacent crests is much smaller than the period between zero upward crossings, and hence $\varepsilon \to 1$.

Actual measurements of swell and storm waves as given by Silvester (1974) are as follows:

	Swell waves	Storm waves
ε	$\simeq 0.3$	$\simeq 0.6$-0.8
H_s	$\simeq 1.42H_{rms}$	$\simeq 1.48H_{rms}$
$H_{1/10}$	$\simeq 1.8H_{rms}$	$\simeq 2.0H_{rms}$

Figure 8.20(b) illustrates the histogram of wave heights (derived from the wave trace shown in Figure 8.20(a)) and shows the fitted Rayleigh distribution. As a matter of interest, these data were recorded in very shallow water, for which the Rayleigh distribution was not expected to be a good fit. Applying a statistical goodness of fit criterion, this proved not to be the case. Further details may be found in Chadwick (1989).

Time domain analysis has traditionally been carried out using analogue data. A rapid method was developed by Tucker to find H_{rms}, from which other wave parameters can be derived by assuming a Rayleigh distribution. More recently, digital data has become available, and Goda (1985) gives details of how to derive time domain parameters directly.

Example 8.3

Using the time series data given in Table 8.3:

Table 8.3 Wave heights and periods.

Wave number	Wave height, H (m)	Wave period, T (s)	Wave number	Wave height, H (m)	Wave period, T (s)
1	0.54	4.2	11	1.03	6.1
2	2.05	8.0	12	1.95	8.0
3	4.52	6.9	13	1.97	7.6
4	2.58	11.9	14	1.62	7.0
5	3.20	7.3	15	4.08	8.2
6	1.87	5.4	16	4.89	8.0
7	1.90	4.4	17	2.43	9.0
8	1.00	5.2	18	2.83	9.2
9	2.05	6.3	19	2.94	7.9
10	2.37	4.3	20	2.23	5.3
			21	2.98	6.9

(a) Determine H_{max}, T_{max}, H_s, T_s, H_z, T_z.
(b) Plot a histogram of the wave heights using a class interval of 1 m.
(c) Determine H_{max}, H_s and H_{rms} from H_z, assuming a Rayleigh distribution.
(d) Calculate the value of $f(h)$ at the centre of each class interval, and hence superimpose the pdf on the histogram (assume that the scale equivalence is $f(h) \equiv n/N\Delta h$).
(e) Suggest reasons for the anomalies between the results in (a) and (c).

Solution

(a) From Table 8.3:
16th wave gives $H_{max} = 4.89$, $T_{max} = 8.0$
For $\boldsymbol{H_s}$, 16th, 3rd, 15th, 5th, 21st, 19th, 18th waves (21/3 = 7 waves) are the highest 1/3 of the waves.
Average to obtain

$$H_s = 3.6\,\text{m} \quad T_s = 7.8\,\text{s}$$

For H_z, T_z average all 21:

$$H_z = 2.4\,\text{m} \quad T_z = 7.0\,\text{s}$$

(b)

Class interval of wave height (m)	No. of waves
0–1	1
1–2	7
2–3	9
3–4	1
4–5	3

The histogram is shown in Figure 8.22.

(c) From part (a):

$$H_z = 2.4\,\text{m}$$

$$\therefore H_{rms} = 1.13 \times 2.4 = 2.71\,\text{m}$$

$$\therefore H_s = 1.414 H_{rms} = 3.83\,\text{m}$$

$$\therefore H_{max} = H_s(1/2 \ln N)^{1/2}$$

$$= 3.83(1/2 \ln 21)^{1/2}$$

$$= 4.73\,\text{m}$$

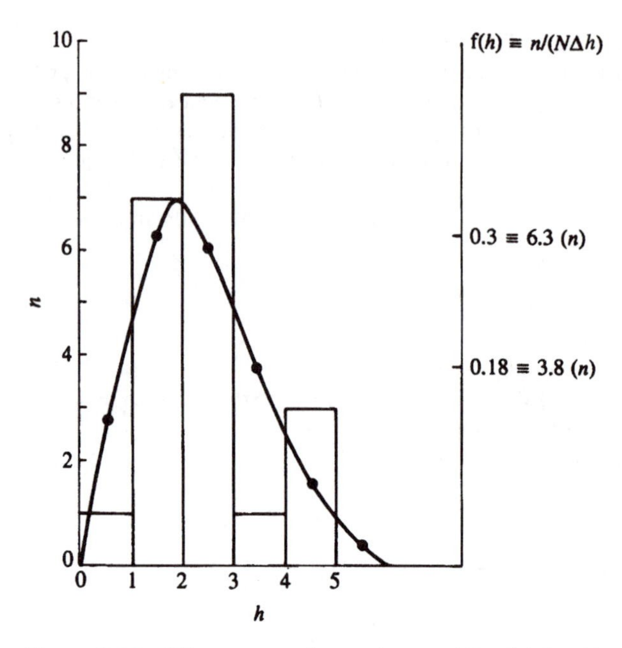

Figure 8.22 Histogram and superimposed Rayleigh pdf.

(d) Using (8.53):
$f(h) = (2h/H_{rms}^2)\exp[-(h/H_{rms})^2]$

h	$f(h)$	$n \equiv f(h)N\Delta h$
0.5	0.13	2.8
1.5	0.3	6.3
2.5	0.29	6.1
3.5	0.18	3.8
4.5	0.078	1.6
5.5	0.0022	0.05

These results are plotted in Figure 8.22.

(e) The Rayleigh distribution is apparently not a good fit. (For statistical reliability we require $N > 100$.)

Frequency domain analysis. The wave trace shown in Figure 8.20(a) can also be analysed in the frequency domain. This is made possible by application of the Fourier series representation. In essence, any sea state can be described mathematically as being composed of an infinite series of sine waves of varying amplitude and frequency. Thus the surface excursion $\eta(t)$ at any time (defined in Figure 8.3) may be represented as

$$\eta(t) = \sum_{n=1}^{\infty} [a_n \cos \omega nt + b_n \sin \omega nt] \tag{8.55}$$

where ω is the frequency ($2\pi/T$) and $t=0$ to $t=\mathrm{T}$; a_n and b_n are amplitudes.

Equation (8.55) may be equivalently written as

$$\eta(t) = \sum_{n=1}^{\infty} c_n \cos(\omega nt + \phi_n)$$

where

$$c_n^2 = a_n^2 + b_n^2$$
$$\tan \phi_n = -b_n/a_n$$

(This is shown graphically in Figure 8.23.)

Noting that the equation for wave energy is $E = \rho g H^2/8$, then wave energy is proportional to (amplitude)$^2/2$ (with units of m^2). Thus the spectral energy density curve $S(f)$ (with units of $\mathrm{m}^2\mathrm{s}$) may be found from

$$S(f)\Delta f = \sum_{f}^{f+\Delta f} \frac{1}{2} c_n^2 \tag{8.56}$$

To accomplish this, values of c_n must be found from equation (8.56). The technique commonly used for doing this is termed the fast Fourier transform (FFT). A description of the FFT techniques is beyond the scope of this chapter, but the reader is directed to Carter *et al.* (1986) for a description of its application to sea waves, and to Broch (1981) for details of the principles of digital frequency analysis.

Suffice to say here that a given wave trace record may be analysed using FFT techniques to produce the spectral density histogram. An example is shown in Figure 8.20(c). Having obtained the spectral density histogram, then the **frequency domain wave parameters** may be found from the following equations:

$$H_{m0} = 4(m_0)^{0.5}$$
$$T_{m01} = m_0/m_1$$
$$T_{m02} = (m_0/m_2)^{0.5}$$
$$T_{\mathrm{p}} = 1/f_{\mathrm{p}} \text{ where } f_{\mathrm{p}} = \text{frequency at the maximum value of } S(f)$$

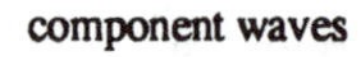

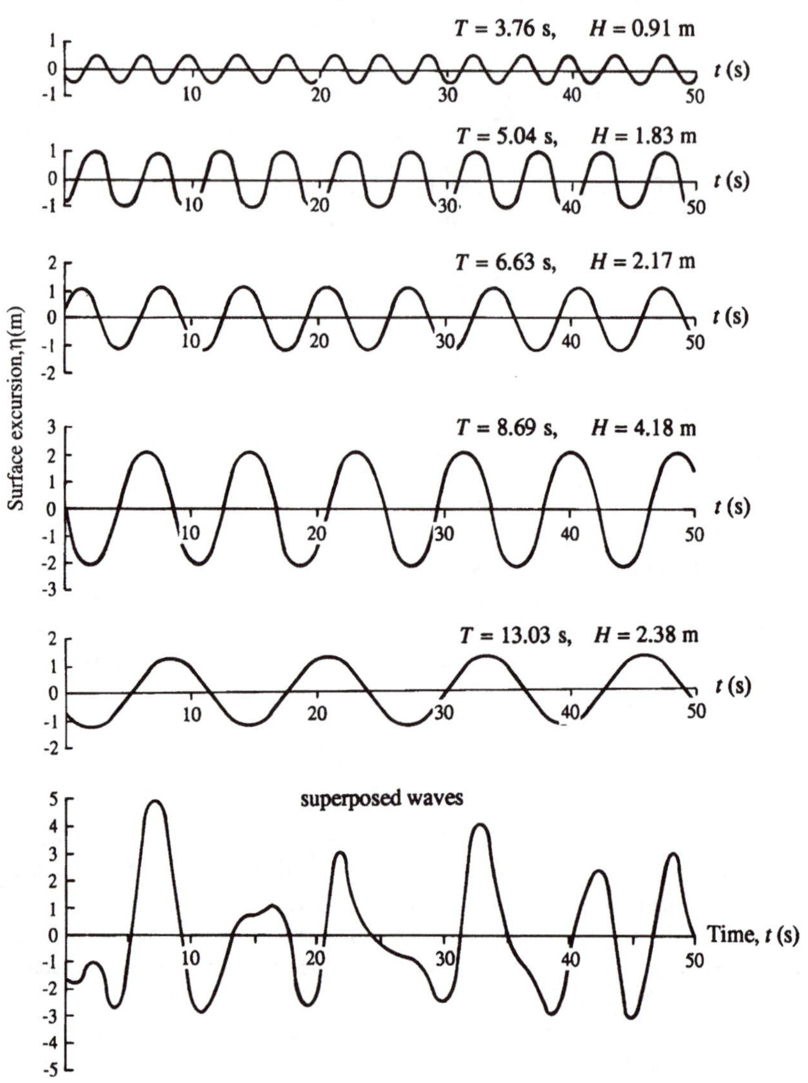

Figure 8.23 Graphical representation of a Fourier series.

$$Q_p = \left(\frac{2}{m_0^2}\right)\int_0^{\infty} f\,S(f)^2\mathrm{d}f \qquad \text{(spectral peakedness)}$$

$$\varepsilon = \left[1 - m_2^2/(\mathrm{m}_0/\mathrm{m}_4)\right]^{0.5} \qquad \text{(spectral width)}$$

$$\sigma^2 = \int_0^{\infty} S(f)\mathrm{d}f = m_0 \qquad \text{(spectral variance)}$$

where

$$m_n = \int_0^\infty S(f) f^n \mathrm{d}f \qquad (n\text{th spectral moment})$$

Frequency domain wave parameters do not have direct equivalent parameters in the time domain. However, as a useful guide, the following parameters have been found to be roughly equivalent:

Time domain parameter	Equivalent frequency domain parameter
H_s	H_{m_0} (approximate)
η_{rms}	$\sqrt{m_0}$ (exact)
T_z	T_{m_02} (approximate)
T_s	$0.95T_p$ (approximate)

Because of the proliferation of wave parameters in both the time and frequency domains, there is confusion in the literature as to the precise definition of some of those parameters. For example, H_{m_0} and H_s are often confused. For this reason (and others) a standard set of sea state parameters was proposed by the International Association for Hydraulic Research. Details may be found in Darras (1987).

Directional wave spectra

The sea state observed at any particular point consists of component waves not only of various heights and periods but also from different directions. Therefore a complete description of the sea state needs to include directional information. Mathematically this may be expressed as

$$\eta(x, y, t) = \sum_{n=1}^{\infty} \sum_{m=0}^{2\pi} a_{n,m} \cos(k_n x \cos\theta_m + k_n y \sin\theta_m - 2\pi f_n t + \phi_{n,m}) \quad (8.57)$$

where a is amplitude, k is wave number $= 2\pi/L$, f is frequency, θ is wave direction, ϕ is phase angle, n is frequency counter, and m is direction counter.

Equivalently, extending the concept of spectral density $S(f)$ to include direction, the directional spectral density $S(f, \theta)$ can be defined as

$$S(f, \theta) = S(f)\, G(f, \theta)$$

where

$$\sum_{f}^{f+\Delta f}\sum_{\theta}^{\theta+\Delta\theta}\frac{1}{2}a_n^2 = S(f,\theta)\Delta f\Delta\theta \tag{8.58}$$

and $G(f, \theta)$ is the directional spreading function, where

$$\int_{-\pi}^{\pi} G(f,\theta)\mathrm{d}\theta = 1$$

An idealized directional spectrum is shown in Figure 8.24. Recently direct measurements of directional spectra have been measured by arrays of wave recorders of various forms (see for example Chadwick *et al.* 1995a,b). The analysis of such records is complex (refer to Goda 1985, or Dean and Dalrymple, 1991), and the analysis techniques do not always work in real sea states, particularly when wave reflections are present. A parametric form

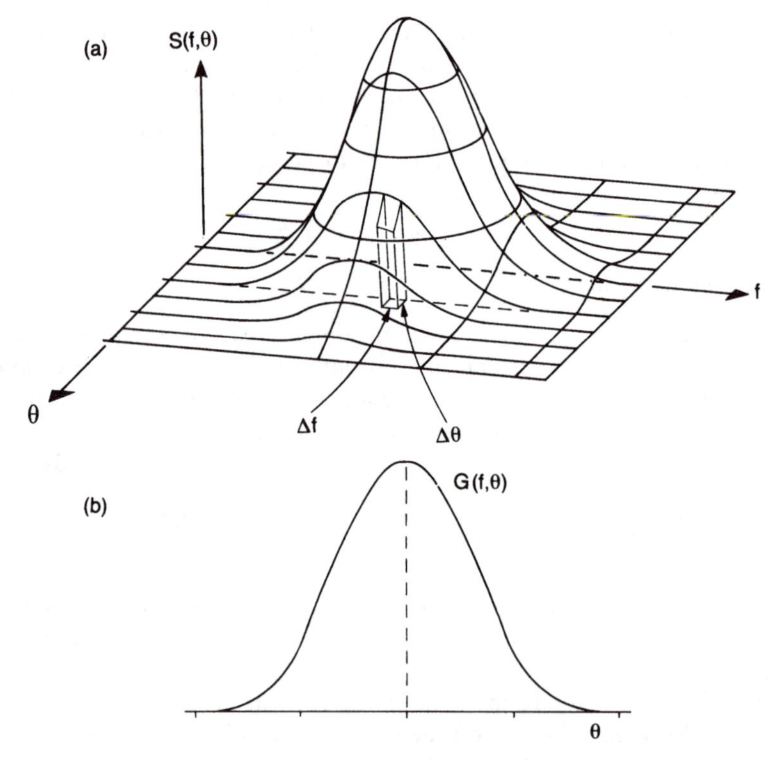

Figure 8.24 Idealized directional spectrum: (a) directional spectral density; (b) directional spreading function.

of the directional spreading function (as given in Goda, 1985) and due to Mitsuyasu is

$$G(f, \theta) = N \cos^{2s}\left(\frac{\theta}{2}\right) \tag{8.59}$$

where N is a normalizing factor, given by

$$N = \frac{1}{\int_{-\pi}^{\pi} \cos^{2s}\left(\frac{\theta}{2}\right)}$$

and $s = s_m(f/f_p)^{\mu}$; $s_m = 10$ for wind waves, 25 to 75 for swell waves; $\mu = -2.5$ for $f \geq f_p$, 5 for $f < f_p$; and f_p is the peak frequency.

Shoaling and refraction of directional wave spectra. In section 8.3, discussion of shoaling and refraction was restricted to considering waves of single period, height and direction. However, as already shown in this section, a real sea state is more realistically represented as being composed of a large number of components of differing periods, heights and directions. Therefore, in determining an inshore sea state due account should be taken of the offshore directional spectrum.

This can be achieved in a relatively straightforward way, provided the principle of linear superposition can be applied. This implies that non-linear processes such as seabed friction and higher-order wave theories are excluded.

The principle of the method is to carry out a refraction and shoaling analysis for every individual component frequency and direction and then to sum the resultant inshore energies at the new inshore directions at each frequency and hence assemble an inshore directional spectrum.

Mathematically, this may be expressed in the following way. For each component frequency and direction

$$S(f, \theta_i) = S(f, \theta_o) K_r^2(f, \theta_o) K_s^2(f) \tag{8.60}$$

where subscript o refers to offshore and subscript i refers to inshore.

The inshore wave direction for each component is given by

$$\theta_i = \cos^{-1}\left(\frac{\cos\theta_o}{K_r^2(f, \theta_o)}\right) \tag{8.61}$$

Summing over all f, θ:

$$m_{oi} = \int_0^{\infty} \int_{\theta \min}^{\theta \max} S_i(f, \theta)\, d\theta\, df = \int_0^{\infty} \int_{\theta \min}^{\theta \max} S_o(f, \theta) K_r^2(f, \theta_o) K_s^2(f) d\theta\, df \tag{8.62}$$

Here it is convenient to use the same directional increments inshore as used offshore, and therefore it is necessary to reallocate the offshore energies to the relevant directional sector inshore. Goda (2000) presents a set of design charts for the effective refraction coefficient and predominant wave direction over parallel contours using the Betchneider–Mitsuyasu frequency spectrum and Mitsuyasu spreading function, which facilitate the ready application of the method described above.

Diffraction of directional wave spectra. It was also Goda (2000) who pioneered the use of directional spectra in the determination of wave diffraction. He defined the effective diffraction coefficient $(K_d)_{eff}$ as

$$(K_d)_{eff} = \left[\frac{1}{m_0}\int_0^{\infty}\int_{\theta\min}^{\theta\max} S(f,\theta)K_d^2(f,\theta)\mathrm{d}\theta\,\mathrm{d}f\right]^{1/2} \tag{8.63}$$

and hence constructed a new set of diffraction diagrams.

These diagrams show that the diffraction of a directional random sea state differs quite markedly from the case of a monochromatic sea. At the edge of the shadow zone for a semi-infinite barrier K_d is approximately 0.7 (cf. $K_d = 0.5$ for a monochromatic wave), and waves of greater height penetrate the shadow zone at equivalent points. For the case of a barrier gap, the wave height variations are smoothed compared with the monochromatic case, with smaller heights in the area of direct penetration and larger heights in the shadow regions.

Note also that there is a shift in the spectral peak wave period. This is because at any particular physical point in the diffraction zone the K_d value will vary with wavelength and frequency. These results have recently been verified by Briggs *et al.* (1995) by a physical model study in the CERC directional spectral wave basin.

8.6 Wave prediction from wind records

Storm waves

For a given wind speed, the waves produced will depend on the duration D and fetch F: the longer the fetch or duration, the bigger the waves produced. However, as the wind contains only a given amount of energy, the wave heights will approach some limiting value for particular values of fetch and duration (when the rate of transfer of energy to the waves equals the energy dissipation by wave breaking and friction). This sea state is referred to as the **fully arisen sea** (FAS). Measurements of the FAS for various wind speeds indicate that the wave energy spectrum is similar for all FASs, independent of location or wind speed.

Methods of predicting waves from wind records

Wind is recorded for meteorological purposes at many sites, and is therefore far more frequently recorded than are waves. Given that storm waves are dependent solely on wind speed, duration and fetch, the most common method used to predict wave climate is based on the use of wind records. This technique is generally referred to as **hindcasting**. As no purely theoretical means of predicting waves from wind has yet been devised, empirical relationships have been derived. The simplest of these provide a prediction in the form of a single wave height and period for a given wind speed, rather than a wave spectrum. The characteristic wave height most commonly used is that of the significant wave height, H_s.

Tables and charts relating H_s to wind speed, fetch and duration have been devised by various people after considerable and painstaking research. Notable examples of these include the Darbyshire and Draper (1963) charts for oceanic and coastal waters around the UK, and the SMB method given in the US *Shore Protection Manual* (Sverdrup *et al.*, 1975). Figure 8.25 presents the Darbyshire and Draper charts. To use these charts, the required wind speed is selected, and the corresponding wave height (and period) is found at the first intersection of this wind speed with either the known fetch or the duration. Thus the chart also determines whether the sea state is fetch or duration limited.

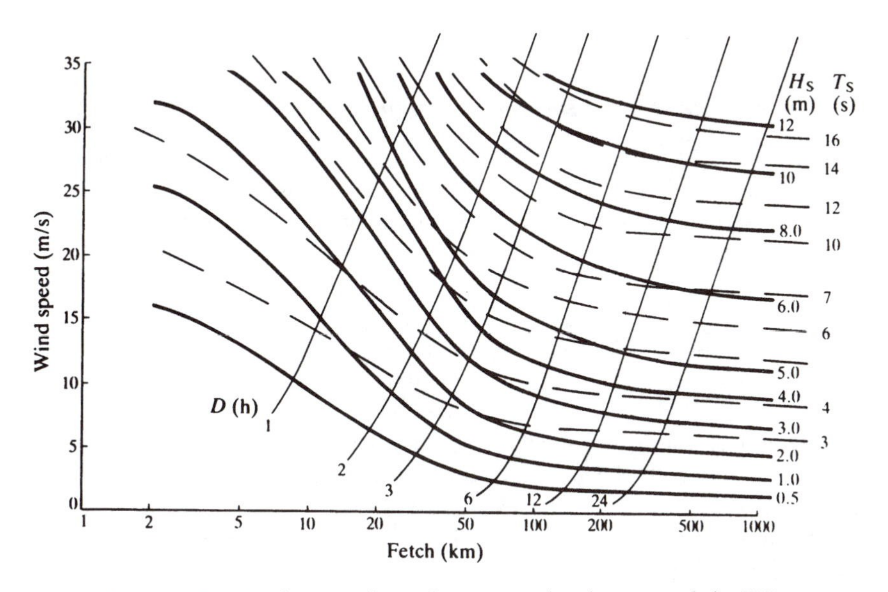

Figure 8.25 Prediction of significant wave heights around the UK.

Example 8.4 Use of the Darbyshire and Draper charts

Using Figure 8.25:

(a) Find the significant wave height H_S and period T_S, and state whether the fully arisen sea has been reached under the following conditions:
wind speed $U = 20\,\text{m/s}$ (i.e. Beaufort force 8)
fetch $F = 100\,\text{km}$
storm duration $D = 6\,\text{h}$.

(b) Find the new significant wave height if the fetch is now increased to 500 km (that is, corresponding to a storm from a different direction).

Solution

(a) The intersection of the two lines representing $U = 20\,\text{m/s}$ and $F = 100\,\text{km}$ on Figure 8.25 gives

$$H_s \simeq 5.75\,\text{m} \quad T_S \simeq 7\,\text{s} \quad D \simeq 4\,\text{h}$$

Hence, as the storm has a duration of 6 h, the fully arisen sea condition has been reached, and the waves are fetch limited.

(b) In this case, intersection of the line representing $U = 20\,\text{m/}s$ with that representing $D = 6\,\text{h}$ occurs before that with the line representing $F = 500\,\text{km}$. Hence the sea state is not fully arisen and the waves are duration limited, giving $H_S \simeq 6\,\text{m}$.

Parametric forms for the spectral energy density curve

In cases where the full wave spectrum is required (for example, physical model testing of breakwaters) then the Pierson–Moskowitz (PM) spectrum is often used. For a fully arisen sea, this may be stated in terms of wave frequency as

$$S(f) = \frac{K_p g^2}{(2\pi)^4 f^5} \exp\left[-\frac{5}{4}\left(\frac{f_p}{f}\right)^4\right] \tag{8.64}$$

where $S(f)$ is the spectral energy density ($\text{m}^2\,\text{s}$); $K_p = 0.0081$; f_p is the frequency at which the peak occurs in the spectrum $= 0.8772g/2\pi u_{19.5}$; and $u_{19.5}$ is the wind speed at 19.5 m above the sea surface (m/s). The PM spectrum was first presented in 1964, and was derived from measurements of ocean waves taken by weather ships in the north Atlantic. It does not describe conditions in fetch-limited seas.

More recent work on wave energy spectra (Hasselmann *et al.*, 1973) has been based on observations in the North Sea (Europe), resulting in the JONSWAP spectrum. This spectrum is given by

$$S(f) = \frac{K_j g^2}{(2\pi)^4 f^5} \exp\left[-\frac{5}{4}\left(\frac{f_p}{f}\right)^4\right] \gamma^a \tag{8.65}$$

where

$$a = \exp\left[-\frac{(f-f_p)^2}{2W^2 f_p^2}\right]$$

and $K_j = 0.076/X_{10}^{0.22}$; $X_{10} = gF/U_{10}^2$; U_{10} is the wind speed at 10 m above the sea surface, F is fetch length in m; $f_p = 3.5g/(U_{10}X_{10}^{0.33})$; $\gamma = 3.3$; and $W = 0.07$ for $f \leq f_p$ or 0.09 for $f > f_p$.

The equation for the JONSWAP spectrum is more complicated than the equation for the PM spectrum because it is a function of both wind speed and fetch. Additionally, the parameter γ (known as the **peak enhancement factor**) is introduced (not to be confused with $\gamma = H/d$). Otherwise the two spectra are of similar form.

Estimates of wave height H_S and period T_S may be derived from these two spectra given (in the *Shore Protection Manual*) by

$$\left.\begin{aligned} H_s &\approx 0.025U_{10}^2 (\text{m}) \\ T_s &\approx 0.79U_{10} (\text{s}) H_s \end{aligned}\right\} \quad \text{for the PM spectrum}$$

$$\left.\begin{aligned} H_s &\approx 0.00051U_{10}F^{0.5} (\text{m}) \\ T_s &\approx 0.059(U_{10}F)^{0.33} (\text{s}) \end{aligned}\right\} \quad \text{for the JONSWAP spectrum}$$

where U_{10} is in m/s and F is in m.

The PM and JONSWAP spectra are deep water spectra. Where a spectrum is required in transitional water depths the TMA spectrum (after Hughes, 1984) may be used. It is a modified JONSWAP spectrum in which the JONSWAP spectrum is multiplied by a function $\phi(f, d)$ that is depth and frequency dependent. Hence

$$S(f)_{\text{TMA}} = S(f)_{\text{JONSWAP}} \phi(f, d)$$

where to within 4%

$$\begin{aligned} \phi(f, d) &= 2\pi f^2 d/g \quad \text{for} \quad f < (2\pi d/g)^{0.5} \\ &= 1 - \frac{1}{2}\left[2 - 2\pi f(d/g)^{0.5}\right]^2 \quad \text{for} \quad f > (2\pi d/g)^{0.5} \end{aligned}$$

Also, the factors K_j and γ used in the JONSWAP spectrum have to be modified for water depth, and are given by

$$K_j = 0.0078\left(\frac{2\pi U^2}{gL_p}\right)^{0.49}$$

$$\gamma = 2.47\left(\frac{2\pi U^2}{gL_p}\right)^{0.39}$$

where L_p is the wavelength corresponding to f_p in depth d.

Sorensen (1993) provides further details and references for these spectra.

Example 8.5

Given a wind speed of 22.66 m/s at 19.5 m above sea level and assuming an oceanic sea state that is not fetch limited:

(a) Calculate $S(f)$ at frequencies 0.05, 0.07, 0.09 and 0.11 Hz.
(b) Hence estimate m_0, H_{mo}, T_p, H_s, T_s.

Solution

(a) In this case the PM spectrum may be applied. Hence using (8.52)

$$S(f) = \frac{0.0081}{(2\pi)^4 f^5} g^2 \exp\left[-\frac{5}{4}\left(\frac{f_p}{f}\right)^4\right]$$

where

$$f_p = \frac{0.8772g}{2\pi U_{19.5}}$$

for $U_{19.5} = 22.66\,\text{m/s}$, $f_p = 0.06\,\text{Hz}$, and hence

f (Hz)	$S(f)$ ($\text{m}^2\,\text{s}$)
0.05	120
0.07	152
0.09	66
0.11	27.7

(b) As $m_0 = \int_0^\infty S(f)\,df$, this may be approximated as $m_0 = \Sigma S(f)\Delta f$.
In this case, taking $\Delta f = 0.02$ and assuming that $S(f)$ only has a value between $f = 0.04$ and $f = 0.12$ (this may be checked by calculating $S(f)$ outside these limits), then for

f	S(f)	Δf	$S(f)\Delta f$
0.05	120	0.02	2.4
0.07	152	0.02	3.04
0.09	66	0.02	1.32
0.11	27.7	0.02	0.55
			Σ7.31

Therefore

$$\mathrm{m}_0 \approx 7.31\,\mathrm{m}^2$$

$$H_0 = 4\sqrt{m_0} = 10.8\,\mathrm{m}$$

$$H_s \approx H_{m0}$$

or

$$H_s \approx 0.025U_{10}^2 = 10.61\,\mathrm{m}$$

$$T_s \approx 0.79U_{10} = 16.3\,\mathrm{s}$$

or

$$T_s \approx 0.95T_p = \frac{0.95}{0.06} = 15.8\,\mathrm{s}$$

$$T_p = \frac{1}{f_p} = 16.7\,\mathrm{s}$$

Effective fetch

In cases where the fetch width is small in comparison with the fetch length, the problem arises of what fetch length (the effective fetch) should be used in predicting wave heights and frequencies for varying wind directions. Three principal approaches have been developed by Saville in 1962, Seymour in 1977 and Donelan in 1980.

Saville's effective fetch concept is based on two assumptions. First, waves are generated over a 45° range either side of the wind direction with energy transfer from the wind to the waves being proportional to the cosine of the angle between the wind and waves. Second, wave growth is proportional to fetch length. Hence

$$F_{\text{eff}} = \frac{\Sigma F_i \cos^2 \theta_i}{\Sigma \cos \theta_i}$$

where F_i, θ_i are calculated at 6° intervals.

This method was first used with the SMB wave prediction method, which itself was found to overpredict wave heights for small fetches. It has subsequently also been used in conjunction with the JONSWAP spectrum.

Seymour's method involves a more complex set of calculations, but is better grounded in wave generation mechanics. It assumes that wave energy E is distributed according to a cosine^2 function over a 180° arc, and that the energy along each direction, E_i, is given by the JONSWAP formulae (noting that E is proportional to $H^2/8$). Hence

$$E = (2/\pi) \sum_i E_i \cos^2(\theta_i - \theta_w)\Delta\theta$$

Donelan's method also uses the JONSWAP formulae, and assumes that the fetch length should be measured along the wave direction θ_w rather than the wind direction ϕ, and that the wind speed used for wave prediction should therefore be the component along the wave direction. Donelan assumed that the predominant wave direction was that which produces the maximum value of wave period. The resulting equations may be found in the CIRIA/CUR Manual (1991).

8.7 Long-term wave statistics

If wave records are collected for an extended period (say for a season or several years), then the computed values of relevant parameters for each wave recording period (such as H_s, T_s) may be presented in the form of scatter diagrams (H_s versus T_s). These graphs can provide very useful information for contractors undertaking marine construction works, or for designers. For some purposes it may be useful to extend these records to estimate the one-in-T-year event (for example, the maximum wave height expected once in 100 years). This requires the application of frequency analysis, which may be considered as the study of rare events. The principles and techniques are fully described in Chapter 10, and here consideration is restricted to their application to wave data.

Application of frequency analysis to wave data

For wave data, the techniques of frequency analysis can be applied directly provided the record is sufficiently long (for example, at least 25 years of record to estimate the 50 year event). Unfortunately, this is rarely the case. More typically, wave records either are unavailable or are of relatively short duration, ranging from a season to a few years. In such cases recourse must be made to using what in statistical terms is referred to as a partial duration series. In this case all events greater than a threshold value are abstracted from the record and subjected to a modified form of frequency analysis. For wave data typically recorded at 3 hour intervals BS 6349 suggests that the first stage is to abstract the maximum wave heights from each storm event in an attempt to preserve the independence of each event. These data are then ranked and assigned a probability. Next, a suitable probability density function (pdf) is chosen, and the relevant wave parameter is plotted against the reduced variate. Finally, a straight line is fitted to the data, and is then extrapolated to the relevant probability to obtain the wave height at the desired return period. Table 8.4 provides the details of the method together with five pdfs recommended in BS 6349. A more recent publication (Offshore Technology Report 89300; Department of Energy, 1990) recommends the use of the extreme value type 1 distribution (otherwise referred to as the Gumbel or Fisher–Tippet type 1 distribution), with the data being fitted by the method of moments using the observed 0.5 m class intervals for cumulative probabilities from 3 hourly data sets. This report also recommends that the wave period to be associated with the predicted wave height should be calculated on the basis of wave steepness $s = H/L = 1/18$. In deep water this gives

$$\frac{1}{18} = \frac{2\pi H_s}{gT_z^2} \quad \text{or} \quad T_z = 3.4H_s^{1/2}$$

Table 8.4 Choice of probability density function (pdf).

Pdf name	Reduced variate	Wave parameter
Weibull	$\log_e \log_e(1/p_n)$	$\log_e(H_n - H_L)$
Fisher–Tippet	$-\log_e \log_e[1/(1-p_n)]$	$-\log_e(H_L - H_n)$
Frechet	$-\log_e \log_e[1/(1-p_n)]$	$\log_e(H_n - H_L)$
Gumbel	$-\log_e \log_e[1/(1-p_n)]$	H_n
Gompertz	$\log_e \log_e(1/p_n)$	H_n

$p_n = 1 - (n/(n_x + 1))$ (exceedence probability)
n = rank no. (highest wave given highest rank)
n_x = total no. of data points
H_n = wave height for rank no. n
H_L = lower (or upper) limiting wave height (chosen by trial)

Several points concerning this procedure need to be explained and emphasized.

Determination of T_R year event probability. The event-based probability needs to be related to the probability for the event with return period T_R. This is achieved as follows.

Given n_x data points in time T_0 (years) and H_{nx} maximum wave height in time T_0, then T_0 is the return period of H_{nx}.

As

$$P_{n_x} = 1 - \left(\frac{n_x}{n_x + 1}\right) \text{(event based exceedence probability)}$$

$$= \frac{1}{n_x + 1}$$

then for a return period T_R the corresponding probability is

$$P(H \geqslant H_{\text{design}}) = \frac{T_0}{T_R}\left(\frac{1}{n_x + 1}\right)$$

For example, if 39 storms in one year are given, and the 50 year wave height is required, then

$$P(H \geqslant H_{\text{design}}) = \frac{1}{50}\frac{1}{(39+1)} = \frac{1}{2000}$$

Plotting position formulae. As stated in Chapter 10, the general plotting position formula is given by

$$P(H \geqslant H_{\text{design}}) = \frac{i - a}{N + b}$$

where a and b are constants depending on the selected pdf. It would appear from the literature, for example BS 6349 and Sorensen (1993), that most wave data have been analysed using the Weibull formula,

$$P(H \geqslant H_{\text{design}}) = \frac{i}{N + 1}$$

whereas hydrological data sets are typically analysed using an unbiased plotting position formula. The use of a biased plotting position appears to the author to be unjustified.

Outliers. Any data set may contain a rare event or events within a relatively short record. For example, the 100 year event may just happen to have occurred in the one season of recording. This is known as an **outlier.** Outliers will not fit any pdf because their assigned exceedence probability will be too high (that is, not rare enough). Where these are known to have occurred, a pdf can be fitted to the remaining data and the return period of the outlier estimated using the fitted pdf.

Fitting a pdf. It may prove necessary to try several pdfs before a good fit is found. Some pdfs contain two parameters to be fitted, and others three. Where three parameters are to be fitted, one must be assumed a priori (H_L, for example) before plotting. This may therefore necessitate an iterative procedure to obtain the best fit.

The 'best fit' straight line (e.g. pdf) can be obtained in three ways mathematically (least squares, method of moments, method of maximum likelihood) or simply by eye. Additionally, the χ^2 goodness of fit test can also be used to determine statistically (that is, objectively) which pdf is the most appropriate. All of these techniques have been used. None of them provides a uniquely optimal approach, and it is very important to plot the data to appreciate visually how the data look in comparison with the fitted pdf.

Suitability of records. Where records are short, the question 'how typical is this record?' must be posed. Any extrapolation to large return periods based on one year of data is likely to over- or underestimate longer-term trends. To provide an indication of how reliable short-term records are, the results of an analysis of this problem are presented in Offshore Technology Report 89 300, Table 18 (Department of Energy, 1990). This shows that if the 50 year return period wave height is estimated from one year's data the 95% confidence level is ±20%. This reduces to 9% with 5 years' data and 6% with 10 years' data. There is no substitute for long-term records. Short records of wave can, however, be extended by the use of hindcasting of longer-term wind records.

Climate change

The possible effects of global warming are now well documented. These are succinctly summarized in *Coastal Defence and the Environment: A Guide to Good Practice* (MAFF, 1993). One possible consequence is an increase in storminess for UK waters. Where long-term wave records exist, this may be discernible by plotting the mean annual value of H_s against time, and extrapolating the trend. The original data set can then be de-trended before estimating the wave height of any return period, and the trend can be added back when estimating the same event at some future time. However, this technique may not be statistically significant in any particular data set, which is not to say that the effect is absent! Climate is also subject to other medium- and long-term changes, which will reduce the reliability of any estimates of rare events.

Encounter probability

This is the probability that an event with return period T_R will occur within a given time of N years. It provides an insight into the probability of a rare event occurring, say, in a particular record, and is also used in planning construction work. It may be derived as follows.

In any one year the occurrence probability is

$$P(X \geqslant x) = 1/T \qquad \text{(by definition)}$$

Conversely, the non-occurrence probability is

$$P'(X \leqslant x) = (1 - P)$$

For two consecutive years:

$$P'(X \leqslant x)_2 = (1 - P)(1 - P)$$

For N years:

$$P'(X \leqslant x)_N = (1 - P)^N$$

Hence

$$P(X \geqslant x)_N = 1 - P'_N = 1 - (1 - P)^N$$

or

$$P(X \geqslant x)_N = 1 - \left(1 - \frac{1}{T}\right)^N$$

Table 8.5 shows an example of the probability that an event with a return period of 100 years will occur within various design lives.

Table 8.5

Design life, N (years)	$P(X \geqslant 100$ year event) (%)
1	1
5	5
50	39.5
100	63.4
200	86.6
500	99.3

8.8 Prediction of extreme still water levels

Principal components

In the design of coastal structures or coastal defence schemes consideration must be given to determining the highest water levels that are likely to occur. Tidal variations are the most obvious source of water level variation, but there are several other sources that may need to be considered. These include storm surges, wind set-up, wave set-up, tsunami, seiches and sea level rise.

Tides are caused by the gravitational and centrifugal forces of the earth/moon/sun system. These forces can be accurately predicted by astronomical calculations. However, calculating tidal level variations at any particular place from a knowledge of these forces is not so easy. As these forces contain many periodic components an alternative method of tidal prediction, known as **harmonic analysis**, is to determine the frequencies and amplitudes of these components by Fourier series analysis of recorded tidal levels. Thus astronomical tidal predictions can be made at any particular location with a tide gauge record.

Tides can be considered as long waves (that is, their frequency is much less than that of short – wind – waves). Hence they are also shallow water waves even in the oceans (because of their immense wavelength). Shallow water waves are strongly affected by shoaling, and therefore significant tidal water level rises may be expected in coastal seas and estuaries. A good first reference to understanding tides may be found in Open University (1989).

The coriolis force is produced by the earth's rotation, and in the northern hemisphere has the effect of a force that always turns a flow to the right. In the English Channel, this results in larger tidal ranges on the French coast than on the English coast.

Storm surges are caused by atmospheric pressure variations. The sea surface under a low-pressure system rises because of the local reduction of pressure, and this locally increased water level tends to move with the depression, again acting as a long wave.

Wind set-up is the result of shear force between wind and water, which is balanced by an increasing water level in the downwind direction.

Wave set-up occurs only inside the surf zone, and is caused by radiation stress gradients (refer to section 8.4 for further details). Wave set down occurs outside the breaker zone and is a maximum at the onset of breaking.

Tsunami are seismically induced long waves, which can dramatically increase in height because of shoaling on the continental shelf.

Seiches are also long waves; they are caused by resonant excitation of any enclosed (lakes) or semi-enclosed (estuaries and harbours) body of water.

Sea level rise can be caused either by movement of the earth's crust (isostatic uplift) or by climate change resulting in sea level rise.

Design extreme still water level

Ideally, each of the possible components contributing to the extreme still water level should be analysed separately and subjected to probability analysis where appropriate. However, the combined probability of, for instance, storm surge and wind set-up will depend upon their joint probabilities. If these are independent then the joint probability is the product of the individual probabilities. In contrast, if one is dependent upon the other, their joint probability reduces to their individual probability. At many locations, storm surge, wind set-up, wave set-up and wave height are likely to be correlated, thus making a joint probability analysis complex.

A useful approach to the problem is to regard any tide gauge record as being composed of astronomical tide and storm surge (which incorporates any wind or wave set-up). Thus surge heights can be abstracted by subtracting the astronomical tide and then applying a probability analysis. Alternatively, the annual maxima of water level can be analysed by probability analysis directly. Such analyses have been carried out for many locations around the UK, and may be found in Graff (1981) and Blackman (1985).

For an ungauged site, an estimate of ESWL may be found from the ESWL of the nearest gauged site by assuming that

$$\frac{\text{ESWL} - \text{MHWS}}{(\text{MHWS} - \text{MLWS})} = \text{constant}$$

Estimated 50 year return period storm surges around the UK may be found in Pugh (1987). They range from about 1 to 1.5 m except for East Anglia and the Thames Estuary, where surges of up to 2.75 m are predicted.

Sea level rise

Considerable effort has been expended in recent years to predict global sea level rise due to global warming. Predictions have been prepared for the Intergovernmental Panel on Climate Change (IPCC), with their best estimate with being 300 mm rise by the year 2050.

Around the UK, sea levels are also affected by crustal movements, and thus currently in the UK the Ministry of Agriculture, Fisheries and Food (MAFF) recommends the following allowances for sea level rise:

NRA region	Allowance
Anglian, Thames, Southern	6 mm/year
North West, Northumbria	4 mm/year
Remainder	5 mm/year

These allowances must be added to the ESWL to determine the future ESWL at any future date.

References and further reading

Battjes, J. A. (1968) Refraction of water waves. *J. Waterways and Harbours Div. ASCE*, **WW4**, 437–57.

Blackman, D. L. (1985) New estimates of annual sea level maxima in the Bristol Channel. *Estuarine, Coastal and Shelf Science*, **20**, 229–32.

Briggs, J. B., Thompson, E. F. and Vincent, C. L. (1995) Wave diffraction around a breakwater. *J. Waterways, Port Coastal and Ocean*, **121**, 23–35.

Broch, J. T. (1981) *Principles of Analog and Digital Frequency Analysis*, Tapir, Norway.

BSI (1984) BS 6349 *British Standard Code of Practice for Maritime Structures, Part 1, general criteria*, British Standards Institution.

Carter, D. J. T. *et al.* (1986) *Estimating Wave Climate Parameters for Engineering Applications*, Offshore Technology Report no. 86, HMSO, London.

Chadwick, A. J. (1989) Measurement and analysis of inshore wave climate. *Proc. Inst. Civ. Eng. Part 2*, **87**, Mar, 23–8.

Chadwick, A. J., Pope, D. J., Borges, J. and Ilic, S. (1995a) Shoreline directional wave spectra, Part 1: an investigation of spectral and directional analysis techniques. *Proc. Inst. Civ. Eng., Water Maritime & Energy*, **112**, 198–208.

Chadwick, A. J., Pope, D. J., Borges, J. and Ilic, S. (1995b) Shoreline directional wave spectra, Part 2: instrumentation and field measurements. *Proc. Inst. Civ. Eng., Water Maritime & Energy*, **112**, 209–14.

CIRIA and CUR (1991) *Manual on the Use of Rock in Coastal and Shoreline Engineering*, CIRIA Special Publication 83/CUR Report 154, London.

Darbyshire, M. and Draper, L. (1963) Forecasting wind-generated sea waves. *Engineering*, **195** (April).

Darras, M. (1987) IAHR list of sea state parameters: a presentation. *IAHR Seminar Wave Analysis and Generation in Laboratory Basins*. XXII Congress, Lausanne, 1–4 Sept 1987, pp. 11–74.

Davidson, M. A., Bird, P. A. D., Bullock, G. N. and Huntley, D. A. (1996) A new non-dimensional number for the analysis of wave reflection from rubble mound breakwaters. *Coastal Engineering*, **28**, 93–120.

Dean, R. G. and Dalrymple, R. A. (1991) *Water Wave Mechanics for Engineers and Scientists*, World Scientific, Singapore.

Department of Energy (1990) *Metocean Parameters – Wave Parameters*, Offshore Technology Report 893000, HMSO, London.

Dodd, N. and Brampton, A. H. (1995) *Wave Transformation Models: A Project Definition Study*, H R Wallingford.

Dyer, K. D. (1986) *Coastal and Estuarine Sediment Dynamics*, Wiley, Chichester.

Fredsoe, J. and Deigaard, R. (1992) *Mechanics of Coastal Sediment Transport*, Advanced Series on Ocean Engineering 3, World Scientific, Singapore.

Goda, Y. (2000) *Random Seas and Design of Maritime Structures*, Advanced Series on Ocean Engineering, Volume 15. World Scientific, Singapore.

Graff, J. (1981) An investigation of the frequency distributions of annual sea level maxima at ports around Great Britain. *Estuarine, Coastal and Shelf Science*, **12**, 389–449.

Hardisty, J. (1990) *Beaches Form and Process*, Unwin Hyman, London.

Hasselmann, K., Barnett, T. P., Bouws, E., Carlsen, H., Cartwright, D. E., Enkee, K., Ewing, J. A., Gienapp, H., Hasselmann, D. E., Kruseman, P., Meerburg, A., Müller, P., Olbers, D. J., Richter, K., Sell, W. and Walden, H. (1973) Measurements of wind-wave growth and swell decay during the joint North Sea wave project (JONSWAP). *Deutsches Hydrographisches Zeitschrift*, **8** (12), 95.

Horikawa, K. (1978) *Coastal Engineering*, University of Tokyo Press, Tokyo.

Horikawa, K. (ed.) (1988) *Nearshore Dynamics and Coastal Processes, Theory Measurement and Predictive Models*, University of Tokyo Press, Tokyo.

Hughes, S. A. (1984) *The TMA Shallow Water Spectrum Descriptions and Applications*, Technical Report 84-7, US Army Corps of Engineers.

Hunt, J. N. (1979) Direction solution of wave dispersion equation. *Journal of Waterway, Port, Coastal, and Ocean Engineering (ASCF)*, **105** (WW4), 457–9.

Huntley, D. A., Davidson, M., Russell, P., Foote, Y. and Hardisty, J. (1993) Long waves and sediment movement on beaches: recent observations and implications for modelling. *Journal of Coastal Research Special Issue*, **15**, 215–29.

Ilic, S. and Chadwick, A. J. (1995) Evaluation and validation of the mild slope evolution equation model for combined refraction–diffraction using field data, *Coastal Dynamics 95*, Gdansk, Poland, pp. 149–60.

Komar, P. D. (1976) *Beach Processes and Sedimentation*, Prentice-Hall, Englewood Cliffs, NJ.

Koutitas, G. K. (1988) *Mathematical Models in Coastal Engineering*, Pentech Press, London.

Li, B. (1994) An evolution equation for water waves. *Coastal Engineering*, **23**, 227–42.

Longuet-Higgins, M. S. (1970) Longshore currents generated by obliquely incident sea waves. *Journal of Geophysical Research*, **75**, 6778–89.

Longuet-Higgins, M. S. and Stewart, R. W. (1964) Radiation stress in water waves, a physical discussion with applications. *Deep-Sea Research*, **75**, 6790–801.

McDowell, D. M. (1988) The interface between estuaries and seas, in *Developments in Hydraulic Engineering* (ed. Novak, P.), Elsevier, Amsterdam, pp. 139–94.

MAFF (1993) *Coastal Defence and the Environment: A Guide to Good Practice*, Ministry of Agriculture, Fisheries and Food, London.

Open University (1989) *Waves, Tides and Shallow Water Processes*, Pergamon Press, Oxford.

Penney, W. and Price, A. (1952) The diffraction of sea waves and shelter afforded by breakwaters. *Philos. Trans Royal Society (London) Series A*, **244**, 236–53.

Pugh, D. T. (1987) *Tides, Surges and Mean Sea Level*, Wiley, Chichester.

Silvester, R. (1974) *Coastal Engineering*, Elsevier, Oxford.

Sorensen, R. M. (1993) *Basic Wave Mechanics for Coastal and Ocean Engineers*, John Wiley & Sons, New York.

Sverdrup, H. U., Munk, W. H. and Bretschneider, C. L. (1975) SMB method for predicting waves in deep water, in *Shore Protection Manual*, US Army Coastal Engineering Research Centre, Washington.

9

Sediment transport

9.1 Introduction

Sediment transport governs or influences many situations that are of importance to mankind. Silt deposition reduces the capacity of reservoirs, interferes with harbour operation and closes or modifies the path of watercourses. Erosion or scour may undermine structures. In rivers and on coastlines, sediment movements form a part of the long term pattern of geological processes.

Clearly, sediment transport occurs only if there is an interface between a moving fluid and an erodible boundary. The activity at this interface is extremely complex. Once sediment is being transported, the flow is no longer a simple fluid flow, since two materials are involved. Sediment transport may be conceived of as occurring in one of two modes:

1. by rolling or sliding along the floor (bed) of the river or sea – sediment thus transported constitutes the bedload;
2. by suspension in the moving fluid (this is usually applicable to finer particles) which is the suspended load.

9.2 The threshold of movement

Description of threshold of movement

If a perfectly round object (a cylinder or sphere) is placed on a smooth horizontal surface, it will readily roll on application of a small horizontal force. In the case of an erodible boundary, of course, the particles are not perfectly round, and they lie on a surface which is inherently rough and may not be flat or horizontal. Thus, the application of a force will only cause motion when it is sufficient to overcome the natural resistance to motion of the particle. The particles will probably be non-uniform in size. At the interface, a moving fluid will apply a shear force (Fig. 9.1(a)), which

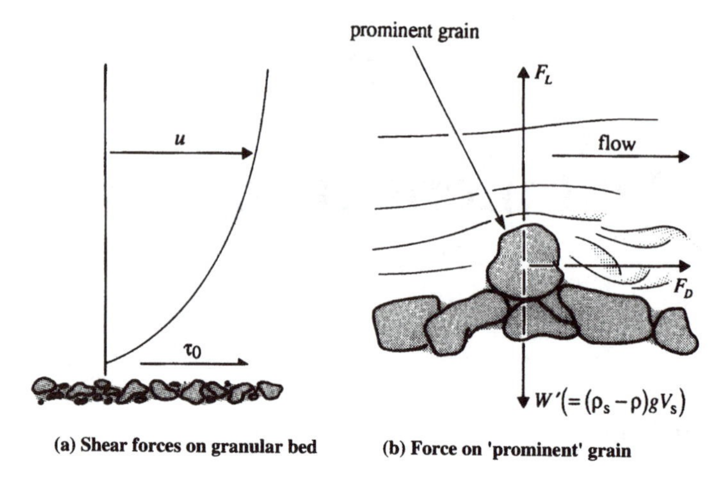

Figure 9.1 Fluid forces causing sediment movement.

implies that a proportionate force will be applied to the exposed surface of a particle. Observations by many experimenters have confirmed that if the shear force is gradually increased from zero, a point is reached at which particle movements can be observed at a number of small areas over the bed. A further small increase in τ_0 (and therefore u) is usually sufficient to generate a widespread sediment motion (of the bedload type). This describes the 'threshold of motion'. After further increments in τ_0, another point is reached at which the finer particles begin to be swept up into the fluid. This defines the inception of a suspended load.

Parameters of sediment transport

Some idea of the problems which face the engineer may be gained by considering the case of a channel flow over a sandy bed which is initially level. Once the shear stress is sufficient to cause transport, 'ripples' will form in the bed (Fig. 9.2). These ripples may grow into larger 'dunes'. In flows having quite moderate Froude Numbers, the dunes will migrate downstream. This is due to sand being driven from the dune crests and then being deposited just downstream on the lee side. Once the flow is sufficient to bring about a suspended load, major changes occur at the bed as the dunes will be 'washed out'. A relationship probably exists between the major parameters of the transport process (Froude Number, sediment properties, fluid properties, shear stress, bed roughness or dune size, and rate of sediment transport). A multitude of attempts have been made to develop a rational theory, so far with limited success. Most of the equations in current use have been developed on the basis of a combination of dimensional analysis, experimentation and simplified theoretical models.

Figure 9.2 Sediment ripple formation.

The entrainment function

A close inspection of an erodible granular boundary would reveal that some of the surface particles were more 'prominent' or 'exposed' (and therefore more prone to move) than others (Fig. 9.1(b)). The external force on this particle is due to the separated flow pattern. The other force acting on the particle is related to its submerged self-weight, W' (where $W' = \pi D^3 g(\rho_S - \rho)/6$ for a spherical particle) and to the angle of repose, ϕ. The number of prominent grains in a given surface area is related to the areal grain packing (= area of grains/total area = A_p). As the area of a particle is proportional to the square of the typical particle size (D^2), the number of exposed grains is a function of A_p/D^2. The shear stress at the interface, τ_0, is presumably the sum of the forces on the individual particles, with the contribution due to prominent grains dominating, so the total force on each prominent grain in unit area may be expressed as

$$F_D \propto \tau_0 D^2/A_p$$

At the threshold of movement $\tau_0 = \tau_{CR}$, so

$$\tau_{CR}\frac{D^2}{A_p} \propto (\rho_s - \rho)g\frac{\pi D^3}{6}\tan\phi$$

This can be rearranged to give a dimensionless relationship

$$\frac{\tau_{CR}}{(\rho_s - \rho)gD} \propto \frac{\pi A_p}{6} \tan \phi \tag{9.1a}$$

In 1936, an American engineer published the results of some pioneering research into sediment transport (Shields, 1936). He showed that the particle entrainment was related to a form of Reynolds' Number, based on the friction velocity u_* (see section 3.4), i.e. $\mathrm{Re}_* = \rho u_* D/\mu$. The left-hand side of (9.1a) is the ratio of shear force to gravity force, and is known as the entrainment function, F_S. Shields plotted the results of his experiments in the form of F_S against Re_*, and proved that there was a well defined band of results indicating the threshold of motion (Fig. 9.3). $\mathrm{Re}_* < 2$ (approximately) corresponds to laminar flow and $\mathrm{Re}_* > 400$ (approximately) to rough turbulent conditions. It will be noted that F_S is constant $(= 0.056)$ for rough turbulent flows. The Shields threshold line has been expressed in a convenient explicit form (van Rijn, 1984, part 1), based on the use of a dimensionless particle size parameter, $D_{gr} = D(g[(\rho_s/\rho) - 1]/\nu^2)^{1/3}$, as follows:

$$\left.\begin{array}{l} \text{For } D_{gr} < 4, F_S = 0.24/D_{gr}, \\ 4 < D_{gr} < 10, F_S = 0.14/D_{gr}^{0.64}, \\ 10 < D_{gr} < 20, F_S = 0.04/D_{gr}^{0.1}, \\ 20 < D_{gr} < 150, F_S = 0.013/D_{gr}^{0.29} \\ \text{and for } D_{gr} < 150, F_S = 0.056. \end{array}\right\} \tag{9.1b}$$

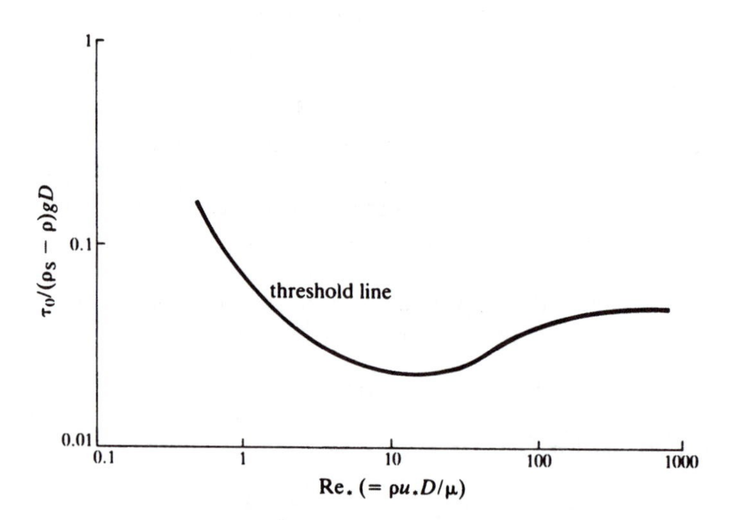

Figure 9.3 Shields' diagram.

Taking $F_S = 0.056$ and substituting in (9.1a)

$$\tau_{CR}/(\rho_S - \rho)gD = 0.056 \qquad (9.1c)$$

Also, from the Chézy formula,

$$\tau_0 = \rho g R S_0$$

where R is the hydraulic radius. Taking ρ_S as $2650\,kg/m^3$, then

$$\frac{1000 \times 9.81\, R \times S_0}{(2650 - 1000) \times 9.81 \times D} = 0.056$$

$$\therefore RS_0/D = 0.0924 \qquad (9.1d)$$

This may be used to estimate (a) the minimum stable particle size for a given channel or (b) the critical shear stress for a given particle size. This treatment is, naturally, only as good as the Chézy formula and the other approximations made in this development. Nevertheless, many channels have been successfully designed with it. The Chézy formula assumes a generally turbulent flow, however the effects of turbulence at the boundary itself are more difficult to analyse than is suggested by the above treatment.

9.3 A general description of the mechanics of sediment transport

Remarks on the approach used in the rest of this chapter

It should already be obvious that the study of sediment transport involves many difficulties. This is why the approach adopted in this chapter is a little different from that used up to now. It is not always possible to provide rigorous proofs of equations. Even where this is possible, it may not be helpful to someone studying the subject for the first time, since some proofs are long and difficult. In general, proofs will be given only if they are reasonably simple. Where the development of an equation involves a complicated mathematical/empirical development, the equation will simply be stated with a brief outline of principles, the appropriate reference(s) and examples of its application. Only a limited selection of sediment transport equations can be given here. No special merit is claimed for this selection, but it is hoped that at least a path will have been cleared through the 'jungle' which will enable the reader to explore some of the more advanced texts.

Conditions at the interface between a flowing fluid and a particulate boundary

For most practical cases, channel flows are turbulent. This means that the flow incorporates the irregular eddying motion, as discussed in Chapter 3.

A close look in the region of the granular boundary at the bed would reveal the existence of a sub-layer comprising 'pools' of stationary or slowly moving fluid in the interstices. This sub-layer zone is not stable, since eddies (with high momentum) from the turbulent zone periodically penetrate the sub-layer and eject the (low momentum) fluid from the 'pools'. The momentum difference between the fluid from the two zones generates a shearing action, which in turn generates more eddies, and so on. Grains are thus subjected by the fluid to a fluctuating impulsive force. Once the force is sufficient to dislodge the more prominent grains (i.e. when $\tau_0 \geq \tau_{CR}$) they will roll over the neighbouring grain(s). As sediment movement becomes more widespread, the pattern of forces becomes more complex as moving particles collide with each other and with stationary particles. As τ_0 increases further, granular movement penetrates more deeply into the bed. Bed movement may most simply be represented as a series of layers in relative sliding motion (Fig. 9.4), with a linear velocity distribution (note the analogy with laminar flow). This model forms the basis of the 'tractive force' equations of Du Boys, Shields, etc. (see section 9.4). In reality, the pattern of movement is much more irregular.

The mechanics of particle suspension

If sediment grains are drawn upward from the channel bed and into suspension, it must follow that some vertical (upward) force is being applied to the grains. The force must be sufficient to overcome the immersed self-weight of the particles. Consider a particle suspended in a vertical flask (Fig. 9.5). If the fluid is stationary, then the particle will fall due to its self-weight (assuming $\rho_s > \rho$), accelerating up to a limiting (or 'terminal') velocity v_{FS} at which the self-weight will be equal in magnitude to the drag force, F_D, acting on the particle. Since drag force $= C_D \times (\frac{1}{2}\rho A U_\infty^2)$ (see section 3.6),

$$(\rho_s - \rho) g V_s = C_D \times \frac{1}{2}\rho A_s v_{FS}^2 = F_D \tag{9.2}$$

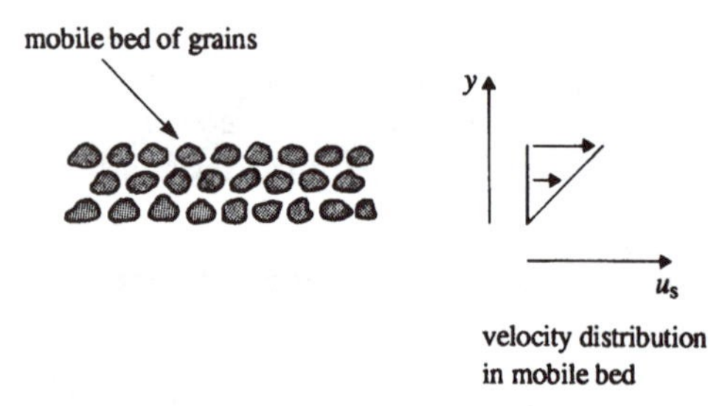

Figure 9.4 Linear velocity distribution for idealized bed motion.

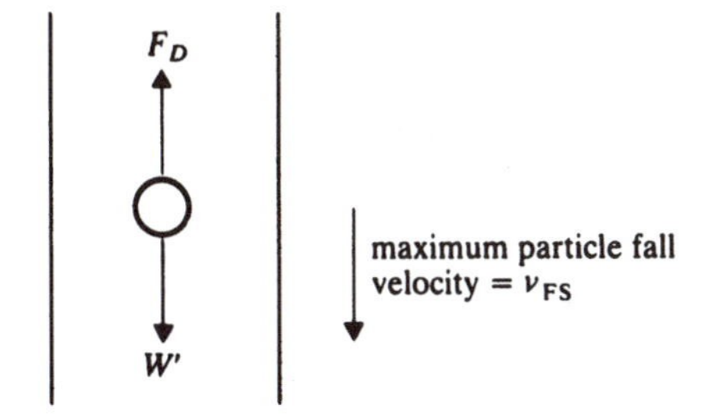

Figure 9.5 Forces acting on a falling particle.

where V_S and A_S are the volume and cross-sectional area of the particle. If a discharge is now admitted at the base of the flask, the fluid is given a vertical upward velocity v. As $v \to v_{FS}$, the particle will cease to fall and will appear to be stationary. If $v > v_{FS}$ then the particle can be made to travel upward.

From this argument, it must follow that the suspension of sediment in a channel flow implies the existence of an upward velocity component. In fact, this should not come as a complete surprise (as a review of section 3.4 will reveal). Fluctuating vertical (and horizontal) components of velocity are an integral part of a turbulent flow. Flow separation over the top of a particle provides an initial lift force (Fig. 9.1(b)) which tends to draw it upwards. Providing that eddy activity is sufficiently intense, then the mixing action in the flow above the bed will sweep particles along and up into the body of the flow (Fig. 9.6(a)). Naturally, the finer particles will be most readily suspended (like dust on a windy day).

The Prandtl model of turbulence (section 3.4) can be used as the basis for a sediment concentration model, assuming (a) that sediment concentration, C (= volume of sediment /(volume of sediment + fluid)), varies as shown in

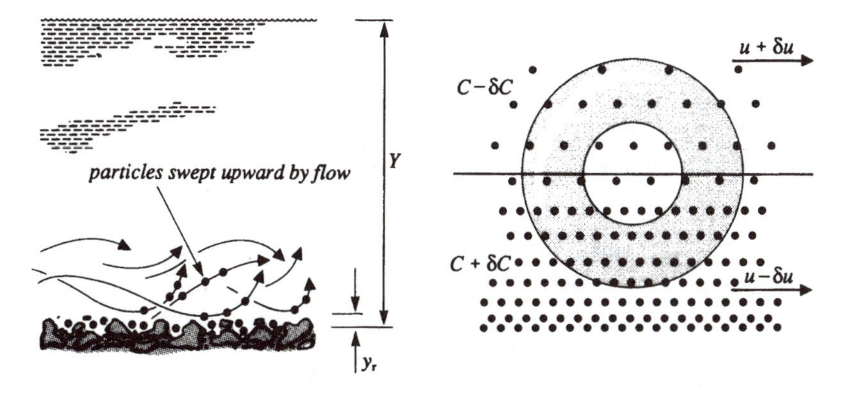

Figure 9.6 Suspension of sediment.

Figure 9.6(b), and (b) that this is an equilibrium condition. The forces acting on the particles are gravity and fluid drag. If the relative velocity between the particle and the fluid in the vertical plane is assumed to be v_{FS} then the upward rate of transport is

$$(u' - v_{FS})(C - \delta y\,(dC/dy)) \tag{9.3a}$$

and the downward transport rate is

$$(u' + v_{FS})(C - \delta y\,(dC/dy)) \tag{9.3b}$$

For equilibrium the sum of the upward and downward transport rates must be zero. Hence, the net transport rate is

$$\left(u'\delta y\frac{dC}{dy} + v_{FS}C\right) = 0 \tag{9.4}$$

From equations 3.4 and 3.6a, $u'\delta y = \tau/(\rho(du/dy))$ and $(du/dy) = u_*/Ky$; also, for steady, two-dimensional channel flows the vertical variation of shear stress is often taken as $\tau = \tau_0(1-(y/Y))$. Combining these with (9.4) gives the result

$$u_*Ky(1-(y/Y))dC + Cv_{FS} = 0 \tag{9.5}$$

This may be integrated to yield

$$\frac{C}{C_r} = \left(\frac{y_r(Y-y)}{y(Y-y_r)}\right)^{(v_{FS}/Ku_*)} \tag{9.6}$$

where C_r is a reference concentration at height y_r.

Equation (9.6) is a simple mathematical model of suspended sediment transport, but it cannot be solved without using experimental results to evaluate the unknown constants.

(a) The value for K is often taken as 0.4, but this is for a clear fluid. There is no general agreement as to the effect of suspended sediment on the value of K, though some experimental results are illustrated in Chang (1988) which indicate that K is not a constant.

(b) A value for C_r is also required, together with the corresponding reference height. Strictly speaking, the concentration is a function of the sediment size and the turbulent flow conditions. Some of the methods for calculating C_r are quite complex and are beyond the scope of this chapter. Chang (1988), Raudkivi (1998) and van Rijn (1984) may be consulted. The formula suggested by van Rijn has the merit of being relatively simple.

The model does appear to fit experimental results quite well, but this should be viewed with caution, since the value of the exponent (v_{FS}/Ku_*) is difficult to estimate with confidence. Also the Prandtl turbulence model, used as the basis for the model, is only a crude approximation. Furthermore, one often has to assume that the turbulent shear is the same with the sediment in suspension as in clear water, which is not true. Nevertheless, a number of solutions to (9.6) have been published, and it is worth illustrating the way in which the model may be applied by an example.

Example 9.1 Suspended sediment transport in a river

A river having a rectangular cross-section 100 m wide and 5 m deep has a bed slope of 1 m in 5 km. The bed of the river consists of fine sand of 0.2 mm diameter, and the bed roughness k_s can be taken as 0.4 mm. The vertical velocity distribution at the vertical centre line of the river is given by $u = 2.5u_* \ln(30y/k_s)$. The drag coefficient for the sediment particles is $C_D = 24/\mathrm{Re}$, sediment density is 2650 kg/m^3 and viscosity of water $\mu = 1.14 \times 10^{-3}$ kg/m s.

Estimate the total suspended sediment discharge based on the relationship $\Delta q_s = Cu\delta y$, where is q_s the transport rate per unit width. Use (9.6) to find C.

Solution

The sediment fall velocity is found from (9.2) with the assumption that sediment particles are spheres, so $A_s = \pi D^2/4$ and $V_s = \pi D^3/6$. Rearranging (9.2) and substituting for $C_D = 24/\mathrm{Re}$:

$$v_{FS} = \frac{(\rho_s - \rho)gD^2}{18\mu} = \frac{(2650 - 1000) \times 9.81 \times (0.2 \times 10^{-3})^2}{18 \times 1.14 \times 10^{-3}} = 0.0316\,\mathrm{m/s}$$

Hydraulic radius $R = A/P = (100 \times 5)/(100 + (2 \times 5)) = 4.55$ m

$$u_* = \sqrt{gRS_0} = \sqrt{9.81} \times 4.55 \times 1/5000 = 0.094\,\mathrm{m/s}$$

The sediment discharge Δq_s must be found for a series of elevations, y, above the bed, and the total sediment discharge can be found by summation. This means that the velocity, u, and concentration C must be evaluated. To evaluate C from (9.6) we need to have values for y_r and C_r; y_r is taken as 5% of the water depth (Chang, 1988, p. 149), and a value for C_r has been calculated as 0.0028, based on Einstein's approach as given by Chang. The river cross-section is divided into five 1 m high horizontal 'slices'. The velocity and concentration are calculated for the mid-height of each slice at the centre of the river and Δq_s evaluated for each slice. Note that because the values of C and u are for the centre of the river, Δq_s is the maximum sediment discharge per unit width through the 'slice'. To find the sediment discharge for each slice over the whole 100 m width of the river it must be remembered that the water velocity and sediment transport will reduce from the maximum at the centre to zero at each of the banks. To allow for this the sediment discharge over the

Table 9.1 Solution of Example 9.1

y	$\bar{y}$	u	C	Δq_s	ΔQ_s
5	4.5	2.992	0.04×10^{-3}	0.12×10^{-3}	8.0×10^{-3}
4	3.5	2.932	0.12×10^{-3}	0.35×10^{-3}	23.35×10^{-3}
3	2.5	2.85	0.24×10^{-3}	0.68×10^{-3}	45.36×10^{-3}
2	1.5	2.733	0.48×10^{-3}	1.31×10^{-3}	87.34×10^{-3}
1	0.5	2.47	1.5×10^{-3}	3.71×10^{-3}	247.12×10^{-3}
0					
					$Q_s = \sum \Delta Q_s = 411.4 \times 10^{-3}\,\mathrm{m^3/s}$

whole width is estimated as $\Delta Q_s = 0.667 \times \Delta q_s \times 100$, which assumes a parabolic distribution of sediment transport across the width of the river. The total sediment discharge is then

$$Q_s = \sum \Delta Q_s$$

Thus, at $\bar{y} = 0.5\,\mathrm{m}$, $u = 2.5u_* \ln(30 \times 0.5/0.0004) = 2.47\,\mathrm{m/s}$.

The concentration is

$$C = C_r \left(\frac{y_r(Y - \bar{y})}{\bar{y}(Y - y_r)} \right)^{(v_{FS}/Ku_*)} = 0.0028 \left(\frac{0.25(5 - 0.5)}{0.5(5 - 0.25)} \right)^{(0.0316/0.4 \times 0.094)} = 1.5 \times 10^{-3}$$

Hence the sediment discharge at the centre of the first 'slice' (between 0 and 1 m above the bed) is

$$\Delta q_s = Cu\,\mathrm{d}y = 1.5 \times 10^{-3} \times 2.47 \times 1 = 3.71 \times 10^{-3}\,\mathrm{m^3/ms}$$

therefore

$$\Delta Q_s = 0.667 \times 100 \times 3.71 \times 10^{-3} = 247.12 \times 10^{-3}\,\mathrm{m^3/s}$$

The solution is completed in tabular form in Table 9.1.

9.4 Sediment transport equations

In practice, virtually all sediment transport occurs either as bedload or as a combination of bedload and suspended load (suspended load rarely occurs in isolation, except for certain cases involving very fine silts). The combined load is known as a total load.

Bedload formulae

Tractive force ('Du Boys type') equations. In 1879, Du Boys proposed an equation which related sediment transport to shear stress:

$$q_s = f(\tau_0)$$

where q_s is the volume of sediment transported per second per unit channel width. Subsequent researchers proposed that the function should be in the form of a power series:

$$q_s = K_1 + K_2\tau_0 + K_3\tau_0^2 + \ldots$$

If higher order terms are neglected, then the constants can be evaluated, since

$$q_s = 0 \text{ if } \tau_0 < \tau_{CR}, \quad \text{therefore } K_1 = 0$$

$$q_s = 0 \text{ if } \tau_0 = \tau_{CR}, \quad \text{therefore } K_2 = -K_3\tau_{CR}$$

Therefore

$$q_s = K_3\tau_0(\tau_0 - \tau_{CR}) \tag{9.7}$$

The problem of evaluating K_3 now arises. One very simple equation, which is based only on limited data, (see Raudkivi (1998)) is

$$K_3 = 3.874 \times 10^{-8}/D^{0.75} \quad \text{(metric units)} \tag{9.8}$$

A more rational formulation was derived by Shields (1936) for a level bed:

$$q_s = \frac{10qS_0\rho^2}{\rho_s}\frac{(\tau_0 - \tau_{CR})}{(\rho_s - \rho)^2 gD}\ \text{m}^3/\text{s} \tag{9.9}$$

Here, again, the equation offers a solution for K_3.

Probabilistic equations. It has already been pointed out that grain movement is brought about by the impulsive force of turbulent eddies. Eddy action does not occur uniformly with time or space. It might therefore be thought that the incidence of an eddy capable of transporting a particular grain is some statistical function of time. H. A. Einstein (1942) proposed just such a probabilistic model of bedload for the case of a level bed of grains.

The basic ideas underlying Einstein's equation are the following:

(a) For an individual grain, migration will take place in a series of jumps (Fig. 9.7) of length $L = K_L D$. During a time T a series of n such jumps will occur, so that the particle will travel a total distance nL.

(b) The probability, p, that a grain will be eroded during the typical time scale, T, must be some function of the immersed self-weight of the particle and the fluid lift force acting on the particle. The immersed self-weight is $(\rho_s - \rho)g(K_V D^3)$, and the lift force is $C_L \rho (K_A D^2) u^2/2$, where grain area $A_s = K_A D^2$ and grain volume $V_s = K_V D^3$. Therefore,

$$p = f\left\{\frac{(\rho_s - \rho)g(K_V D^3)}{C_L \rho (K_A D^2) u^2/2}\right\} \tag{9.10}$$

u is a 'typical' velocity at the sub-layer. Researchers have proposed that

$$u \simeq 11.6 u_* \simeq 11.6\sqrt{gR' S_0}$$

where R' is that proportion of the hydraulic radius appropriate to sediment transport. Equation (9.10) is usually expressed as

$$p = f\{\mathbf{B}_* \cdot \Psi\} \tag{9.11}$$

where

$$\mathbf{B}_* = \frac{K_V}{C_L K_A 135/2} \qquad \Psi = \frac{(\rho_s - \rho)D}{\rho R' S_0} \tag{9.12}$$

(c) The number of grains of a given size in area $A(= K_L D \times 1)$ is $K_L D / K_A D^2$, therefore the number of grains dislodged during time T will

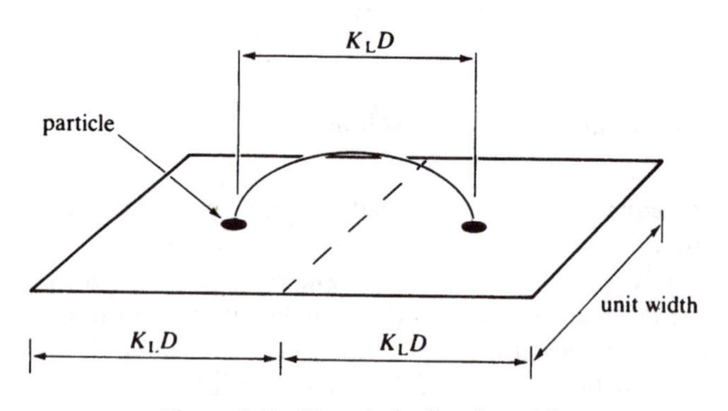

Figure 9.7 Einstein bedload model.

be $pK_L D/K_A D^2$. The volume of grains crossing a given boundary must therefore be

$$\frac{pK_L D}{K_A D^2} K_V D^3 = \frac{pK_L K_V D^2}{K_A} \tag{9.13}$$

The volume must also be given by $q_s T$. If the time T is some function of particle size and fall velocity, say $T = K_T D/v_{FS}$, then

$$q_s T = q_s \frac{K_T D}{v_{FS}} \tag{9.14}$$

Equating (9.13) and (9.14),

$$\frac{q_s K_T D}{v_{FS}} = \frac{pK_L K_V D^2}{K_A}$$

Therefore

$$p = \frac{q_s K_T K_A}{v_{FS} K_L K_V D} \tag{9.15}$$

Equating (9.11) and (9.15), and evaluating v_{FS} from (9.2), leads, with some rearrangement, to

$$\Phi = q_s \sqrt{\frac{\rho}{(\rho_s - \rho) g D^3}} = f\{\mathbf{B}_*, \Psi\} \tag{9.16}$$

where Φ is a dimensionless bedload function and $\mathbf{B}_*$ and Ψ have been defined above.

(d) Following Einstein, a number of researchers investigated the relationship between Φ and Ψ. A typical result is due to Brown (in Rouse, 1950)

$$\Phi = 40(1/\Psi)^3 \tag{9.17}$$

which is valid for $\Phi > 0.04$. As Φ (and therefore q_s) $\rightarrow 0$, $1/\Psi \rightarrow 0.056$, which corresponds to the Shields threshold condition.

Example 9.2 River bedload

A European river has the following hydraulic characteristics: $Q = 450\,\text{m}^3/\text{s}$, width 50 m, depth 6 m, bed slope (S_0), 3×10^{-4}. The sediment has a typical diameter $D_{50} = 0.01$ m. Estimate the bedload transport, using the Shields and Einstein–Brown formulae.

Solution

From the data given,

$$R = \frac{A}{P} = \frac{50 \times 6}{50 + (2 \times 6)} = 4.84\,\text{m}$$

From (9.1c),

$$\tau_{CR} = 0.056(\rho_s - \rho)gD_{50} = 0.056(2650 - 1000) \times 9.81 \times 0.01$$
$$= 9.064\,\text{N/m}^2$$

Also,

$$\tau_0 = \rho g R S_0 = 1000 \times 9.81 \times 4.84 \times 3 \times 10^{-4}$$
$$= 14.245\,\text{N/m}^2$$

Using the Shields formula (equation (9.9)) with $q = 450/50 = 9\,\text{m}^3/\text{s m}$

$$q_s = \frac{10 \times 9 \times 3 \times 10^{-4} \times 1000^2}{2650} \frac{(14.245 - 9.064)}{(2650 - 1000)^2 \times 9.81 \times 0.01}$$
$$= 1.976 \times 10^{-4}\,\text{m}^3/\text{s m}$$

Therefore

$$Q_s (= 50 \times q_s) = 0.01\,\text{m}^3/\text{s}$$

Using the Einstein–Brown formula, from (9.12),

$$\Psi = \frac{(\rho_s - \rho)D}{\rho R' S_0} = \frac{(2650 - 1000) \times 0.01}{1000 \times 4.84 \times 3 \times 10^{-4}} = 11.364$$

Notice that, in the absence of other information, it has been assumed that $R' = R$. From (9.17),

$$\Phi = 40(1/11.364)^3 = 0.0273$$

and, using (9.16),

$$\Phi = q_s \sqrt{\frac{\rho}{(\rho_s - \rho)} \frac{1}{gD^3}} = q_s \sqrt{\frac{1000}{(2650 - 1000)} \frac{1}{9.81 \times 0.01^3}}$$
$$= q_s \times 248.6$$

Therefore

$$248.6 q_s = 0.0273$$

i.e.

$$q_s = 1.096 \times 10^{-4}\,\mathrm{m^3/s}$$

or

$$Q_s = 0.0055\,\mathrm{m^3/s}$$

The substantial difference in the two estimates is not surprising. Of the two, the Einstein–Brown method is probably the more reliable. The answers obtained are the maximum probable values, the actual load may be less. It should also be emphasized that most bedload formulae apply primarily to coarse sand, and perhaps to some gravels.

Total load formulae. It is possible to calculate the total load from the sum of the bedload and suspended load. Separate equations are available for bed-load (e.g. those in the preceding section) and for suspended load. However, experimental data are still rather sparse, and it is very difficult to separate bed and suspended load from these data. For this reason, some researchers have tackled directly the problem of total load. Two examples of total load formulae are outlined below.

Energy (stream power) formula (Bagnold, 1966)

The equation is based on the immersed weight of sediment per unit bed area:

$$\text{total weight transport per unit channel width}(W_T'U) = W_b'U_b + W_s'U_s \tag{9.18}$$

where b refers to bedload and s to suspended load, and the U terms refer to 'typical' sediment transport velocities.

The bedload component. From a consideration of the forces acting on a particle (Fig. 9.1(b)), movement will occur when $F_D >$ resistance. The resistance is assumed to be a function of W' (the immersed weight of the particles) and ϕ:

$$F_D = W' \tan\phi$$

So if W_b' is the immersed weight of particles per unit bed area being transported as bedload, then the bedload work rate is

$$W_b'U_b \tan\phi$$

The power, P, required to maintain the bed movement is provided by the flow. For uniform flow in a channel of rectangular cross-section:

$$P = \rho g V b Y S_0 \text{ or } P/b = \rho g V Y S_0 \tag{9.19}$$

where $V = Q/A$. Only a fraction of the stream power is absorbed in bed movement. Bagnold used an 'efficiency' e_b to estimate the bedload power:

$$e_b P/b = e_b \rho g V Y S_0 = W_b' U_b \tan\phi \tag{9.20a}$$

The suspended load component. If a suspended load exists, then the fluid must supply an effective upward velocity, which must be equal and opposite to v_{FS}. The suspended load work rate is given by

$$W_s' v_{FS} = W_s' U_s \frac{v_{FS}}{U_s} = W_s' U_s \tan\phi_s$$

As some power has been absorbed in bed movement, the power remaining is $(1 - e_b)P/b$. Power absorbed in suspended load transport is

$$(e_s P/b)(1 - e_b) = W_s' v_{FS} = W_s' U_s \tan\phi_s \tag{9.20b}$$

where e_s is a suspended load 'efficiency'.

Total load. This is obtained by adding the two components

$$W_T' U = W_b' U_b + W_s' U_s = \frac{P}{b}\left\{\frac{e_b}{\tan\phi} + \frac{e_s(1 - e_b)}{\tan\phi_s}\right\} \tag{9.21}$$

($W_T' U = (\rho_s - \rho) g q_s$, where q_s is the total sediment transport per unit width.)

This leaves the problem of evaluating e_b, e_s, $\tan\phi$, etc. Bagnold assumed a universal constant value for $e_s (= 0.015)$ and hence estimated that $e_s(1 - e_b) \simeq 0.01$, both values being for fully turbulent flow. Some inevitable uncertainties were ignored in obtaining these values. The two remaining values have been reduced to analytical form.

tan ϕ.

$$\text{If } G^2 = \frac{\rho_s D^2 u_*^2}{\rho \times 14 \nu^2}$$

where G is analogous to the Reynolds Number, Bagnold's data lead to the following:
for $G^2 < 150$,

$$\tan\phi = 0.75 \tag{9.22a}$$

for $150 < G^2 < 6000$,

$$\tan\phi = -0.236\log G^2 + 1.25 \quad (9.22b)$$

for $G^2 > 6000$,

$$\tan\phi = 0.374 \quad (9.22c)$$

e_b. If $\rho_s = 2650\,kg/m^3$ and $0.3 < V < 3.0\,m/s$, then
for $0.015 < D$ mm < 0.06,

$$e_b = -0.012\log 3.28V + 0.15 \quad (9.23a)$$

for $0.06 < D$ mm < 0.2,

$$e_b = -0.013\log 3.28V + 0.145 \quad (9.23b)$$

for $0.2 < D$ mm < 0.7,

$$e_b = -0.016\log 3.28V + 0.139 \quad (9.23c)$$

for D mm > 0.7,

$$e_b = -0.028\log 3.28V + 0.135 \quad (9.23d)$$

The Bagnold total load equation must thus be regarded as primarily a 'sand in water' transport equation for water depths $Y > 150$ mm and particle sizes limited to the range given above.

Ackers and White (A & W) formula (White, 1972) and in revised form in Ackers (1993)

The A & W formula is one of the more recent developments in this field. Initially the underlying theoretical work was developed by considering the transport of coarse material (bedload) and fine material (suspended load) separately. Ackers and White then sought to establish 'transitional' relationships to account for the intermediate grain sizes. The functions which emerged are based upon three dimensionless quantities, G_{gr}, F_{gr} and D_{gr}: G_{gr} is the sediment transport parameter, which is based on the stream power concept. For bedload, the effective stream power is related to the velocity of flow and to the net shear force acting on the grains. Suspended load is assumed to be a function of total stream power, P. The particle mobility number, F_{gr}, is a function of shear stress/immersed weight of grains. The critical value of F_{gr} (i.e. the magnitude representing inception of motion) is denoted by A_{gr}. Finally, a dimensionless particle size number, D_{gr}, expresses the relationship between immersed weight of grains and viscous forces.

The equations are then as follows:

$$G_{gr} = \frac{q_S D_m}{qD}\left[\frac{u_*}{V}\right]^n = C\left[\frac{F_{gr}}{A_{gr}} - 1\right]^m \tag{9.24a}$$

$$F_{gr} = \frac{u_*^n}{\sqrt{gD[(\rho_s/\rho) - 1]}}\left(\frac{V}{\sqrt{32}\log(10D_m/D)}\right)^{1-n} \tag{9.24b}$$

$$D_{gr} = D\left(\frac{g[(\rho_s/\rho) - 1]}{\nu^2}\right)^{1/3} \tag{9.25}$$

(note that $V = Q/A$). The index n does have a physical significance, since its magnitude is related to D_{gr}. For fine grains $n = 1$, for coarse grains $n = 0$, and for transitional sizes $n = f(\log D_{gr})$.

The values for n, m, A_{gr} and C are as follows:
for $D_{gr} > 60$ (coarse sediment with $D_{50} > 2\,\text{mm}$):

$$n = 0, m = 1.78, A_{gr} = 0.17, C = 0.025 \tag{9.26a}$$

for $1 < D_{gr} < 60$ (transitional and fine sediment, with D_{50} in the range 0.06–2 mm):

$$n = 1 - 0.56\log D_{gr} \tag{9.26b}$$

$$m = 1.67 + 6.83/D_{gr} \tag{9.26c}$$

$$A_{gr} = 0.14 + 0.23/D_{gr}^{1/2} \tag{9.26d}$$

$$\log C = 2.79\log D_{gr} - 0.98(\log D_{gr})^2 - 3.46 \tag{9.26e}$$

Equation (9.24a) may also be expressed in the form $F_{gr} = A_{gr} + A_{gr}(G_{gr}/C)^{1/m}$, which can be regarded as equating a function of transport mobility (F_{gr}) to the other terms.

The equations have been calibrated by reference to a wide range of data, and good results are claimed – 'good results' in this context meaning that for 50% or more of the results,

$$\frac{1}{2} < \left(\frac{\text{estimated } q_s}{\text{measured } q_s}\right) < 2$$

Example 9.3 Siltation of reservoir

A reservoir having a capacity of $20 \times 10^6\,m^3$ is to be sited in a river valley. The river has the following characteristics: width 10 m; bed slope 1 in 3000; discharge $87\,m^3/s$ (assumed to be constant); depth 5 m. The river boundary is alluvial ($D_{50} = 0.3\,mm$, $\rho_s = 2650\,kg/m^3$). Estimate the time which would elapse before the reservoir capacity is reduced to half its original capacity. Assume a rectangular channel section.

Solution

Using Bagnold's method

Estimation of tan ϕ_s. The natural rate of fall v_{FS} may be obtained from the relationships

$$C_D = F/\frac{1}{2}\rho A v_{FS}^2$$

and

$$F = (\rho_s - \rho) g K_V D^3$$

(i.e. assuming that the only forces acting on the body are the immersed self-weight and the drag). For high Reynolds Numbers (and turbulent wake) $C_D = 0.44$ for a sphere.

If the sediment is assumed to be spherical, then the area $A = K_A D^2 = (\pi/4)D_{50}^2$, and the volume $= K_V D^3 = (\pi/6)D_{50}^3$. Therefore

$$0.44 = \frac{(2650 - 1000) \times 9.81 \times \pi \times (0.3 \times 10^{-3})^3 \times 4}{\frac{1}{2} \times 1000 \times \pi \times (0.3 \times 10^{-3})^2 \times v_{FS}^2 \times 6}$$

Hence

$$v_{FS} = 0.121\,m/s$$

Assuming that $U_s \to$ velocity of flow, then $U_s \simeq 87/(10 \times 5) = 1.74\,m/s = V$. Therefore

$$\tan\phi_s = 0.121/1.74$$

Estimation of e_b *and tan* ϕ. For $D_{50} = 0.3\,mm$, use (9.23c):

$$e_b = -0.016 \times \log(3.28 \times 1.74) + 0.139 = 0.127$$

$$R = 2.5\,m$$

$$u_*^2 = gRS_0 = 9.81 \times 2.5 \times 1/3000$$

Therefore

$$G^2 = \frac{2650 \times (0.3 \times 10^{-3})^2 \times 9.81 \times 2.5 \times 1}{14 \times 1000 \times (1.14 \times 10^{-6})^2 \times 3000} = 107.2$$

Therefore, from (9.22a), $\tan\phi = 0.75$.

Estimation of stream power. From (9.19),
$P/b = \rho g V Y S_0 = 1000 \times 9.81 \times 1.74 \times 5 \times 1/3000 = 28.45\,\mathrm{W/m}$. Thus, from (9.21),

$$W_T' U = 28.45 \left\{ \frac{0.127}{0.75} + \left(0.01 \times \frac{1.74}{0.121} \right) \right\} = 8.91\,\mathrm{N/ms}$$

Therefore, over 10 m width flux $= 10 \times 8.91 = 89.1\,\mathrm{N/s}$. This represents a volume

$$Q_s = \frac{89.1}{(2650 - 1000) \times 9.81 \times (1 - p_s)}\,\mathrm{m^3/s}$$

p_s is the porosity or voidage of the grains when packed closely together, and is usually approximately 0.3. Therefore, the rate at which the reservoir will fill is given by

$$\frac{89.1}{(2650 - 1000) \times 9.81 \times (1 - 0.3)} = 7.864 \times 10^{-3}\,\mathrm{m^3/s}$$

Therefore, the annual sediment volume $= 247\,320\,\mathrm{m^3}$. So the sediment will have reduced the capacity to half its original value in

$$\frac{20 \times 10^6}{2 \times 247\,320} \simeq 40 \text{ years}$$

Using the A & W equations

$$u_* = \sqrt{\tau_0/\rho} = \sqrt{gRS_0} = \sqrt{9.81 \times 2.5 \times 1/3000}$$
$$= 0.0904\,\mathrm{m/s}$$

From (9.25),

$$D_{gr} = D \left(\frac{g[(\rho s/\rho) - 1]}{\nu^2} \right)^{1/3} = 0.3 \times 10^{-3} \left(\frac{9.81[(2650/1000) - 1]}{(1.1 \times 10^{-6})^2} \right)^{1/3}$$
$$= 7.12$$

Since D_{gr} is in the transitional range, (9.26b)–(9.26e) are used to calculate n, m, A and C:
from (9.26b)

$$n = 1 - 0.56 \log D_{gr} = 1 - 0.56(\log 7.12) = 0.5226$$

from (9.26c)

$$m = 1.67 + 6.83/D_{gr} = 1.67 + 6.83/7.12 = 2.629$$

from (9.26d)

$$A_{gr} = 0.14 + 0.23/D_{gr}^{1/2} = 0.14 + 0.23/7.12^{1/2} = 0.2262$$

from (9.26e)

$$\begin{aligned}\log C &= 2.79 \log D_{gr} - 0.98(\log D_{gr})^2 - 3.46 \\ &= 2.79(\log 7.12) - 0.98(\log 7.12)^2 - 3.46 = -1.7938\end{aligned}$$

Therefore

$$C = 0.016$$

Using (9.24b),

$$\begin{aligned}F_{gr} &= \frac{u_*^n}{\sqrt{gD[(\rho_s/\rho) - 1]}} \left(\frac{V}{\sqrt{32} \log(10D_m/D)} \right)^{1-n} \\ &= \frac{0.0904^{0.5226}}{\sqrt{9.81 \times 0.3 \times 10^{-3}[(2650/1000) - 1]}} \left(\frac{1.74}{\sqrt{32} \log[10 \times 5/0.3 \times 10^{-3}]} \right)^{1-0.5226} \\ &= 1.0574\end{aligned}$$

Therefore, from (9.24a),

$$\begin{aligned}G_{gr} &= \frac{q_s D_m}{qD} \left(\frac{u_*}{V} \right)^n = C \left(\frac{F_{gr}}{A_{gr}} - 1 \right)^m \\ &= \frac{q_s \times 5}{8.7 \times 0.3 \times 10^{-3}} \left(\frac{0.0904}{1.74} \right)^{0.5226} = 0.016 \left(\frac{1.0574}{0.2262} - 1 \right)^{2.629}\end{aligned}$$

Hence

$$q_s = 1.2 \times 10^{-3}\ \mathrm{m^3/ms}$$

Therefore if the porosity, $p_s = 0.3$, the reservoir will fill at a rate:

$$1.2 \times 10^{-3} \times 10/(1 - 0.3) = 17.133 \times 10^{-3}\ \mathrm{m^3/s}$$

Therefore annual volume of sediment deposited is 521 087 m^3. Hence the reservoir capacity will be reduced by half in

$$\frac{20 \times 10^6}{2 \times 521\,087} = 19 \text{ years}$$

9.5 Concluding notes on sediment transport

Limitations of transport equations

Sediment transport processes are complex and the above treatment is brief and introductory. Many important issues could not be covered. For example, it has been assumed that estimates of sediment transport rates may be based on one 'typical' particle size. (D_{50}, say), but this is not realistic since an actual river boundary consists of a range of particle sizes. It is possible to use a transport formula to estimate the transport rate for each of a series of particle size fractions and then add the rates to form a total. Even this is not correct since:

1. The exposed particles of a given size will constitute only a fraction of the area of the river boundary.
2. Some particles of a given size will be wholly or partly sheltered by surrounding (larger) particles, which will affect their mobility.

An attempt to incorporate the effect of these conditions was made by Einstein *et al.*, details of which may be found in Graf (1971) or Yang (1996). Yang also provides examples of comparative tests on a range of transport formulae.

The sediment transport formulae given here pre-suppose equilibrium conditions i.e. rates of erosion are balanced by rates of deposition, that the discharge and boundary roughness (and hence shear stress) do not vary and that sediments are non-cohesive. For natural rivers changes can occur over fairly short periods (e.g. the passage of a flood). Furthermore, in some cases, the finer fractions of sediment may be eroded from the bed and banks and not be replaced by deposition. The remaining (coarser) fractions are less easily eroded, so the sediment transport rate reduces and the channel becomes more stable; a process which is sometimes known as 'armouring'.

Sediment transport in estuaries

Sediment which is transported down a river and into a sheltered estuary is often fine and silty in nature. In the estuary the water will be saline, due to mixing with seawater. The current system in the estuary will be the resultant of the river discharge and the tidal cycle, the tides will also cause the depth of the water to vary. The sediment in the estuary is subjected to a number of changes:

(a) electro-chemical changes due to the salt in the water (increasing salinity often causes flocculation which tends to increase deposition of the suspended sediment).

(b) possible human impacts due to industrial effluents, shipping movements etc.
(c) repeated cycles of erosion, transport and deposition due to the tides.
(d) changes in the magnitudes of tidal currents (due to the range of conditions between spring and neap tides).
(e) flushing of the estuary due to fluvial flood events.

The outcome is a change in the nature and behaviour of the sediment, which becomes transformed into estuarial mud. A chemical analysis would usually reveal a complex cocktail of silt, metals and other chemicals and clay. The mud is 'sticky' in consistency because there is a degree of inter-particulate adhesion (or cohesion), and it is therefore known as a cohesive sediment or cohesive mud. It is often found that there is a thin surface layer of mud, which is fairly liquid, below which, the mud is consolidated and is less easily eroded. The cohesiveness has an effect on the threshold of motion and on sediment mobility, and may significantly attenuate wave action.

Cohesive muds act as a sink for organic materials and other pollutants. This can have the beneficial effect of providing a food supply, for invertebrate organisms and hence for other wildlife such as wading birds. On the other hand high concentrations of suspended sediment can produce a biochemical oxygen demand (BOD) and hence a deleterious effect on aquatic life. An understanding of cohesive sediments must encompass their biological, chemical, physical and ecological attributes.

Some estuarial sediments may be a combination of cohesive and non-cohesive sediments, adding to the complexity of the transport processes. More details about cohesive sediments may be found in Dyer (1997), Raudkivi (1998) and Whitehouse *et al.* (2000).

Marine sediment transport

The flow field in a marine environment is complex, since there are tidal and other currents, waves and variations in depth, due to the tides. In intermediate or shallow waters, waves induce an oscillatory motion in the water over the seabed, and waves may break, producing intense local turbulence. Sediments may be sandy or stony (gravel/shingle). For some situations sediment transport formulae originally developed for river flows have been adapted for maritime applications. However, where wave breaking occurs, it is normal practice to use transport formulae which have been developed and calibrated specifically for such conditions. A brief introduction to coastal sediment transport is included in Chapter 16. For further information see Dyer (1986), Raudkivi (1998) or Reeve, Chadwick and Fleming (2004).

Research into many aspects of sediment transport is ongoing. The general availability of personal computers means that computational models are more commonly used. These have the potential to permit better simulation of the turbulent flow field and hence of the distribution of sediment transport in the water. However, calibration of such models and the accurate establishment of boundary conditions is still a problem due to limitations in the available data.

References and further reading

Ackers, P. (1993) Sediment transport in open channels: Ackers and White update. *Proc. Instn. Civil Engrs; Water, Maritime and Energy*, **101**, 247–9.

Bagnold, R. A. (1966) *An Approach to the Sediment Transport Problem from General Physics*. Professional Paper, 422-I. US Geological Survey, Washington, DC.

Bagnold, R. A. (1980) An empirical correlation of bedload transport rates in flumes and natural rivers. *Proc. Roy. Soc.*, **A372**, 453–73.

Bagnold, R. A. (1986) Transport of solids by natural water flow: evidence for a worldwide correlation. *Proc. Roy. Soc.*, **A405**, 369–74.

Chang, H. H. (1988) *Fluvial Processes in River Engineering*, Wiley, New York.

Cheong, H. F. and Shen, H. W. (1983) Statistical properties of sediment movement. *Am. Soc. Civ. Engrs, J. Hydraulic Engng*, **109**(12), 1577–88.

Dyer, K. R. (1986) *Coastal and Estuarine Sediment Dynamics*, Wiley, Chichester.

Dyer, K. R. (1997) *Estuaries, a Physical Introduction*, Wiley, Chichester.

Einstein, H. A. (1942) Formulas for the transportation of bedload. *Trans. Am. Soc. Civ. Engrs*, **107**, 561–77.

Graf, W. H. (1971) *Hydraulics of Sediment Transport*, McGraw-Hill, New York.

McDowell, D. M. (1989) A general formula for estimation of the rate of transport of bed load by water. *J. Hydraulic Res.*, **27**(3), 355–61.

McDowell, D. M. and O'Connor, B. A. (1977) *Hydraulic Behaviour of Estuaries*, Macmillan, London.

Nnadi, F. N. and Wilson, K. K. (1996) Bed load motion at high shear stress: dune washout and plane-bed flow. *Am. Soc. Civ. Engrs, J. Hydraulic Engng*. **121**(3), 267–73.

Raudkivi, A. J. (1998) *Loose Boundary Hydraulics*, 4th edn, Balkema, Rotterdam.

Reeve, D., Chadwick A. and Fleming, C. (2004). *Coastal Engineering: Processes, Theory and Design Practice*, E & FN Spon, London.

Rijn, L. van (1984) Sediment transport (in 3 parts). *Am. Soc. Civ. Engrs, J. Hydraulic Engng: Part I*, **110**(10), 1431–56; *Part II*, **110**(11), 1613–41; *Part III*, **110**(12), 1733–54.

Rouse, H. (ed.) (1950) *Engineering Hydraulics*, Wiley, New York.

Shields, A. (1936) Anwendung der Ahnlichkeitsmechanik und der Turbulenzforschung auf die Geschiebebewegung, Heft **26**. *Preuss. Vers. für Wasserbau und Schiffbau*, Berlin.

White, W. R. (1972) *Sediment Transport in Channels, a General Function*. Rep. Int. 102, Hydraulics Research, Wallingford.

White, W. R., Milli, H. and Crabbe, A. D. (1973) *Sediment Transport, an Appraisal of Existing Methods*, Rep. Int. 119. Hydraulics Research, Wallingford.

Whitehouse, R., Soulsby, R., Roberts, W., Mitchener, H. (2000) *Dynamics of Estuarine Muds*, Thomas Telford, London.

Yalin, M. S. (1977) *Mechanics of Sediment Transport*, 2nd edn, Pergamon, Oxford.

Yang, C. T. (1996) *Sediment Transport: Theory and Practice*, McGraw-Hill, New York.

10
Flood hydrology

10.1 Classifications

Hydrology has been defined as the study of the occurrence, circulation and distribution of water over the world's surface. As such, it covers a vast area of endeavour and is not the exclusive preserve of civil engineers. Engineering hydrology is concerned with the quantitative relationship between rainfall and 'runoff' (i.e. passage of water on the surface of the earth) and, in particular, with the magnitude and time variations of runoff. This is because all water resource schemes require such estimates to be made before design of the relevant structures may proceed. Examples include reservoir design and flood risk management including flood alleviation schemes and land drainage. Each of these examples involves different aspects of engineering hydrology, and all involve subsequent hydraulic analysis before safe and economical structures can be constructed.

Engineering hydrology is conveniently subdivided into two main areas of interest, namely, surface water hydrology and groundwater hydrology. The first of these is further subdivided into rural hydrology and urban hydrology, since the runoff response of these catchment types to rainfall is very different. A catchment is an area of the earth's surface which drains into a particular river or underground storage.

The most common use of engineering hydrology is the prediction of 'design' events. This may be considered analogous to the estimation of 'design' loads on structures. Design events do not mimic nature, but are merely a convenient way of designing safe and economical structures for water resources schemes. As civil engineers are principally concerned with the extremes of nature, design events may be either floods or droughts. The design of hydraulic structures will normally require the estimation of a suitable design flood (e.g. for spillway sizing) and sometimes a design drought (e.g. for reservoir capacity).

The purpose of this chapter is to introduce the reader to some of the concepts of engineering hydrology. The treatment is limited to the estimation of design floods for rural and urban catchments. This limitation is necessary because a complete introduction to hydrology would occupy a textbook in its own right. However, the chapter is considered useful, since hydrological design is the precursor to many hydraulic designs. Consequently, civil engineers should have an overall understanding of both subjects and their interactions.

10.2 Methods of flood prediction for rural catchments

Historically, civil engineers were faced with the problem of flood prediction long before the current methods of analysis were available. Two techniques in common use in the 19th and early 20th centuries were those of using the largest recorded 'historical' flood and the use of empirical formulae relating rainfall to runoff.

The former was generally more accurate, but as runoff records were sparse, the latter was often used in practice. This was possible because rainfall records have been collected for much longer periods than have runoff records.

The occurrence of a series of catastrophic floods in the 1960s in the UK prompted the Institution of Civil Engineers to instigate a comprehensive research programme into methods of flood prediction. This was carried out at the Institute of Hydrology (IH), and culminated in the publication of the *Flood Studies Report* (NERC, 1975) usually abbreviated as the FSR. The report represented a milestone in British hydrology, assimilating previous knowledge with new techniques which were comprehensively tested against an enormous data set of hydrological information. The result was the formulation of a new set of design methods which could be applied with greater confidence to a wide range of conditions.

Since the publication of the FSR, a further 18 supplementary reports have been written refining its methods. In order to consolidate these reports and in the light of more recent research and greater access to hydrological data the Institute of Hydrology (now the Centre for Ecology and Hydrology) undertook a review of the FSR methods in the 1990s. The resulting *Flood Estimation Handbook* (IH, 1999), usually abbreviated as the FEH, largely supersedes the *Flood Studies Report*.

There are fundamentally two types of flood prediction technique recommended in the FEH. These are statistical methods (e.g. frequency analysis) and the unit hydrograph rainfall-runoff model. In addition, there are two types of catchment, those which are gauged (i.e. have recorded rainfall and runoff records) and those which are ungauged. One of the major aspects of the FSR was the derivation of techniques which allow flood prediction for ungauged catchments. This involved finding quantitative relationships between catchment descriptors and flood magnitudes for large numbers of gauged catchments, and the application of these results to ungauged catchments by the use of multiple regression techniques. The FEH has changed the emphasis from relying on generalised regression equations to the application of techniques for transferring hydrological data from gauged to ungauged catchments.

In the following sections, the basic ideas of catchment descriptors, frequency analysis and unit hydrographs are introduced, and the application of these methods to gauged and ungauged catchments is discussed.

10.3 Catchment descriptors

A good starting point for a quantitative assessment of runoff is to consider the physical processes occurring in the hydrological cycle and within the catchment, as shown in Figures 10.1(a) and (b).

Circulation of water takes place from the ocean to the atmosphere by evaporation, and this water is deposited on a catchment mainly as rainfall. From there, it may follow several routes, but eventually the water is returned to the sea via the rivers.

Within the catchment, several circulation routes are possible. Rainfall is initially intercepted by vegetation and may be re-evaporated. Secondly, infiltration into the soil or overland flow to a stream channel or river may occur. Water entering the soil layer may remain in storage (in the unsaturated zone) or may percolate to the groundwater table (the saturated zone). All subsurface water may move laterally and eventually enter a stream channel. The whole system may be viewed as a series of linked storage processes with inflows and outflows, as shown in Figure 10.1(c). Such a representation is referred to as a conceptual model. If equations defining the storages and flows can be found, a mathematical catchment model can be constructed (see Beven, 2000).

Using this qualitative picture, a set of descriptors may be proposed which determine the response of the catchment to rainfall. These might include the following:

1. catchment area;
2. soil type(s) and depth(s);
3. vegetation cover;
4. stream slopes and surface slopes;
5. rock type(s) and area(s);
6. drainage network (natural and artificial);
7. lakes and reservoirs;
8. impermeable areas (e.g. roads, buildings, etc.).

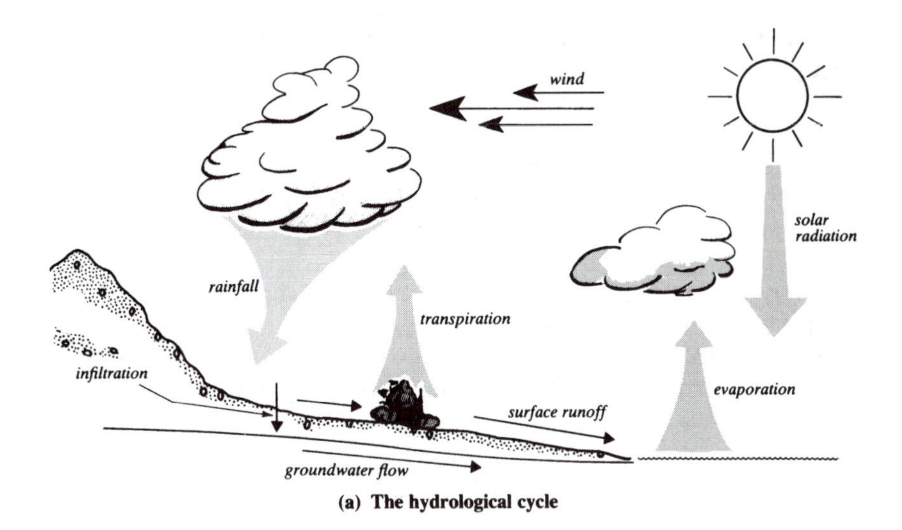

(a) The hydrological cycle

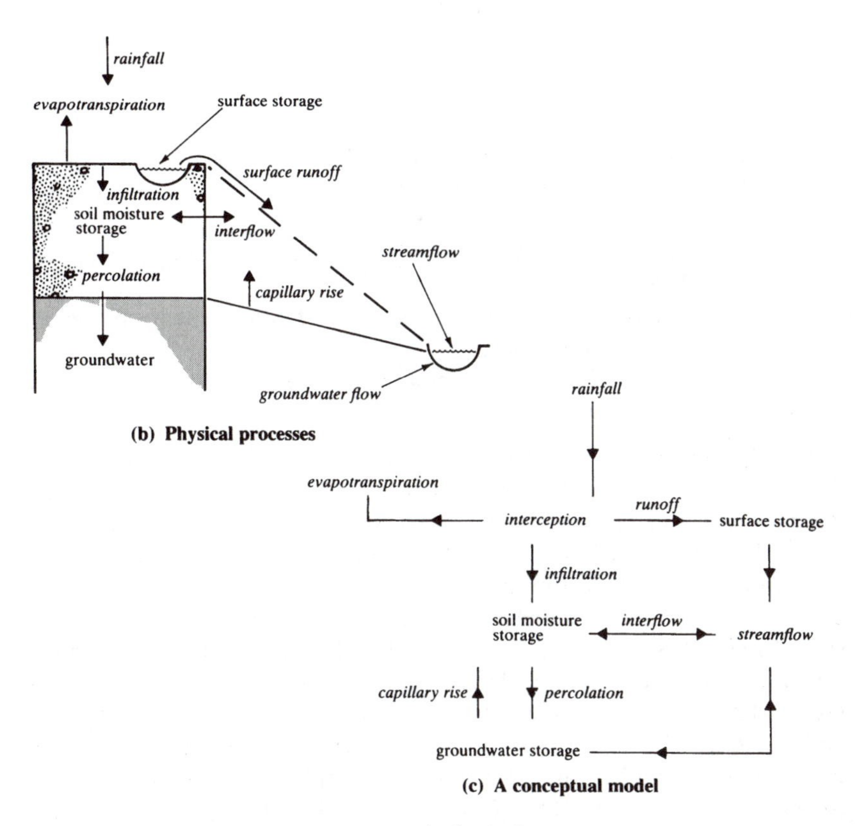

(b) Physical processes

(c) A conceptual model

Figure 10.1 Hydrological process.

In addition, different catchments will experience different climates, and hence the response of the catchment to rainfall will depend also on the prevailing climate. This may be represented by:

(a) rainfall (depth, duration and intensity);
(b) evaporation potential (derived from temperature, humidity, windspeed and solar radiation measurements or from evaporation pan records).

However, from an engineering viewpoint, qualitative measures of catchment descriptors are inadequate in themselves, and quantitative measures are necessary to predict flood magnitudes. This was one of the tasks performed by the FEH team which led, for example, to the following equation for rural catchments:

$$QMED = 1.172\, AREA^{AE} \left(\frac{SAAR}{1000}\right)^{1.560} \times FARL^{2.642} \left(\frac{SPRHOST}{100}\right)^{1.211} 0.0198^{RESHOST} \qquad (10.1)$$

where

$QMED =$ the median annual maximum flood (m^3/s);
$AREA =$ the catchment area (km^2);
$AE =$ the exponent of the $AREA$ term;
$SAAR =$ the average annual rainfall over the standard period 1961–1990 (mm);
$FARL =$ a number indicating the degree of flood attenuation due to reservoirs and lakes;
$SPRHOST =$ the standard percentage runoff depending on soil type (%);
$RESHOST =$ a number dependent on soil type.

An adjustment to the *QMED* for urbanisation may be undertaken. Full details are given in the FEH and values for the descriptors are available in digital form for catchments with an area of 0.5 km^2 or greater in the UK on the FEH CD-ROM (CEH, 1999a). This equation contains all the catchment descriptors which were found to be statistically significant, and may be applied to ungauged catchments. However, it should only be used for UK rural catchments, and as a last resort for the design of minor works if no suitable data can be transferred from another gauged catchment (see section 10.4). The equation will only give an approximate value for *QMED*, and this reflects the difficulty of predicting natural events with any certainty. Furthermore it should be noted that the topographic descriptors (e.g. *AREA*) provided in digital form are derived from a digital terrain model with a 50 m×50 m grid size. Consequently the definition of these descriptors for small, urbanised or relatively flat catchments may require checking against local information (e.g. maps and site surveys).

10.4 Frequency analysis

For gauged catchments with records of sufficient length the techniques of frequency analysis may be applied directly to determine the magnitude of a flood event with a specified return period. The concept of return period is an important one because it enables the determination of risk (economic or otherwise) associated with a given flood magnitude. The return period is usually expressed in years. It may be formally defined as the average time interval between flood events which are greater than a specified value. The qualifier 'on average' is often misunderstood. For example, although a 100-year flood event will occur, on average, once every 100 years, it may occur at any time (i.e. today or in several years' time). Also, within any particular 100-year period, floods of greater magnitude may occur.

The annual maxima series

The return period is often estimated from a frequency analysis of an annual maxima series, in which the largest flood event from each year of records at a site is abstracted. The annual frequency with which a given flood magnitude is exceeded is the reciprocal of return period, T. For sufficiently large T this frequency represents the annual probability that a particular flood will exceed a specified value, Q, which may be written as

$$P(Q) = \frac{1}{T} \tag{10.2}$$

The confusion surrounding the interpretation of return period may be alleviated by the use of the term annual probability of flooding, particularly when dealing with people who do not have a technical background (Fleming, 2002). For example the 1 in 100-year flood event is easier to understand as the 1% annual probability of flooding (i.e. there is a 1 in 100 chance of a particular flood magnitude being exceeded in any year). For consistency with the terminology used in the FEH and other technical literature, return period will be used throughout this chapter. However, the reader's attention is drawn to promoting the use of the term annual probability of flooding.

The annual maxima series requires a relatively long duration record (typically more than 13 years). In order to capture the seasonal effects of climate for a region it is usual to divide the record into water years (e.g. 1 October–30 September in the UK). The resulting series, in statistical terms, is considered to be an independent series and constitutes a random sample from an unknown population. The series may be plotted as a histogram, as shown in Figure 10.2(a). Taking, as an example, a 31-year record, the annual maxima are divided into n class intervals of equal size, ΔQ. The probability that the discharge will exceed, say, $60\,m^3/s$ is equal to the number of events greater than $60\,m^3/s$ divided by the total number of events, N:

$$P(60) = (4+3+2)/31 = 0.29$$

and the corresponding return period is

$$T = 1/P(60) = 3.4 \text{ years}$$

If the histogram is now replaced by a smooth curve, as shown in Figure 10.2(b), then

$$P(Q) = \int_Q^\infty f(Q)\mathrm{d}Q$$

The function $f(Q)$ is known as a probability density function (pdf) and, by definition,

$$\int_0^\infty f(Q)\mathrm{d}Q = 1$$

(i.e. $P(Q \geqslant 0) = 1$). Hence, the scale equivalent to the histogram ordinate is given by $f_i/N\Delta Q$ as

$$\sum_{i=1}^{n} \left(\frac{f_i}{N\Delta Q} \Delta Q \right) = 1$$

where f_i is the number of annual maxima in class i.

The point of this analysis is that it makes it possible to estimate the probability that the discharge will exceed any given value greater than the maximum value in the data set ($90\,m^3/s$ in this case). Replacing the histogram with the pdf allows such estimates to be made.

The technique which is used in practice looks rather different from the histogram, so the method is now extended. If, instead of drawing a pdf, the cumulative probability of non-exceedence, $F(Q)$ is drawn, i.e.

$$F(Q) = 1 - P(Q) = \int_0^Q f(Q)\mathrm{d}Q$$

then Fig 10.2(c) is the result. $F(Q)$ is also known as the cumulative distribution function, F. It is usual to plot the discharge axis, Q, as the ordinate against a transformed version of the F scale known as the reduced variate, y, of the distribution as shown in Figure 10.2(d). This is known as probability paper. The probability scale is non-linear and depends on the shape

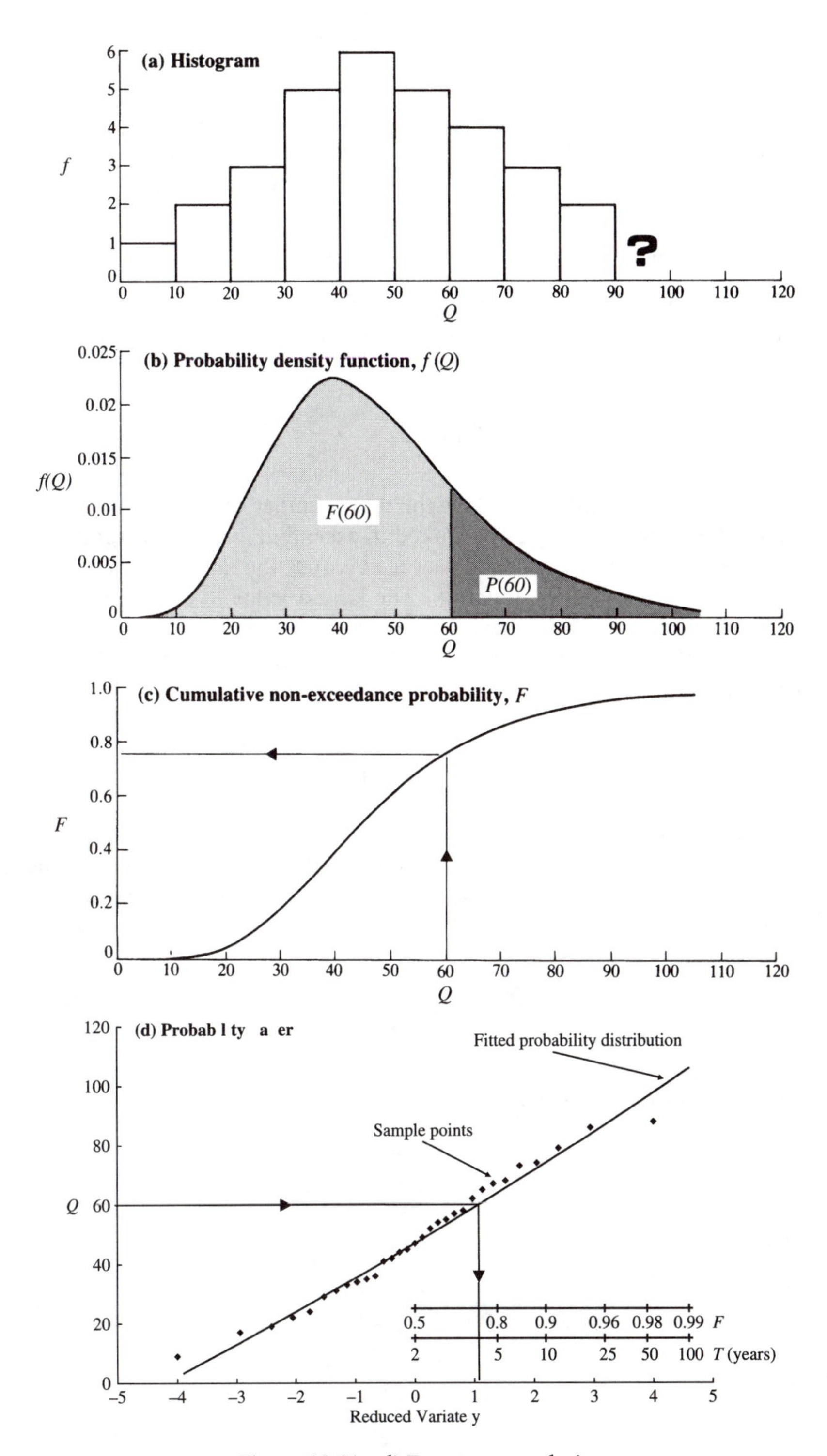

Figure 10.2(a–d) Frequency analysis.

of the original pdf. The convention is to calculate the reduced variate scale so that the two-parameter case of the distribution plots as a straight line. (*Note*: this is only true if the pdf is an exact fit to the sample data!)

Plotting positions

To plot the sample data on suitable probability paper requires the estimation of the cumulative probability of non-exceedence of each event. As an initial estimate, one could say that

$$F(Q_i) = \frac{i}{N}$$

where i is the rank number and N is the total number of events, and where the annual maxima data, Q_i, are ranked in ascending order of magnitude.

However, this formula is unsatisfactory because the data set is a sample drawn from an unknown population. The largest value in the sample may have a considerably higher return period than that suggested by this simple formula which exhibits bias at the extremes of the distribution. A more general formula is

$$F(Q_i) = \frac{i - \mathrm{a}}{\mathrm{N} + \mathrm{b}}$$

where a and b are constants for particular pdfs.

For the case of the generalised logistic distribution (discussed in the following section) a more unbiased plotting position (due to Gringorten) is given by

$$F(Q_i) = \frac{i - 0.44}{N + 0.12} \qquad (10.3)$$

For example, again using the data set with $N = 31$, for the largest discharge ($90\,\mathrm{m^3/s}$) $i = 31$ so

$$F(90) = \frac{31 - 0.44}{31 + 0.12} = 0.982$$

Since

$$T = \frac{1}{P(Q)} = \frac{1}{1 - F(Q)} \qquad (10.4)$$

then

$$T = \frac{1}{1 - 0.982} = 55.6 \text{ years}$$

In other words, the expected return period of the largest discharge from the record is 55.6 years.

It is sometimes useful to be able to estimate the probability, or risk, r, that an event of return period T will be exceeded during a particular time interval of M years. Since $F(Q)$ is the probability of Q not being exceeded in a year then the probability of annual events not exceeding the given magnitude Q in a period of M years is

$$[F(Q)]^M$$

since the events are independent. Consequently, the probability of Q being exceeded at least once in M years must be

$$r = 1 - [F(Q)]^M$$

or

$$r = 1 - [1 - P(Q)]^M$$

where $P(Q)$ is the probability of Q being exceeded in one year.

From equation (10.4) then this risk, r, is given by

$$r = 1 - \left(1 - \frac{1}{T}\right)^M \tag{10.5}$$

The generalised logistic distribution

Various pdfs have been tested to see if they fit hydrological data sets, for example the log-normal, Gumbel, general extreme value, and log-Pearson type III. A general result is that such data tends to be asymmetrical and that no single pdf is universally applicable. It should be noted that different analysis techniques and distributions will assign different return periods to observed events in a record. The FEH recommends fitting the generalised logistic distribution (GL) to annual maximum flow series in the UK for estimating return periods in the range 2–200 years.

For the GL (Hosking and Wallis, 1997), the pdf is

$$f(Q) = \frac{\alpha^{-1} e^{-(1-k)w}}{(1 + e^{-w})^2} \tag{10.6}$$

and the cumulative distribution function is

$$F = \frac{1}{1 + e^{-w}} \tag{10.7}$$

where

$$w = \begin{cases} -k^{-1} \ln\{1 - k(Q - \xi)/\alpha\}, & k \neq 0 \\ (Q - \xi)/\alpha, & k = 0 \end{cases} \tag{10.8}$$

and the quantile function (flood frequency curve) is

$$Q(F) = \begin{cases} \xi + \alpha\left[1 - \{(1 - F)/F\}^k\right]/k, & k \neq 0 \\ \xi - \alpha \ln\{(1 - F)/F\}, & k = 0 \end{cases} \tag{10.9}$$

The three parameters ξ, α and k are known as location, scale and shape parameters, respectively. For a GL they define the shape of the pdf and may be estimated from the L-moments of the data set (explained later).

Pooled (regional) frequency analysis

The FEH advocates the use of pooling groups whereby the combined annual maximum series of a group of hydrologically similar gauged catchments is analysed. The reasoning behind the use of pooling groups is to generate a large enough data sample to which the probability distribution can be fitted and extrapolated to predict the magnitude of long return period events at the site of interest – called the subject site. This is important since most river gauging records are short by comparison with the return periods used for flood risk analysis. The use of pooling groups (sometimes referred to as regional frequency analysis) is based on the following basic assumptions (Hosking and Wallis, 1997):

1. the annual maximum series at the different gauging stations are independent of each other; and
2. the data at each gauging station follow the same underlying frequency distribution scaled by a site-specific index flood.

For UK catchments the median annual maximum flood, *QMED*, is used as the index flood. By definition the median has a probability of non-exceedence of 0.5. Substituting into equation (10.9) gives $\xi = QMED$. The growth curve is defined by

$$x = \frac{Q(F)}{QMED} \tag{10.10}$$

which can be expressed in shorthand form as

$$x = \frac{Q}{QMED} \tag{10.11}$$

where x is the growth curve factor. Hence the GL growth curve for $k \neq 0$ becomes

$$x = 1 + \frac{\beta}{k}\left(1 - \left(\frac{1-F}{F}\right)^{k}\right) \quad (10.12)$$

where

$$\beta = \frac{\alpha}{\xi} \quad (10.13)$$

Since ξ is fixed at *QMED* this leaves only the k and β parameters to be estimated.

In the case where $k = 0$ the GL reduces to the two-parameter logistic distribution and the growth curve is given by

$$x = 1 - \beta \ln\left(\frac{1-F}{F}\right) \quad (10.14)$$

The logistic reduced variate, y, form of equation (10.14) is given by

$$x = 1 + \beta y \quad (10.15)$$

where

$$y = -\ln\left(\frac{1-F}{F}\right) \quad (10.16)$$

The GL and logistic growth curves may be plotted on logistic probability paper as shown in Figure 10.3. It should be noted that where the fitted distribution is bounded above (for cases where $k > 0$, because of the mathematical properties of equation (10.12)) the user must assess whether the upper bound adversely affects design events derived from the curve.

Fitting the GL growth curve using L-moments

The remaining parameters, k and β, of the GL growth curve may be estimated from the sample L-moment ratios for each gauging station in the pooling group. A full discussion of the method of L-moments can be found in Hosking and Wallis (1997) and in summary form in the FEH. An outline of the procedure for estimating k and β is given below, and Example 10.1 illustrates the method for a single site.

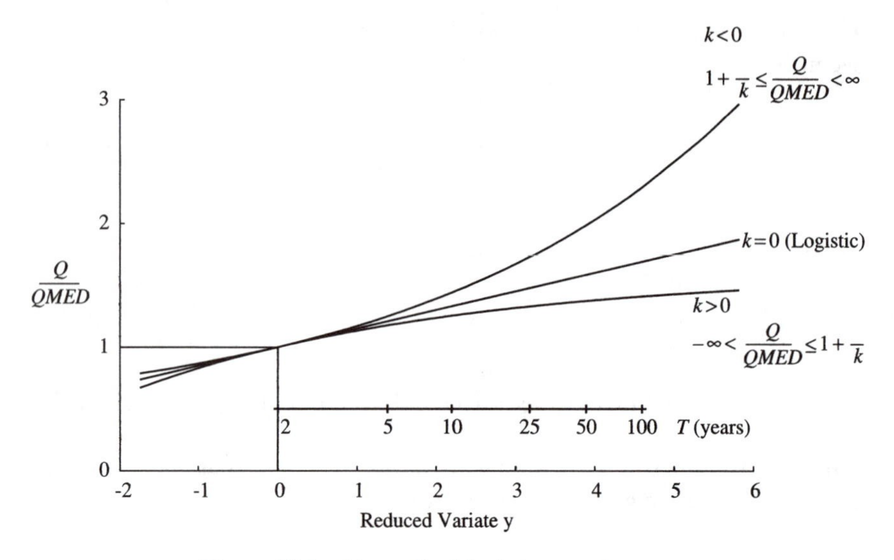

Figure 10.3 Generalised logistic growth curves.

1. The probability weighted moments, b, are estimated for a given gauge record using

$$b_0 = \frac{1}{N}\sum_{j=1}^{N} Q_j \qquad \text{(i.e. the sample mean)} \tag{10.17a}$$

$$b_1 = \frac{1}{N}\sum_{j=2}^{N} \frac{(j-1)}{(N-1)} Q_j \tag{10.17b}$$

$$b_2 = \frac{1}{N}\sum_{j=3}^{N} \frac{(j-1)(j-2)}{(N-1)(N-2)} Q_j \tag{10.17c}$$

$$b_3 = \frac{1}{N}\sum_{j=4}^{N} \frac{(j-1)(j-2)(j-3)}{(N-1)(N-2)(N-3)} Q_j \tag{10.17d}$$

where Q_j is the jth element of a sample of annual maximum flows arranged in ascending order; and N is the sample size (i.e. the number of annual maxima in the record).

2. The sample L-moments are defined as

$$l_1 = b_0 \tag{10.18a}$$

$$l_2 = 2b_1 - b_0 \tag{10.18b}$$

$$l_3 = 6b_2 - 6b_1 + b_0 \qquad (10.18c)$$

$$l_4 = 20b_3 - 30b_2 + 12b_1 - b_0 \qquad (10.18d)$$

3. The sample L-moment ratios are defined as

$$\mathrm{L-CV} = t_2 = \frac{l_2}{l_1} \qquad (10.19a)$$

$$\mathrm{L-skewness} = t_3 = \frac{l_3}{l_2} \qquad (10.19b)$$

$$\mathrm{L-kurtosis} = t_4 = \frac{l_4}{l_2} \qquad (10.19c)$$

   These ratios are calculated for each gauge record in the pooling group. A weighted average (with the weightings based on the hydrological similarity of each gauging station to the subject site) is used to obtain the L-moment ratios for the pooling group.

4. Finally the GL growth curve parameters are estimated from:

$$k = -t_3 \qquad (10.20a)$$

$$\beta = \frac{t_2 k \sin \pi k}{k\pi (k + t_2) - t_2 \sin \pi k} \qquad (10.20b)$$

where t_2 and t_3 are the L-moment ratios for the pooling group.

The gauging stations making up the pooling group are selected on the basis of hydrological similarity to the subject site (in terms of the *AREA*, *BFIHOST* and *SAAR* descriptors) and so that the combined record length is approximately $5T$. The FEH gives detailed guidance on assessing the suitability of a pooling group. The methods involved and data requirements require computer analysis using the WINFAP-FEH software package (CEH, 1999b) for UK gauging stations and, as such, are outside the scope of this chapter.

Example 10.1 illustrates the fitting procedure for a single site which may be carried out using a spreadsheet (available at the supporting website at www.sponpress.com/supportmaterial/0415306094.html). It should be noted that fitting the GL to the annual maximum flows for a single site (i.e. in the absence of a pooling group) is acceptable provided the record length, $N > 2T$, where T is the maximum return period of interest.

Example 10.1 Flood frequency analysis

Estimate the 20-year flood event for the River Dart at Austin's Bridge using frequency analysis of the 43-year river flow record.

Data. Table 10.1 lists the annual maximum discharges for water years 1958/59 to 2000/01 (Data copyright © Environment Agency).

Solution

Table 10.2 lists the discharges from lowest to highest, and the associated calculations for the probability weighted moments (using equation (10.17)). The cumulative non-exceedence probabilities are calculated using the Gringorten plotting position formula (equation 10.3). The logistic-reduced variate values are obtained from equation (10.16). Table 10.2 also lists the *QMED* of the series, the L-moments, L-moment ratios, and the k and β sample estimators of the GL distribution (calculated from equations 10.18, 10.19 and 10.20, respectively). Table 10.3 shows the results

Table 10.1 Annual maximum discharges for the River Dart at Austin's Bridge. (Data copyright Environment Agency.)

Water year (from Oct.)	Peak discharge (m^3/s)	Water year (from Oct.)	Peak discharge (m^3/s)
1958	196.886	1980	250.105
1959	327.612	1981	183.630
1960	243.110	1982	317.843
1961	226.810	1983	269.781
1962	218.298	1984	223.727
1963	295.526	1985	222.161
1964	209.387	1986	261.088
1965	227.526	1987	207.150
1966	225.380	1988	179.819
1967	252.966	1989	210.063
1968	206.502	1990	237.144
1969	154.929	1991	92.551
1970	213.542	1992	302.603
1971	140.779	1993	161.489
1972	274.012	1994	215.800
1973	309.446	1995	158.365
1974	284.042	1996	224.299
1975	120.397	1997	175.910
1976	116.637	1998	246.980
1977	172.219	1999	328.887
1978	180.563	2000	371.035
1979	549.735		

Table 10.2 Single site frequency analysis for the River Dart at Austin's Bridge using sample data from annual maxima record.

Water year	Annual maximum flow Q_j (m^3/s)	j	$\frac{(j-1)}{(n-1)}Q_j$	$\frac{(j-1)(j-2)}{(n-1)(n-2)}Q_j$	$\frac{(j-1)(j-2)(j-3)}{(n-1)(n-2)(n-3)}Q_j$	$\frac{Q_j}{QMED}$	Cumulative non-exceedence probability $F(Q)$	Logistic reduced variate y
1991	92.551	1				0.414	0.013	−4.331
1976	116.637	2	2.777			0.521	0.036	−3.282
1975	120.397	3	5.733	0.140		0.538	0.059	−2.763
1971	140.779	4	10.056	0.491	0.012	0.629	0.083	−2.408
1969	154.929	5	14.755	1.080	0.054	0.692	0.106	−2.135
1995	158.365	6	18.853	1.839	0.138	0.708	0.129	−1.910
1993	161.489	7	23.070	2.813	0.281	0.722	0.152	−1.718
1977	172.219	8	28.703	4.200	0.525	0.770	0.175	−1.548
1997	175.910	9	33.507	5.721	0.858	0.786	0.199	−1.396
1988	179.819	10	38.533	7.519	1.316	0.804	0.222	−1.256
1978	180.563	11	42.991	9.437	1.887	0.807	0.245	−1.126
1981	183.630	12	48.094	11.730	2.639	0.821	0.268	−1.004
1958	196.886	13	56.253	15.092	3.773	0.880	0.291	−0.889
1968	206.502	14	63.917	18.707	5.145	0.923	0.314	−0.779
1987	207.150	15	69.050	21.894	6.568	0.926	0.338	−0.674
1964	209.387	16	74.781	25.535	8.299	0.936	0.361	−0.572
1989	210.063	17	80.024	29.277	10.247	0.939	0.384	−0.472
1970	213.542	18	86.434	33.730	12.649	0.954	0.407	−0.375
1994	215.800	19	92.486	38.348	15.339	0.965	0.430	−0.280

Table 10.2 (*Contd.*)

Water year	Annual maximum flow Q_j (m^3/s)	j	$\frac{(j-1)}{(n-1)}Q_j$	$\frac{(j-1)(j-2)}{(n-1)(n-2)}Q_j$	$\frac{(j-1)(j-2)(j-3)}{(n-1)(n-2)(n-3)}Q_j$	$\frac{Q_j}{QMED}$	Cumulative non-exceedence probability $F(Q)$	Logistic reduced variate y
1962	218.298	20	98.754	43.355	18.426	0.976	0.454	−0.186
1985	222.161	21	105.791	49.025	22.061	0.993	0.477	−0.093
1984	223.727	22	111.864	54.568	25.920	1.000	0.500	0.000
1996	224.299	23	117.490	60.178	30.089	1.003	0.523	0.093
1966	225.380	24	123.422	66.227	34.769	1.007	0.546	0.186
1961	226.810	25	129.606	72.706	39.988	1.014	0.570	0.280
1965	227.526	26	135.432	79.277	45.584	1.017	0.593	0.375
1990	237.144	27	146.803	89.514	53.709	1.060	0.616	0.472
1960	243.110	28	156.285	99.108	61.942	1.087	0.639	0.572
1998	246.980	29	164.653	108.430	70.480	1.104	0.662	0.674
1980	250.105	30	172.692	117.936	79.607	1.118	0.686	0.779
1967	252.966	31	180.690	127.805	89.464	1.131	0.709	0.889
1986	261.088	32	192.708	141.006	102.229	1.167	0.732	1.004
1983	269.781	33	205.547	155.414	116.560	1.206	0.755	1.126
1972	274.012	34	215.295	168.035	130.227	1.225	0.778	1.256
1974	284.042	35	229.939	185.073	148.058	1.270	0.801	1.396
1963	295.526	36	246.272	204.225	168.486	1.321	0.825	1.548
1992	302.603	37	259.374	221.417	188.204	1.353	0.848	1.718
1973	309.446	38	272.607	239.362	209.442	1.383	0.871	1.910
1982	317.843	39	287.572	259.516	233.565	1.421	0.894	2.135

1959	327.612	40	304.211	281.952	260.805	1.464	0.917	2.408
1999	328.887	41	313.226	297.946	283.049	1.470	0.941	2.763
2000	371.035	42	362.201	353.367	344.533	1.658	0.964	3.282
1979	549.735	43	549.735	549.735	549.735	2.457	0.987	4.331
Total	9986.734		5872.185	4252.730	3376.663			

Probability weighted moments: $b_0 = 232.250$ $b_1 = 136.562$ $b_2 = 98.901$ $b_3 = 78.527$

	Sample statistics	
Median	*QMED*	223.727
L-moments:	l_1	232.250
	l_2	40.875
	l_3	6.279
	l_4	10.020
L-CV	t_2	0.176
L-skewness	t_3	0.154
L-kurtosis	t_4	0.245

Generalised Logistic distribution sample estimators:

	k	−0.154
	β	0.177

Table 10.3 GL distribution fitted using sample estimators derived from annual maxima series for the River Dart at Austin's Bridge.

Cumulative distribution function F	GL growth curve factor x (equation (10.12))	Logistic reduced variate y (equation (10.16))	GL peak flow Q (m^3/s) (equation (10.10))	Return period T (years) (equation (10.4))
0.05	0.581	−2.944	129.920	1.053
0.10	0.670	−2.197	149.869	1.111
0.15	0.730	−1.735	163.419	1.176
0.20	0.779	−1.386	174.275	1.250
0.25	0.821	−1.099	183.689	1.333
0.30	0.859	−0.847	192.262	1.429
0.35	0.895	−0.619	200.339	1.538
0.40	0.930	−0.405	208.158	1.667
0.45	0.965	−0.201	215.901	1.818
0.50	1.000	0.000	223.727	2.000
0.55	1.036	0.201	231.798	2.222
0.60	1.074	0.405	240.296	2.500
0.65	1.115	0.619	249.448	2.857
0.70	1.160	0.847	259.567	3.333
0.75	1.212	1.099	271.125	4.000
0.80	1.273	1.386	284.916	5.000
0.85	1.352	1.735	302.450	6.667
0.900	1.463	2.197	327.237	10.000
0.905	1.477	2.254	330.406	10.526
0.910	1.492	2.314	333.757	11.111
0.915	1.508	2.376	337.314	11.765
0.920	1.525	2.442	341.103	12.500
0.925	1.543	2.512	345.157	13.333
0.930	1.562	2.587	349.515	14.286
0.935	1.583	2.666	354.227	15.385
0.940	1.606	2.752	359.353	16.667
0.945	1.631	2.844	364.973	18.182
0.950	1.659	2.944	371.187	20.000

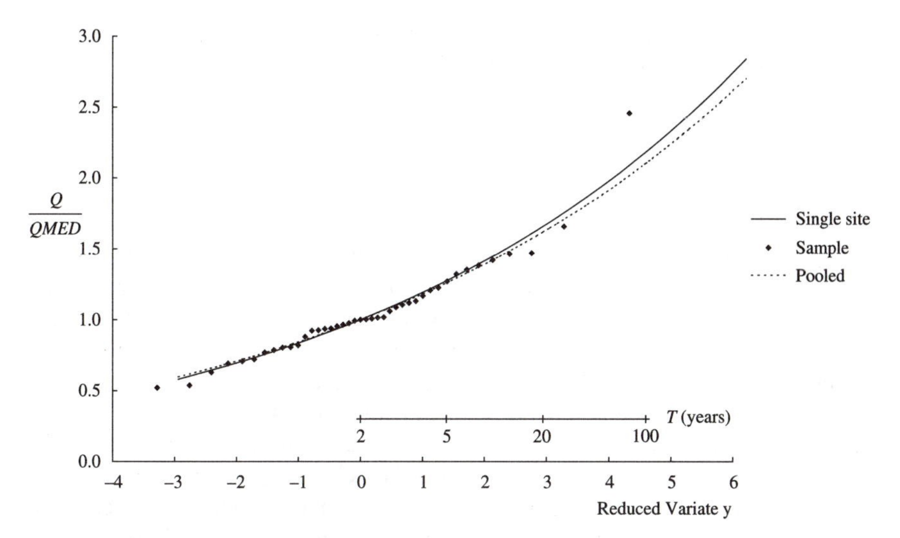

Figure 10.4 GL growth curve for the River Dart at Austin's Bridge. (Sample data copyright Environment Agency; pooled data courtesy of the centre for Ecology and Hydrology, Wallingford, UK.)

of fitting the GL distribution using the sample k and β estimates over a range of F values. The sample data and the fitted distribution are plotted in Fig. 10.4. In addition the pooled growth curve derived from the WINFAP-FEH software (CEH, 1999b) is shown for comparison. Figure 10.5 shows the histogram of the data series and the fitted pdf.

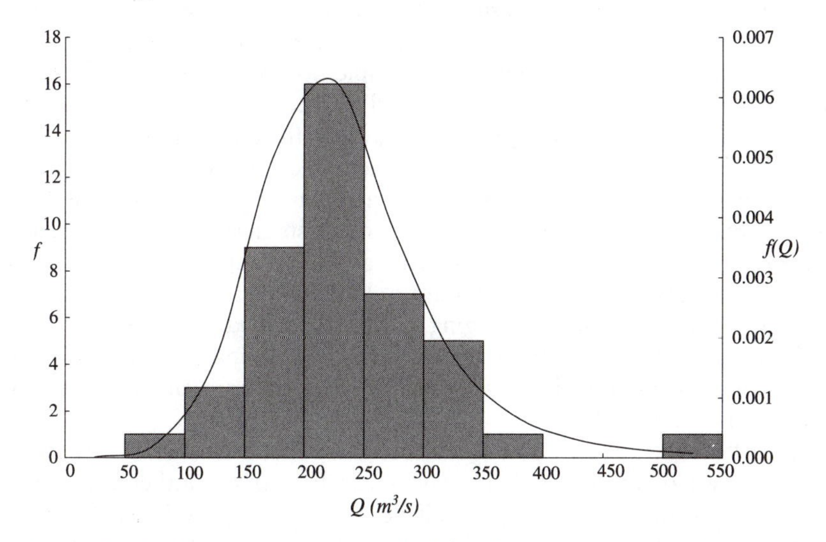

Figure 10.5 Histogram and GL probability density function for the annual maxima series for the River Dart at Austin's Bridge. (Data copyright Environment Agency.)

From the GL distribution fitted to the single site record the growth curve factor, x, for the 20-year event is 1.659 giving a flood of 371.187 m^3/s (the pooled growth curve gives a flood magnitude of 362.212 m^3/s).

Methods for estimating QMED

The FEH recommends that for gauged catchments with a record length greater than 13 years *QMED* can be estimated from the median of the annual maxima series. For records between 2 and 13 years *QMED* can be derived from the series of peak flows that exceed a given threshold value (called the peaks over a threshold or POT series).

For ungauged catchments, or where the record is less than 2 years in length, the FEH recommends transferring data (in the form of the GL growth curve) from a nearby donor gauged catchment (usually at a different gauge on the same river) or from analogue gauged catchments (i.e. with similar catchment descriptors to the subject site – the pooling group). The estimated *QMED* at the subject site (e.g. from catchment descriptors) can be adjusted using the ratio of the observed to predicted *QMED* at the gauged site.

Improving the validity of flood frequency analysis

Since a gauge record is a sample of the time series of river flow for a catchment, the record should be representative of the entire series (population), i.e. the sample statistics should be stationary – whereby the statistical properties of the sample used to estimate the population parameters are not affected by the start time of the sample nor the sample length. Non-stationarity in a record may be exhibited as trend (underlying increase or decrease in values over time), periodicity (cyclic fluctuation), or as a sudden change. The causes of non-stationarity include climate change, changes to the catchment (such as land use), and errors and changes in the methods of data measurement. The FEH recommends tests that can be employed to screen observed data for non-stationarity as well as adjustments that can be made to account for a limited amount of land use change. In addition each flood peak data set for UK gauging stations provided with the WINFAP-FEH software (CEH, 1999b) contains notes concerning the features of the catchment and its record. An approach for dealing with climate change is presented in section 10.10.

River flow records may contain observations that are substantially larger than the rest of the series. Such values can be termed outliers, however, this does not mean that they should be automatically excluded from the annual maxima series. One of the reasons for the use of pooling groups, the generalised logistic distribution and fitting the distribution using L-moments is the relative robustness of the analysis against outliers (Hosking and Wallis, 1997; Ahmad *et al.*, 1988). In addition, where historic records exist (e.g. historic flood marks and written accounts of extreme events) predating river gauging at a site this information can be used to review the flood frequency estimates made using gauged data (see Bayliss and Reed, 2001).

10.5 Unit hydrograph theory

Components of a natural hydrograph

Figure 10.6(a) shows a flood hydrograph and the causative rainfall. The hydrograph is composed of two parts, the surface (or quick response) runoff, which is formed directly from the rainfall, and the baseflow. The latter is supplied from groundwater sources which do not generally respond quickly to rainfall. The rainfall may also be considered to be composed of two parts. The net or effective rainfall is that part which forms the surface runoff while the rainfall losses constitute the remaining rainfall (this is either evaporated or enters soil moisture and groundwater storages). Simple techniques for separating runoff and rainfall have been developed, and are described in detail elsewhere (see Shaw, 1994; Wilson, 1990). Here, only a brief introduction to baseflow and rainfall separation techniques is given.

Baseflow and rainfall separation techniques

Given that baseflow represents the groundwater contribution to the total hydrograph, a simple method for its approximate determination is that shown in Figure 10.6(a). There the baseflow is assumed constant throughout the storm with a value equal to the total flow before the onset of storm runoff. A more sophisticated method is recommended in the FSR/FEH, which is shown in Figure 10.7. Here the point at which surface runoff begins is denoted by point x. The baseflow recession curve is extended to point A (at the time of peak runoff). From A to B a straight line is drawn. Point B is the point at which the surface (or quick response) runoff ends. The position of point B is found by first calculating the *LAG* (time from the centroid of rainfall to peak discharge) and then setting point B at a time of 4 × *LAG* from the end of the rainfall. This method was found to work well for UK catchments, but does not have a theoretical justification.

Once the baseflow separation line has been drawn the volume of surface runoff (V_{SR}) may be calculated (it is the area under the hydrograph above the baseflow line). The net or effective rainfall depth (P_{net}) may then be determined as

$$V_{SR} = P_{net} \cdot A$$

where A is the catchment area.

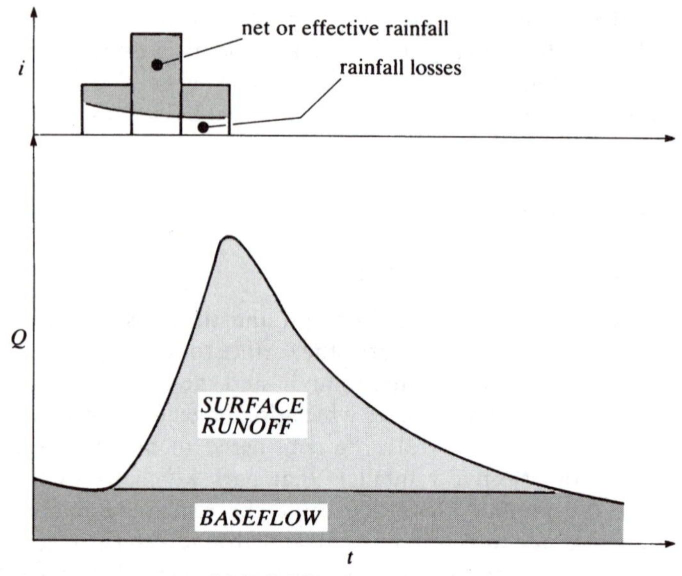

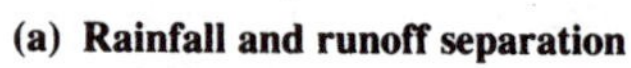

(a) Rainfall and runoff separation

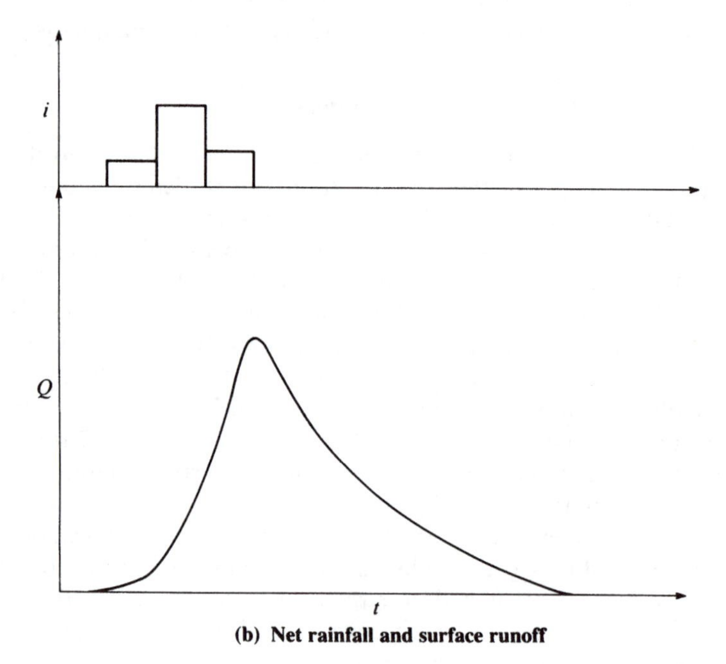

(b) Net rainfall and surface runoff

Figure 10.6 Hydrograph analysis.

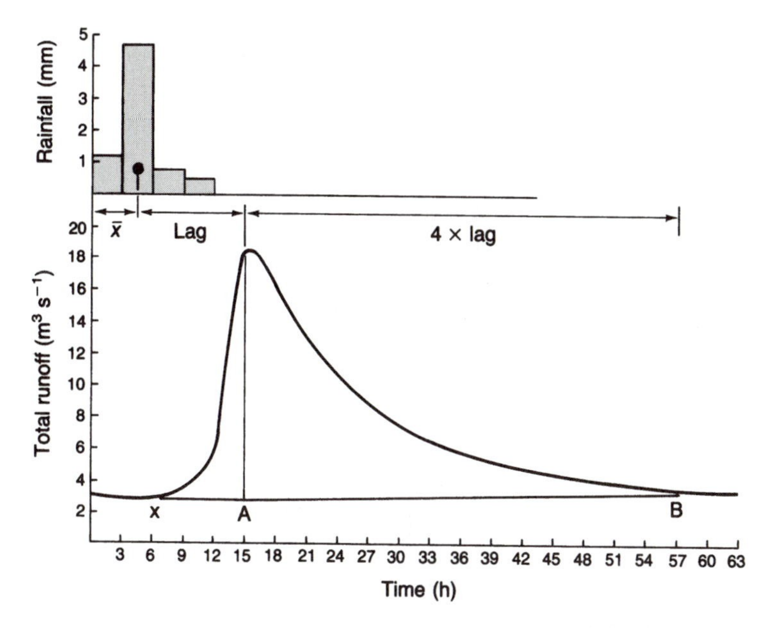

Figure 10.7 FSR/FEH method for baseflow and rainfall separation.

All that now remains is to find the time distribution of the net rainfall. The traditional method for doing this is to employ the concept of the ϕ index. The ϕ index assumes a constant loss rate (in mm rain/hour). This may be conceived as being equivalent to a constant value of infiltration. In practice the value of the ϕ index is set such that over the whole storm P_{net} (mm) of rainfall is contained above the ϕ index line as shown in Figure 10.8(a). This method may be criticised for its lack of realism in representing the physical processes occurring during interception, evaporation and infiltration. A more realistic method might be to use a suitable infiltration curve (as illustrated in Figure 10.6(a)) rather than a constant rate and such a method has also been employed in practice. When the *flood studies team* investigated this problem, they considered both the ϕ index method and the use of infiltration curves, but decided for various reasons to use the concept of percentage runoff. This is illustrated in Figure 10.8(b). The concept here is that in each block of rain a given percentage (PR) of storm rainfall forms the net rainfall.

Where both storm rainfall and total discharge have been measured the net rainfall and ϕ index or percentage runoff can all be found directly once the baseflow has been separated. For ungauged UK catchments the FEH developed a regression equation for PR for rural catchments given by

$$PR_{RURAL} = SPR + DPR_{CWI} + DPR_{RAIN} \tag{10.21a}$$

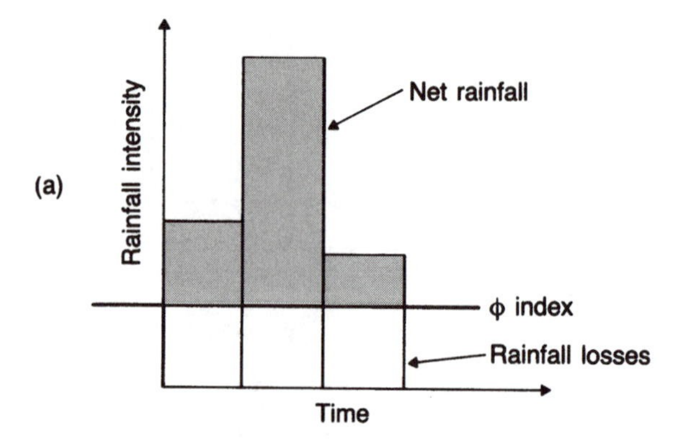

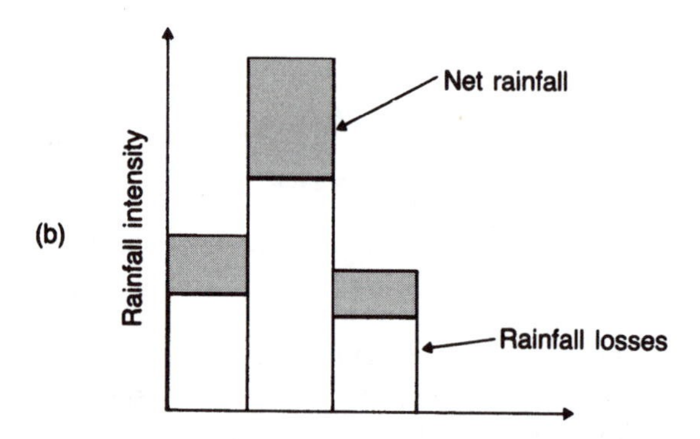

Figure 10.8 Rainfall separation technique.

where

$$DPR_{CWI} = 0.25(CWI - 125) \tag{10.21b}$$

$$DPR_{RAIN} = \begin{cases} 0 & \text{for } P \leqslant 40\,\text{mm} \\ 0.45\,(P-40)^{0.7} & \text{for } P > 40\,\text{mm} \end{cases} \tag{10.21c}$$

and the PR adjusted for urbanisation is

$$PR = PR_{RURAL}\,(1.0 - 0.615URBEXT) + 70\,(0.615URBEXT) \tag{10.21d}$$

The base flow, BF, is given by

$$BF = (33(CWI - 125) + 3.0SAAR + 5.5)10^{-5}AREA \tag{10.22}$$

where

AREA is the catchment area (km^2);
BF is baseflow (m^3/s);
CWI is a catchment wetness index (mm);
DPR_{CWI} is the dynamic contribution to percentage runoff from *CWI*;
DPR_{RAIN} is the dynamic contribution to percentage runoff from rainfall;
P is the storm rainfall depth (mm);
PR is the percentage runoff from both rural and urban areas;
PR_{RURAL} is the percentage runoff from rural areas;
SAAR is the average annual rainfall over the standard period 1961–1990 (mm);
SPR is the standard percentage runoff depending on soil type (*SPRHOST*);
URBEXT is the extent of urban land cover.

The calculation procedures for baseflow and rainfall separation are illustrated in Example 10.2.

Example 10.2 Hydrograph analysis

For the storm runoff event given in Table 10.4

(a) Draw the rainfall hyteograph and hydrograph to a common time axis.
(b) Separate the baseflow component using the FSR/FEH method and determine the volume of surface runoff and lag time.
(c) Separate the rainfall using the ϕ index method.
(d) Explain why must step (b) be carried out before step (c).
(e) Suggest why the ϕ index method is unsatisfactory and suggest other more appropriate techniques.

Solution

(a) As shown in Figure 10.7
(b) First determine the position of the rainfall centroid by taking moments. Referring to Figure 10.7

$$\bar{x} = \frac{(1.08 \times 1.5) + (4.7 \times 4.5) + (0.72 \times 7.5) + (0.54 \times 10.5)}{(1.08 + 4.7 + 0.72 + 0.54)}$$

$$\bar{x} = 4.807\,\text{h}$$

hence determine the *LAG* time

$$LAG = 15 - \bar{x} \approx 10.2\,\text{h}$$

$$\text{as } 4 \times LAG \approx 41\text{h}$$

then the position of point B is as shown in Figure 10.7.

Table 10.4 North Tyne at Tarset – catchment area 285 km^2.

Date	Time (h)	Mean areal rainfall (Observed) (mm)	Runoff at Tarset (Observed) (m^3/s)
14.4.75	09.00		3.28
		1.08	
	12.00		3.14
		4.70	
	15.00		3.09
		0.72	
	18.00		3.53
		0.54	
	21.00		5.76
	24.00		18.59
15.4.75	03.00		16.03
	06.00		12.63
	09.00		10.31
	12.00		8.64
	15.00		7.52
	18.00		6.65
	21.00		6.03
	24.00		5.43
16.4.75	03.00		4.97
	06.00		4.59
	09.00		4.22
	12.00		3.92
	15.00		3.74
	18.00		3.50
	21.00		3.34
	24.00		3.19

Next find the volume of surface runoff (V_{SR}) as the area under the hydrograph above the baseflow line

$$V_{SR} \approx 0.7 \times 10^6 \text{ m}^3$$

(c) First find the total depth of effective rainfall

$$P_{net} = \frac{V_{SR}}{A} = \frac{0.7 \times 10^6}{285 \times 10^6}$$
$$= 2.46 \text{ mm}$$

As the total storm rainfall is 7.04 mm and the net rainfall is 2.46 mm then the total 'losses' are $7.04 - 2.46 = 4.58$ mm. If this is expressed as a constant loss rate then the ϕ index is given by $4.58/4 = 1.145$ mm/3 h.

(d) Step (c) requires a value for P_{net} which is found from V_{SR} determined in (b). Hence (b) must be carried out before (c).

(e) For ϕ index $= 1.145$ mm/3 h, 3 of the 4 rainfall blocks contain less rainfall than the ϕ index. Hence a constant loss rate of 1.145 mm cannot be sustained. One method of overcoming this problem is to reset the ϕ index line in such a position that P_{net} is contained above it. In this case setting the ϕ index $= 4.7 - 2.46 = 2.24$ (mm/3 h) gives the solution. However the concept of a constant loss rate has now been lost. An alternative is to use the concept of percentage runoff (PR). In this case

$$PR = \frac{P_{net}}{P_{storm}} \times 100 = \frac{2.46}{7.04} \times 100 = 35\%$$

and each block of storm rainfall contributes 35% of its value to net rainfall, i.e. 0.38, 1.64, 0.25 and 0.19 mm.

Unit hydrograph rainfall-runoff modelling principles

After baseflow and rainfall separation the net rainfall and corresponding surface runoff are shown in Figure 10.6(b). The purpose of unit hydrograph theory is to be able to predict the relationships between the two for any storm event. A unit hydrograph is thus a simple model of the response of a catchment to rainfall.

This concept was first introduced by Sherman in 1932, and rests on three assumptions as shown in Figures 10.9(a–c). These are:

1. any uniform net rainfall having a given duration will produce runoff of specific duration, regardless of intensity;
2. the ratios of runoff equal the ratios of net rainfall intensities, provided that the rainfalls are of equal duration;
3. the hydrograph representing a combination of several runoff events is the sum of the individual contributory events, i.e. the principle of superposition may be applied.

These assumptions all imply that the response of a catchment to rainfall is linear, which is not true. However, unit hydrographs have been found

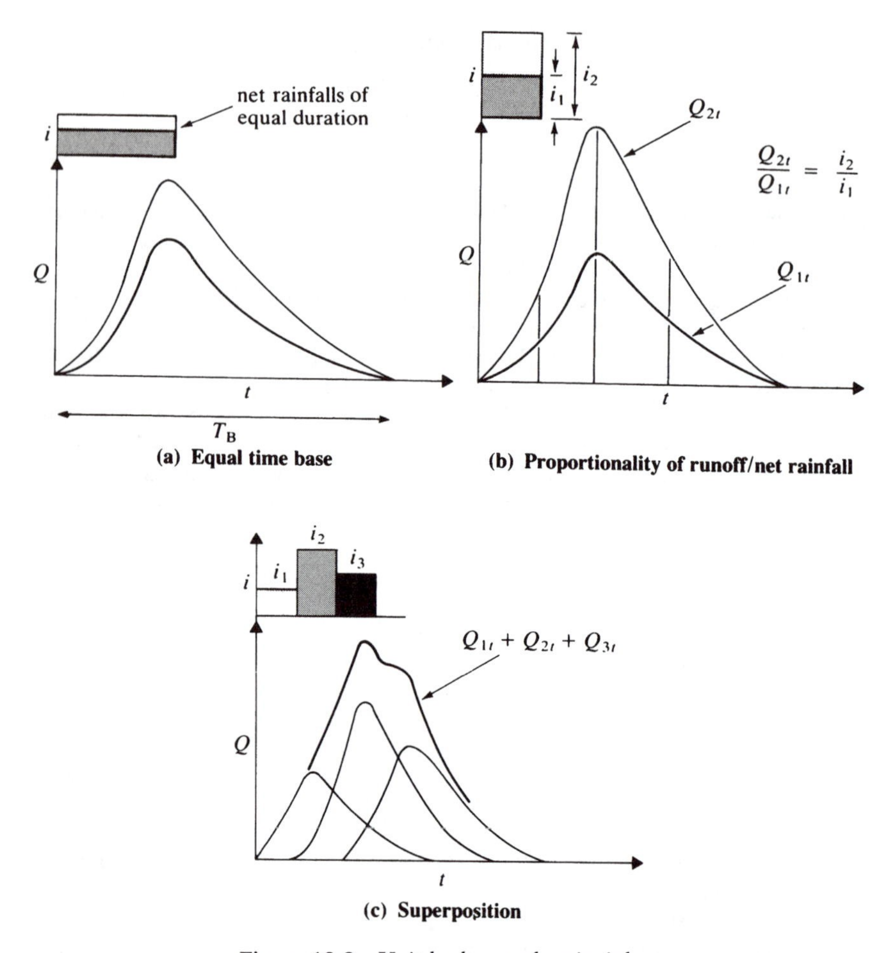

Figure 10.9 Unit hydrograph principles.

to work reasonably well for a wide range of conditions. Unit hydrograph theory has, therefore, been used extensively for design flood prediction. For a full discussion of the various approaches to rainfall-runoff modelling see Beven (2000).

Unit hydrograph definition and convolution

The foregoing principles are embodied in the definition of the unit hydrograph and its application.

The P mm, D h unit hydrograph is the hydrograph of surface runoff produced by P mm of net rainfall in D hours, provided the net rainfall falls uniformly over the catchment in both space and time. Both P and D may have any values, but commonly P is taken as 10 mm and D as 1 h.

Once a unit hydrograph has been derived for a catchment, it may be used to predict the surface runoff for any storm event by the process of convolution. This is shown diagrammatically in Figure 10.10(b), along with the unit hydrograph definition (Fig. 10.10(a)).

The process may also be expressed in matrix form as

$$\mathbf{P}\cdot\mathbf{U}=\mathbf{q} \tag{10.23}$$

where **P** is the matrix of net rainfalls, **U** is the matrix of unit hydrograph ordinates and **q** is the matrix of surface runoff ordinates.

Alternatively, the process may be laid out in tabular form as shown in Table 10.5. A spreadsheet for the matrix method is available at the supporting website (at www.sponpress.com/supportmaterial/0415306094.html) and an example follows to illustrate the process of convolution.

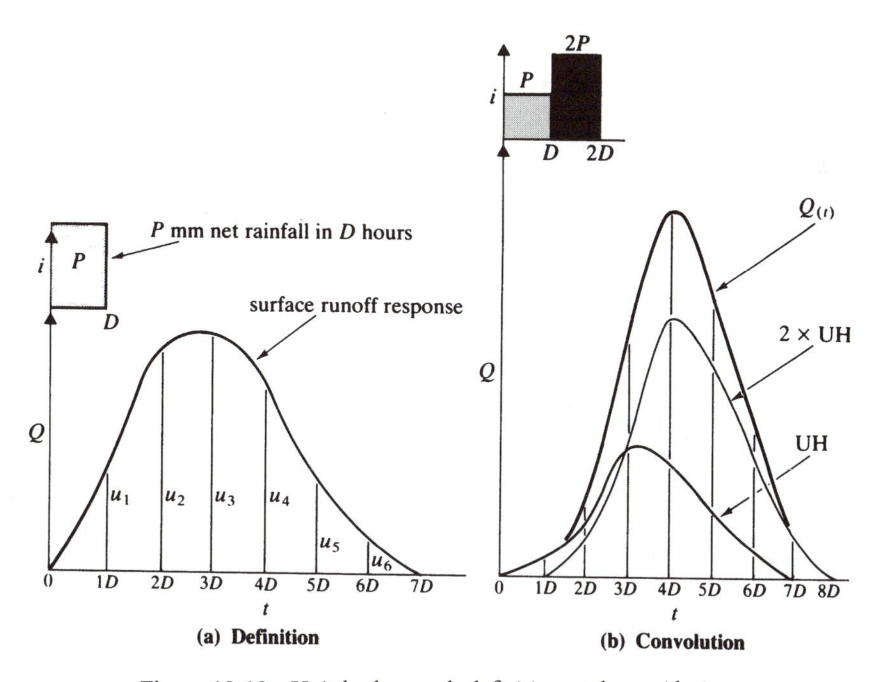

Figure 10.10 Unit hydrograph definition and convolution.

Table 10.5 Tabular and matrix methods of convolution.

Time	Rainfall	Unit hydrograph ordinates						Surface runoff
		u_1	u_2	u_3	u_4	u_5	u_6	
1	P_1	p_1u_1						$=q_1$
2	P_1	$p_2u_1+p_1u_2$						q_2
3			$p_2u_2+p_1u_3$					q_3
4				$p_2u_3+p_1u_4$				q_4
5					$p_2u_4+p_1u_5$			q_5
6						$p_2u_5+p_1u_6$		q_6
7							p_2u_6	q_7

Equivalent matrix form

$$\begin{vmatrix} p_1 & 0 & 0 & 0 & 0 & 0 \\ p_2 & p_1 & 0 & 0 & 0 & 0 \\ 0 & p_2 & p_1 & 0 & 0 & 0 \\ 0 & 0 & p_2 & p_1 & 0 & 0 \\ 0 & 0 & 0 & p_2 & p_1 & 0 \\ 0 & 0 & 0 & 0 & p_2 & p_1 \\ 0 & 0 & 0 & 0 & 0 & p_2 \end{vmatrix} \quad \begin{vmatrix} u_1 \\ u_2 \\ u_3 \\ u_4 \\ u_5 \\ u_6 \\ u_7 \end{vmatrix} = \begin{vmatrix} q_1 \\ q_2 \\ q_3 \\ q_4 \\ q_5 \\ q_6 \\ q_7 \end{vmatrix}$$

Note: $p_i = P_i / P$ for a P mm, D h unit hydrograph; i.e. if $P_1 = 30$ mm and the unit hydrograph is a 10 mm, D h unit hydrograph, then $p_1 = 30/10 = 3$.

Example 10.3 Convolution

Given the following 10 mm 1 h unit hydrograph and net rainfall profile, determine the surface runoff hydrograph.

Unit hydrograph data

Time (h)	0	1	2	3	4	5	6	7
Discharge (m^3/s)	0	1	4	6	5	3	1	0

Net rainfall profile

Time (h)	0		1		2		3
Net rainfall (mm)		10		30		20	

Solution

The solution may be found using the tabular method given in Table 10.5 as follows:

		Unit hydrograph ordinates (m^3/s per 10 mm)								Surface runoff
Time	P_{net}	0	1	4	6	5	3	1	0	in m^3/s
0	10	1×0								0
1	30	3×0	$+1\times1$							1
2	20	2×0	$+3\times1$	$+1\times4$						7
3			2×1	$+3\times4$	$+1\times6$					20
4				2×4	$+3\times6$	$+1\times5$				31
5					2×6	$+3\times5$	$+1\times3$			30
6						2×5	$+3\times3$	$+1\times1$		20
7							2×3	$+3\times1$	$+1\times0$	9
8								2×1	$+3\times0$	2
9									2×0	0

Derivation of unit hydrographs

For gauged catchments, unit hydrographs are derived by hydrograph analysis of measured storm events. In general, storm events are complex (i.e. varying intensity and duration), and the unit hydrograph must be 'unearthed' from such events. As surface runoff can be predicted by convoluting the unit hydrograph with net rainfall, the converse must be true.

The matrix equation (10.23) may be solved for **U** by inversion. First **P** is converted into a square matrix by pre-multiplying by its transpose $|\mathbf{P}^T|$, hence:

$$\mathbf{P}^T \cdot \mathbf{P} \cdot \mathbf{U} = \mathbf{P}^T \cdot \mathbf{q}$$

This equation may now be inverted to give

$$\mathbf{U} = [\mathbf{P}^T \cdot \mathbf{P}]^{-1} \cdot \mathbf{P}^T \cdot \mathbf{q}$$

In practice, the method suffers from complications due to instability which arises because the unit hydrograph is only an approximate model and because of data errors. Full details of this method and its application are given in the Institute of Hydrology Report No. 71 (1981).

Synthetic unit hydrographs

For ungauged catchments, unit hydrographs cannot be derived directly. However, measures of catchment descriptors may be used to estimate a unit hydrograph. Such unit hydrographs are termed 'synthetic', and examples of two are shown in Figure 10.11. They have a simple triangular form, whose shape is determined by three parameters – the time to peak (T_P), the peak runoff (U_P) and the time base (TB).

In the FEH the recommended equations for determining a synthetic unit hydrograph are based on the so called instantaneous unit hydrograph (IUH). An IUH is a theoretical concept in which the unit hydrograph is the runoff response to P mm of rainfall falling instantaneously rather than in ΔT h. The IUH can however be easily converted to any ΔT h period as given in the equations below. The parameters of the synthetic unit hydrograph in the FEH were derived using multiple regression techniques on data from gauged (rural) catchments.

First the time to peak $T_P(0)$ of the 10 mm IUH is calculated from

$$T_P(0) = 4.270 DPSBAR^{-0.35} PROPWET^{-0.80} DPLBAR^{0.54} (1 + URBEXT)^{-5.77} \tag{10.24a}$$

This is then converted to the time to peak of any required duration ΔT h using the equation

$$T_P = T_P(\Delta T) = T_P(0) + \frac{\Delta T}{2} \tag{10.24b}$$

where ΔT is usually taken as $\leqslant 0.2 T_P(0)$.

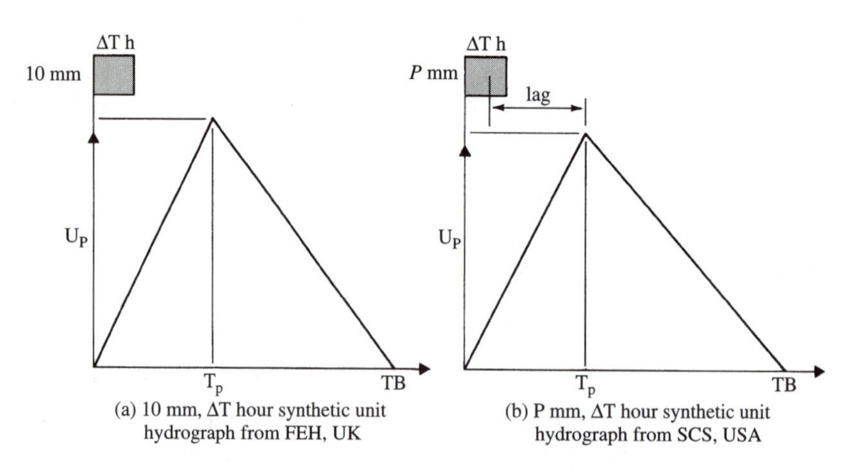

Figure 10.11 Synthetic unit hydrographs.

The remaining parameters are given by

$$U_P = 2.2AREA/T_P \tag{10.24c}$$

$$TB = 2.52T_P \tag{10.24d}$$

where

DPSBAR is the catchment slope (m/km);
PROPWET is the proportion of time during the standard period 1961–1990 that the $SMD \leqslant 6$ mm;
DPLBAR is a measure of the mean drainage length of the catchment;
SMD is the soil moisture deficit (mm);
URBEXT is the extent of urban land cover;
AREA is the catchment area (km^2);
and where T_P and TB are in hours and U_P is in m^3/s.

The P mm, ΔT h synthetic unit hydrograph derived by the US Soil Conservation Service is given by

$$T_P = \text{lag} + \Delta T/2 \qquad U_P = 0.208\text{AP}/T_P \qquad \text{TB} = 2.67T_P$$

where lag $\cong 0.6t_c$, t_c is the time of concentration (h), A is the catchment area (km^2) and P is the net rainfall (mm), and where T_P and TB are in hours and U_P is in m^3/s.

These two unit hydrographs have a very similar shape. The main difference lies in the estimation of T_P, American practice being to calculate t_c rather than T_P from empirical formulae.

Design flood estimation using the unit hydrograph rainfall-runoff model

Figure 10.12 shows, in outline, the procedure for estimating a design flood event for any given return period. It is universally applicable. An important feature of the method is the estimation of the full flow hydrograph, which is necessary in flood routing studies. The central component of this procedure is the determination of the appropriate storm duration and associated rainfall depth and profile. Underestimation of the peak runoff will occur if the selected duration is either too short or too long, but the peak runoff is more sensitive to underestimation of the storm duration. Wherever possible the parameters of the method (T_P, *SPR*, *BF*, etc.) should be derived from observed rainfall and river flow events.

For the UK, full details of each component of the rainfall-runoff method are given in the FEH, Volume 4 (IH, 1999). It should be noted that the FEH

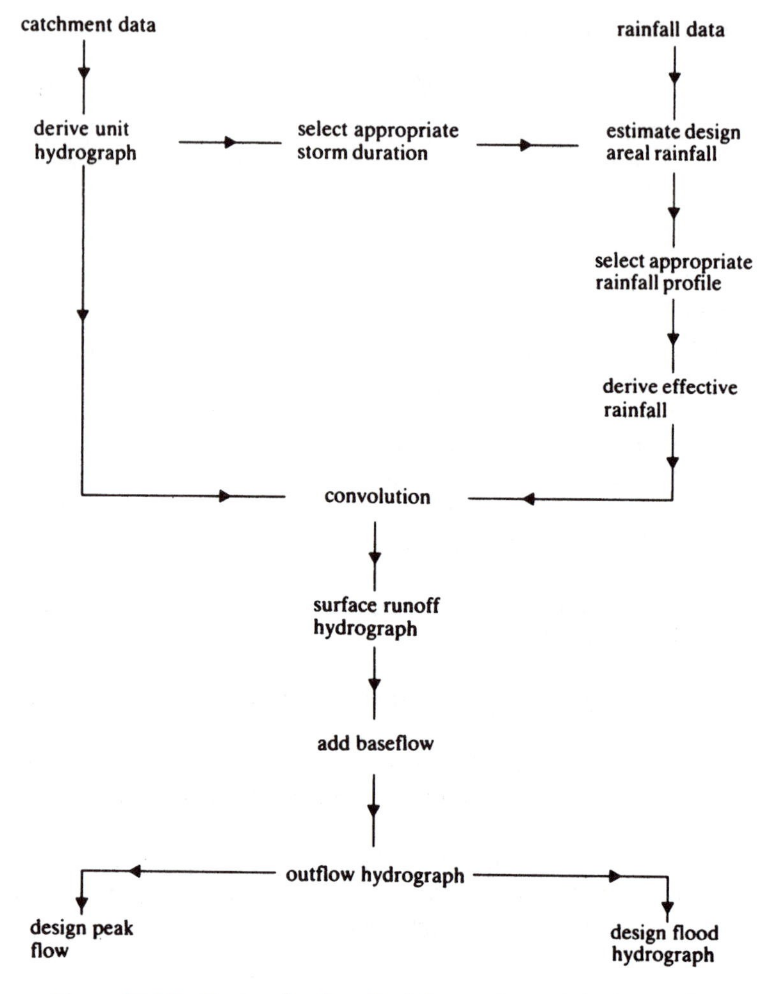

Figure 10.12 Design flood methodology using unit hydrographs.

rainfall-runoff method is largely the same as that of the FSR (NERC, 1975). The differences are in the use of the FEH digital catchment descriptors and the refined rainfall depth-duration-frequency analysis. Consequently, estimates of peak discharge using the FEH method are, on average, higher than those made with the FSR (Ashfaq and Webster, 2002). At the time of writing, the Centre for Ecology and Hydrology in the UK is reviewing the FEH rainfall-runoff method. The calculations and extraction of values from tables and figures in the FEH are both time-consuming and liable to user error. Consequently, the FEH rainfall-runoff model is available as part of commercial river modelling software packages, such as ISIS (Wallingford

Software Ltd and Halcrow Group Ltd, 2003) and MIKE 11 FEH (DHI Software, 2003a).

For the USA, details may be found in the *National Engineering Handbook* (US Soil Conservation Service, 1972), and a full discussion of the methods with examples is given in McCuen (1998).

Example 10.4 Design hydrograph estimation

Estimate the design hydrograph with a 100-year peak flow for the River Dart at Austin's Bridge using the FEH rainfall-runoff method

Data. Table 10.6 lists the catchment descriptors.

Solution

Full details of the method can be found in the FEH volume 4(IH, 1999) and only a summary is given here. Furthermore, the user needs access to the FEH CD-ROM (CEH, 1999a).

First calculate the parameters of the synthetic unit hydrograph:

$$T_P(0) = 4.270\,(121.4)^{-0.35}\,(0.47)^{-0.80}\,(19.79)^{0.54}\,(1+0.004)^{-5.77} \quad \text{(from 10.24a)}$$

$$T_P(0) = 7.134\,\text{h}$$

The required time interval, ΔT, is 1 h (since $\Delta T \leqslant 0.2T_P(0)$ and rounded down).

$$T_P = T_P(\Delta T) = 7.134 + \frac{1}{2} \qquad \text{(from 10.24b)}$$

$$T_P = 7.634\,\text{h}$$

Table 10.6 Catchment descriptors for the River Dart at Austin's Bridge. (Data derived from the FEH CD-ROM courtesy of the Centre for Ecology and Hydrology, UK).

AREA (km^2)	248.88
DPLBAR (km)	19.79
DPSBAR (m/km)	121.4
PROPWET	0.47
SAAR (mm)	1771.0
SPRHOST (%)	32.9
URBEXT	0.004
Grid reference	SX 274950, 066050

$$U_P = 2.2 \times 248.88/7.634 \qquad \text{(from 10.24c)}$$

$$U_P = 71.723\,\text{m}^3/\text{s per} 10\,\text{mm}$$

$$TB = 2.52 \times 7.634 \qquad \text{(from 10.24d)}$$

$$TB = 19.238\,\text{h}$$

Next, a storm duration (D) must be selected. The FEH recommends

$$D = T_P\left(1 + \frac{SAAR}{1000}\right)$$

and

$$D = \text{odd multiple integer of}\,\Delta T$$

In this case,

$$D = 7.634\left(1 + \frac{1771}{1000}\right) = 21.154\,\text{h}$$

Take $D = 21 \times \Delta T = 21\,\text{h}$. Using this duration, an areal rainfall depth may be calculated which will produce a 100-year flood event.

Since *URBEXT* is $0.004 < 0.125$ (i.e. rural), then The FEH recommends a 140-year return period rainfall is required to generate the 100-year return period flood flow (FEH volume 4, Fig. 3.2/Table 3.1). The point rainfall depths with associated return periods between 1 and 1000 years and durations in the range 1 h to 8 days are available at 1 km grid intervals for the UK on the FEH CD-ROM (CEH, 1999a). In this case the M140-21h (the rainfall of 140-year return period and 21 h duration) point rainfall depth is 176.9 mm (reproduced from the FEH CD-ROM courtesy of the Centre for Ecology and Hydrology, UK).

Finally this point rainfall value is reduced by an areal reduction factor (*ARF*) to allow for the effect of catchment area. In this case

$$ARF \approx 0.924 \quad \text{(using FEH Volume 4, Fig. 3.4)}$$

Therefore $P = 0.924 \times 176.9 = 163.5\,\text{mm}$.

This rainfall depth is next distributed according to a selected profile. The FEH recommend the use of the so-called 75% winter profile for rural catchments (FEH volume 4, Fig. 3.5). Finally the net rainfall is found by multiplying by the percentage runoff (PR) equations (10.21a–d). In this case:

$$DPR_{CWI} = 0.25\,(CWI - 125) \quad \text{(from 10.21b)}$$

where CWI is obtained from the $SAAR$ (FEH volume 4, Fig. 3.7), as 126 mm, giving

$$DPR_{CWI} = 0.25(126 - 125) = 0.25\%$$

Since $P > 40\,\text{mm}$, then

$$DPR_{RAIN} = 0.45(163.5 - 40)^{0.7} = 13.1\% \quad \text{(from 10.21c)}$$

So

$$PR_{RURAL} = 32.9 + 0.25 + 13.1 = 46.3\% \quad \text{(from 10.21a)}$$

and the PR adjusted for urbanisation is

$$PR = 46.3(1.0 - 0.615 \times 0.004) + 70(0.615 \times 0.004) = 46.4\% \quad \text{(from 10.21d)}$$

Thus the net rainfall profile given in Table 10.7 is produced.

Finally the net rainfall profile is convoluted with the unit hydrograph and a baseflow allowance added.

The baseflow is given by

$$BF = (33\,(126 - 125) + 3.0 \times 1771 + 5.5)\,10^{-5} \times 248.88 \text{ (from 10.22)}$$

$$BF = 13.32\,\text{m}^3/\text{s}$$

The results are summarised in Table 10.7 and plotted in Figure 10.13.

It is worth emphasising the need to review flood predictions in the light of other sources of information (e.g. the use of local recorded events to derive the parameter values). This type of hydrological model is limited by the range of observed data used in its parameter formulation (Wheater, 2002). In fact, the river Dart at Austin's Bridge is an example of a catchment for which the FSR rainfall-runoff model results

Table 10.7 Synthetic unit hydrograph analysis for the River Dart at Austin's Bridge – Results of convolution.

Time (h)	Net rain (cm)	Unit hydrograph (m^3/s per cm net rain)	Surface runoff (m^3/s)	Runoff + baseflow (m^3/s)
0	0.077	0.000	0.000	13.3
1	0.100	9.395	0.723	14.0
2	0.132	18.790	2.386	15.7
3	0.172	28.186	5.289	18.6
4	0.225	37.581	9.809	23.1
5	0.293	46.976	16.442	29.8
6	0.381	56.371	25.827	39.1
7	0.494	65.766	38.793	52.1

Continued

Table 10.7 (*Contd.*)

Time (h)	Net rain (cm)	Unit hydrograph (m^3/s per cm net rain)	Surface runoff (m^3/s)	Runoff + baseflow (m^3/s)
8	0.637	69.461	55.960	69.3
9	0.812	63.280	77.782	91.1
10	0.940	57.099	105.493	118.8
11	0.812	50.918	139.751	153.1
12	0.637	44.737	178.657	192.0
13	0.494	38.556	219.655	233.0
14	0.381	32.375	260.229	273.5
15	0.293	26.195	297.804	311.1
16	0.225	20.014	329.622	342.9
17	0.172	13.833	352.634	366.0
18	0.132	7.652	363.885	377.2
19	0.100	1.471	362.464	375.8
20	0.077	0.000	350.695	364.0
21			331.127	344.4
22			305.278	318.6
23			275.000	288.3
24			241.859	255.2
25			207.226	220.5
26			172.368	185.7
27			138.523	151.8
28			106.978	120.3
29			79.434	92.8
30			57.512	70.8
31			40.798	54.1
32			28.278	41.6
33			19.021	32.3
34			12.286	25.6
35			7.491	20.8
36			4.187	17.5
37			2.025	15.3
38			0.736	14.1
39			0.113	13.4
40			0.000	13.3

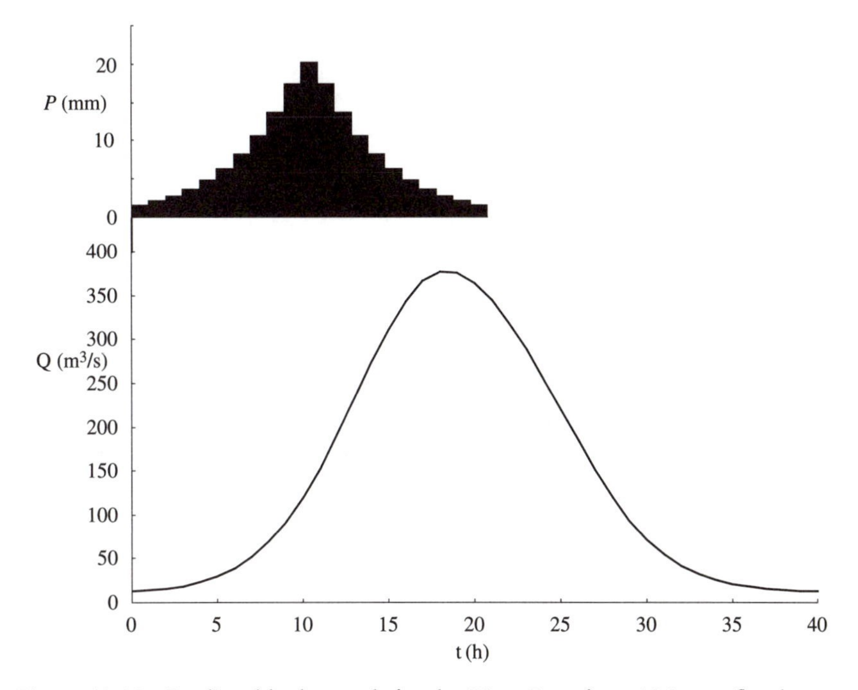

Figure 10.13 Predicted hydrograph for the River Dart for a 100-year flood event.

do not compare closely with a frequency analysis of the gauge records for long return period events (Boorman *et al.*, 1990).

10.6 Summary of design flood procedures for rural catchments

This is illustrated in Figure 10.14. The two basic approaches of frequency analysis and unit hydrograph rainfall-runoff models are complementary. The accuracy of each method depends on the amount and quality of available data. Estimates from gauged catchments are more accurate than those from ungauged catchments. Detailed advice on the choice of which method to use for different applications is available in the FEH, volume 1 (IH, 1999).

10.7 Flood routing

General principles

So far, the discussion has centred on methods of estimating flood events at a given location. However, the engineer requires estimates of both the stage and discharge along a watercourse resulting from the passage of a flood wave. The technique of flood routing is used for this purpose.

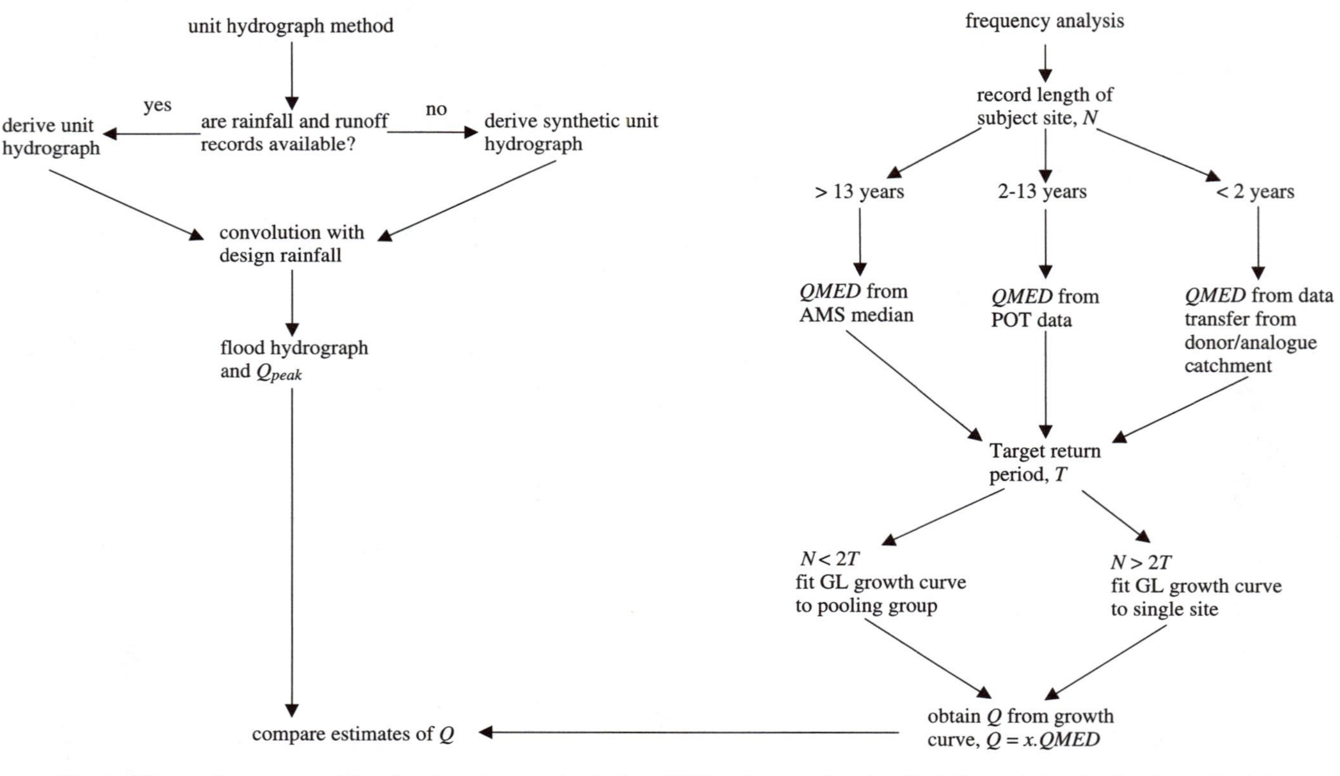

Figure 10.14 Summary of flood estimation methods (see FEH volume 1 for detailed discussioin of selection criteria).

There are two distinct kinds of problem:

1. reservoir routing: to find the outflow hydrograph over the spillway from the inflow hydrograph;
2. channel routing: to find the outflow hydrograph from a river reach (particular length of river channel) from the inflow hydrograph.

In each case the peak flow of the outflow hydrograph is less than and later than that of the inflow hydrograph. These processes are referred to as 'attenuation' and 'translation', respectively.

To determine the outflow hydrograph from the inflow hydrograph requires the application of the continuity equation in the form:

$$I - O = \frac{dV}{dt} \tag{10.25}$$

where I is the inflow rate, O the outflow rate, V is the volume and t is time. This is shown diagrammatically in Figure 10.15. Expressing equation (10.25) in a finite difference form gives

$$\frac{(I_t + I_{t+\Delta t})}{2}\Delta t - \frac{(O_t + O_{t+\Delta t})}{2}\Delta t = V_{t+\Delta t} - V_t \tag{10.26}$$

where I_t and O_t are inflow and outflow rates at time t and $I_{t+\Delta t}$ and $O_{t+\Delta t}$ are inflow and outflow rates at time $t + \Delta t$. This equation may be solved successively through time for a known inflow hydrograph if the storage volume can be related to outflow (reservoir case) or channel properties.

Reservoir routing

This case is shown in Figure 10.16. The outflow is governed by the height (stage h) of water above the spillway crest level, and the volume of live storage is also governed by this height. Hence, for a given reservoir, both the volume and outflow can be expressed as functions of stage. This is achieved by a topographical survey and application of a suitable weir equation, respectively. Equation (10.25) may be solved by a number of numerical methods, including the Newton–Raphson method which is demonstrated below.

Rearranging equation (10.26) in terms of unknown and known values

$$\underbrace{\frac{V_{t+\Delta t}}{\Delta t} + \frac{O_{t+\Delta t}}{2}}_{\text{i.e unknown}} - \underbrace{\left(\frac{I_t + I_{t+\Delta t}}{2} + \frac{V_t}{\Delta t} - \frac{O_t}{2}\right)}_{\text{known}} = 0 \tag{10.27}$$

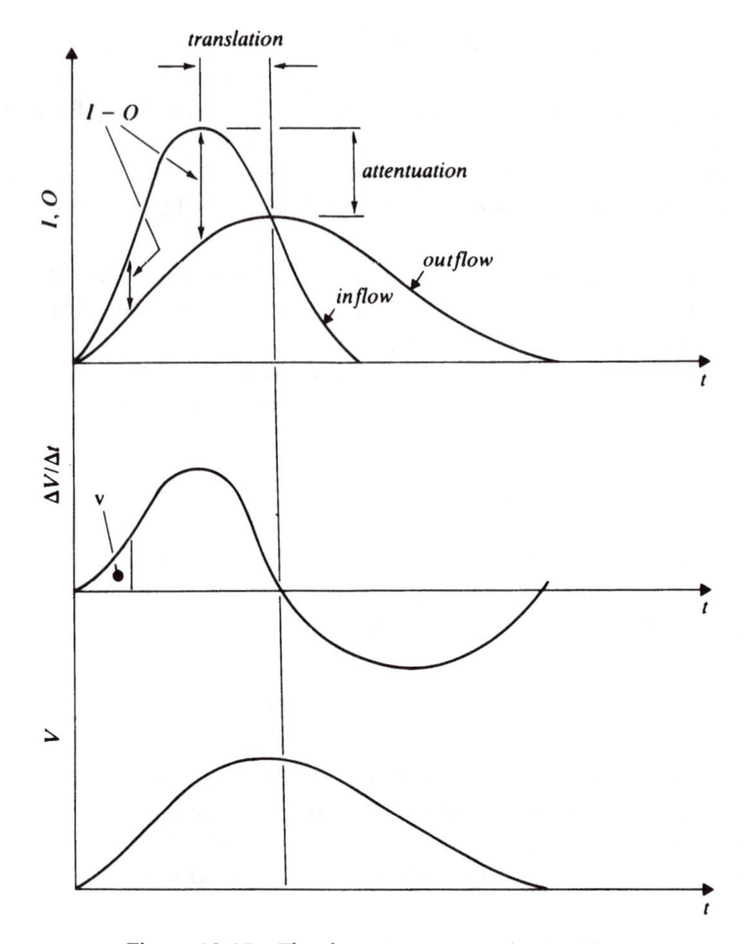

Figure 10.15 Flood routing – general principles.

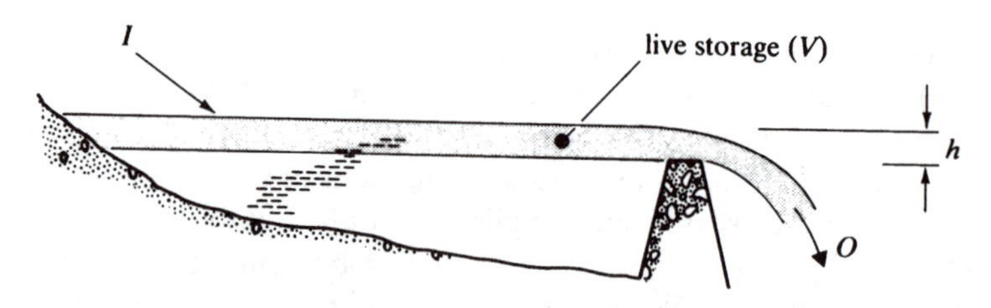

Figure 10.16 Reservoir routing.

The storage-stage relationship may be expressed as

$$V = V(h) \tag{10.28}$$

where $V(h)$ represents a function of stage, h, at time $t + \Delta t$.

The outflow-stage relationship may be expressed as

$$O = Kh^n \tag{10.29}$$

where K and n may be assumed constant and their values depend upon the type of spillway structure (refer to chapter 13 for further details).

Substituting equations (10.28) and (10.29) into equation(10.27) yields a non-linear equation in one unknown, h

$$F(h) = \frac{V(h)}{\Delta t} + \frac{Kh^n}{2} - \left(\frac{I_t + I_{t+\Delta t}}{2} + \frac{V_t}{\Delta t} - \frac{O_t}{2}\right) = 0 \tag{10.30}$$

This equation can be solved using a root finding method (i.e. find h for $F(h) = 0$) such as the Newton–Raphson method (Chapra and Canale, 1998), where a better estimate of the root is h_{i+1} and is given by

$$h_{i+1} = h_i - \frac{F(h_i)}{F'(h_i)} \qquad i = 0, 1, 2\ldots \tag{10.31}$$

where h_i is the previous estimate of the root. $F'(h)$ is obtained by differentiating equation (10.30) with respect to h and using the fact that volume differentiates to area, giving

$$F'(h) = \frac{A(h)}{\Delta t} + \frac{nKh^{n-1}}{2} \tag{10.32}$$

where $A(h)$ represents the surface area of the reservoir as a function of stage, h, at time $t + \Delta t$.

The calculation of equation (10.31) is repeated until there is no significant change in the estimate of h. A suitable starting estimate for h_i is to use the stage at the previous time step, t (provided that $h_i > 0$ to avoid a division by zero).

Care must be exercised in the selection of a suitable time step, Δt, to ensure that the hydrograph shape and peak are adequately represented and that the solution converges. Convergence can be checked by re-running the calculations using smaller time steps until there is no noticeable change between runs. Because of the number of calculations involved, the method is suited to solution by spreadsheet (an example is available at the supporting website at www.sponpress.com/supportmaterial/0415306094.html) or computer program (e.g. the FORTRAN code accompanying the FEH volume 4, IH, 1999).

Example 10.5 Reservoir flood routing

Determine the outflow hydrograph resulting from the probable maximum flood (PMF) at Ardingly Reservoir from the following data:

Reservoir plan area at spillway crest level = $0.8\,\text{km}^2$;
Reservoir plan area 3 m above spillway crest level = $1.0\,\text{km}^2$;
spillway type = circular shaft;
discharge equation: $O = 64h^{3/2}$;
discharge preceding occurrence of PMF = $5\,\text{m}^3/\text{s}$.

PMF

time (h)	0	2	4	6	8	10	12	14	16	18	20
inflow (m³/s)	5	8	15	30	85	160	140	95	45	15	10

Solution

The approximate equations for V and A in terms of h may be derived by assuming a linear variation of area with stage:

$$A(h) = \left(0.8 + h\frac{(1-0.8)}{3}\right) \times 10^6$$

$$A(h) = \left(0.8 + \frac{h}{15}\right) \times 10^6\,\text{m}^2$$

where $A(h)$ is the reservoir area at stage h. Hence

$$V(h) = \left(\frac{0.8 + A(h)}{2}\right) h \times 10^6$$

$$V(h) = \left(0.8 + \frac{h}{30}\right) h \times 10^6\,\text{m}^3$$

The outflow is related to stage through the weir equation $O = 64h^{3/2}$
Choosing $\Delta t = 2\,\text{h}$, i.e. 7200 s to correspond with the inflow data then

$$F(h) = \frac{V(h)}{7200} + \frac{64h^{3/2}}{2} - \left(\frac{I_t + I_{t+\Delta t}}{2} + \frac{V_t}{7200} - \frac{O_t}{2}\right) = 0 \qquad \text{(from 10.30)}$$

The values of the variables in the bracketed term are all known at the start of a given time step.

$$F'(h) = \frac{A(h)}{7200} + 48h^{1/2} \qquad \text{(from 10.32)}$$

The solution may now proceed by using the tabular method shown in Table 10.8, once the initial conditions have been specified. At time zero, the outflow is $5\,\text{m}^3/\text{s}$, the corresponding stage is 0.183 m (from the spillway discharge equation) and the storage volume is $0.148 \times 10^6\,\text{m}^3$. Hence, the first row of Table 10.8 may be

Table 10.8 Solution for reservoir routing (Newton-Raphson method calculations have been carried out using a spreadsheet and are not shown).

Time, t (h)	Inflow, I (m^3/s)	Stage, h (m)	Volume, V (m^3)	Area, A (m^2)	Outflow, O (m^3/s)
0	5	0.183	147314.2	812183.4	5.00
2	8	0.194	156431.3	812931.4	5.47
4	15	0.239	192709.7	815901.1	7.46
6	30	0.346	280941.9	823078.9	13.04
8	85	0.644	528961.1	842928.3	33.07
10	160	1.185	994733.5	878994.4	82.55
12	140	1.560	1328793.2	903975.9	124.66
14	95	1.521	1293862.1	901396.1	120.05
16	45	1.243	1046250.4	882893.0	88.73
18	15	0.899	746338.5	859948.7	54.57
20	10	0.637	522831.6	842443.4	32.51

completed. The stage at the next time step is obtained by applying equation 10.31 until $F(h) \approx 0$. The next row of Table 10.8 may now be completed. The inflow and outflow hydrographs are shown in Figure 10.17. It is left as an exercise for the reader to repeat the calculations for different time steps in order to assess convergence.

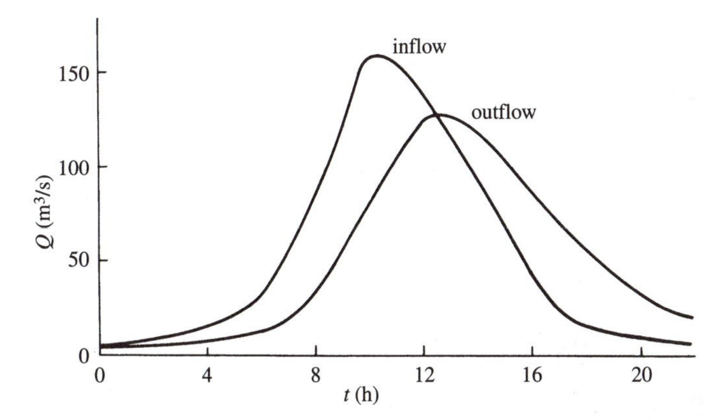

Figure 10.17 Reservoir flood routing – solution to Example 10.5.

Channel routing

In this case the storage volume is not a simple function of stage, and therefore solution of the continuity equation is more complex. It is commonly solved using the full equations of gradually varied unsteady flow as described in Chapter 14. There are a number of commercial software packages available for one-dimensional river modelling, for example ISIS (Wallingford Software Ltd and Halcrow Group Ltd, 2003), MIKE 11 (DHI Software, 2003b) and HEC-RAS (US Army Corps of Engineers, 2002). However, there are simpler techniques which can be applied if previous inflow and outflow hydrographs have been recorded. These are referred to as hydrological routing methods and are used in some conceptual models for hydrological forecasting. These routing methods are based on the so-called Muskingum method (after McCarthy in 1938) described below.

Channel storage may be considered to consist of two parts, prism and wedge storage, as shown in Figure 10.18. Assuming no sudden change of cross-section within the reach, then approximate expressions for inflow, outflow and storage are

$$I = ay_i^n \quad O = ay_o^n$$

where a and n are constants. Now

$$\text{Prism storage} = by_o^m$$

$$\text{Wedge storage} = c\,(y_i^m - y_o^m)$$

where b and c are constants. So

$$\text{Total storage}\, V = \text{prism storage} + \text{wedge storage}$$

$$= by_o^m - cy_o^m + cy_i^m$$

Substituting for y_i and y_o and assuming $m = n$ (approximately correct for natural channels),

$$V = \frac{b}{a}O - \frac{c}{a}O + \frac{c}{a}I$$

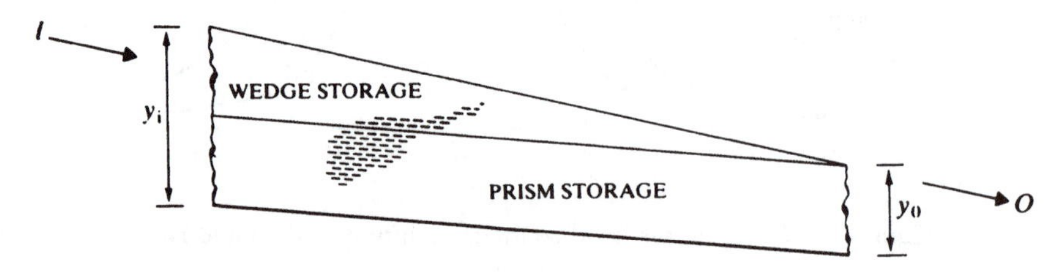

Figure 10.18 Channel routing.

or

$$V = \frac{b}{a}\left(\frac{c}{b}I + O - \frac{c}{b}O\right)$$

Taking $K = b/a$ and $X = c/b$

$$\begin{aligned} V &= K(XI + O - XO) \\ &= K[XI + (1 - X)O] \end{aligned} \tag{10.33}$$

where K is called the storage constant and has dimensions of time and X is a dimensionless weighting factor between 0 and 0.5 (but normally between 0.2 and 0.4).

Equation (10.33) is the Muskingum equation. It is obviously only an approximation but has been used widely with reasonable results.

Substitution of this equation into the continuity equation yields

$$O_{t+\Delta t} = C_0 I_{t+\Delta t} + C_1 I_t + C_2 O_t \tag{10.34}$$

where

$$C_0 = -\frac{KX - 0.5\Delta t}{K - KX + 0.5\Delta t} \tag{10.35a}$$

$$C_1 = \frac{KX + 0.5\Delta t}{K - KX + 0.5\Delta t} \tag{10.35b}$$

$$C_2 = \frac{K - KX - 0.5\Delta t}{K - KX + 0.5\Delta t} \tag{10.35c}$$

and

$$C_0 + C_1 + C_2 = 1$$

Hence, the outflow may be determined through time from the inflow hydrograph, the known values of K and X and a starting value for the outflow.

In the Muskingum method, both K and X may be determined by a graphical technique from a previously recorded event. A value for X is assumed, and a plot of V is drawn, derived from the known inflows and outflows against $[XI + (1 - X)O]$. If the assumed value of X is correct, then a straight-line plot with gradient K should result (cf. equation (10.33)). If this is not the case, a new value of X is chosen and the procedure repeated.

Example 10.6 Channel routing using the Muskingum method

Given the inflow and outflow hydrograph for a river reach in Table 10.9(a),

(a) assuming that the storage V can be written as in equation (10.33) find K and X for the flood;
(b) taking the outflow hydrograph from (a) as inflow to the next river reach with $K = 27$ and $X = 0.2$ find the peak outflow.

Solution

(a) The first stage of the solution is to find the change in reach storage (ΔV) as a function of time. This may be found from the finite difference form of the continuity equation (10.26). Second the cumulative volume of reach storage (V) may be found by summing the ΔV s. Third a value of X is chosen and V plotted against $[XI + (1 - X)O]$. In this case three values of X (0.2, 0.25 and 0.3) have been tried. The results are tabulated in Table 10.9(b) and plotted in Figure 10.19. The best result appears to be for $X = 0.25$, for which $K = 17.66$ h has been estimated from Figure 10.19.

Table 10.9(a) Data.

	Hour		Inflow	Outflow
Day	midnight (m)	noon (n)	(m^3/s)	(m^3/s)
1		m	1.05	1.05
2	n		3.54	1.47
		m	9.62	3.68
3	n		16.27	8.12
		m	20.43	13.36
4	n		20.94	17.66
		m	19.04	19.13
5	n		12.9	18.05
		m	9.06	16.24
6	n		6.93	11.15
		m	5.43	8.69
7	n		4.07	6.65
		m	3.34	5.09
8	n		2.69	4.02
		m	2.27	3.23

Table 10.9(b) Tabular solution to Example 10.6(a).

Time (h)	Inflow (m^3/s)	Outflow (m^3/s)	Change in storage ($m^3 \times 10^3$)	Storage volumes ($m^3 \times 10^3$)	$[XI + (1-X)O]$		
					$X = 0.2$	$X = 0.25$	$X = 0.3$
0.00	1.05	1.05		0.00	1.05	1.05	1.05
			44.71				
12.00	3.54	1.47		44.71	1.88	1.99	2.09
			173.02				
24.00	9.62	3.68		217.73	4.87	5.17	5.46
			304.34				
36.00	16.27	8.12		522.07	9.75	10.16	10.57
			328.75				
48.00	20.43	13.36		850.82	14.77	15.13	15.48
			223.56				
60.00	20.94	17.66		1074.38	18.32	18.48	18.64
			68.90				
72.00	19.04	19.13		1143.29	19.11	19.11	19.10
			−113.18				
84.00	12.90	18.05		1030.10	17.02	16.76	16.50
			−266.33				
96.00	9.06	16.24		763.78	14.80	14.45	14.09
			−246.24				
108.00	6.93	11.15		517.54	10.31	10.10	9.88
			−161.57				
120.00	5.43	8.69		355.97	8.04	7.87	7.71
			−126.14				
132.00	4.07	6.65		229.82	6.13	6.01	5.88
			−93.53				
144.00	3.34	5.09		136.30	4.74	4.65	4.56
			−66.53				
156.00	2.69	4.02		69.77	3.75	3.69	3.62
			−49.46				
168.00	2.27	3.23		20.30	3.04	2.99	2.94

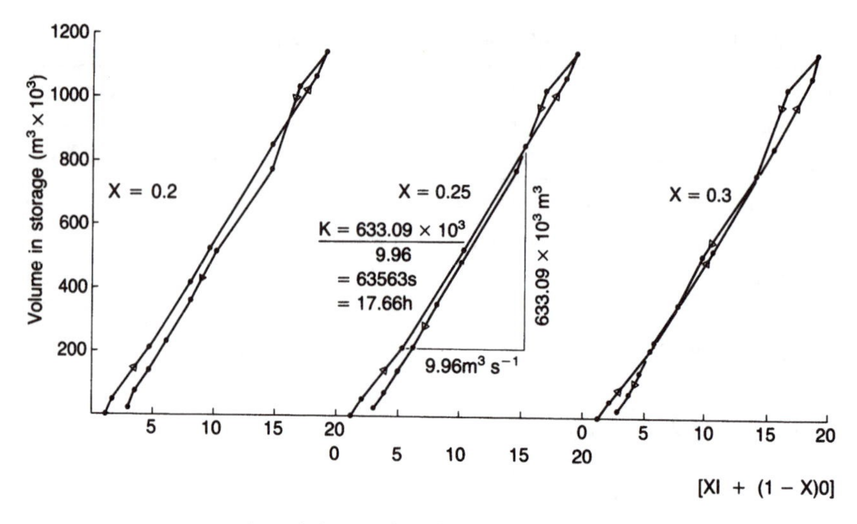

Figure 10.19 Plot of the results of Table 10.9(b) for Example 10.6.

(b) Here equation (10.34) is applied after solution of the equations 10.35(a–c) for C_0, C_1 and C_2 given by

$$C_0 = -\frac{(27 \times 0.2) - (0.5 \times 12)}{27 - (27 \times 0.2) + (0.5 \times 12)} = 0.02$$

(note $\Delta t = 12\,\text{h}$ for the inflow hydrograph and thus K must have the same units)

$$C_1 = \frac{(27 \times 0.2) + (0.5 \times 12)}{27 - (27 \times 0.2) + (0.5 \times 12)} = 0.41$$

$$C_2 = \frac{27 - (27 \times 0.2) - (0.5 \times 12)}{27 - (27 \times 0.2) + (0.5 \times 12)} = 0.57$$

as a check on the calculation

$$C_0 + C_1 + C_2 = 0.02 + 0.41 + 0.57 = 1$$

which is correct.

The calculation of the new outflow hydrograph is shown in tabular form in Table 10.10. The peak outflow is seen to be $16.48\,\text{m}^3/\text{s}$ at $t = 96\,\text{h}$. As with Example 10.5 (reservoir flood routing) an initial value for the outflow is required (here set to equal the inflow) and the tabular solution is easier to follow if the relevant equation (10.34) is written above the table.

This graphical technique is tedious. Cunge (1969) presented a simpler alternative to this approach, known as the constant-parameter Muskingum-Cunge method. He demonstrated that K is approximately equal to the time of travel of the flood wave, i.e.

$$K \cong \frac{\Delta L}{c} \tag{10.36}$$

Table 10.10 Tabular solution to Example 10.6(b).

Time (h)	Inflow (m^3/s)	C_0I_t (m^3/s)	C_1I_t (m^3/s)	C_2O_t (m^3/s)	Outflow (m^3/s)
0.00	1.05	+	0.43 +	0.60 =	1.05
12.00	1.47	0.03	0.60	0.60	1.06
24.00	3.68	0.07	1.51	0.73	1.28
36.00	8.12	0.16	3.33	1.37	2.40
48.00	13.36	0.27	5.48	2.83	4.96
60.00	17.66	0.35	7.24	4.94	8.66
72.00	19.13	0.38	7.84	7.16	12.56
84.00	18.05	0.36	7.40	8.76	15.36
96.00	16.24	0.32	6.66	9.39	16.48
108.00	11.15	0.22			16.28

where ΔL is the length of the river reach and c is the flood wave celerity ($c \cong \sqrt{gy}$) and

$$X \cong 0.5 - \frac{\overline{Q}_P}{2S_0\bar{B}c\Delta L} \tag{10.37}$$

where $\overline{Q}_P$ is the mean flood peak and $\overline{B}$ is the mean surface width of the channel.

Using these equations allows rapid calculation of K and X for each reach, and the constant-parameter Muskingum–Cunge method may also be applied to rivers without recorded outflow hydrographs (since $\overline{Q}_P$ is a reference flow determined from previous records and remains constant throughout each time step). Recalculating the parameters at each time step for each reach – the variable-parameter Muskingum–Cunge (VPMC) method – provides an improved simulation of the routed hydrograph shape (Ponce and Yevjevich, 1978; Price, 1978). However, the VPMC method is non-linear and requires a numerical solution using a computational model. Furthermore the VPMC method results in a small loss of volume (Tang *et al.*, 1999).

Many of the commercially available one-dimensional river modelling software packages provide a facility for flood routing calculations.

10.8 Design floods for reservoir safety

The safe design of reservoirs entails, amongst other things, the provision of a spillway which will prevent flood waters overtopping the dam, causing subsequent collapse and the release of a flood wave. In the past, many such failures have occurred, causing loss of life and property on a disastrous scale (for details, see Binnie, 1981). As a result of such failures in the UK, the Reservoirs (Safety Provisions) Act of 1930 was introduced at the instigation of the Institution of Civil Engineers (ICE). Subsequently, the Reservoirs (Safety Provisions) Act 1975 was introduced, though this was only brought into force in 1986. The 1975 Act applies to reservoirs with a capacity in excess of 25 000 m^3.

The main purpose of these acts is to ensure that new reservoirs are designed by competent engineers (known as Panel engineers, appointed through ICE) and that existing reservoirs are regularly inspected and repaired in accordance with the recommendations of the Panel engineers. In 1986 the Building Research Establishment set up a National Dams Database to record the construction details and historic performance of UK dams.

The design philosophy behind spillway sizing is to make them sufficiently large to pass design floods safely. UK practice is to categorise reservoirs in terms of the potential hazard to life and property should the dam be overtopped (ICE, 1996). In most cases the design flood is the 'probable maximum flood' (PMF). For lesser hazard reservoirs the 1 in 10 000-year flood is used.

The estimation of a PMF is problematic, because, theoretically, such events should never occur. The interim report on reservoir safety (ICE, 1933), re-published with additions in 1960, tackled this problem by gathering information on the largest recorded historical floods. Using this information, they produced an 'envelope curve' of specific peak discharge (Q/A) against catchment area (A). Floods lying on the envelope curve were termed 'normal maximum floods', and it was suggested that reservoirs be designed for a 'catastrophic flood' equal to twice the normal maximum flood.

With the advent of the FSR (NERC, 1975), the Institution of Civil Engineers produced a new guide, *Floods and Reservoir Safety* (1996). In this guide, the concept of the PMF was introduced, and details were given of how it may be calculated. The essence of the method is to estimate an inflow hydrograph based on the unit hydrograph rainfall-runoff model. The storm rainfall (depth and profile) to be used is the estimated maximum rainfall, as calculated by the Meteorological Office. This is convoluted with a unit hydrograph in which all the parameters are set to maximise runoff. Finally, the PMF is routed through the reservoir to find the outflow hydrograph, assuming a maximum initial water level with an allowance for wind induced wave action. At the time of writing, research is being undertaken to validate the FEH (IH, 1999) method for estimating the PMF both in terms of the extreme rainfall values and the rainfall-runoff model, and, as an interim measure, the original FSR method is continuing to be used in the UK (Babtie Group, 2000).

Since the introduction of the 1930 Act, there have been no major dam failures in the UK due to overtopping. However, many reservoirs are now over 100 years old, and are suffering from the effects of age. It is to be expected that the inspection and repair of old reservoirs will assume more prominence than the construction of new ones (in the UK) for the foreseeable future. Guidance for the design of reservoirs that do not fall under the 1975 Act is available in the UK (Hall *et al.*, 1993; Kennard *et al.*, 1996). Furthermore, there is ongoing research into the development of a risk-based approach to reservoir safety where all the potential causes of dam failure and the consequences can be taken into account using defined probabilities (Brown and Root, and BRE, 2002).

10.9 Methods of flood prediction for urban catchments

The runoff response of urban catchments to rainfall is different from that of rural catchments. For a given rainfall, flood rise times are quicker, flood peaks higher and flood volumes larger. This is due to two main physical differences between urban and rural catchments; the large proportion of impermeable areas (roads, roofs, pavements, etc.) and the existence of artificial piped drainage systems. The percentage runoff from impermeable areas is typically 60–90%. The drainage systems convey this increased volume to the outflow point of the catchment much more quickly than natural drainage would. The design of drainage systems calls for the application of both hydrological and hydraulic principles.

Historically, such designs have been based on the rational method (attributed to Lloyd-Davies, 1906). Some of the shortcomings of this method were overcome in the Transport and Road Research Laboratory (TRRL) hydrograph method introduced in 1963 (refer to TRRL, 1976). The *Wallingford Procedure* (National Water Council, 1981) refined the hydrological principles resulting in a modified rational method and a completely new method based on a conceptual model of rainfall-runoff processes. Much of the hydrological work is derived from the methods in the FSR with additional research carried out specifically for urban catchments. The *Wallingford Procedure* is commonly used in UK practice for generating the design storms and the consequent runoff that enters the drainage system. Since the publication of the FEH (IH, 1999), the revised rainfall depth-duration-frequency model for the UK is increasingly being used to estimate the design storm events.

The following two sections describe, in outline, the rational method and the *Wallingford Procedure*. Complete details are available in the references.

The rational method

The basis of this method is a simplistic relationship between rainfall and runoff of the form

$$Q_P = CiA \qquad (10.38)$$

where Q_P is the peak runoff rate, i is the rainfall intensity, A is the catchment area (normally impermeable area only) and C is the runoff coefficient. In the modified rational method this equation becomes

$$Q_P = \frac{C_V C_R i A}{0.36} \qquad (10.39)$$

with the units: Q_P in l/s, i in mm/h, A in hectares, and where C_V is the volumetric runoff coefficient, C_R is the routing coefficient and A is the total catchment area. The recommended equation for determining C_V as given in the *Wallingford Procedure*, Volume 1, is

$$C_V = \frac{PR}{100} \qquad (10.40a)$$

where PR is the (urban) percentage runoff which is found from

$$PR = 0.829\,PIMP + 25.0\,SOIL + 0.078\,UCWI - 20.7 \qquad (10.40b)$$

where

$SOIL$ is a number depending on soil type
$PIMP$ is percentage impermeable area to total catchment area
$UCWI$ is the urban catchment wetness index (mm) (related to $SAAR$)

The recommended value for C_R is a fixed value of $C_R = 1.3$ for all systems.

To apply these equations to urban storm water drainage design requires a knowledge of the critical storm duration to estimate i. The assumption made is that this storm duration is equal to the time of concentration of the catchment (t_c), given by

$$t_c = t_e + t_f \qquad (10.41)$$

where t_e is the time of entry into the drainage system (between 3 and 8 min) and t_f is the time of flow through the drainage system.

The value of i for any given return period and duration may be found using the methodology and data given in the *Wallingford Procedure*, Volumes 3 and 4, for UK catchments. First, values of Jenkinson's r ($M5-60$ min/$M5$–2-day rainfall) and $M5-60$ min (rainfall of 5-year return period and 60-min duration) are read from mapped values for the UK. Next the value of $M5-D/M5-60$ min (where D is the required duration) is read from plotted data using the value of r to obtain the required value of $M5-D$. The value of $MT-D$ (where T is the required return period) is then found from tabulated data relating $M5-D$ to return period T. This value of $MT-D$ (the point rainfall of required return period and duration) is next reduced by multiplying by an areal reduction factor (ARF), which is plotted as a function of duration and area, to obtain the design catchment rainfall depth. Finally the design rainfall intensity (i) is found simply from

$$i = \frac{MT - D}{D} \tag{10.42}$$

To calculate t_f, a pipe size (D), length (L) and gradient (S_0) must be chosen, and the full bore velocity (V) calculated (using the HRS tables or charts). Thus, for a single pipe $t_f = L/V$, t_c is found from equation (10.41) and Q_P, from equation (10.39).

If this value of Q_P exceeds the pipefull discharge for the pipe initially chosen, then the procedure is repeated with a larger pipe size. For a complete drainage system this analysis is carried out sequentially in the downstream direction. t_c and A will, of course, increase, and consequently i will decrease, in the downstream direction.

Example 10.7 Application of the modified rational method

The modified rational method is to be used to design a small storm water drainage system. The following design information has already been established:

Pipe No	Length (m)	Gradient	Area (ha)	*PIMP* (%)
1.000	40	1/133	0.81	35
1.010	40	1/542	0.53	35
2.000	45	1/75	0.66	30
1.020	35	1/32	0.32	30

Time of entry = 4 min
SOIL = 0.3
k_s = 1.5 mm
UCWI = 72 mm
C_R = 1.3

Available pipe sizes (mm) 150, 225, 300, 375, 450
Design storm intensities (2-year return period)

Time (min)	1	2	3	4	5	6
Rainfall intensity (mm/h)	86.1	76	68.4	62.4	57.6	53.5

(a) Sketch a key plan of the system.
(b) Determine the necessary pipe diameters using the modified rational method.

Notes:

1. The Hydraulic Research Station design tables relating pipe gradient, diameter, pipe-full discharge and velocity are required for simple application of the method.
2. The rainfall intensity table will require interpolation, preferably by plotting the intensity curve, for accurate results.

Solution

(a) A key plan of the system is shown in Figure 10.20. The notation system gives each pipe a reference decimal number. The pipe at the upstream end of the system is given the reference number 1.000 and the sequential system of pipes having the largest cumulative value of t_c should all be labelled 1.xxx. All other pipes in the system will form branches into the main pipe run and should be numbered sequentially from upstream to downstream as pipes 2.000, 2.010... and 3.000, 3010, etc.

(b) The solution is laid out in tabular form in Table 10.11, which is convenient for hand calculation. Several points may be noted concerning this solution as follows:

1. Calculations start with the upstream pipe 1.000. Initially the smallest available pipe size (150 mm) is tried. In this instance a 225 mm diameter pipe is required.
2. For pipe 1.010 the cumulative area and time of concentration includes the upstream pipe 1.000. Because of the very shallow gradient a 375 mm diameter pipe is required.
3. For pipe 2.000 the calculations follow the same pattern as for pipe 1.000, as it is another upstream pipe.
4. For pipe 1.020 the cumulative area includes those for all preceding pipes and the time of concentration is the time of flow of pipes 1.000, 1.010 and 1.020 plus the time of entry. The minimum permitted diameter is that of pipe 1.010 (e.g. 375 mm) as in design pipe diameters are not allowed to decrease downstream to

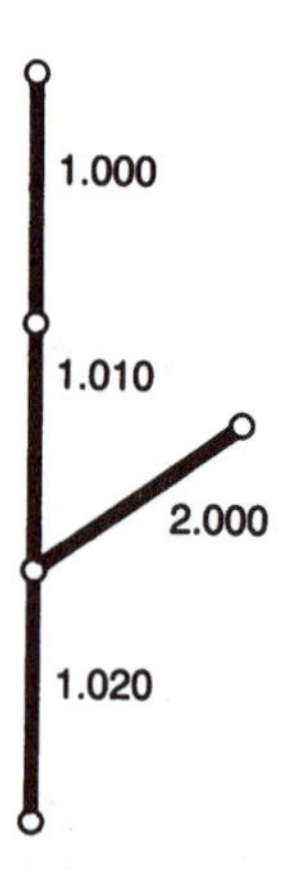

Figure 10.20 Key plan of the sewer system in Example 10.7.

Table 10.11 Tabular solution for the modified rational method.

Pipe number	Pipe length (m)	Pipe gradient	Assumed diameter (mm)	Pipefull velocity (m/s)	t_f (min)	t_c (min)	i (mm/h)	Area (ha)	Cumm Area (ha)	PIMP (%)	C_V	Q_{full} (l/s)	Q_p (l/s)	Comments
1.000	40	1/133	150	0.76	0.88	4.88	58.0	0.81	0.81	35	0.21	35.6	13.4	Diameter too small
			225	0.99	0.67	4.67	59.0					36.2	39.5	Diameter correct
1.010	40	1/541	225	0.49	1.36	6.03	53.5	0.53	1.34	35	0.21	54.4	19.5	Diameter too small
			300	0.59	1.13	5.8	54.5					55.4	41.9	Diameter too small
			375	0.68	0.98	5.65	55.0					55.9	76.0	Diameter correct
2.000	45	1/75	150	1.01	0.74	4.74	58.5	0.66	0.66	30	0.17	23.7	17.9	Diameter too small
			225	1.33	0.56	4.56	59.5					24.1	53.0	Diameter correct
1.020	35	1/32	375	2.84	0.2	5.85	53.0	0.32	2.32	30	0.19	84.4	313.0	Diameter too large but necessary as upstream pipe has 375 mm diameter

prevent potential blockage problems. Finally, the volumetric runoff coefficient (C_v) is found as a weighted average of the contributions from all upstream contributing areas (because *PIMP* varies with each contributing area) i.e.

$$\text{(Cumulative) } C_{V1.020} = \frac{(C_{V1.000} \times 0.81 + C_{V1.010} \times 0.53 + C_{V2.000} \times 0.66 + C_{V1.020} \times 0.32)}{2.32}$$

Whilst the rational method is often satisfactory from a design viewpoint, it can be criticised for its inability to simulate a real storm event for the following reasons:

(a) the inaccuracy of the runoff coefficient C_V – note that this is improved in the modified rational method;
(b) the rainfall intensity varies continuously down the system: this does not correspond to a real storm event;
(c) the assumption of full bore conditions to calculate t_f – in reality, each pipe will be running partially full most of the time.

For these reasons the modified rational method is only recommended for the design of drainage systems in catchments which do not exceed 150 ha in area.

The Wallingford Procedure surface runoff model

This is based on a conceptual model of the physical processes occurring in an urban catchment. In the model the net rainfall depth is calculated from a percentage runoff equation derived from regression analysis (e.g. the Wallingford Procedure constant *PR* equation 10.40b or the UK variable *PR* model (Osborne, 1993)), and an allowance is made for depression storage. The time distribution of this net rainfall is calculated according to the type of surface the rain falls upon. The runoff response of each subcatchment area is derived by routing the net rainfall through two non-linear reservoirs to produce a hydrograph of surface runoff. The runoff hydrograph represents the inflow at a particular point in the drainage system.

Drainage network analysis

The remaining aspect of the analysis is to predict the discharge and depth in each pipe (sewer) in the drainage system during a storm event. This may be undertaken by means of one of the channel routing methods described earlier. However, particularly where surcharging and surface flooding are likely (e.g. in the assessment of the flood capacity of existing drainage systems), the full dynamic equations for one-dimensional, unsteady, gradually varied free surface flow need to be solved (see Chapter 14). In addition, the equations need to be modified to represent pressurised pipe full flow. These solution of the equations concerned require a computational model.

There are a number of commercial software packages for the design of drainage systems for new developments, such as WinDes (Micro Drainage Limited, 2003) in the UK; and for the analysis of large sewerage networks, such as InfoWorks CS (Wallingford Software Ltd, 2003) and MOUSE (DHI Software, 2003c). These packages incorporate the *Wallingford Procedure* and FEH storm generation methods and offer a range of surface runoff models that can be applied in different countries.

Supplementary information regarding urban drainage analysis

Since the introduction and widespread adoption of the *Wallingford Procedure* in the UK, the Wastewater Planning User Group (WaPUG) has been formed to both disseminate knowledge of the techniques and to highlight and solve problems in applying the method. A series of user notes have been published which the reader is recommended to read if and when involved in storm water drainage design. Guidance on the selection of the method of analysis and the design parameters for urban storm drainage systems can be found in the European standard EN 752 (BSI, 1998) and in *Sewers for Adoption* (WRc, 2001) in the UK.

Sustainable urban drainage systems

One of the consequences of traditional piped urban drainage systems is the reduced infiltration of rainwater into the soil. Consequently, the increased runoff during extreme storm events must be conveyed in the drainage system otherwise flooding occurs. Furthermore, if the system outfalls into a river there is a risk of increased erosion leading to channelisation and degradation of the stream. As a result, there is a need for sustainable urban drainage systems which maximise the attenuation storage capacity of the system. This may be achieved through the use of natural storage features such as soakaways, ponds, filter drains, etc. Further details and guidance can be obtained from the ongoing work of the CIRIA organisation in the UK (CIRIA, 2000a,b, 2001).

10.10 Climate change impacts in flood hydrology

There has been growing concern for the potential impacts of human-induced global climate change on flood risk assessment. One of the issues facing the hydrologist is that the use of historic records of river flow and rainfall do not cover a sufficient period of time to capture the perceived effects of increased greenhouse gas emissions. For example, the FEH (IH, 1999) findings using the available historic records for the UK could not demonstrate conclusively that climate change has (or, indeed, has not) affected flood behaviour. Consequently, the hydrologist needs to rely on computational model predictions of what if scenarios of potential climate change in order to take a precautionary approach to flood risk assessment. Current guidance in the UK is to make an allowance of an increase of 20% in peak flows or volumes over the next 50 years, and to test the sensitivity of these estimates wherever possible through running computational models of the hydrological and/or flow processes (DEFRA and EA, 2003). The input data for such models is available from the outputs of the global climate model of the UK Meteorological Office at the Hadley Centre run for four representative climate change scenarios. Detailed guidance is available in the DEFRA/EA report (DEFRA and EA, 2003).

It should be noted that scaling down results from global climate models which do not simulate all the atmospheric processes fully to use for river catchment flood estimation is an active area of research (Wheater, 2002). As climate models and the available data improve then, inevitably, the scenario estimates will require updating. It is likely that the next generation of flood prediction tools will be increasingly based on running rainfall-runoff models with long time series of rainfall as input (modified to allow for climate change). The resulting synthetic river flow sequences can then be subjected to frequency analysis in order to estimate the return periods of flood events for design. However, there are issues over the calibration of these models by varying the parameters so that the model simulates observed flows from observed rainfall to an acceptable degree of accuracy (Beven, 2000).

References and further reading

Ahmad, M. I., Sinclair, C. D. and Werritty, A. (1988) Log-logistic flood frequency analysis, *J. Hydrology*, **98**, 205–24.

Ashfaq, A. and Webster, P. (2002) Evaluation of the FEH rainfall-runoff method for catchments in the UK, *J. CIWEM*, **16**, 223–8.

Babtie Group (2000) *Floods and Reservoir Safety*, Research Report for DETR, Babtie Group.

Bayliss, A. C. and Reed, D. W. (2001) *The Use of Historical Data in Flood Frequency Estimation*, Report to MAFF, CEH, Wallingford, UK.

Beven, K. J. (2000) *Rainfall-Runoff Modelling: The Primer*, Wiley, London.

Binnie, G. M. (1981) *Early Victorian Water Engineers*, Thomas Telford, London.

Boorman, D. B., Acreman, M. C. and Packman, J. C. (1990) *A Review of Design Flood Estimation using the FSR Rainfall-Runoff Method*, Institute of Hydrology Report No. 111, IH, Wallingford, UK.

British Standards Institution (1998) *BS-EN 752 Drain and Sewer Systems Outside Buildings*, BSI, London.

Brown and Root, and BRE (2002) *Reservoir Safety – Floods and Reservoir Safety Integration*, Report Ref. XU0168, DEFRA, UK.

Butler, D. and Davies, J. W. (2000) *Urban Drainage*, E & FN Spon, London.

Centre for Ecology and Hydrology (CEH) (1999a) *FEH CD-ROM 1999*, CEH, Wallingford, UK.

Centre for Ecology and Hydrology(CEH) (1999b) *WINFAP-FEH Version* 1.0, 1999, CEH, Wallingford, UK.

Chapra, S. C. and Canale, R. P. (1998) *Numerical Methods for Engineers, with Programming and Software Applications*, 3rd edn, McGraw-Hill, Singapore.

CIRIA (2000a) *Sustainable Urban Drainage Systems – Design Manual for Scotland and Northern Ireland*, Report No. C521, CIRIA, UK.

CIRIA (2000b) *Sustainable Urban Drainage Systems – Design Manual for England and Wales*, Report No. C522, CIRIA, UK.

CIRIA (2001) *Sustainable Urban Drainage Systems – Best Practice Manual*, Report No. C523, CIRIA, UK.

Cunge, J. A. (1969) On the subject of a flood propagation method, *J. Hydraulics Res*, IAHR 7, 205–30.

DEFRA and EA (2003) *UK Climate Impacts Programme 2002 Climate Change Scenarios: Implementation for Flood and Coastal Defence – Guidance for Users*, R & D Technical Report W5B-029/TR, Environment Agency, UK.

DHI Software (2003a) *MIKE 11 FEH*, DHI Water and Environment, UK.

DHI Software (2003b) *MIKE 11*, DHI Water and Environment, UK.

DHI Software (2003c) *MOUSE*, DHI Water and Environment, UK.

Fleming, G. (2002) *Flood Risk Management: Learning to Live with Rivers*, Thomas Telford, London.

Hall, M. J., Hockin, D. L. and Ellis, J. B. (1993) *The Design of Flood Storage Reservoirs*, CIRIA Report No. RP393, CIRIA and Butterworth-Heinemann Ltd, London.

Hosking, J. R. M. and Wallis, J. R. (1997) *Regional Frequency Analysis: an Approach based on L-moments*, Cambridge University Press, UK.

Institute of Hydrology (IH) (1981) *Derivation of a Catchment Average Unit Hydrograph*, Report No. 71, NERC.

Institute of Hydrology (IH) (1999) *Flood Estimation Handbook*. Vol. 1 *Overview*. Vol. 2 *Rainfall Frequency Estimation*. Vol. 3 *Statistical Procedures for Flood Frequency Estimation*. Vol. 4 *Restatement and Application of the Flood Studies Report Rainfall-Runoff Method*. Vol. 5 *Catchment Descriptors*, IH, Wallingford, UK.

Institution of Civil Engineers (ICE) (1960) *Floods in Relation to Reservoir Practice*, London.

Institution of Civil Engineers (ICE) (1996) *Floods and Reservoir Safety*, 3rd edn, Thomas Telford, London.

Kennard, M. F., Hosking, C. G. and Fletcher, M. (1996) *Small Embankment Reservoirs*, CIRIA Report No. 161, CIRIA, London.

Linsley, R. K., Kohler, M. A. and Paulhus, J. L. H. (1982) *Hydrology for Engineers*, 3rd edn, McGraw-Hill, London.

Lloyd-Davies, D. E. (1906) The elimination of storm water from sewerage systems. *Proc. Instn. Civ. Engrs.*, **164**(2), 41–67.

McCuen, R. H. (1998) *Hydrologic Design and Analysis*, 2nd edn, Prentice-Hall, New Jersey.

Micro Drainage Limited (2003) *WinDes*, Micro Drainage Limited, UK.

National Water Council (1981) *Design and Analysis of Urban Storm Drainage. The Wallingford Procedure.* Vol. 1 *Principles, Methods and Practice.* Vol. 2 *Program User's Guide.* Vol. 3 *Maps.* Vol. 4 *Modified Rational Method.* Vol. 5 *Programmers Manual*, Hydraulics Research Ltd. Wallingford, UK.

Natural Environment Research Council (NERC) (1975) *Flood Studies Report.* Vol. I *Hydrological Studies.* Vol. II *Meteorological Studies.* Vol. III *Flood Routing Studies.* Vol. IV *Hydrological Data.* Vol. V *Maps*, London.

Osborne, M. P. (1993) *A New Runoff Volume Model*, WaPUG User Note No. 28, WaPUG, UK.

Ponce, V. M. and Yevjevich, V. (1978) Muskingum-Cunge methods with variable parameters, *J. Hydr. Div.*, ASCE, 104(12), 1663–7.

Price, R. K. (1978) A river catchment flood model. *Proc. ICE.*, **65**(2), 655–68.

Shaw, E. M. (1994) *Hydrology in Practice*, 3rd edn, Chapman & Hall, London.

Tang, X., Knight, D. W. and Samuels, P. G. (1999) Volume conservation in variable-parameter Muskingum-Cunge method, *J. Hydraulic Engrg.*, ASCE, **125**(6), 610–20.

Transport and Road Research Laboratory (TRRL) (1976) *A Guide for Engineers to the Design of Storm Sewer Systems*, 2nd edn., Road Note 35.

US Army Corps of Engineers (2002) *HEC-RAS River Analysis System, version 3.1*, Hydrologic Engineering Center, Davis, CA.

US Soil Conservation Service (1972) *National Engineering Handbook* (Section 4: Hydrology) US Government Printing Office, Washington, DC.

US Soil Conservation Service (1986) *Urban Hydrology for Small Watersheds*, Technical Report 55, USDA, Springfield, VA.

Wallingford Software Ltd (2003) *InfoWorks CS, version 5.0*, Wallingford Software Ltd, UK.

Wallingford Software Ltd and Halcrow Group Ltd (2003) *ISIS Flow User Manual: version 2.2*, Wallingford Software Ltd, UK.

Wastewater Planning User Group (WaPUG) *User notes*, Various dates. WaPUG, UK.

Wheater, H. S. (2002) Progress in and prospects for fluvial flood modelling, *Phil. Trans. R. Soc. Lond. A.*, **360**, 1409–31.

Wilson, E. M. (1990) *Engineering Hydrology*, 4th edn, Macmillan, London.

WRc (2001) *Sewers for Adoption: a Design and Construction Guide for Developers*, 5th edn, WRc, Swindon, UK.

11

Dimensional analysis and the theory of physical models

11.1 Introduction

The preceding chapters have explored various aspects of hydraulics. In each case a fundamental concept has been described and, where possible, that concept has been translated into an algebraic expression which has then been used as the basis of a mathematical model. Mathematical models are functions which represent the behaviour of a physical system, and which can be solved on a computer or calculator. A mathematical model is very convenient, since it is available whenever the engineer needs to use it. However, it may have occurred to the reader that some problems could be so complex that no adequate mathematical model could be formulated. If such a problem is encountered, what is the engineer to do? To deal with such problems it is necessary to find an alternative to mathematical models. One alternative which is frequently adopted is the use of scale model experiments. However, this approach also raises questions. For example, even when the experimental results have been obtained, there may be no self-evident (e.g. geometrical) relationship between the model behaviour and the behaviour of the full scale prototype. Thus, if an engineer wishes to employ model tests, two problems must be faced:

1. the design of the model and of the experimental procedure;
2. the correct interpretation of the results.

To this end it is necessary to identify physical laws which apply equally to the behaviour of model and prototype. Our understanding of such laws has developed progressively over the last century or so.

11.2 The idea of 'similarity'

Ideas about basic forms of similarity are often gained early in life from simple shapes. These ideas are formalized mathematically as 'geometrical similarity'. Geometrical similarity requires:

1. that all the corresponding lengths of two figures or objects are in one ratio;
2. that all corresponding angles are the same for both figures or objects.

Now, (1) may be expressed algebraically as

$$L''/L' = \lambda_L$$

where λ_L is a 'scale factor' of length (Fig. 11.1(a)). Similar statements may be made about other characteristics of two systems, where such systems exhibit some form of 'similarity', i.e. where they have certain features in common. For example, measurements of the velocity patterns in two systems may reveal that at corresponding co-ordinates there is a relationship between the velocity U' in one system and the velocity U'' in the other system. If that relationship is in the form $U''/U' = \lambda_U$, then the two systems are 'kinematically' similar (Fig. 11.1(b)). This could equally well be restated in terms of the fundamental dimensions as

$$\lambda_T = T''/T' \quad \lambda_L = L''/L'$$

since velocity has dimensions of LT^{-1}.

There may also be systems which exhibit similarity in their force patterns (Fig. 11.1(c)) so that

$$\lambda_F = F''/F'$$

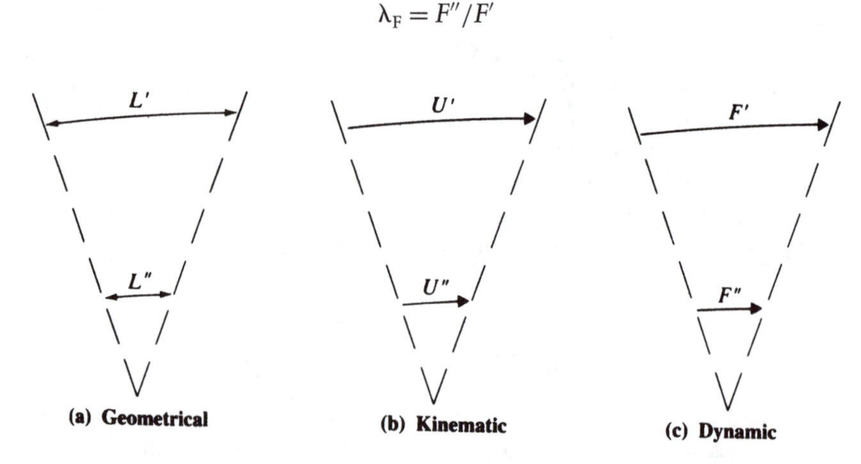

Figure 11.1 Similarity.

and since force = mass × acceleration, force may be said to possess the dimensions MLT^{-2}, therefore, for systems with similar forces,

$$\lambda_T = T''/T' \qquad \lambda_L = L''/L' \qquad \lambda_M = M''/M'$$

must all be true. Systems exhibiting similarity of this nature are referred to as being 'dynamically' similar.

These statements indicate the sorts of scales which may be of interest. They do not, however, help us to determine scale magnitudes. For this purpose, we must develop equations which reflect the appropriate dimensional considerations.

11.3 Dimensional homogeneity and its implications

Consider the relationship

$$3\,\text{eggs} + 2\,\text{bananas} = 4\,\text{Rolls-Royces}$$

This will immediately be rejected as nonsense. There is no link, numerical, dimensional or of any other sort, between the left-hand and the right-hand sides. The point of this statement of the obvious is to impress the fact that most people intuitively accept the idea of homogeneity or harmony within an equation. Dimensional homogeneity must be true for any equation which purports to describe a set of physical events.

If any physical quantity, J, is considered, it will be possible to reduce it to some function of the three fundamental dimensions, mass, length and time: i.e.

$$J = f\,[M, L, T]$$

Furthermore, if the magnitude of J is compared for two similar systems, then

$$\frac{J''}{J'} = \frac{f\,[M, L, T]''}{f\,[M, L, T]'} = \lambda_J$$

Self-evidently, this ratio must be dimensionless. This is true if the function is in the form of a product, and therefore

$$J = K[M^a L^b T^c] \tag{11.1}$$

where K is a numeric, and a, b and c are powers or indices whose magnitudes have to be determined. As an example, if J is a velocity, then in (11.1), $a = 0, b = 1$ and $c = -1$, since velocity has dimensions LT^{-1}.

In one sense it can be argued that dimensional relationships are arbitrary, since magnitudes depend on the choice of units (feet, metres, pounds, kilogrammes, etc.). For this reason, an equation which is a statement of a physical law is often used in a dimensionless form. Dimensionless equations are completely general, and are therefore frequently the basis for the representation of experimental data.

11.4 Dimensional analysis

Dimensional analysis is a powerful tool for deriving the dimensionless relationships referred to above. The methodology has long been used by engineers and scientists, and the techniques have been progressively refined over the years. Three approaches will be mentioned here. They are (1) the indicial method, (2) Buckingham's method and (3) the matrix method. However, it must be emphasized that all methods are absolutely dependent on the correct identification of all the factors which govern the physical events being analysed. The omission of a single factor may give quite misleading results. The procedure is best explained through the medium of a worked example. To illustrate the reason for using this procedure, the reader should imagine that we have been presented with a set of data relating to some unfamiliar or new type of engineering system. The indicial method will be used here.

Example 11.1 Dimensional analysis procedure

As part of a development programme, scale model tests have been carried out on a new hydraulic machine. The experimental team have presented the following data. Thrust force F, the flow velocity u, the viscosity μ and density ρ of the fluid. A typical size of the system, L, is also given. Two questions must be posed, namely (a) how to analyse or plot the data in the most informative way and (b) how to relate the performance of the model to that of the working prototype.

Solution

It seems reasonable to postulate that the force F is related to the other given quantities, so one might say

$$F = f[\rho, u, \mu, L]$$

The form of the function is completely unknown, but it has been proposed above:

1. that the function must be in the form of a power product;
2. that there must be a dimensional balance between both sides of the equation.

From 1, the equation may be rewritten as

$$F = \mathrm{K}[\rho^a u^b \mu^c L^d] \tag{11.2}$$

To meet the requirement of 2, each quantity must be reduced to its fundamental dimensions.

Force, F, is measured in newtons, but newtons are not fundamental dimensions. However, by definition, force is equated to the product of mass (M) and acceleration (which has dimensions LT^{-2}). The fundamental dimensions of F are therefore MLT^{-2}. The dimensions of the other terms are: $\rho = ML^{-3}$, $u = LT^{-1}$, $\mu = ML^{-1}T^{-1}$ and $L = L$. These dimensional relationships may be conveniently tabulated as a matrix:

$$\begin{array}{c} \\ M \\ L \\ T \end{array} \begin{array}{c} \begin{array}{ccccc} F & \rho & u & \mu & L \end{array} \\ \begin{bmatrix} 1 & 1 & 0 & 1 & 0 \\ 1 & -3 & 1 & -1 & 1 \\ -2 & 0 & -1 & -1 & 0 \end{bmatrix} \end{array}$$

Expressing each quantity in (11.2) in terms of its dimensions,

$$MLT^{-2} = \mathrm{K}[(ML^{-3})^a (LT^{-1})^b (ML^{-1}T^{-1})^c L^d] \tag{11.3}$$

For dimensional homogeneity, the function has the dimensions MLT^{-2}, and since a product relationship exists, the sum of the indices (or powers) of each dimension in the function may be equated to the index of the same dimension on the left-hand side of the equation.

Thus, on the left-hand side, M has the index 1. In the function bracket, M has the indices a and c. Therefore,

$$1 = a + c \tag{11.4}$$

Similarly, T has the index -2 on the left-hand side, and the indices $-b$ and $-c$ in the function, therefore

$$2 = b + c \tag{11.5}$$

The corresponding indicial equation for L is

$$1 = -3a + b - c + d \tag{11.6}$$

There are thus three equations (one for each fundamental dimension) but four unknown indices, so that a complete solution is unattainable. However, a partial solution is worthwhile. To obtain the partial solution, it is necessary to select (or guess) three 'governing variables'. Suppose that ρ, u and L are selected. The indices of ρ, u and L are a, b and d, respectively. So (11.4), (11.5) and (11.6) must be rearranged in terms of a, b and d.

From (11.4),

$$a = 1 - c \tag{11.7}$$

From (11.5),

$$b = 2 - c \tag{11.8}$$

Substituting (11.7) and (11.8) into (11.6)

$$1 = -3(1 - c) + (2 - c) - c + d$$

and therefore

$$d = 2 - c \tag{11.9}$$

Substituting for a, b and d in (11.2),

$$F = \mathrm{K}[\rho^{1-c} u^{2-c} \mu^{c} L^{2-c}]$$

$$= \mathrm{K}\left[\rho u^2 L^2 \left(\frac{\mu}{\rho u L}\right)^{c}\right]$$

Since the function represents a product, it may be restated as

$$F = \rho u^2 L^2 \mathrm{K}\left(\frac{\rho u L}{\mu}\right)^{c}$$

or as

$$\frac{F}{\rho u^2 L^2} = \mathrm{K}\left(\frac{\rho u L}{\mu}\right)^{-c} \tag{11.10}$$

where K and c are unknown.

There are a number of important points to be made about (11.10).

(a) Two groups have emerged from the analysis, $F/\rho u^2 L^2$ and $\rho u L/\mu$. If the reader cares to check, it will be found that both groups are dimensionless. For conciseness, dimensionless groups are referred to as 'Π' groups. Thus we might state

$$\Pi_1 = F/\rho u^2 L^2$$

and

$$\Pi_2 = \rho u L/\mu$$

Π_2 is, of course, the Reynolds Number, which has been encountered before.

(b) Dimensionless groups are independent of units and of scale. Π_1 and Π_2 are therefore equally applicable to the model or to the prototype.

(c) Both Π groups represent ratios of forces, as will now be shown:

$$\text{the 'inertia force' of a body, } F = \text{mass} \times \text{acceleration}$$

$$\text{the mass of a body} = \rho \times \text{volume} = \rho L^3$$

$$\text{acceleration} = du/dt, \text{ which has the dimensions of (velocity/time)} = LT^{-2}$$

Therefore

$$\text{mass} \times \text{acceleration} = \rho L^3 L T^{-2} = \rho L^4 T^{-2} (= \rho u^2 L^2)$$

Therefore Π_1 represents the ratio of the thrust force to the inertia force:

$$\text{the inertia force per unit area} = \rho L^4 T^{-2} L^{-2} = \rho L^2 T^{-2}$$

Now $\rho L^2 T^{-2}$ may be rewritten as ρu^2, so looking now at Π_2, the ratio (inertia force/viscous force) can be written as

$$\frac{\rho u^2}{\text{viscous force}}$$

Viscous force is represented by the viscous shear stress, $\tau = \mu\, du/dy$. Dimensionally, this is the same as $\mu u/y$. Therefore,

$$\frac{\rho u^2}{\mu(u/y)} = \frac{\rho u y}{\mu}$$

which is Π_2.

(d) All three fundamental dimensions are present in (11.10). Therefore, if the model is to truly represent the prototype, then both model and prototype must conform to the law of dynamic similarity. For this to be so, the magnitude of each dimensionless group must be the same for the model as for the prototype:

$$\Pi_1' = \Pi_1''$$

$$\Pi_2' = \Pi_2''$$

where Π' refers to the prototype and Π'' to the model.

If the above statement (regarding equality of Π groups) is not fulfilled, it must follow that ratios of forces in the model do not correspond to the ratios of forces in the prototype, and hence that the model and prototype are not dynamically similar.

(e) It would, of course, be possible to choose alternatives to ρ, u and L as governing variables. Different Π groups might then emerge. However, because the product $\rho u L$ represents the inertia force (which is a parameter relevant to any flow), these three quantities are very frequently selected. The groups which are ultimately the most helpful for experimental analysis have emerged largely by trial and error.

The reader is strongly advised to review Example 11.1 carefully, since it contains all the main ideas and processes which form the basis of dimensional analysis.

The indicial method of dimensional analysis is perfectly satisfactory as long as the number of variables involved is small. For problems involving larger numbers of variables, a more orderly process is helpful. This leads us to the use of the Buckingham or the matrix methods.

11.5 Dimensional analysis involving more variables

Buckingham's method

Buckingham's ideas were published in a paper in 1915. In outline, he proposed:

1. that if a physical phenomenon was a function of m quantities and n fundamental dimensions, dimensional analysis would produce $(m-n)\Pi$ groups;
2. that each Π should be a function of n governing variables plus one more quantity;
3. the governing quantities must include all fundamental dimensions;
4. the governing quantities must not combine among themselves to form a dimensionless group;
5. as each Π is dimensionless, the final function must be dimensionless, and therefore dimensionally

$$f\,[\Pi_1, \Pi_2, \ldots, \Pi_{(m-n)}] = M^0 L^0 T^0$$

Referring back to Example 11.1, there were five quantities (F, ρ, u, μ, L) and three dimensions (M, L, T), from which we derived two groups. To obtain Π_1, using Buckingham's approach, with ρ, u, and L as governing variables,

$$\Pi_1 = \rho^a u^b L^c F = M^0 L^0 T^0$$

Therefore

$$(ML^{-3})^a (LT^{-1})^b L^c (MLT^{-2}) = M^0 L^0 T^0$$

Equating indices,

$$M \quad a+1=0 \qquad \text{therefore } a=-1$$

$$T \quad -b-2=0 \qquad \text{therefore } b=-2$$

$$L \quad -3a+b+c+1=0$$

substituting for a and b,

$$+3-2+c+1=0 \qquad \text{therefore } c=-2$$

so

$$\Pi_1 = F/\rho u^2 L^2$$

Any remaining Π groups are produced in the same way, and thus $\Pi_2 = \mu/\rho u L$, as before.

The matrix method

This is based on the work of a number of investigators, notably Langhaar (1980) and Barr (1983). The method used here is described in detail in Matthews and Morfett (1986), and is a development of the previous work. It can readily be programmed for computer application, to speed up the analysis. The procedure entails:

1. setting up a dimensional matrix with the governing variables represented by the first n columns;
2. partitioning the matrix to form an '$n \times n$' matrix and an additional matrix;
3. applying Gauss–Jordan elimination to produce a leading diagonal of unity in the $n \times n$ matrix;
4. abstracting the requisite II groups from the final index matrix.

All of the remaining examples in this chapter employ this method.

11.6 Applications of dynamic similarity

Pipe flow

The history of the development of the various pipe flow equations has already been outlined in Chapter 4. It is noteworthy that dimensional procedures underlie a number of these developments, such as Nikuradse's diagram. An example illustrating the use of these procedures follows.

Example 11.2 Analysis of pipe flow

The hydraulic head loss, h_f, in a simple pipeline is assumed to depend on the following quantities:

the density (ρ) and viscosity (μ) of the fluid;
the diameter (D), length (L) and roughness (k_S) of the pipe;
a typical flow velocity (usually the mean velocity) V.

(a) Develop the appropriate dimensionless groups to describe the flow.

(b) A 10 km pipeline, 750 mm diameter, is to be used to convey oil ($\rho = 850\,\text{kg/m}^3, \mu = 0.008\,\text{kg/m s}$). The design discharge is 450 l/s. The pipeline will incorporate booster pumping stations at suitable intervals. As part of the design procedure, a model study is to be carried out. A model scale $\lambda_L = 1/50$ has been selected, and air is to be used as the model fluid. The air has a density of $1.2\,\text{kg/m}^3$ and viscosity of $1.8 \times 10^{-5}\,\text{kg/m s}$.

At what mean air velocity will the model be correctly simulating the flow of oil?

If the head loss in the model is 10 m for the full pipe length, what will be the head loss in the prototype?

Solution

(a) If the matrix method is employed, the dimensional matrix is first produced. The principal quantities of interest are:

1. the characteristics of the fluids: ρ, μ;
2. the characteristics of the pipeline: length L, diameter D and roughness k_S;
3. the characteristics of the flow: typical (usually mean) velocity V, head loss h_f.

The dimensional matrix ($\mathbf{M}_D$) is then

$$\begin{array}{c} \\ M \\ L \\ T \end{array} \begin{array}{c} \begin{array}{cccccccc} V & D & \rho & | & h_f & \mu & k_S & L \end{array} \\ \left[\begin{array}{ccc|cccc} 0 & 0 & 1 & 0 & 1 & 0 & 0 \\ 1 & 1 & -3 & 1 & -1 & 1 & 1 \\ -1 & 0 & 0 & 0 & -1 & 0 & 0 \end{array}\right] \end{array} \tag{11.11}$$

The governing variables are assumed to be V, D and ρ.

Applying the Gauss–Jordan elimination procedure,

$$\begin{array}{c} \\ \rho \\ D \\ V \end{array} \begin{array}{c} \begin{array}{cccccccc} \rho & D & V & | & h_f & \mu & k_S & L \end{array} \\ \left[\begin{array}{ccc|cccc} 1 & 0 & 0 & 0 & 1 & 0 & 0 \\ 0 & 1 & 0 & 1 & 1 & 1 & 1 \\ 0 & 0 & 1 & 0 & 1 & 0 & 0 \end{array}\right] \end{array} \tag{11.12}$$

and hence $\mathbf{M}_k^*$, the index matrix is

$$\begin{array}{c} \\ \rho \\ D \\ V \\ h_f \\ \mu \\ k_S \\ L \end{array}\begin{array}{c} \begin{array}{cccc} \Pi_1 & \Pi_2 & \Pi_3 & \Pi_4 \end{array} \\ \begin{bmatrix} 0 & -1 & 0 & 0 \\ -1 & -1 & -1 & -1 \\ 0 & -1 & 0 & 0 \\ 1 & 0 & 0 & 0 \\ 0 & 1 & 0 & 0 \\ 0 & 0 & 1 & 0 \\ 0 & 0 & 0 & 1 \end{bmatrix} \end{array} \tag{11.13}$$

Therefore

$$\Pi_1 = \frac{h_f}{D} \qquad \Pi_2 = \left(\frac{\rho DV}{\mu}\right)^{-1}$$

$$\Pi_3 = \frac{k_S}{D} \qquad \Pi_4 = \frac{L}{D}$$

Because D was used as a governing variable, Π_1 emerged as $h_f D^{-1}$. It is conventional to use $h_f L^{-1}$ in preference to $h_f D^{-1}$, since $h_f L^{-1}$ represents the hydraulic gradient. This makes no practical difference, providing the model is geometrically similar to the prototype, i.e.

$$\frac{D''}{D'} = \frac{L''}{L'}$$

in which case,

$$\frac{(h_f D^{-1})''}{(h_f D^{-1})'} = \frac{(h_f L^{-1})''}{(h_f L^{-1})'}$$

(b) Proceeding to the numerical solution, the prototype velocity is obtained, as usual:

$$V' = \frac{0.45}{(\pi/4) \times 0.75^2} = 1.019\,\text{m/s}$$

For dynamic similarity, the ratios of forces in the model must equal the corresponding ratios in the prototype. In pipe flow the force ratio is represented by the Reynolds Number, so

$$\Pi_2'' = \Pi_2'$$

For the prototype,

$$\Pi_2' = \frac{\rho' V' D'}{\mu'} = \frac{850 \times 1.019 \times 0.75}{0.008} = 81\,202$$

Therefore

$$\Pi_2'' = 81\,202 = \frac{1.2 \times V'' \times (0.75/50)}{1.8 \times 10^{-5}}$$

Therefore the velocity in the model $V'' = 81.2\,\text{m/s}$. The velocity which is required if the model is to truly represent the prototype is known as the 'corresponding velocity'.

To estimate the head loss in the prototype, the modified Π_1 equation ($\Pi_1 = h_f L^{-1}$) is used:

$$\Pi_1' = \Pi_1''$$

i.e.

$$\left(\frac{h_f}{L}\right)' = \left(\frac{h_f}{L}\right)''$$

In the model, $h_f'' = 10\,\text{m}$ and $L'' = 200\,\text{m}$. Therefore

$$\frac{h_f'}{10\,000} = \frac{10}{200}$$

So $h_f' = 500\,\text{m}$, which will be the total head required from the booster pumps.

Free surface flows

In Example 11.2, the evaluation of V'' presented no problem. This was so due to the dominance of one pair of forces (momentum and shear), which are both represented by the Reynolds Number. Hydraulics problems are by no means always so straightforward. A case in point is the family of free surface flows. These flows are controlled by momentum, shear and gravity forces, and this combination highlights a difficulty which has to be overcome through a fundamental grasp of the mechanics of fluids.

Example 11.3 Open channel flow

State the physical quantities which relate to steady open channel flow. Hence derive the dimensionless groups which describe such a flow. Why is it theoretically impossible to produce a valid scale model? How is this problem usually overcome?

Solution

The principal characteristics of a steady channel flow are:

a typical velocity (usually the mean velocity), V;
the frictional resistance, F;
the fluid characteristics, i.e. density (ρ) and viscosity (μ);
the geometrical characteristics of the channel, i.e. the depth, y (or hydraulic radius, R);

bed slope S_0, and surface roughness, k_S;
the gravitational acceleration, g.

The dimensional matrix is

$$\begin{array}{c} \\ M \\ L \\ T \end{array} \begin{array}{c} \begin{array}{cccccccc} V & R & \rho & \mu & g & k_S & S_0 & F \end{array} \\ \left[\begin{array}{cccccccc} 0 & 0 & 1 & 1 & 0 & 0 & 0 & 1 \\ 1 & 1 & -3 & -1 & 1 & 1 & 0 & 1 \\ -1 & 0 & 0 & -1 & -2 & 0 & 0 & -2 \end{array}\right] \end{array} \quad (11.14)$$

Following Gauss-Jordan elimination,

$$\begin{array}{c} \\ \rho \\ R \\ V \end{array} \begin{array}{c} \begin{array}{cccccccc} \rho & R & V & \mu & g & k_S & S_0 & F \end{array} \\ \left[\begin{array}{cccccccc} 1 & 0 & 0 & 1 & 0 & 0 & 0 & 1 \\ 0 & 1 & 0 & 1 & -1 & 1 & 0 & 2 \\ 0 & 0 & 1 & 1 & 2 & 0 & 0 & 2 \end{array}\right] \end{array} \quad (11.15)$$

The index matrix is then

$$\begin{array}{c} \\ \rho \\ R \\ V \\ \mu \\ g \\ k_S \\ S_0 \\ F \end{array} \begin{array}{c} \begin{array}{ccccc} \Pi_1 & \Pi_2 & \Pi_3 & \Pi_4 & \Pi_5 \end{array} \\ \left[\begin{array}{ccccc} -1 & 0 & 0 & 0 & -1 \\ -1 & 1 & -1 & 0 & -2 \\ -1 & -2 & 0 & 0 & -2 \\ 1 & 0 & 0 & 0 & 0 \\ 0 & 1 & 0 & 0 & 0 \\ 0 & 0 & 1 & 0 & 0 \\ 0 & 0 & 0 & 1 & 0 \\ 0 & 0 & 0 & 0 & 1 \end{array}\right] \end{array} \quad (11.16)$$

Therefore

$$\Pi_1 = \frac{\rho R V}{\mu} \qquad \Pi_2 = \frac{V^2}{gR}$$

$$\Pi_3 = \frac{k_S}{R} \qquad \Pi_4 = S_0 \qquad \Pi_5 = \frac{F}{\rho R^2 V^2}$$

In Example 11.2, the Reynolds Number was the sole criterion for dynamic similarity. In channel flow there are two criteria: Reynolds Number (Π_1) and Froude Number ($\Pi_2 = \mathrm{Fr}^2$). Furthermore, in modelling channel flows, it is almost universal practice to use water as the 'model fluid'. This means that $\rho' = \rho''$ and $\mu' = \mu''$.

From Π_1

$$\frac{\rho V' R'}{\mu} = \frac{\rho V'' R''}{\mu}$$

therefore $V'R' = V''R''$, or

$$\frac{V''}{V'} (= \lambda_V) = \frac{R'}{R''} \quad (11.17)$$

From Π_2,

$$\left(\frac{V^2}{Rg}\right)' = \left(\frac{V^2}{Rg}\right)''$$

Therefore, since g must be the same for model as for prototype,

$$\frac{V''^2}{V'^2} = \frac{R''}{R'}$$

Therefore

$$\frac{V''}{V'}(= \lambda_V) = \left(\frac{R''}{R'}\right)^{1/2} \tag{11.18}$$

It is clear that (11.17) and (11.18) are incompatible unless $R'' = R'$, i.e. unless the model and the prototype are identical in size. This is impracticable. It is possible to overcome this problem by considering the nature of the flow. In channels, as in pipes, frictional resistance is a function of the viscosity (and therefore of the Reynolds Number) for laminar and transitional turbulent flows. Once the flow is completely turbulent, resistance is independent of viscosity. Therefore, if the magnitude of the Reynolds Number representing the model flow is sufficiently great to indicate 'complete turbulence', (11.17) may be ignored. Model scaling is then based on (11.18) only. In practice, the design of hydraulic models involves rather more complexities than have emerged here, as will be seen in section 11.7.

Hydraulic machines

The quantities which are usually considered in a dimensional analysis of hydraulic machines are:

1. the power (P) and rotational speed (N) of the machine;
2. the pressure head (H) generated by the machine;
3. the corresponding discharge (Q);
4. the typical machine size (D) and roughness (k_s);
5. the fluid characteristics (ρ and μ).

(Note that it is conventional for the pressure head to be represented as gH rather than simply as H.)

If ρ, N and D are used as governing variables, the following groups emerge:

$$\Pi_1 = \frac{P}{\rho N^3 D^5} \qquad \Pi_2 = \frac{Q}{ND^3} \qquad \Pi_3 = \frac{gH}{N^2D^2}$$

$$\Pi_4 = \frac{\rho ND^2}{\mu} \qquad \Pi_5 = \frac{k_s}{D}$$

Most hydraulic machines operate in the completely turbulent zone of flow, so that the Reynolds Number term (Π_4) is neglected. Since geometrical similarity must apply if two machines are to be compared, the geometrical term (Π_5) is also usually ignored. Thus, Π_1, Π_2 and Π_3 are the groups used in analysis. However, one further group, known as the 'specific speed' of a machine, is derived by combining groups as follows:

Pumps. For pumps, the interest centres on the discharge and head of a particular machine. The product

$$\Pi_2^{1/2}\Pi_3^{-3/4} = \frac{NQ^{1/2}}{(gH)^{3/4}} = K_N \tag{11.19}$$

is a dimensionless specific speed. Historically the 'g' term was disregarded, and so conventionally the specific speed becomes

$$N_S = NQ^{1/2}/H^{3/4} \tag{11.20}$$

This is not dimensionless.

Turbines. For turbines the emphasis lies with the power output and head terms. The product

$$\Pi_1^{1/2}\Pi_3^{-5/4} = \frac{NP^{1/2}}{\rho^{1/2}(gH)^{5/4}} = K_N \tag{11.21}$$

is again a dimensionless specific speed. However, since both the ρ and g terms are constant, they are usually ignored, and therefore

$$N_S = NP^{1/2}/H^{5/4} \tag{11.22}$$

Again this is not dimensionless.

The name 'specific speed' is an unfortunate term since neither K_N nor N_S represent a meaningful 'speed' anywhere in the machine. The name owes its origins to a concept which arose early in the development of hydraulic machines. It is best to think of it as a numeric whose value is related to the geometrical form and designed duty of a given machine. This value will be a constant for a particular 'family' of similar machines, and will be calculated for the designed optimum operating conditions.

Example 11.4 Pump similarity

A 100 mm diameter pump, which is driven by a synchronous electric motor, has been tested with the following results:

power supplied to pump = 3.2 kW;
pump speed = 1450 rev/min, head = 12.9 m, discharge = 20 l/s;
efficiency = 79%; fluid: water at 18° C.

A geometrically similar pump is required to deliver 100 l/s against a 20 m head. The new pump will be driven by a synchronous motor. The speeds may be either 960 rev/min or 1450 rev/min. Calculate the principal characteristics (size, speed and power) of the new machine. Calculate also the specific speed, N_S.

Solution

(Note that in the solution Π'' terms refer to the 100 mm pump). Π_3 is constant, i.e. $\Pi_3'' = \Pi_3'$. Therefore

$$\frac{9.81 \times 12.9}{1450^2 \times 0.1^2} = \frac{9.81 \times 20}{N'^2 D'^2}$$

so

$$N'D' = 180.5$$

or

$$N' = 180.5/D'$$

Π_2 is constant, i.e. $\Pi_2'' = \Pi_2'$. Therefore

$$\frac{0.02}{1450 \times 0.1^3} = \frac{0.1}{N'D'^3}$$

so

$$N'D'^3 = 7.25$$

or

$$N' = 7.25/D'^3$$

Thus, combining these two equations

$$N' = \frac{180.5}{D'} = \frac{7.25}{D'^3}$$

therefore $D'^2 = 0.0402$, i.e. $D' = 0.2$m. Hence

$$N' = 180.5/0.2 = 902.5 \text{ rev/min}$$

Therefore, the speed must be 960 rev/min.

Check diameter:

$$960 = 180.5/D'$$

therefore $D' = 0.188$ m. In practice, D' would probably be rounded up to 0.2 m.

Check discharge from Π_2:

$$\frac{0.02}{1450 \times 0.1^3} = \frac{Q'}{960 \times 0.2^3}$$

therefore $Q' = 0.106\,\text{m}^3/\text{s}$.

To estimate the power of the new pump, use Π_1:

$$\frac{3.2}{1000 \times 1450^3 \times 0.1^5} = \frac{P'}{1000 \times 960^3 \times 0.2^5}$$

therefore $P' = 29.72$ kW.

The specific speed is obtained from (11.20), so for the new pump

$$N_S' = \frac{960 \times 0.1^{1/2}}{20^{3/4}} = 32.1$$

Check against the original pump:

$$N_S'' = \frac{1450 \times 0.02^{1/2}}{12.9^{3/4}} = 30.1$$

The difference between N_S'' and N_S' arises due to rounding up the speed of the new pump from 902.5 to 960 rev/min.

11.7 Hydraulic models

The design of a major new hydraulic system (e.g. a hydroelectric scheme, a dock or a hydraulic structure) may be approached in one of three ways:

1. by theoretical reasoning and the use of numerical models;
2. on the basis of previous experience of similar systems;
3. by scale model experiments.

It has already been pointed out that, even with modern computing facilities, many complex problems still defy complete theoretical analysis, especially where flows are two- or three-dimensional (e.g. in rivers or estuaries of complex geometry). In such cases physical scale models are an important research or design tool. Ideally, a scale model of a river will reproduce the prototype system such that conditions of dynamic similarity are attained. However, as illustrated by Example 11.3, a number of criteria of dynamic similarity may be applicable and it is usually impossible to design a model which complies with all of them, so one criterion has to be adopted as representing the predominant force system. By ignoring the effect of other criteria, a difference between scale model and prototype is introduced which implies differences in behaviour; this error is known as scale effect. The magnitude of such errors depends on the relative magnitudes of the predominant force and the other force(s) (e.g. the relative importance of Reynolds' and Froude Numbers). To ensure that the behaviour of a hydraulic model is sufficiently accurate it is important to undertake checks, by reference either to past experience of similar systems or to field measurements of the system under consideration. Adjustments may have to be made to the model to achieve acceptable performance (see Example 11.5).

River models

For present purposes, a river will be defined as a channel in which flow is unidirectional. River models may be constructed:

(a) to study flow patterns only. The model then has a 'rigid' or 'fixed' bed (e.g. of wood or concrete); or
(b) to study flow and sediment movement. The model must then be constructed with a particulate 'mobile' bed.

River models may be 'undistorted' or 'distorted'. An undistorted model is geometrically similar to the prototype, and therefore has the same scale (λ_L) for the vertical and horizontal dimensions.

A distorted model has differing horizontal and vertical scales (λ_x and λ_y, respectively, with $\lambda_x < \lambda_y$) to save space and cost (but note that distorted mobile bed models may not simulate prototype sediment transport behaviour accurately). The possibility of distortion is based on studies of river flows. The flow pattern may be envisaged as two distinct zones (Fig. 11.2(a)):

1. The zone near to the bank in which velocity must vary in the x-direction (due to frictional shear at the sides) and in the y-direction (due to friction at the bed). According to Keulegan (1938) this zone extends approximately $2.5y$ from each bank (and therefore occupies $2 \times 2.5y = 5y$ of the total width of the river).

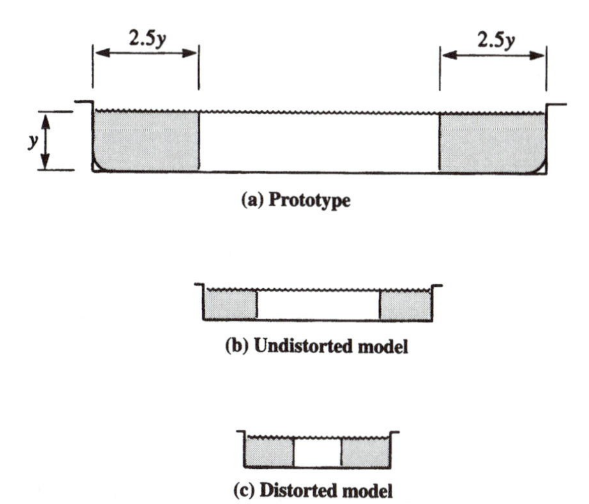

Figure 11.2 River models.

2. The central zone, in which velocity varies in the y-direction, but varies very little in the x-direction. The flow is therefore practically two-dimensional. The properties of a two-dimensional flow may be represented by a 'sample', it is unnecessary to reproduce the whole flow. Therefore if the model channel width $B'' > 5y''$, a sample of the central zone will be reproduced (Fig. 11.2(c)). Distortion may help in achieving a sufficiently high Reynolds' Number to achieve completely turbulent flow, though this is a function of both the Reynolds' Number and surface roughness (see Fig. 4.5, and a corresponding version of the Moody diagram for open channels in Henderson (1966), p. 93). Also it is often impossible to achieve the correct magnitudes for the local losses at changes of section shape or direction in distorted models. A guide value of $\mathrm{Re} > 1000$ may be taken as indicating turbulent flow, or if friction velocity is used $\mathrm{Re}_* > 70$, though it is desirable to exceed these values by a comfortable margin (see section 5.5).

For mobile-bed models further quantities (e.g. boundary shear, sediment diameter, immersed sediment density) now have to be scaled and additional boundary conditions may be required. There are still only two degrees of freedom for the scaling of the model. This generally means the adoption of a separate sediment transport scale. One approach to designing such models is based on the two parameters used for the Shields Diagram (F_s and Re_*, see section 9.2). Thus, if F_s and Re_* are the same for the model and prototype it is assumed that the model will produce the same bed conditions, and that the ratio of particle size to roughness will be the same. It is often the case that the sediment density in the model differs from that in the prototype to achieve the correct scaling. Where suspended sediment transport is modelled, the ratio of shear velocity to fall velocity, υ_{FS} becomes an important parameter, which may involve scaling the sediment density. A range of granular materials is available for modelling, e.g. coal dust ($\rho_s = 1300\,\text{kg/m}^3$), araldite ($\rho_s = 1120\,\text{kg/m}^3$), polystyrene ($\rho_s = 1040\,\text{kg/m}^3$).

Example 11.5 Model of a large river

Explain the difference between distorted and undistorted models, and justify the use of distortion. Illustrate the application of distortion by reference to a model study of a lowland river having the following prototype dimensions: depth = 5 m, width = 190 m, slope = 0.0001, mean velocity = 1.15 m/s, $k_s = 0.02\,\text{m}$.

Solution

The first part of the question has been covered in the preceding text.

The preliminary stage in finding the model scale is to calculate the value of Re_*. To calculate u_* recall that

$$u_* = \left(\frac{\tau_0}{\rho}\right)^{1/2}$$

$$\text{friction factor, } \lambda = \frac{8\tau_0}{\rho V^2} = 8\left(\frac{u_*}{V}\right)^2$$

$$V = C(RS_0)^{1/2}$$

where

$$C = \left(\frac{8g}{\lambda}\right)^{1/2} = \left(\frac{V}{u_*}\right)g^{1/2}$$

Therefore, substituting for C in the Chézy formula and rearranging,

$$u_* = (gRS_0)^{1/2} \simeq (gyS_0)^{1/2}$$

The data for the lowland river will therefore lead to

$$u'_* = (9.81 \times 5 \times 0.0001)^{1/2} = 0.07\,\text{m/s}$$

$$\text{Re}'_* = \frac{\rho u'_* k'_s}{\mu} = \frac{1000 \times 0.07 \times 0.02}{1.14 \times 10^{-3}} = 1228$$

Now $u''_* = \lambda_V u'_*$, and from (11.18) $\lambda_V = \lambda_R^{1/2} \simeq \lambda_y^{1/2}$. Also $k''_S = \lambda_L k'_S = \lambda_y k'_S$. Therefore

$$\text{Re}''_* = \rho(u'_*\lambda_y^{1/2})(k'_S\lambda_y)/\mu = \text{Re}'_*\lambda_y^{3/2} = 1228\lambda_y^{3/2}$$

Since $\text{Re}''_* \geqslant 70$ for fully turbulent flow,

$$1228\lambda_y^{3/2} \geqslant 70$$

i.e.

$$\lambda_y \geqslant \left(\frac{70}{1228}\right)^{2/3} \geqslant \frac{1}{6.752}$$

For an undistorted model, horizontal length scale and vertical length scale are the same. Therefore $B'' = B' \times \lambda_x$, i.e.

$$B'' = 190 \times \frac{1}{6.752} = 28.14\,\text{m}$$

and $y'' = y' \times \lambda_y$, i.e.

$$y'' = 5 \times \frac{1}{6.752} = 0.74\,\text{m}$$

The model is thus the size of a substantial river! This is completely impracticable in terms of space and cost.

Turning to the concept of the 'distorted' model, $y'' = 0.74$. Therefore the zones of three-dimensional flow near the two banks account for $2 \times (2.5 \times 0.74) = 3.7\,\text{m}$. Therefore, if $B > 3.7\,\text{m}$, a sample of the central zone will be reproduced, say $\lambda_x = 1/40$, which gives

$$B'' = 190 \times (1/40) = 4.75\,\text{m}$$

Other scales are determined as follows:

$$S_0 \propto y/x$$

so

$$\lambda_S = \frac{\lambda_y}{\lambda_x} = \frac{1/6.75}{1/40} = 5.93$$

therefore

$$S_0'' = S_0'\lambda_S = 0.0001 \times 5.93 = 0.000593$$

$Q = V \times \text{area} = VBy$, therefore

$$\lambda_Q = \lambda_V\lambda_B\lambda_y = \lambda_y^{1/2}\lambda_x\lambda_y = \lambda_x\lambda_y^{3/2}$$

so

$$\lambda_Q = \frac{1}{40}\left(\frac{1}{6.75}\right)^{3/2} = \frac{1}{701}$$

Now $Q' = 1.15 \times 190 \times 5 = 1092.5\,\mathrm{m^3/s}$, therefore

$$Q'' = 1092.5 \times \frac{1}{701} = 1.56\,\mathrm{m^3/s}$$

(which would still strain the pumping resources of most laboratories!).

Now that the model has been distorted, the variation of depth with distance may not accurately reproduce the prototype characteristics. It is usual to carry out adjustments to the surface roughness in the model (k''_S) until an acceptable degree of accuracy is attained. This is known as 'calibration' of the model.

Models of estuarial and coastal hydraulics

Transport phenomena on coasts and in estuaries represent some of the most challenging problems in physical scale modelling. Flows are often complex, and scale effects may become significant.

Estuarial models. Estuaries have been well described as meeting places (of salt and fresh water, of river flows and tidal flows, of different flora and fauna, etc.). The co-existence of salt and fresh water introduces the problem of differential density effects (considered further in section 17.7). The large areas of water may mean that Coriolis Forces (due to the earth's rotation) are significant, especially for mobile-bed models. Furthermore, the flow velocities and turbulence are usually three-dimensional. It is not possible to reproduce all of these effects in a scale model.

Acceptable results can often be achieved by making a number of simplifying assumptions:

1. the horizontal shear stresses are usually greater by an order of magnitude than those on vertical planes;
2. if $\lambda_\rho = 1$, then Froudian scaling is used for current velocity;
3. Coriolis Forces are ignored except where bed forms due to sediment transport are important;
4. oscillatory flows are scaled using the Strouhal Number (see section 3.6).

To reproduce the correct horizontal shear forces and water surface profile it is often necessary to exaggerate the bed roughness in the model. However, this inevitably introduces a scale effect which may be significant where dispersion effects are being modelled. For example, the rate of dispersion of a plume from a sewage treatment works into a wide estuary is often overestimated in a model.

Correct interpretation of the results of model tests depend crucially on the skills of the engineers responsible for design and operation of the model, and on the availability of field data for proving and calibration purposes.

Coastal models. Problems relating to tides and waves are encountered, for example, in the design of docks, coastal defences and sea outfalls (discharge pipes from storm sewers or sewage treatment works).

Physical models (Fig. 11.3) are still widely used for these studies, though recent progress in the application of finite element analysis is having an increasing impact. Since tides are a particular type of wave, this section is, in effect, all about waves.

Figure 11.3 Hydraulic model of Port Belawan, Sumatra. (Courtesty of HR Wallingford Ltd.)

Waves may be simply (and roughly) divided into two categories, namely 'long' waves and 'short' waves. The parameter of wave length is L/y, and it is generally held that for long waves $L/y > 20$ and for short waves $L/y < 2$. Short waves are primarily wind-generated in deep waters. Long waves may be swell waves or may be tides. The principal characteristics of a wave (see Fig. 11.4) are the geometrical terms (height H, length L and mean water depth y) and the temporal terms (periodic time T and celerity c). Dimensional analysis shows that the main two parameters of the flow are the Froude and Strouhal Numbers:

$$\mathrm{Fr} = c/(gy)^{1/2} \quad \text{(for long waves)}$$

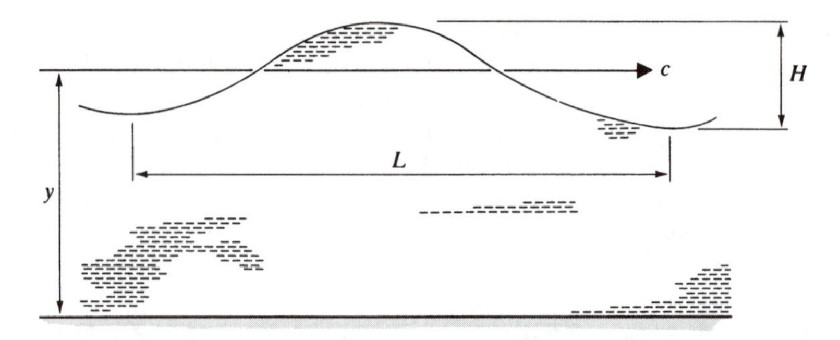

Figure 11.4 Wave profile.

$$\mathrm{Fr} = c/(gL)^{1/2} \quad \text{(for short waves)}$$

$$\mathrm{St} = cT/L$$

It should be noted that, to avoid surface tension effects on the wave form, $L \gg 0.017\,\mathrm{m}$ (in practice $L > 0.17\,\mathrm{m}$).

None of the above equations inherently excludes the possibility of some distortion. It must be recalled, however, that waves are a function of gravitation and of the fluid. A wave shape cannot be arbitrarily distorted. On the other hand, current patterns can be reproduced by distorted models. Thus, for example, wave diffraction patterns around a breakwater are not accurately reproduced in a distorted model, though in practice experienced researchers can often interpret the results obtained from distorted models with sufficient accuracy.

Example 11.6 Harbour model

A major and developing port is subject to excessive wave action during storms. This leads on occasion to damage to vessels at their moorings. A model test is to be carried out to optimize the design of a new breakwater. The harbour occupies $0.7 \times 2\,\mathrm{km}$, and the maximum space in the hydraulics laboratory is $10 \times 20\,\mathrm{m}$. Select the model scales.

Solution

The absolute maximum length scale would be

$$\lambda_L = \frac{20}{2000} = \frac{1}{100}$$

However, allowances must be made for access and for the wave- and tide-producing mechanisms, so say $\lambda_L = 1/120$. (It is being assumed here the model will be undistorted, since wave diffraction is the centre of interest.)

Other scales are as follows:

$$\text{wave height } \lambda_H (= H''/H') = 1/120$$

If St is the same for the model and the prototype then

$$\text{St}'' = \text{St}' \quad \text{and} \quad c = (gy)^{1/2}$$

then rearranging we obtain

$$\frac{T''}{T'} = \frac{c'L''}{c''L'}$$

i.e.

$$\lambda_{T_w} = \frac{\lambda_L}{\lambda_{c_w}} = \frac{1/120}{\sqrt{1/120}} = \frac{1}{10.95}$$

say, $\lambda_{T_w} = 1/11$

$$\text{tidal period } T_T = L/c_T$$

therefore

$$\lambda_{T_T} = \frac{\lambda_L}{\lambda_{c_T}} = \frac{\lambda_L}{\sqrt{\lambda_L}} \simeq \frac{1}{11}$$

Therefore tidal period in model $= 12/11 = 1.09\,\text{h}$.

Models of hydraulic structures

Hydraulic structures are small compared with rivers or estuaries. In consequence, hydraulic models of such structures can usually be constructed to a relatively large scale (say $1/10 < \lambda_L < 1/50$) and need not be distorted. Velocity scales are usually based on the Froude Number (for flows with a free surface) or the Reynolds Number (for ducted flows). However, there are some exceptions to this general rule. For example, to simulate cavitation, the pressures in the model should be the same as those for the prototype. This is not always possible. Taking the case of a spillway model, the velocities in the model will be lower than those in the prototype, so the lowest model pressures will probably not be as low as those in the prototype. Nevertheless, an experienced researcher will usually be able to use the results from the model to predict conditions on the prototype. Models can also be used to investigate local scour problems, or the current patterns set up by the structure.

Example 11.7 Spillway model

A spillway system is to be designed for a dam (Fig. 11.5). The design discharge over the spillway is $15\,\text{m}^3/\text{s}$ per metre width at a design head of 3 m. A hydraulic model is to be used to confirm the estimates of the spillway performance, and also to help in the design of the scour bed just downstream of the stilling basin. Estimate the scales of such a model if the scour bed is armoured with stones having a representative size $D_{50} = 35\,\text{mm}$.

Solution

For fully turbulent flow, $\text{Re}_* > 70$. To obtain Re_*, u_* must first be evaluated. From (3.7),

$$\frac{V}{u_*} = \frac{1}{0.4}\ln\left(\frac{y}{k_S}\right)$$

where k_S represents the roughness, which will be taken as being equal to $D_{50}/2$. The magnitude of the velocity, V, over the scour bed is obtained from the continuity equation, $V = Q/A = 15/6 = 2.5\,\text{m/s}$. Therefore

$$\frac{2.5}{u_*} = \frac{1}{0.4}\ln\left(\frac{6}{0.035/2}\right)$$

Therefore $u_* = 0.171\,\text{m/s}$. Hence,

$$\text{Re}'_* = \frac{1000 \times 0.171 \times 0.035}{0.001} = 5985$$

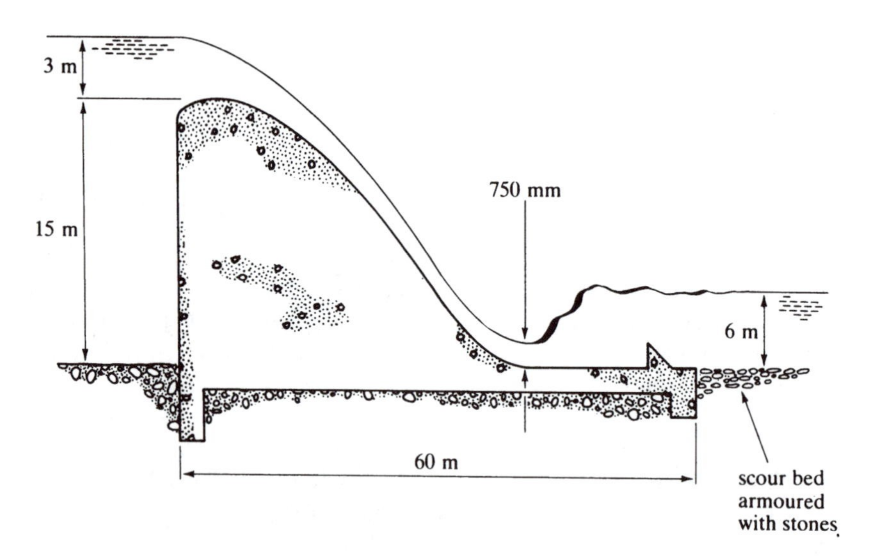

Figure 11.5 Spillway section (Example 11.7).

and $\lambda_L^{1.5} = 70/5985$, so $\lambda_L = 1/19.4$ (say scale = 1/20). Now, $\lambda_Q = \lambda_L^{2.5} = 1/1789$, so $Q'' = 8.385$ l/s and $\lambda_V = \lambda_L^{1/2} = 1/4.472$. The densimetric Froude Numbers are, for the prototype,

$$\Pi_2' = \frac{1000 \times 0.171^2}{(2560 - 1000) \times 9.81 \times 0.035} = 0.0516$$

and for the model (assuming the same sediment density),

$$\Pi_2'' = \frac{1000 \times (0.171/4.472)^2}{(2650 - 1000) \times 9.81 \times (0.035/20)} = 0.0516$$

so the sediment would be of 1.75 mm diameter in the model. The sediment in the model can be of the same density as that in the prototype.

References and further reading

Barr, D. I. H. (1983) A survey of procedures for dimensional analysis. *Int. J. Mech. Engng Education*, **11**(3), 147–59.

Henderson, F. M. (1966) *Open Channel Flow*, Macmillan, New York.

Hughes, S. (1993) *Physical Models and Laboratory Techniques in Coastal Engineering*, World Scientific, London.

Keulegan, G. H. (1938) Laws of turbulent flows in open channels. *J. Res.*, **21**, Paper No. 1151, US National Bureau of Standards, Washington, DC.

Langhaar, H. L. (1980) *Dimensional Analysis and Theory of Models*, Robert E. Krieger, Florida.

Matthews, F. W. and Morfett, J. C. (1986) A method of dimensional analysis for computer application. *Int. J. Mech. Engng Education*, **14**(1), 65–73.

Novak, P. and Cabelka, J. (1981) *Models in Hydraulic Engineering*, Pitman, London.

Sharp, J. J. (1981) *Hydraulic Modelling*, Butterworths, London.

Yalin, M. S. (1971) *Theory of Hydraulic Models*, Macmillan, London.

Part Two

Aspects of Hydraulic Engineering

12

Pipeline systems

12.1 Introduction

In Part One, the principles of hydraulics were explained. Part Two considers how these principles may be applied to practical design problems. In this chapter, various aspects of pipelines are considered. The starting point is the design of simple (one pipe) pipelines. This is followed by a discussion of series, parallel and branched pipelines, leading to the analysis of distribution systems. Finally, the steady flow design of pumping mains is discussed, and the important topic of surge protection for pumping mains and turbine installations is introduced.

12.2 Design of a simple pipe system

Aspects of design

Consider Figure 12.1, which shows a typical pipeline between two storage tanks. Broadly, there are two aspects to the design of this system:

(a) hydraulic calculations;
(b) detail design.

Under (a), a suitable pipe diameter must be determined for the available head and required discharge. This aspect has already been covered in Chapter 4. However, it should be noted that to estimate the local head losses, the proposed valves and fittings must be known. In addition, the maximum and minimum pressures within the pipeline must be found, to ensure that the pressure rating of the pipe is sufficient and to check that sub-atmospheric pressures do not occur. This is most easily visualized by drawing the energy line and hydraulic gradient on the plan of the pipe longitudinal section (refer to the following section).

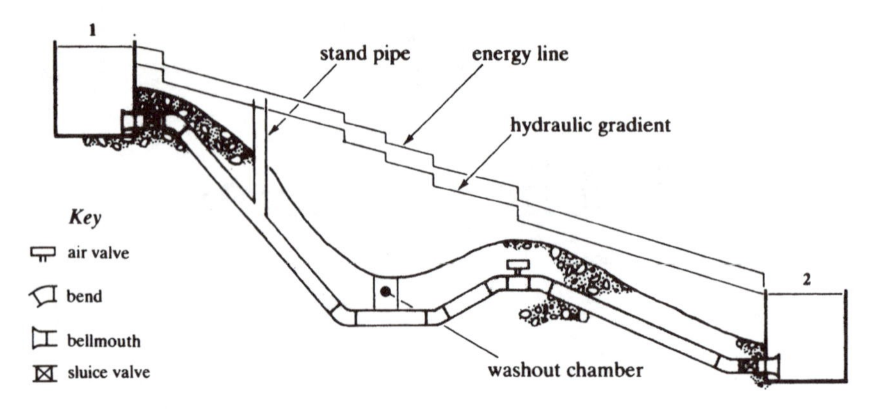

Figure 12.1 Design of a simple pipe system.

Under (b), the following must be considered:

(i) pipeline material, pressure rating and jointing system;
(ii) provision of valves, bends, fittings and thrust blocks;
(iii) locations of any necessary air valves and washouts.

The energy line and hydraulic gradient

Applying Bernoulli's energy equation in Figure 12.1,

$$\frac{p_1}{\rho g}+\frac{V_1^2}{2g}+z_1=\frac{p_2}{\rho g}+\frac{V_2^2}{2g}+z_2+h_f+h_L$$

where h_f is the frictional head loss and h_L is the local head loss. As

$$p_1=p_2=0 \qquad V_1=V_2=0$$

then

$$z_1-z_2=H=h_f+h_L$$

This equation is used to find the required diameter for the given discharge (cf. Example 4.5).

At any point between the reservoirs,

$$\frac{p}{\rho g}+\frac{V^2}{2g}+z=\text{height of the energy line}$$

and

$$\frac{p}{\rho g}+z=\text{height of the hydraulic gradient}$$

Both of these lines have a slope of $S_f(h_f/L)$, and local head losses are represented by a step change. They are both shown in Figure 12.1. The energy line begins and ends at the water level in the upper and lower tanks, and the hydraulic gradient is always a distance $V^2/2g$ below the energy line.

The usefulness of the hydraulic gradient lies in the fact that it represents the height to which water would rise in a standpipe (piezometer). Hence, the location of the maximum and minimum pressures may be found by finding the maximum and minimum heights between the pipe and the hydraulic gradient.

If the hydraulic gradient is below the pipe, then there is sub-atmospheric pressure at that point. This condition is to be avoided since cavitation may occur (if $p/\rho g < -7.0\,\text{m}$), and if there are any leaks in the pipeline matter will be sucked into the pipe, possibly causing pollution of the water supply.

Pipe materials and jointing systems

Table 12.1 lists typical pipe materials, associated linings and jointing systems for water supply pipelines.

The choice of material will depend on relative cost and ground conditions. All of the materials in Table 12.1 are in common usage. More detailed guidance is given in Twort *et al.* (1985). Pipe joints, as shown in Figure 12.2, are normally of the spigot and socket type for underground pipes, where there are no lateral forces to resist. Flanged joints are used in pumping stations, service reservoirs, etc., where lateral forces must be resisted and where easy removal of pipe sections is required.

Table 12.1 Pipe materials and joints for water supply pipes.

Material	Protective lining	Joint type(s)
cast iron (old water mains)	bitumen	spigot and socket with run lead sealer, flanged
ductile iron	bitumen, spun concrete	spigot and socket with rubber ring sealer, flanged
steel	epoxy resin	welded (large pipes), screwed (small pipes)
uPVC	none	spigot and socket with rubber ring sealer, sleeved with chemical sealer, flanged
asbestos cement	bitumen	sleeved with rubber ring sealer

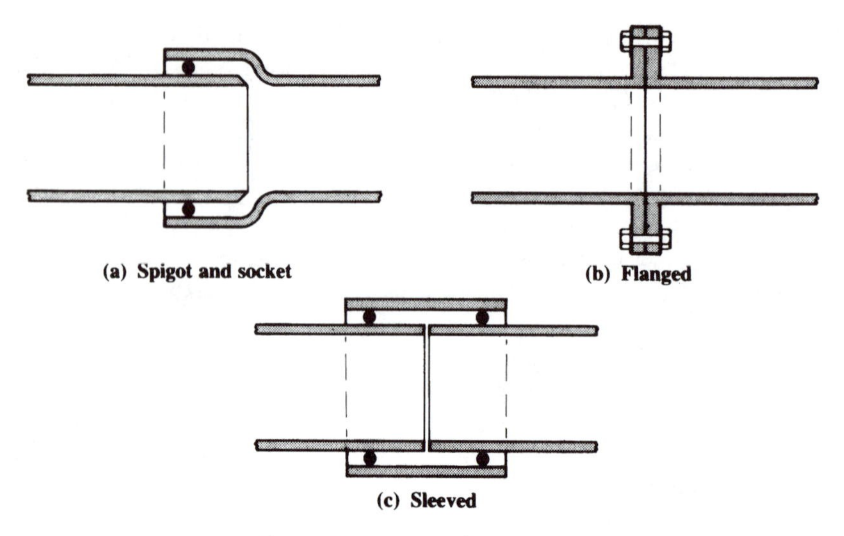

Figure 12.2 Typical pipe joints.

Thrust blocks

Thrust blocks should be provided at all fittings where a change of velocity or flow direction occurs. The forces acting may be calculated by applying the momentum equation (refer to Chapter 2). Thrust blocks are normally designed to withstand either the static head or the pipe test pressure, whichever is greater.

Air valves and washouts

Air valves should be provided at all high points in a water main, so that entrained air is removed during normal operation and air is evacuated during filling. To fulfil this dual role, double orifice air valves are commonly used. It should be noted, however, that air valves should not be placed in any regions of sub-atmospheric pressure, because air would then enter the pipe, rather than be expelled from it.

Washouts are normally placed at all low points so that the water main may be completely emptied for repair or inspection.

12.3 Series, parallel and branched pipe systems

Introduction

Figure 12.3 shows all three cases. To determine the heads and discharges is more complex than in simple pipe problems, and requires the use of the continuity equation in addition to the energy and frictional head loss equations.

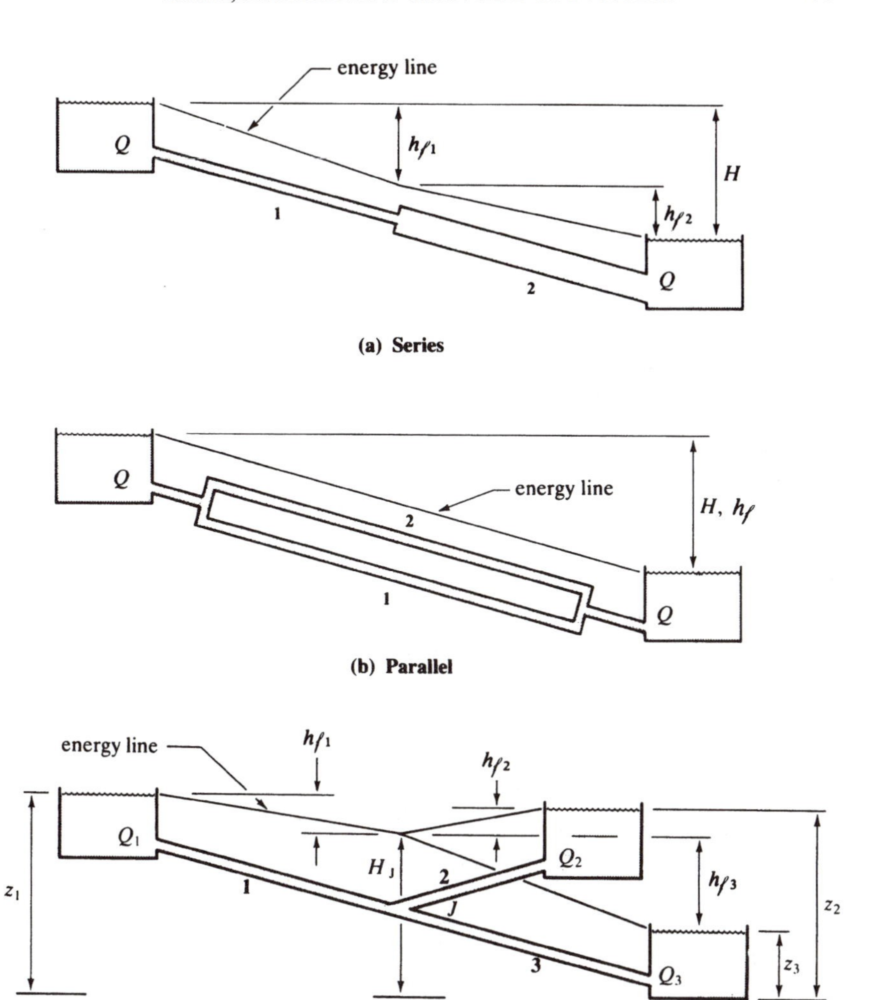

Figure 12.3 Series, parallel and branched pipes.

Series solution

Energy $H = h_{f1} + h_{f2}$
Continuity $Q = Q_1 = Q_2$
As h_{f1}, h_{f2}, are initially unknown, a solution method is as follows:

1. guess h_f;
2. calculate Q_1 and Q_2;
3. if $Q_1 = Q_2$, then the solution is correct;
4. if $Q_1 \neq Q_2$, then return to (1).

Parallel solution

Energy $H = h_{f1} = h_{f2}$
Continuity $Q = Q_1 + Q_2$
This problem can be solved directly for Q_1 and Q_2

Branched solution

Energy $h_{f1} = z_1 - H_J$
$h_{f2} = z_2 - H_J$
$h_{f3} = H_J - z_3$

where H_J is the energy head at junction J.
Continuity $Q_3 = Q_1 + Q_2$
As H_J is initially unknown, a method of solution is as follows:

1. guess H_J;
2. calculate Q_1, Q_2 and Q_3;
3. if $Q_1 + Q_2 = Q_3$, then the solution is correct;
4. if $Q_1 + Q_2 \neq Q_3$, then return to (1).

Example 12.1 Branched pipe system

Given the following data for the system shown in Figure 12.3(c), calculate the discharge and the pressure head at the junction J:

Pipe	Length (km)	Diameter (mm)	Roughness, k_s (mm)
1	5	300	0.03
2	2	150	0.03
3	4	350	0.03

Item	Elevation (m above datum)
Reservoir 1	800
Reservoir 2	780
Reservoir 3	700
Junction J	720

Solution

The method is that given in the previous section, using Figure 4.5 (HRS chart) to find the discharges. Try $H_J = 750\,\text{m}$ above datum.

Then

$$h_{f1} = 800 - 750 = 50\,\text{m}$$

$$\frac{100h_{f1}}{L_1} = \frac{50 \times 100}{5000} = 1$$

therefore

$$Q_1 = 140 \text{ l/s}$$

and

$$h_{f2} = 780 - 750 = 30\,\text{m}$$

$$\frac{100h_{f2}}{L_2} = \frac{100 \times 30}{2000} = 1.5$$

therefore

$$Q_2 = 30 \text{ l/s}$$

and

$$h_{f3} = 750 - 700 = 50\,\text{m}$$

$$\frac{100h_{f3}}{L_3} = \frac{100 \times 50}{4000} = 1.25$$

therefore

$$Q_3 = 250 \text{ l/s}$$

Hence, for

$$H_J = 750\,\text{m}$$

$$Q_1 + Q_2 = 170 \text{ l/s}$$

$$Q_3 = 250 \text{ l/s}$$

This is not the correct solution. A better guess would be achieved by reducing H_J, hence reducing Q_3 and increasing Q_1 and Q_2. For $H_J = 740\,\text{m}$ above datum,

$$h_{f1} = 60\,\text{m} \qquad 100S_{f1} = 1.2 \qquad Q_1 = 160 \text{ l/s}$$

$$h_{f2} = 40\,\text{m} \qquad 100S_{f2} = 2.0 \qquad Q_2 = 35 \text{ l/s} \qquad Q_1 + Q_2 = 195 \text{ l/s}$$

$$h_{f3} = 40\,\text{m} \qquad 100S_{f3} = 1.0 \qquad Q_3 = 210 \text{ l/s}$$

For $H_J = 735$ m above datum,

$$h_{f1} = 65\,\text{m} \qquad 100S_{f1} = 1.3 \qquad Q_1 = 165\ \text{l/s}$$
$$h_{f2} = 45\,\text{m} \qquad 100S_{f2} = 2.25 \qquad Q_2 = 37\ \text{l/s} \qquad Q_1 + Q_2 = 202\ \text{l/s}$$
$$h_{f3} = 35\,\text{m} \qquad 100S_{f3} = 0.875 \qquad Q_3 = 200\ \text{l/s}$$

This is an acceptable solution. The pressure head at J (neglecting velocity head) is given by

$$\begin{aligned} p_J/\rho g &= H_J - z_J \\ &= 735 - 720 = 15\,\text{m} \end{aligned}$$

It is worth noting that if the junction elevation were greater than 735 m above datum, then negative pressures would result. This problem could be overcome by increasing the diameter of pipe No. 2.

12.4 Distribution systems

General design considerations

A water supply distribution system consists of a complex network of interconnected pipes, service reservoirs and pumps which deliver water from the treatment plant to the consumer. Water demand is highly variable, both by day and season. Supply, by contrast, is normally constant. Consequently, the distribution system must include storage elements, and must be capable of flexible operation. Water pressures within the system are normally kept between a maximum (about 70 m head) and a minimum (about 20 m head) value. This ensures that consumer demand is met, and that undue leakage due to excessive pressure does not occur. The topography of the demand area plays an important part in the design of the distribution system, particularly if there are large variations of ground levels. In this case, several independent networks may be required to keep within pressure limitations. For greater operational flexibility, however, they are usually interconnected through booster pumps or pressure reducing valves.

In addition to new distribution systems a common need is for improvement to existing (often ageing) systems. It is good practice to use a ring main system in preference to a branching system. This prevents the occurrence of 'dead ends' with the consequent risk of stagnant water, and permits more flexible operation, particularly when repairs must be carried out. Many existing systems have very high leakage rates (30–50%). Leaks are often very difficult to locate, and have an important bearing on the accuracy of any hydraulic analysis of the system.

An essential prerequisite to the improvement of an existing system is to have a clear understanding of how that system operates. This is often quite difficult to achieve. Plans of the pipe network, together with elevations, diameters, water levels, etc., are required. In addition, the demands must be estimated on a *per capita* consumption basis, or preferably by simultaneous field measurements of pressures and flows at key points in the system.

The analysis of such systems is generally carried out by computer simulation, in which a numerical model of the system is initially calibrated to the field data before being used in a predictive mode. The model is a simplified version of the real system and, in particular, demands from the system are assumed to be concentrated at pipe ends or junctions. This allows relatively simple models to be used without great loss of accuracy, providing that a judicious use of pipe junctions is made.

Such models have been succesfully used to locate areas of leakage within a system. However, if leakage rates are high and of unknown location, then a computer simulation may give misleading results, and therefore be of little value.

Many computer models have been developed with varying degrees of success and applicability. One successful model developed by the Water Research Centre is called WATNET (see Creasy, 1982), and is currently in use in many UK water companies.

Hydraulic analysis

The solution methods described for the analysis of series, parallel and branched pipes are not very suitable for the more complex case of networks. A network consists of loops and nodes as shown in Figure 12.4. Applying the continuity equation to a node,

$$\sum_{i=1}^{n} q_i = 0 \qquad (12.1)$$

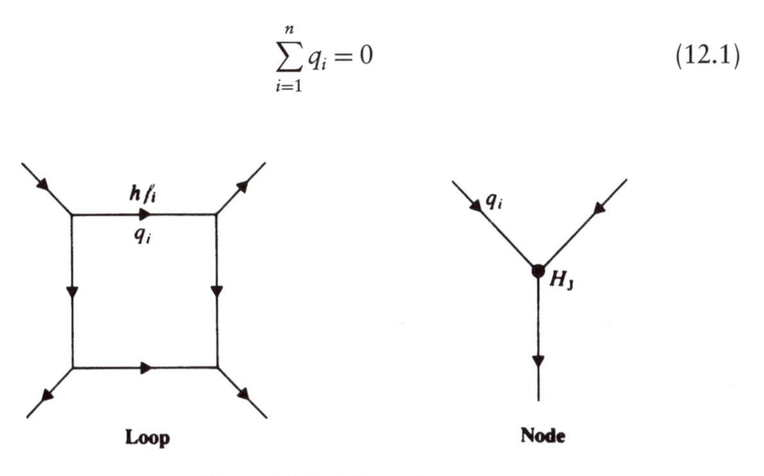

Figure 12.4 Networks.

where n is the number of pipes joined at the node. The sign convention used here sets flows into a junction as positive.

Applying the energy equation to a loop, then

$$\sum_{i=1}^{m} h_{fi} = 0 \quad (12.2)$$

where m is the number of pipes in a loop. The sign convention sets flow and head loss as positive clockwise.

In addition,

$$h_{fi} = f(q_i) \quad (12.3)$$

where $f(q_i)$ represents the Darcy–Weisbach/Colebrook–White equation. Equations (12.1)–(12.3) comprise a set of simultaneous non-linear equations, and an iterative solution is generally adopted. The two standard solution techniques, the loop method and the nodal method, are now discussed.

The loop method

This method, originally proposed by Hardy-Cross in 1936, essentially consists of eliminating the head losses from (12.2) and (12.3) to give a set of equations in discharge only. It may be applied to loops where the external discharges are known and the flows within the loop are required. The basis of the method is as follows:

1. assume values for q_i to satisfy $\Sigma q_i = 0$;
2. calculate h_{fi} from q_i;
3. if $\Sigma h_{fi} = 0$, then the solution is correct;
4. if $\Sigma h_{fi} \neq 0$, then apply a correction factor δq to all q_i and return to (2).

A reasonably efficient value of δq for rapid convergence is given by

$$\delta q = -\frac{\Sigma h_{fi}}{2\Sigma h_{fi}/q_i} \quad (12.4)$$

Step 2 may be carried out using the HRS charts or tables (for hand calculations) or using Barr's explicit formula for λ (for computer solution). Due account must be taken of the sign of q_i and h_{fi}, and of the use of an appropriate convention (i.e. clockwise positive).

Equation (12.4) may be derived as follows. Using the Darcy–Weisbach formula,

$$h_f = \frac{\lambda L V^2}{2gD}$$

or, for a given pipe and assuming that λ is constant,

$$h_f = \mathrm{k}Q^2$$

Taking the true flow to be Q, then

$$Q = (q_i + \delta q)$$

and taking the true head loss to be H_f

$$H_f = \mathrm{k}(q_i + \delta q)^2$$

Expanding by the binomial theorem,

$$H_f = \mathrm{k}q_i^2 \left[1 + 2\frac{\delta q}{q_i} + \frac{2(2-1)}{2!}\left(\frac{\delta q}{q_i}\right)^2 + \ldots \right]$$

ignoring second-order terms and above (for $\delta q \ll q_i$). Then

$$H_f = \mathrm{k}q_i^2[1 + (2\delta q/q_i)]$$

For a loop,

$$\Sigma H_{fi} = 0 = \Sigma \mathrm{k}q_i^2 + 2\delta q\,\Sigma \mathrm{k}q_i^2/q_i$$

or

$$0 = \Sigma h_{fi} + 2\delta q\,\Sigma h_{fi}/q_i$$

Hence

$$\delta q = -\frac{\Sigma h_{fi}}{2\Sigma h_{fi}/q_i}$$

Example 12.2 Flows in a pipe loop

For the square pipe loop shown in Figure 12.5, find:

(a) the discharges in the loop;
(b) the pressure heads at points B, C and D, if the pressure head at A is 70 m and A, B, C and D have the same elevations.

All pipes are 1 km long and 300 mm in diameter, with roughness 0.03 mm.

Solution

(a) It is convenient for hand solution to use a tabular layout in conjunction with the HRS charts or tables for finding h_{fi} from q_i and d.

Initial trial (assume values for q_i)

Pipe	q_i (l/s)	h_{fi} (m)	h_{fi}/q_i
A–B	+60	+2.00	0.0333
B–C	+40	+0.93	0.0233
C–D	0	0	0
A–D	−40	−0.93	0.0233
Σ		+2.00	0.0799

Note the 'positive clockwise' sign convention for q_i and h_{fi}.

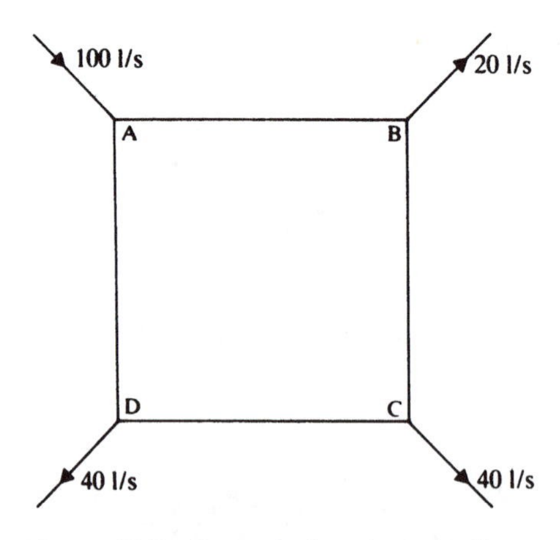

Figure 12.5 Example flow in a pipe loop.

Apply correction factor,

$$\delta_q = -\frac{2}{2(0.0799)} = -12.5 \text{ l/s}$$

Second trial (new discharges $= q_i - 12.5$)

Pipe	q_i (l/s)	h_{fi} (m)	h_{fi}/q_i
A–B	+47.5	+1.3	0.0274
B–C	+27.5	+0.48	0.0175
C–D	−12.5	−0.12	0.0096
A–D	−52.5	−1.58	0.0301
Σ		+0.08	0.0846

$$\delta q = -\frac{0.08}{2(0.0846)} = -0.5 \text{ l/s}$$

As $\delta q = -0.5$ l/s, this solution is sufficiently accurate for most practical purposes.

(b) To find the pressure heads (p/ρ_g) at B, C and D, apply the energy equation. Ignoring velocity heads and recalling that the elevations are the same at A, B, C and D,

$$\frac{p_B}{\rho g} = \frac{p_A}{\rho g} - h_{fA-B} = 70 - 1.3 = 68.7 \text{ m}$$

$$\frac{p_C}{\rho g} = \frac{p_B}{\rho g} - h_{fB-C} = 68.7 - 0.48 = 68.22 \text{ m}$$

$$\frac{p_D}{\rho g} = \frac{p_A}{\rho g} - h_{fA-D} = 70 - 1.58 = 68.42 \text{ m}$$

As a check

$$\frac{p_C}{\rho g} = \frac{p_D}{\rho g} - h_{fD-C} = 68.42 - 0.12 = 68.3$$

Comparing with the previous estimate ($p_c/\rho g = 68.22$ m), the difference is equal to Σh_{fi} (the closing error).

The nodal method

This method, originally proposed by Cornish in 1939, consists of eliminating the discharges from (12.1) and (12.3) to give a set of equations in head losses only. It may be applied to loops or branches where the external heads are known and the heads within the networks are required. The basis of the method is as follows:

1. assume values for the head (H_j) at each junction;
2. calculate q_i from H_j;
3. if $\Sigma q_i = 0$, then the solution is correct;
4. if $\Sigma q_i \neq 0$, then apply a correction factor δH to H_j and return to 2, where

$$\delta H = \frac{2\Sigma q_i}{\Sigma q_i / h_{fi}} \tag{12.5}$$

The derivation of (12.5) is similar to that for (12.4).

Step 2 may be carried out using the HRS charts or tables (for hand calculations) or using the Colebrook–White/Darcy–Weisbach equation for q in terms of h_f (for computer solution). An appropriate sign convention must be used for q_i and h_{fi}, e.g. q_i and h_{fi} positive entering a node.

Example 12.3 Flows in a branched pipe network

Resolve Example 12.1 using the nodal method.

Solution

Assume a trial value of $H_J = 750\,\text{m}$

Pipe	h_{fi} (m)	q_i (l/s)	q_i/h_{fi}
1	+50	+140	+2.8
2	+30	+30	+1.0
3	−50	−250	+5.0
Σ		−80	8.8

Apply correction factor,

$$\delta H = \frac{2(-80)}{8.8} = -18.18\,\text{m}$$

Second trial $H_J = 750 - 18.18 = 731.82\,\text{m}$

Pipe	h_{fi} (m)	q_i (l/s)	q_i/h_{fi}
1	+68.18	+168	+2.5
2	+48.18	+37	+0.8
3	−31.82	−191	+6.0
Σ		+14	9.3

$$\delta H = \frac{2(14)}{9.3} = +3.01\,\text{m}$$

Third trial $H_J = 731.82 + 3.01 = 734.83\,\text{m}$
As the solution to Example 12.1 gave $H_J = 735\,\text{m}$, then convergence has been achieved and the solution is as given in Example 12.1.

Complex networks

For more complex networks (e.g. more than one loop or junction) the loop and nodal methods may both be applied with minor modifications. In the case of the loop method, the correction factor δq at each iteration must be carried over from one loop to the next through any common pipes. For the nodal method, the correction factor δH is applied to successive nodes through the network at each iteration.

The choice of loop or nodal method for the analysis of complex networks depends on the available data. In terms of efficiency of solution and required computer storage space, the loop method requires more storage but converges more quickly, whereas the nodal method requires less storage but convergence is slower and not always achieved. The WATNET program uses a hybrid combined loop–nodal method, which gives the benefits of both methods. For details of other available methods refer to Novak (1983).

12.5 Design of pumping mains

Introduction

Typical applications of pumping mains include river abstractions (low level supply to high level demand), borehole supplies from groundwater and surface water and foul water drainage from low-lying land. In all of these cases there are at least three elements to consider:

1. hydraulic design;
2. economic matching of pump and pipeline;
3. detail design.

These three elements are now considered in turn.

Hydraulic design

The primary requirement is to determine a suitable pump and pipe combination for the required design discharge (Q). Consider Figure 12.6, which shows a simple pumping main. At start-up, the pump is required to deliver the design discharge (Q) against the static head (H). However, as soon as flow commences, frictional losses are introduced (h_{fs} and h_{fd}) which vary with discharge. To attain the design discharge (Q), the head provided by the pump (H_p) must exactly match the static plus friction heads at Q. Hence the discharge is a function of both the pump and the pipeline. For a given system, the head-discharge characteristic curves for the pump may be superimposed on that for the pipeline, as shown in Figure 12.7. The point of intersection of the two characteristic curves locates the one possible combination of head and discharge for the system under steady flow conditions. The intersection point is referred to as the operating point.

To obtain the required discharge (Q), it may be necessary to investigate several pump and pipe combinations. In addition, it is obviously desirable

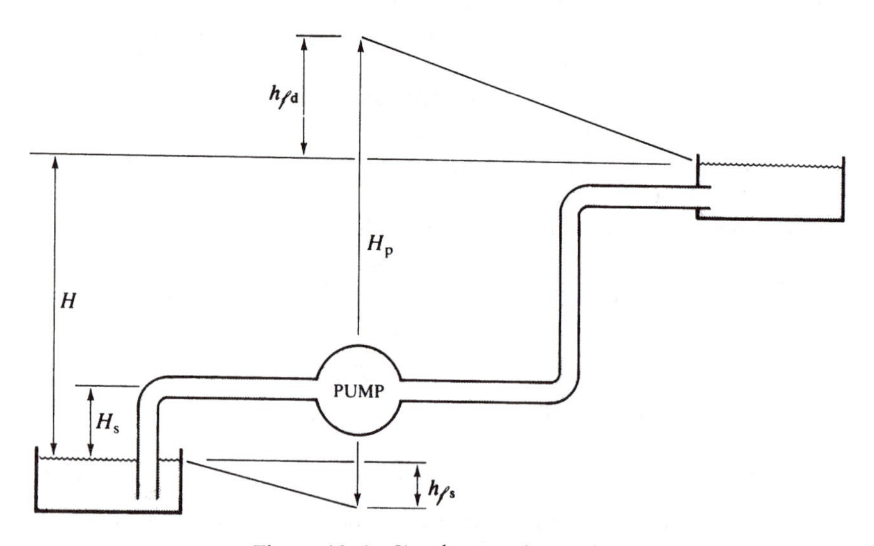

Figure 12.6 Simple pumping main.

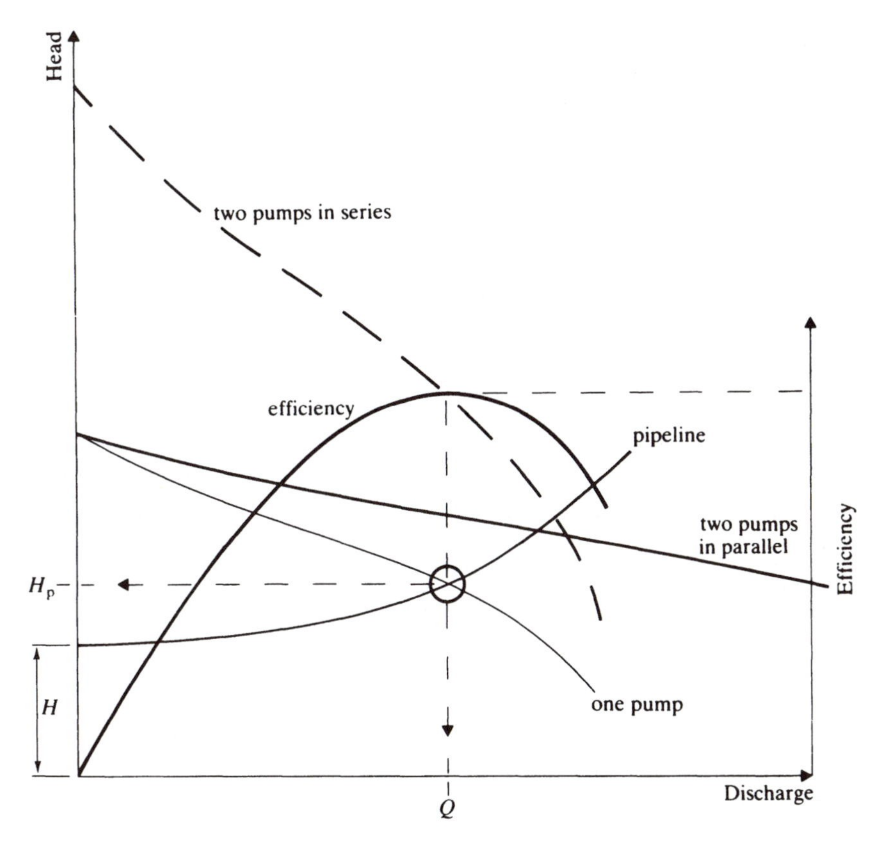

Figure 12.7 Pump and pipeline characteristics for a mixed flow pump.

that the pump should be running at or near peak efficiency at the design discharge (Q). This condition may be checked by drawing the pump efficiency curve (as shown in Fig. 12.7).

Pumps are often arranged in series (for large static heads) or parallel (for varying discharge requirements). To obtain the discharge in these cases requires the estimation of their combined characteristic curves. These are obtained from the characteristic for the single pump case as follows:

For series operation (of n pumps)

$$H_{np} = nH_{\mathrm{p}} \qquad Q_{np} = Q_{\mathrm{p}}$$

where subscript np denotes n pumps and subscript p denotes one pump.

For parallel operation (of n pumps)

$$H_{np} = H_{\mathrm{p}} \qquad Q_{np} = nQ_{\mathrm{p}}$$

The resulting characteristic curves for $n = 2$ are shown in Figure 12.7.

A secondary requirement is to prevent cavitation occurring in the pump. This is likely to occur if the water level in the suction well is several metres below the pump. As cavitation in pumps has already been described in section 7.7, it is not discussed further here.

Example 12.4 Design of a simple pumping main

Given the pump characteristics below, determine the pump efficiency and power requirement for a pipeline of diameter 300 mm, length 5 km, roughness 0.03 mm and a static lift of 10 m. Comment on the suitability of this pump and pipeline combination.

Pump characteristics

Discharge (l/s)	0	10	20	30	40	50	60	70
Head (m)	26.25	24.00	21.75	19.50	17.50	15.00	11.75	6.75
Efficiency (%)	0	28	51	68	80	85	80	64

Solution

(a) Determine the pipeline characteristic by finding the frictional losses at the discharges given in the table above. These are most readily found using the HRS charts or tables.

Pipeline characteristics

Discharge (l/s)	20	30	40	50	60	70
Friction head (m)	1.35	2.8	4.7	7.25	9.75	13.5
Friction + static head (m)	11.35	12.8	14.7	17.25	19.75	23.5

(b) Draw the characteristic curves as shown in Figure 12.8.
(c) The operating point is given by:

$$Q_d = 45.5 \text{ l/s} \quad H_p = 16\,\text{m} \quad \eta = 84\%$$

The power consumption is given by:

$$\begin{aligned} P &= \rho g Q H/\eta \\ &= 10^3 \times 9.81 \times 45.5 \times 10^{-3} \times 16/0.84\,\text{W} \\ &= 8.5\,\text{kW} \end{aligned}$$

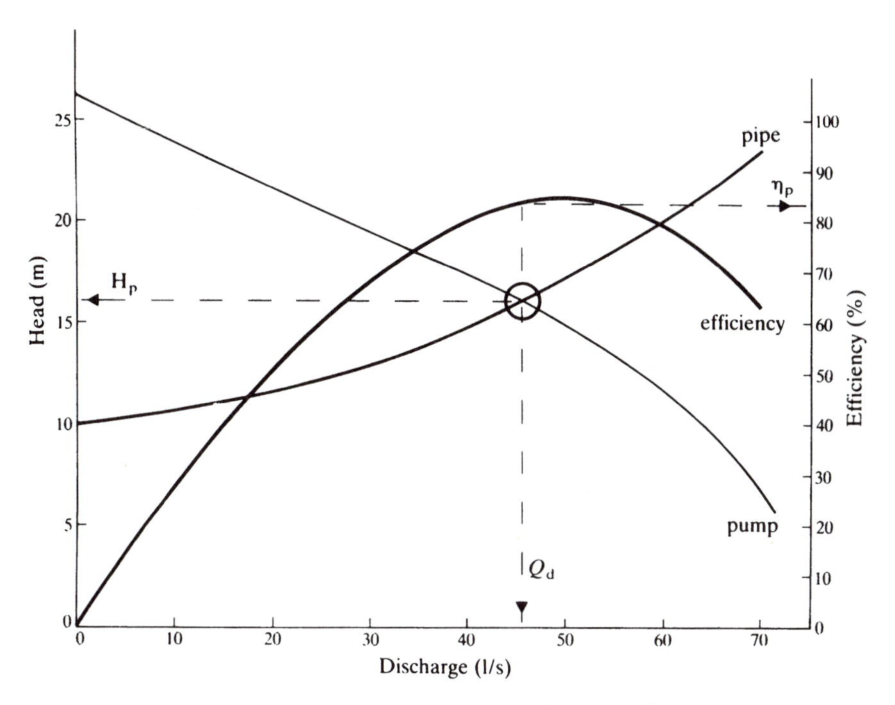

Figure 12.8 Characteristics curves for Example 12.4.

The operating point is satisfactory since the system is operating very close to the peak efficiency point of the pump.

Economics of pumping mains

For the required design discharge, a variety of possible pump and pipe combinations are possible. The final choice is not determined purely on the basis of hydraulic considerations. It is necessary to determine the least-cost solution. There are three basic elements involved:

(a) the pipe cost;
(b) the pump cost;
(c) the running costs.

Pipe costs increase with diameter but, conversely, pump and running costs will decrease with pipe diameter (because of smaller frictional losses). Hence a least-cost solution is possible, as is shown in Figure 12.9.

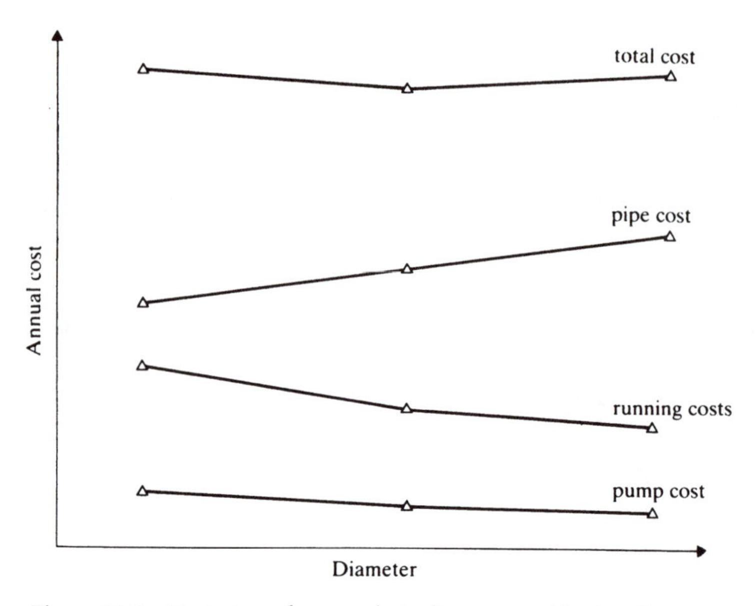

Figure 12.9 Variation of pumped pipeline costs with pipe diameter.

The normal method for finding this solution is:

1. estimate the capital cost of pipes (material and laying) and the associated pumps for various pump diameters;
2. convert capital cost to annual cost by assessing interest charges or by discounted cash flow analyses;
3. estimate the annual power charges and other running costs (maintenance, etc.);
4. tabulate the results and compare.

Details of such calculations are given in Twort *et al.* (1985).

Detail design

A typical centrifugal pump installation is shown in Figure 12.10. The pump is installed between two valves for easy removal in case of repair or maintenance. On the suction side, a combined bellmouth entry and strainer are necessary, together with a non-return valve to ensure self-priming. On the delivery side, a second non-return valve is necessary to prevent damage from possible surge pressures. In addition, an air valve and flow meter (venturi type) are desirable.

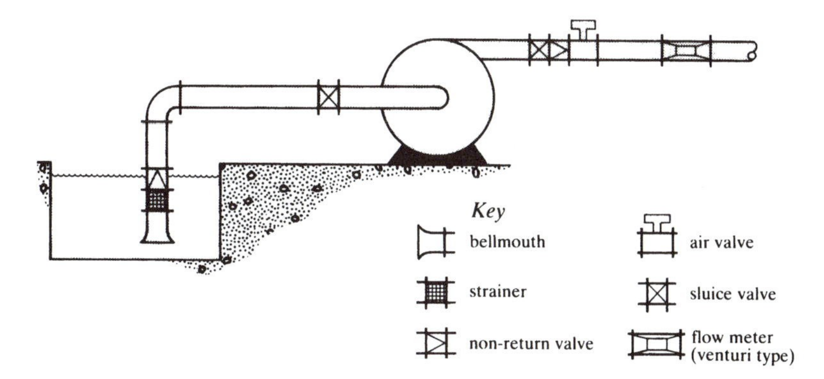

Figure 12.10 Typical centrifugal pump installation.

12.6 Surge protection

General description

The main thrust of this chapter has been the application of principles associated with steady full-pipe flows. However, it would be a serious omission in any design procedure if the possibility of an unsteady flow was not considered. It was asserted in Chapter 6 that changes in conditions at a controlling boundary could produce such flows. The first step, therefore, is to check this aspect of the system for all possible operating conditions, especially those associated with an emergency (pump breakdown, emergency shutdown, etc.). The techniques introduced in Chapter 6 would be suitable for this purpose. If a surge problem is predicted, clearly some countermeasures must be taken. These measures are given the general title of surge protection measures.

The crux of the problem is how to modify the system economically in such a way as to reduce the effects of the unsteady flow to an acceptable level. Two approaches are possible:

1. Surge protection by mechanical means – this might entail limiting the rate of valve closure, for example. However, for some systems, the scale of the system or the nature of emergency shutdown procedures would preclude this approach.
2. Surge protection by hydraulic means – this implies the incorporation of a hydraulic device which automatically limits the pressure rise in the system.

Since this is a hydraulics text, we shall concentrate on the second group of methods. This is not in any way to discount the methods outlined under (1) where they are appropriate.

The simple surge tower

This is one of the simplest devices to analyse and to design. It consists of a large vertical tube connected at its base to the pipeline (Figs 12.11 and 12.12). The top is open to the atmosphere. It is usually sited as close as possible to the controlling boundary which is causing the surge. With the valve open and steady flow in the pipe, the equilibrium water level in the surge tower will correspond to the pressure in the pipeline (i.e. it will lie on the hydraulic gradient). If the control valve is suddenly and completely closed, a rise in pressure is generated as the water decelerates. This causes a rise in the surge tower level. Since the valve is shut, the discharge in the pipe is all diverted into the surge tower. Thus, at time t after valve closure, if V is the mean velocity in the pipeline and V_{ST} the mean velocity in the surge tower, the continuity equation may be written

$$VA = V_{ST}A_{ST} \tag{12.6a}$$

Figure 12.11 Construction of the surge tower for Dinorwig pumped storage scheme, Wales. (Photograph courtesy of CEGB / John Mills.)

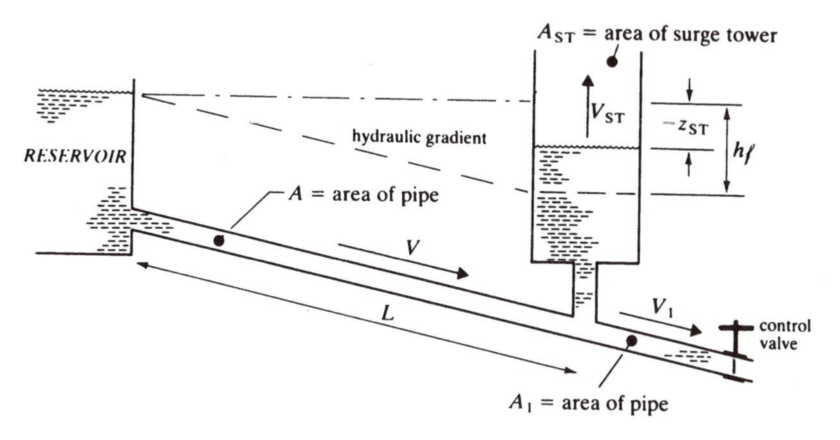

Figure 12.12 Surge tower.

The rise in the water level in the surge tower means that the difference between reservoir level and surge tower level is now z_{ST}. The corresponding pressure difference, δp, is given by

$$\delta p = \rho g z_{ST}$$

The resultant force acting on the fluid in the pipeline is obtained by taking the sum of the forces due to the pressure difference δp, the frictional resistance and any local losses. Taking each force in turn:

$$\text{force due to pressure difference} = \delta p A = \rho g z_{ST} A$$

$$\text{resistance due to pipe friction} = \rho g h_f A = \frac{\rho g \lambda L V^2 A}{2gD}$$

$$\begin{array}{c}\text{resistance due to local losses}\\ \text{(e.g. at the entry to the surge tower)}\end{array} = \rho g h_L A = \frac{\rho g K_L V^2 A}{2g}$$

Since L and D are constant, and λ will be assumed to remain constant, the pipe friction resistance term may be written

$$\rho g h_f A = \rho g K_f V^2 A/2g$$

where $K_f = \lambda L/D$. The total resistance will retard the flow of water in the pipeline upstream of the surge tower. The mass of water in the pipe is ρAL, and its rate of change of momentum is $\rho AL \mathrm{d}V/\mathrm{d}t$. The momentum equation may therefore be written as

$$-\left(-\rho g z_{ST} A + \frac{\rho g K_f V^2 A}{2g} + \frac{\rho g K_L V^2 A}{2g}\right) = \rho AL \frac{\mathrm{d}V}{\mathrm{d}t}$$

Dividing by ρA throughout,

$$-g\left(-z_{\mathrm{ST}}+\frac{K_f V^2}{2g}+\frac{K_{\mathrm{L}} V^2}{2g}\right)=L\frac{\mathrm{d}V}{\mathrm{d}t} \tag{12.7}$$

As it stands, this is not a standard differential equation, but it can be converted into a homogeneous equation as follows. The surge tower velocity, $V_{\mathrm{ST}}=dz_{\mathrm{ST}}/\mathrm{d}t$, so (12.6a) may be written as

$$V=\frac{dz_{\mathrm{ST}}}{\mathrm{d}t}\frac{A_{\mathrm{ST}}}{A} \tag{12.6b}$$

Therefore

$$\frac{\mathrm{d}V}{\mathrm{d}t}=\frac{\mathrm{d}^2 z_{\mathrm{ST}}}{\mathrm{d}t^2}\frac{A_{\mathrm{ST}}}{A}$$

Equation (12.7) may therefore be rewritten as

$$-g\left[-z_{\mathrm{ST}}+\frac{K_f}{2g}\left(\frac{\mathrm{d}z_{\mathrm{ST}}}{\mathrm{d}t}\frac{A_{\mathrm{ST}}}{A}\right)^2+\frac{K_{\mathrm{L}}}{2g}\left(\frac{\mathrm{d}z_{\mathrm{ST}}}{\mathrm{d}t}\frac{A_{\mathrm{ST}}}{A}\right)^2\right]=L\frac{\mathrm{d}^2 z_{\mathrm{ST}}}{\mathrm{d}t^2}\frac{A_{\mathrm{ST}}}{A} \tag{12.8}$$

This is a second-order differential equation, and readers who are familiar with these equations will recognize that it represents a 'damped' harmonic motion. A graph of z_{ST} against t would therefore take the form of a sinusoidal curve whose amplitude decreases with time. Unfortunately, (12.8) is not quite in the standard mathematical form, due to the fact that the turbulent flow leads to resistance in terms of V^2 (or $(\mathrm{d}z/\mathrm{d}t)^2$. It will be shown below that a numerical solution is available. However, some useful information may be obtained if the friction and local loss terms are ignored. This yields

$$-gz_{\mathrm{ST}}=L\frac{\mathrm{d}^2 z_{\mathrm{ST}}}{\mathrm{d}t^2}\frac{A_{\mathrm{ST}}}{A} \tag{12.9}$$

Now this is a standard second-order equation, which represents an undamped harmonic motion. A graph of z_{ST} against t would take the form of a sine curve. It is easily shown that the maximum amplitude of the curve is

$$z_{\mathrm{ST_{max}}}=\pm V_0\sqrt{\frac{L}{g}\frac{A}{A_{\mathrm{ST}}}} \tag{12.10}$$

where V_0 is the original velocity of flow in the pipeline before valve closure.

The periodic time T_p may also be deduced:

$$T_p = 2\pi\sqrt{\frac{L}{g}\frac{A_{ST}}{A}} \tag{12.11}$$

Equations (12.8) and (12.9) both draw broadly the same picture of events. When the valve is shut, the pipe flow is diverted into the surge tower. The water level in the tower accelerates upwards, overshoots the reservoir water level and comes to rest at elevation $z_{ST_{max}}$. The difference in head between the surge tower and the reservoir cannot be sustained, so the surge tower level begins to fall and water flows in the negative direction in the pipeline. The surge tower level falls until it comes to rest at some level ($z_{ST_{min}}$, say) below reservoir level. The flow then reverses and the surge tower level begins to rise again, and so on.

The design of a surge tower for a particular pipeline would be carried out in a number of stages. The effectiveness of the surge tower in suppressing excessive surge pressures is usually checked by modelling the system on a computer. A simple computational model for predicting surge tower behaviour is given in Chapter 14.

Surge protection for hydroelectric schemes

A typical high head hydroelectric scheme (Fig. 12.13) might consist of a reservoir feeding a low pressure tunnel or pipeline. The tunnel might be of considerable length and have a small gradient. At the end of the tunnel the system would be divided into a series of high pressure penstocks laid down the hillside to the turbine house. It might well be convenient to construct the surge tower as a vertical excavation at the downstream end of the tunnel, as

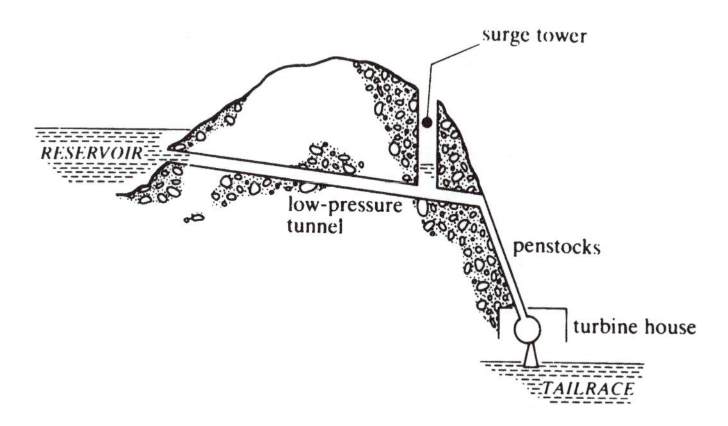

Figure 12.13 Typical hydroelectric scheme.

shown. The tunnel is then protected from excessive surge pressures, which could otherwise cause damage (possibly even a tunnel collapse) that would be extremely expensive to repair. Another important aspect of the surge tower would be the periodic time T_p. The flow to the turbine has to be continuously regulated (or 'governed') so that the output from the turbines matches the electrical power requirement. In a badly designed system, the governor and surge tower can interact in such a way that the system is continuously 'hunting', i.e. a steady flow to the turbines is hardly ever achieved.

A wide variety of surge tower designs has been developed to match the differing requirements of designers. Pickford (1969) gives an excellent summary of these, together with an extensive survey of numerical methods.

The whole of this argument has been based on the assumption that the control valve has been completely closed. For any valve movement other than complete closure, there must be a flow in the pipe between the surge tower and the control valve (Fig. 12.12). The continuity equation thus becomes

$$VA - V_1A_1 = V_{\mathrm{ST}}A_{\mathrm{ST}} = \frac{\mathrm{d}z_{\mathrm{ST}}}{\mathrm{d}t}A_{\mathrm{ST}}$$

which may be converted into finite difference form:

$$\delta z_{\mathrm{ST}} = \left(\frac{VA - V_1A_1}{A_{\mathrm{ST}}}\right)\delta t \tag{12.14}$$

V_1 has to be determined from the hydraulic characteristic of the controlling boundary.

The emphasis here has been placed on the ability of the surge tower to limit the effects of unsteady flows. The surge tower will additionally provide a small local 'reservoir' of liquid which assists in preventing negative pressures by feeding water towards the valve when the valve is opened and the water in the pipeline from the reservoir is still accelerating to meet demand.

The surge tower should, theoretically, be placed as near to the controlling boundary as possible, though different systems may impose a variety of requirements.

Surge protection in pumped mains

Unsteady flow can be generated by pumps when they are started or stopped. During start-up, the pump operating point moves from the zero discharge point to its normal steady-state position. The transient pressures generated during start-up are not usually serious, though they should be checked. If a pump is switched off (or if a power failure occurs), then a negative pressure surge is generated downstream of the pump and a positive one upstream. The initial negative surge may generate a vapour cavity at any high point along the pipeline. The subsequent implosion of the cavity could, theoretically, generate high surge pressures. This can be avoided by following good practice at the design stage.

A variety of approaches are used, including the following:

Mechanical methods. A typical mechanical method would be the use of a flywheel attached to the pump shaft. The increase in the pump inertia means that the rate of deceleration is reduced. A drawback of this method is the increase in starting power, so it is not suitable for all pumps.

Bypassing. This entails the installation of an additional pipeline (Fig. 12.14) around the pump. This pipeline incorporates a non-return valve. During normal operation the pressure downstream of the pipe is higher than that upstream, so the valve remains closed. If the pump fails and the pressure down stream falls below the upstream pressure, then the valve opens and flow is admitted, limiting the fall in pressure. The bypass is also useful in long pipelines, where there may be a number of pumping stations along the pipeline. At times of low demand some of the pumps may be turned off, and the flow will then proceed through the bypass. Bypassing will not work in pipelines where there is negative pressure upstream of the pump.

Air chambers. An air chamber is, in effect, a very compact surge tower with its upper end capped so as to form an air-tight cylinder (Fig. 12.15). Under normal operating conditions (V_0, p_0) the chamber is partially filled with liquid. The pressure of the air trapped above the liquid must be the same as that of the liquid. Consequently, under surge conditions, the liquid level in the chamber will rise or fall to correspond with pressure in the pipeline, in much the same way as for the surge tower. Air chambers can usually be designed to be much more compact than surge towers.

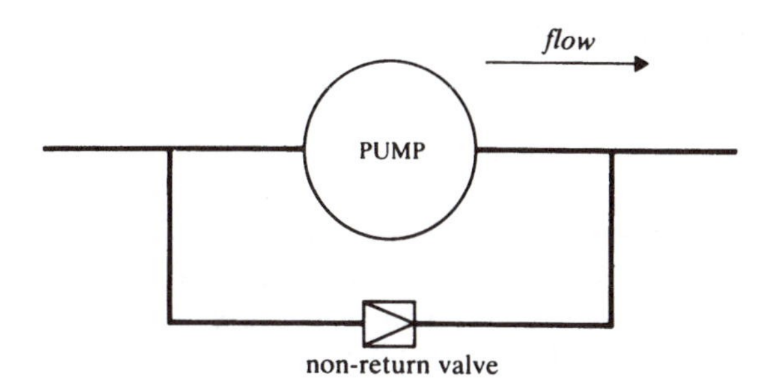

Figure 12.14 Bypass system.

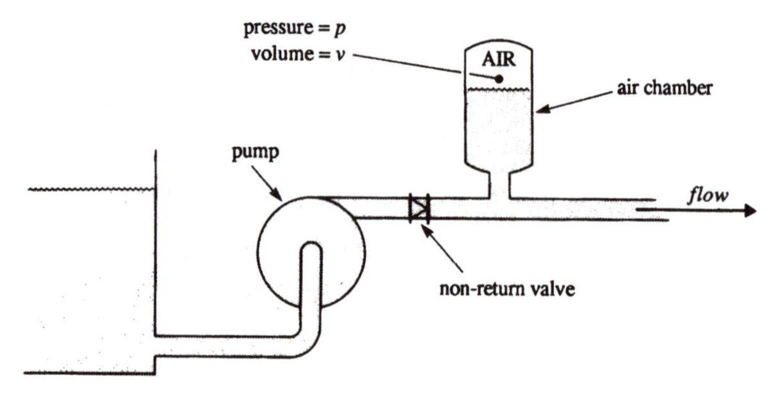

Figure 12.15 Air chamber.

Equations of motion can be developed for a pipeline with an air chamber. These are comparable in form with (14.27) and (12.12). In the momentum equation, an additional term is required to account for the air pressure in the chamber. To estimate the air pressure it is usually assumed that the pressure and volume of the air are related by

$$pv^{1.2} = \text{constant}$$

(readers who are familiar with the gas laws will realize that this equation lies between the equations for isothermal and adiabatic compression). The implication of these remarks is that the instantaneous pressure change in the pipe will almost certainly be greater for an air chamber than for a surge tower. The volume of liquid in the chamber must be adequate to prevent the pressure in the pipe falling to vapour pressure. Conversely, the air volume must be adequate to 'cushion' positive surge pressures. It will be necessary to check the volume of air regularly (air is soluble in water), and to pump air in from time to time to maintain the required level.

A wealth of information is available on various aspects of surge control. The reader is referred in the first instance to Fok (1978), Pickford (1969), Parmakian (1963) and Wood (1970).

References and further reading

Creasy, J. D. (ed.) (1982) *A guide to water network analysis and the WRC computer program WATNET*, Tech. Rep. 177, Water Research Centre, Medmenham.

Fok, A. T. K. (1978) Design charts for air chambers on pump pipe lines. *Am. Soc. Civ. Engrs, J. Hydraulics Divn.*, **104**(HY9), 1289–1303.

Novak, P. (ed.) (1983) *Developments in Hydraulic Engineering–1*, Applied Science Publishers, London.

Parmakian, J. (1963) *Waterhammer Analysis*, Dover, New York.
Pickford, J. (1969) *Analysis of Surge*, Macmillan, London.
Twort, A. C., Law, F. M. and Crowley, F. W. (1985) *Water Supply*, 3rd edn, Edward Arnold, London.
Wood, D. J. (1970) Pressure surge attenuation utilizing an air chamber. *Am. Soc. Civ. Engrs, J. Hydraulics Divn.*, **96**(HY5), 1143–55.

13

Hydraulic structures

13.1 Introduction

Hydraulic structures are devices which are used to regulate or measure flow. Some are of fixed geometrical form, while others may be mechanically adjusted. Hydraulic structures form part of most major water engineering schemes, for irrigation, water supply, drainage, sewage treatment, hydropower, etc. It is convenient to group the structures under three headings:

(a) flow measuring structures, e.g. weirs and flumes;
(b) regulation structures, e.g. gates or valves;
(c) discharge structures, e.g. spillways.

As the reader progresses through the chapter, it will be observed that for most of these structures the depth–discharge relationship is based on the Bernoulli (or specific energy) equation. However, some modifications have to be incorporated to account for the losses of energy which are inevitably incurred in real flows.

An immense amount of experimental and theoretical information has been accumulated during this century, which is reflected in the body of literature now available. The most succinct statements of present knowledge are usually to be found in the relevant British, International or American standards. However, standards may not necessarily be found to cover the more esoteric structures.

13.2 Thin plate (sharp-crested) weirs

This type of device is formed from plastic or metal plate of a suitable gauge. The plate is set vertically and spans the full width of the channel. The weir itself is incorporated into the top of the plate. The geometry of the weir depends on the precise nature of the application. In this section we will concentrate on two basic forms, the rectangular weir and the vee (or triangular) weir. However, other forms are available, such as the compound weir.

It has already been stated that the primary purpose of a weir is to measure discharge. Once the upstream water level exceeds the crest height P_S, water will flow over the weir. As the depth of water above the weir (h_1) increases, the discharge over the weir increases correspondingly. Thus, if there is a known relationship between h_1 and Q, we need only to measure h_1 in order to deduce Q. The 'ideal' relationship between h_1 and Q may readily be derived for each weir shape on the basis of the Bernoulli equation. If these relationships are compared, it is evident that the triangular weir possesses greater sensitivity at low flows, whereas the rectangular weir can be designed to pass a higher flow for a given head and channel width.

Rectangular weirs

There are two types of rectangular weir (see Fig. 13.1).

1. 'Uncontracted' or full-width weirs comprise a plate with a horizontal crest extending from one side of the channel to the other – the crest section is as illustrated in the figure.
2. A 'contracted' weir, by contrast, has a crest width which is less than the channel width.

Since the operation of a weir is based on the use of a gauged depth to estimate the discharge, we must know how these two quantities are related. The actual flow over a weir is quite complex, involving a three-dimensional velocity pattern as well as viscous effects. The simplest method of developing a numerical model which represents a weir is to use the Bernoulli equation as a starting point. An idealized relationship between depth and discharge is obtained. This relationship can then be modified to take account of the differences between ideal and real flows.

The rectangular weir equation. Before developing this equation it should be recalled that, even with an ideal fluid, the velocity distribution at the weir is not uniform (see section 2.8) due to the variation in static pressure with depth. For this reason, the Bernoulli equation is first applied along one streamline, as shown in Figure 13.2. Thus, the total energy on the streamline A–A at Station 1 is

$$H_1 = \frac{p_1}{\rho g} + z_1 + \frac{u_1^2}{2g}$$

Writing $z_1 + p_1/\rho g = y_1$, we obtain $H_1 = y_1 + u_1^2/2g$. At Station 2 the liquid passes over the weir and forms an overspilling 'jet' whose underside

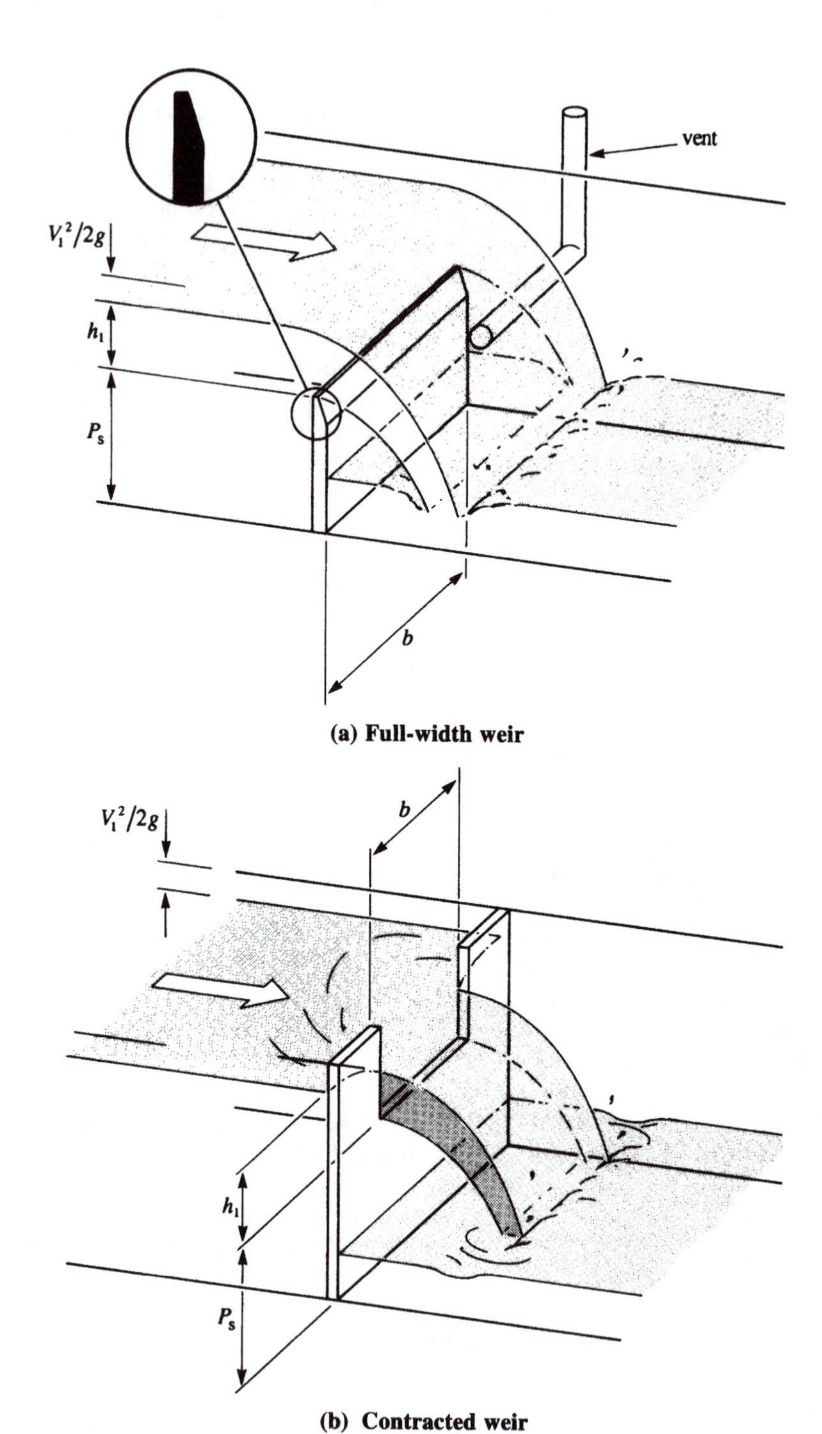

Figure 13.1 Rectangular weirs.

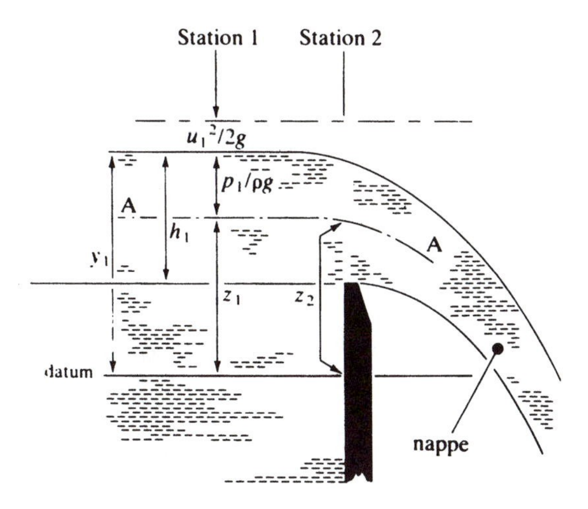

Figure 13.2 Flow over a thin plate weir.

(or 'lower nappe') is exposed to the atmosphere. The pressure distribution here cannot be of the linear hydrostatic form, and it is usual (after Weisbach) to assume that p_2 is atmospheric pressure. Therefore $H_2 = z_2 + u_2^2/2g$ and, assuming that no losses occur,

$$H_1 = H_2$$

i.e.

$$y_1 + u_1^2/2g = z_2 + u_2^2/2g$$

Therefore

$$u_2^2/2g = y_1 - z_2 + u_1^2/2g$$

and therefore

$$u_2 = [2g(y_1 - z_2 + u_1^2/2g)]^{1/2}$$

u_2 is thus a function of y_1 and z_2, i.e. u_2 varies with elevation above the weir crest. At Station 2, the ideal discharge through an elemental strip of width b and depth δz is

$$\delta Q_{\text{ideal}} = u_2 b\ \delta z = [2g(y_1 - z_2 + u_1^2/2g)]^{1/2} b\ \delta z$$

In order to integrate the equation it is necessary to know the limiting values of z_2. If the datum is now raised to the same elevation as the weir crest, the lower limit becomes $z_2 = 0$. A value for the upper limit may be obtained by making the rather drastic assumption that the elevation of the water surface at Station 2 is the same as that at Station 1, and therefore that $y_1 = z_2 = h_1$. (This implies a horizontal water surface, and therefore that streamlines at Station 2 are parallel and horizontal. This is physically impossible.) With these assumptions,

$$Q_{\text{ideal}} = \int_0^{h_1} u_2 b \, dz = \int_0^{h_1} [2g(h_1 - z_2 + u_1^2/2g)]^{1/2} b \, dz$$

$$= \frac{2}{3} b\sqrt{2g} \left[\left(h_1 + \frac{u_1^2}{2g} \right)^{3/2} - \left(\frac{u_1^2}{2g} \right)^{3/2} \right] \tag{13.1}$$

as b is constant for a rectangular weir.

If $u_1^2/2g$ is negligible compared with h_1, then (13.1) reduces to

$$Q_{\text{ideal}} = \frac{2}{3} b\sqrt{2g} h_1^{3/2} \tag{13.2}$$

Modifications to the rectangular weir equation. Due to the converging pattern of streamlines approaching the weir, the cross-sectional area of the jet just downstream of the weir will be significantly less than the cross-section of the flow at the weir itself. This 'contraction' or 'vena contracta' in the flow implies that the actual discharge, Q, will be less than Q_{ideal}.

There are two further discrepancies in the theory:

1. the pressure distribution in the flow passing over the weir is not atmospheric, but is distributed as shown in Figure 13.3;
2. the flow approaching the weir will be subject to viscous forces: this produces two effects – (a) a non-uniform velocity distribution in the channel and (b) a loss of energy between Stations 1 and 2.

It was seen in Chapter 2 that it is conventional for Q and Q_{ideal} to be linked through the experimentally derived coefficient, C_d, i.e. $C_d = Q/Q_{\text{ideal}}$. C_d is not strictly a constant. It can be shown by dimensional analysis that $C_d = f\{\text{Re}, \text{We}, h_1/P_S\}$. However, provided that the overspilling liquid forms a 'jet' whose lower nappe springs clear of the weir plate, the value of C_d will not vary greatly with Q (or Re). The problems of calibration (i.e. evaluating C_d experimentally) become far greater if the nappe clings to the downstream face of the weir plate. Such conditions tend to arise when the discharge is small. Under these conditions, the effects of viscosity and surface tension combine to bring about unstable, fluctuating flow conditions. Since the value of h_1 is small at low flows, an accurate determination

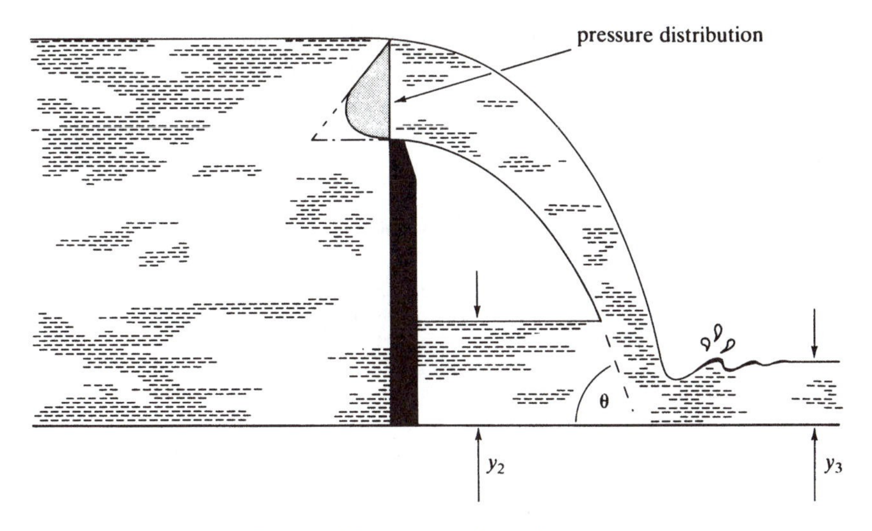

Figure 13.3 Section through nappe.

of C_d becomes impracticable. It will be seen later that these considerations lead to alternative weir shapes.

A vertical section through the overspilling water passing through any weir would clearly show that a vertical contraction occurred. However, in the case of a contracted weir, contraction additionally occurs at the sides of the jet. The jet of an uncontracted weir clings to the channel sides, so air cannot readily gain access to the underside of the nappe. Any air initially trapped beneath the nappe will tend to be entrained in the flow of water. This, in turn, implies the possibility of sub-atmospheric pressure beneath the nappe. For this reason the nappe for an uncontracted weir may tend to be drawn towards the weir more readily than the jet of a contracted weir. In order to offset this, some uncontracted weirs incorporate a vent pipe to admit air to the underside of the nappe (Fig. 13.1(a)). As the falling jet strikes the channel bed downstream of the weir, it must deflect away from the weir and along the channel. This means that there must be a net force in the direction of the acceleration (see Fig. 13.3). To produce this force, the depth y_2 (between the weir and the falling jet) must exceed y_3 downstream of the point of impact, such that

$$(\rho g b/2)(y_2^2 - y_3^2) = \rho Q V_3 - \rho Q V_3 \cos\theta \qquad (13.3)$$

($V = Q/A = Q/by$). This mechanism also tends to reduce the volume of air under the nappe, especially for an uncontracted weir.

A number of empirical discharge formulae have been developed which incorporate C_d. For uncontracted weirs under free discharge conditions two typical equations are:

1. the Rehbock formula:

$$Q = \frac{2}{3}\sqrt{2g}(0.602 + 0.083h_1/P_S)b(h_1 + 0.0012)^{3/2}\,\text{m}^3/\text{s} \qquad (13.4)$$

which is valid for $30\,\text{mm} < h_1 < 750\,\text{mm}, b > 300\,\text{mm}, P_S > 100\,\text{mm}$, $h_1 < P_S$ (in foot-sec units $Q = \frac{2}{3}\sqrt{2g}(0.602 + 0.083h_1/P_S)b(h_1 + 0.004)^{3/2}\,\text{ft}^3/\text{s}$)

2. White's formula:

$$Q = 0.562(1 + 0.153h_1/P_S)b\sqrt{g}(h_1 + 0.001)^{3/2}\,\text{m}^3/\text{s} \qquad (13.5)$$

which is valid for $h_1 > 20\,\text{mm}$, $P_S > 150\,\text{mm}$, $h_1 < 2.2P_S$ (in foot-sec units $Q = 0.562(1 + 0.153h_1/P_S)b\sqrt{g}(h_1 + 0.003)^{3/2}\,\text{ft}^3/\text{s}$)

Ackers *et al.* (1978) suggest that the accuracy of the formulae depends, to a discernible extent, on the siting of the gauging station for measuring the upstream head (h_1). They recommend that this should be at least $2.67P_S$ upstream of the weir, to avoid undue drawdown effects. The British Standard (BS 3680) recommendation is that the station should be between $4h_1$ and $5h_1$ upstream of the weir. For American practice, see the *Water Measurement Manual* (US Bureau of Reclamation, 1967).

For contracted rectangular weirs, (13.1) for Q_{ideal} is applicable. However, the magnitude of the actual discharge, Q, will be affected by the vertical and lateral contraction of the jet. A number of empirical equations have been developed to enable estimates of C_d to be made. Two such formulae are given below.

(a)The Hamilton–Smith formula is

$$C_d = 0.616(1 - h_1/P_S) \qquad (13.6)$$

which is valid for $B > (b + 4h_1), h_1/B < 0.5, h_1/P_S < 0.5, 75\,\text{mm} < h_1 < 600\,\text{mm}, P_S > 300\,\text{mm}, b > 300\,\text{mm}$. This has the advantage of simplicity, and therefore ease and speed in use. (B = channel width)

(b)The Kindsvater–Carter formula makes use of the concept of the 'effective' head and width, h_e and b_e, where

$$h_e = (h + k_h) \quad \text{and} \quad b_e = (b + k_b)$$

k_h and k_b are experimentally determined quantities which allow for the effects of viscosity and surface tension. It has been found that $k_h = \text{constant} = 0.001\,\text{m}$.

The formula is therefore

$$Q = \frac{2}{3} C_e \sqrt{2g} b_e h_e^{3/2} \text{ m}^3/\text{s}$$

where

$$C_e = f\left\{\frac{b}{B}, \frac{h_1}{P_S}\right\}$$

and must be determined empirically. BS 3680 gives charts for the determination of C_e and k_b.

Example 13.1 Determination of discharge for a rectangular weir

A long channel 1.5 m wide terminates in a full-width rectangular weir whose crest height is 300 mm above the stream bed. If the measuring station is recording a depth of 400 mm above the weir crest, estimate the discharge:

(a) assuming a flow with a negligible velocity of approach (take C_d as 0.7);
(b) including velocity of approach (taking the Coriolis coefficient, α, as 1.1 and C_d as 0.7);
(c) using White's formula.

Solution

(a)
$$Q = \frac{2}{3} C_d b \sqrt{2g} h_1^{3/2} \text{ (from (13.2))}$$
$$= \frac{2}{3} \times 0.7 \times 1.5 \times \sqrt{2 \times 9.81} \times 0.4^{3/2} = 0.784 \text{ m}^3/\text{s}$$

(b) If the approach velocity is to be incorporated, (13.1) must be used. However, u_1 is initially unknown. The equation is therefore solved by successive approximations. For the sake of simplicity, it is usual to assume that the velocity at Station 1 is uniform, i.e. $u_1 = Q/A_1 = V_1$.

First trial. If $Q = 0.784 \text{ m}^3/\text{s}$ (from (a)), then

$$V_1 = 0.784/[(0.3 + 0.4) \times 1.5] = 0.747 \text{ m/s}$$

Therefore

$$Q = \frac{2}{3} \times 0.7 \times 1.5 \times \sqrt{2 \times 9.81} \left[\left(0.4 + \frac{1.1 \times 0.747^2}{2 \times 9.81}\right)^{3/2} - \left(\frac{1.1 \times 0.747^2}{2 \times 9.81}\right)^{3/2}\right]$$
$$= 0.861 \text{ m}^3/\text{s}$$

Second trial. $V_1 = 0.861/[(0.3+0.4) \times 1.5] = 0.82\,\text{m/s}$, therefore

$$Q = \frac{2}{3} \times 0.7 \times 1.5 \times \sqrt{2 \times 9.81}\left[\left(0.4 + \frac{1.1 \times 0.82^2}{2 \times 9.81}\right)^{3/2} - \left(\frac{1.1 \times 0.82^2}{2 \times 9.81}\right)^{3/2}\right]$$

$$= 0.875\,\text{m}^3/\text{s}$$

which is sufficiently accurate for practical purposes.

(c) Using White's formula (equation (13.5)),

$$Q = 0.562\left[1 + \left(0.153 \times \frac{0.4}{0.3}\right)\right] \times 1.5\sqrt{9.81}(0.4 + 0.001)^{3/2}$$

$$= 0.807\,\text{m}^3/\text{s}$$

'Submergence' and the modular limit

All of the above equations have rested upon the assumption that the downstream depth, y_3, is too low to impede the free discharge over the weir. There is then a unique relationship between h_1 and Q. This is known as 'modular flow'. Weirs are designed for modular operation. In the field, however, circumstances may arise in which modular operation is not possible. A partial blockage of the channel downstream of a weir might be an example of such circumstances.

Consider what happens to Q if h_1 remains constant but y_3 increases until it rises above the crest. Up to a certain limiting value of y_3, h_1 and Q will be unaffected. The modular limit of operation occurs when this limiting y_3 has been reached. If y_3 increases further, h_1 will have to increase in order to maintain the discharge. The weir is then said to be 'submerged'. Thus, for a given upstream depth the discharge through a submerged weir is less than the free discharge. Furthermore, the discharge is now related to the depths upstream and downstream of the weir (i.e. to h_1 and $h_3 (= y_3 - P_S)$), and not just to h_1. The measurement of h_3 is problematical, since the surface of the water downstream of the weir is highly disturbed. If an estimate of the discharge Q_S of a submerged weir has to be made, Villemonte's formula may be used:

$$\frac{Q_S}{Q} = \left[1 - \left(\frac{h_3}{h_1}\right)^{3/2}\right]^{0.385} \tag{13.7}$$

where Q is the free discharge which corresponds to the upstream depth h_1. Much information is given in the British Standard (BS 3680) and in Ackers *et al.* (1978).

Vee weirs

It has been stated that rectangular weirs suffer from a loss of accuracy at low flows. The vee weir largely overcomes this problem. The variation of b with height, together with the narrow nappe width in the jet, means that for a given increase in Q, the increase in h_1 for a vee weir will be much greater than for a rectangular weir. Conversely, the greater sensitivity limits the range of discharge for which a vee weir can be economically applied. The underlying theory and assumptions are the same as for the rectangular weir excepting, of course, the fact that b is not a constant but a function of z. Thus, the discharge through an elementary strip across the weir (Fig. 13.4) will be

$$\delta Q_{\text{ideal}} = u_2 b\,\delta z = \left[2g\left(h_1 - z_2 + \frac{u_1^2}{2g}\right)\right]^{1/2} b\,\delta z$$

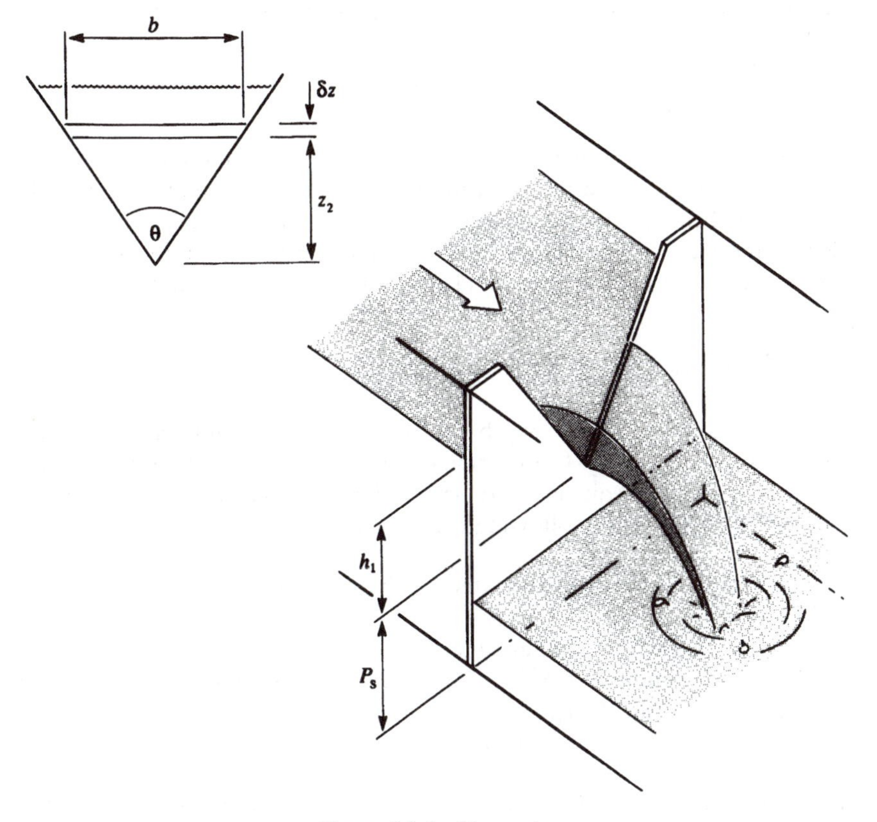

Figure 13.4 Vee weir.

However

$$b = 2z_2 \tan(\theta/2)$$

so

$$Q_{\text{ideal}} = \int_0^{h_1} \left[2g\left(h_1 - z_2 + \frac{u_1^2}{2g}\right)\right]^{1/2} 2z_2 \tan\left(\frac{\theta}{2}\right) dz \qquad (13.8)$$

The approach velocity, u_1, is almost always negligible for vee weirs in view of the smaller discharges for which they are designed. Therefore

$$Q_{\text{ideal}} = \int_0^{h_1} [2g(h_1 - z_2)]^{1/2} 2z_2 \tan(\theta/2)\, dz$$

$$= \frac{8}{15}\sqrt{2g}\tan(\theta/2) h_1^{5/2} \qquad (13.9)$$

The value of C_d is a function of Re, We, θ and h_1, and magnitudes are given in BS 3680 for a wide range of weirs. The 90° weir is probably the most widely used. As a first approximation, a C_d of 0.59 may be used for this angle.

Plate weirs of special form

A wide variety of weir shapes have emerged over the years. One example is the Sutro weir, in which the variation of b with z is such that Q is directly proportional to h_1. Other examples include the Trapezoidal weir and the Compound weir (a rectangular weir with a vee weir sunk in its crest). The reader is directed to Ackers *et al.* (1978) for further information.

13.3 Long-based weirs

In contrast to plate weirs, long-based weirs are larger and generally more heavily constructed (e.g. from concrete). They are usually designed for use in the field, and consequently may have to handle large discharges. An ideal long-based weir has the following characteristics:

1. it is cheaply and easily fabricated (perhaps off-site);
2. it is easily installed;
3. it possesses a wide modular range;
4. it produces a minimum afflux (i.e. increase in upstream depth due to the installation of the weir);
5. it requires a minimum of maintenance.

There are a number of different designs, of which a selection is considered in detail below.

The rectangular ('broad-crested') weir

Rectangular weirs are solid weirs of rectangular cross-section, which span the full width of a channel. They form part of the family of critical depth meters introduced in Chapter 5. However, it is worth recapitulating the following facts:

(a) A hump placed on the bed of the channel results in a local increase in the velocity of flow and a corresponding reduction in the elevation of the water surface.

(b) Given a hump of sufficient height, critical flow will be produced in the flow over the hump. There is then a direct relationship between Q and h_1, i.e. the flow is modular. Long-based weirs are designed for modular flow.

By definition, a rectangular weir is not streamlined. This, in turn, implies that the streamlines at the upstream end of the weir will not be parallel, since the flow will be accelerating. If frictional resistance is ignored, then the streamlines will become parallel and the flow becomes critical given a sufficient length of crest. It is then possible to derive a straightforward expression for Q_{ideal} in terms of H_1 (see Fig. 13.5), as was seen in Chapter 5 (N.B. $H_1 = h_1 + V_1^2/2g$). Thus, as $y_c = \frac{2}{3}H_1$,

$$Q_{\text{ideal}} = \sqrt{g}\, b\left(\frac{2}{3}H_1\right)^{3/2}$$

In SI units this becomes

$$Q_{\text{ideal}} = 1.705bH_1^{3/2}\ \text{m}^3/\text{s} \tag{13.10}$$

(in foot-sec units, $Q_{\text{ideal}} = 3.09bH_1^{3/2}\ \text{ft}^3/\text{s}$). This equation may be modified to account for the differences between real and ideal fluids, using the relationship $Q/Q_{\text{ideal}} = C_d$:

$$Q = C_d \times 1.705bH_1^{3/2} = CbH_1^{3/2}\ \text{m}^3/\text{s} \tag{13.11}$$

In reality, the flow over a long-crested rectangular weir is more complex than ideal flow. Frictional shearing action promotes the growth of a boundary layer, which in turn implies that the water surface profile takes the form shown in Figure 13.5(b). Additionally, as the flow approaches the critical condition, there is a tendency for ripples to form on the surface.

The value of C_d has to be derived empirically. Singer proposed an equation for C_d in the form

$$C_d = 0.848C_F \tag{13.12}$$

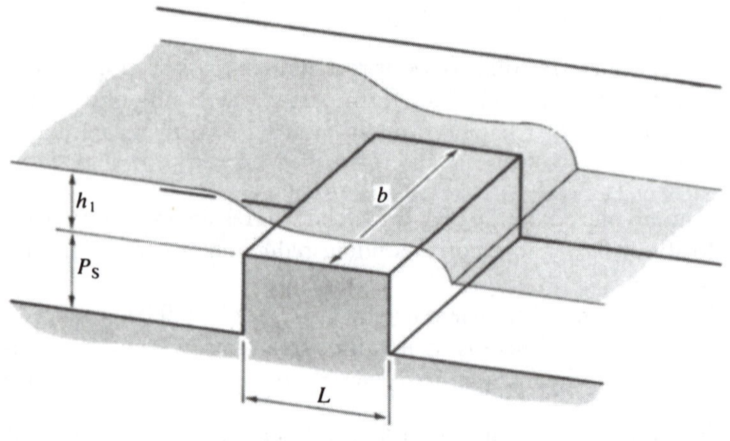

(a) Flow over broadcrested weir

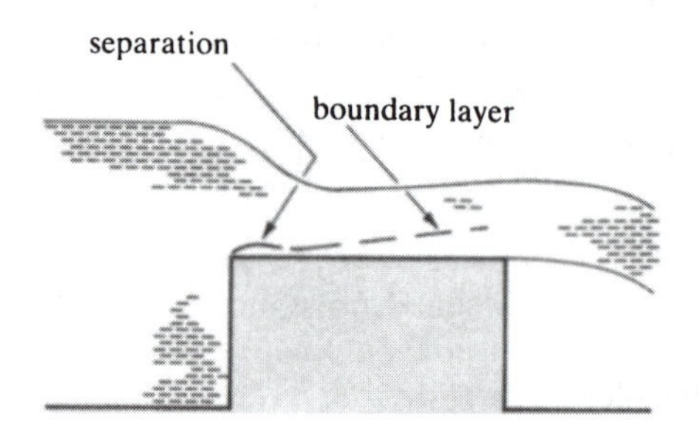

(b) Effect of boundary layer on flow

Figure 13.5 Broad-crested weir.

The correction factor, C_F, is a function of h_1/L and $h_1/(h_1+P_S)$ and is presented in graphical form in Ackers *et al.* However, as a first approximation, where $0.45 < h_1/L < 0.8$ and $0.35 < h_1/(h_1+P_S) < 0.6$,

$$C_F \simeq 0.91 + 0.21\frac{h_1}{L} + 0.24\left(\frac{h_1}{h_1+P_S} - 0.35\right) \tag{13.13}$$

These equations presuppose the establishment of critical flow over the crest. If the stage downstream of the weir rises above the stage which permits critical depth at the weir, then the weir will be submerged. An equation for Q_{ideal} for submerged flow can be developed, based on the energy and continuity equations. However, it is more usual to apply an empirical submergence correction factor ($= Q_s/Q$), which will be a function of both upstream depth (h_1) and depth over the crest (h_2). The modular equation (13.11) can be multiplied by this factor to give the actual discharge.

Round-nosed weirs

These are so named because of the radius formed on the leading corner. In other respects they resemble the broad-crested weir. The incorporation of the radius makes it more robust of the two, and also means that the discharge coefficient is insensitive to minor damage. The magnitude of C_d is higher than that for a rectangular weir. However, very little is known about the performance of such weirs under submerged conditions.

It is worth noting that the region just downstream of any weir can be liable to scouring, especially in natural channels. For this reason, some form of 'stilling' arrangement may have to be incorporated.

Crump weirs

In 1952, E. S. Crump published details of a weir with a triangular profile, which had been developed at the Hydraulics Research Station. This was claimed to give a wide modular range, and also to give a more predictable performance under submerged conditions than other long-based weirs (Fig. 13.6).

Crump proposed upstream and downstream slopes of 1:2 and 1:5, respectively, which were based on sound principles. The upstream slope was designed so that sediment build-up would not reach the crest. The downstream slope was shallow enough to permit a hydraulic jump to form on the weir under modular flow conditions, thus providing an integral energy dissipator. Also, under submerged conditions, losses are not too high and

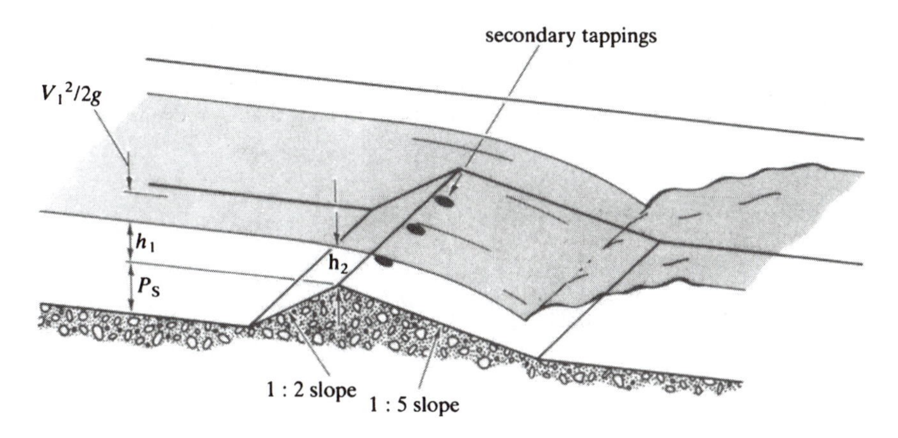

Figure 13.6 Crump weir.

the afflux is minimized. The primary gauging station is upstream of the weir. However, there is a second gauging point on the weir itself, just downstream of the crest. The second reading is used when the weir is submerged. The accuracy of the weir depends on the sharpness of the crest, so some weirs incorporate a metal insert at the crest. The secondary gauging tappings are drilled into the insert. The secondary tappings are sited near the crest in order to be clear of the disturbed flow further downstream.

The discharge equation for a Crump weir is in the form $Q = C_d C_v b g^{1/2} h_1^{3/2}$, which is clearly based on the same concepts as the corresponding equation for a rectangular weir. The value of C_d is about 0.63. The value of C_v varies with the ratio $h_1/(h_1 + P_S)$. For submerged flow,

$$\frac{Q_S}{Q} = 1.04\left[0.945 - \left(\frac{h_p}{H_1}\right)^{3/2}\right]^{0.256} \tag{13.14}$$

where h_p is the head at the second gauging point. The onset of submergence occurs when

$$\text{downstream depth} \geq 0.75 \times (\text{upstream depth})$$

Since 1952, other variants have been developed, e.g. weirs with upstream and downstream slopes both 1:2. However, some of these designs have a less stable C_d, and inferior submerged performance (see Ackers *et al.*, 1978).

Example 13.2 Discharge over a broad-crested weir

A reservoir has a plan area of 100 000 m^2. The discharge from the reservoir passes over a rectangular broad-crested weir, having the following dimensions: $b = 15\,\text{m}$, $P_S = 1\,\text{m}$, $L = 2\,\text{m}$. Estimate the discharge over the weir if the reservoir level is 1 m above the weir crest.

If there is no inflow into the reservoir, estimate the time taken for the reservoir level to fall from 1 m to 0.5 m above the weir crest. Assume that C_d remains constant.

Solution

From (13.13),

$$C_F = 0.91 + \left(0.21 \times \frac{1}{2}\right) + 0.24\left(\frac{1}{1+1} - 0.35\right) = 1.051$$

Therefore

$$C_d = 0.848 \times 1.051 = 0.891$$

Substituting in (13.11),

$$Q = 0.891 \times 1.705 \times 15 \times 1^{3/2} = 22.79\,\mathrm{m}^3/\mathrm{s}$$

If the level in the reservoir falls from 1 m to 0.5 m, then let the upstream depth $= y$ when time $= t$ seconds after commencement of fall in reservoir level:

$$Q = 0.891 \times 1.705 \times 15 \times y^{3/2} = 22.79 y^{3/2}\,\mathrm{m}^3/\mathrm{s}$$

Therefore, during a time interval δt, the outflow will be

$$-Q\,\delta t = -22.79 y^{3/2}\,\delta t\,\mathrm{m}^3$$

and the corresponding reduction in reservoir water level will be $-\delta y$. Thus, outflow will also be given by the product (plan area $\times\ \delta y$) $= -100\,000\ \delta y$. Therefore

$$100\,000\,\delta y = 22.79\, y^{3/2}\,\delta t$$

i.e.

$$\delta t = \frac{100\,000\,\delta y}{22.79\, y^{3/2}}$$

Integrating,

$$t = \frac{100\,000}{22.79}\int_{1.0}^{0.5} \frac{\mathrm{d}y}{y^{3/2}} = \frac{100\,000}{22.79}\left[-\frac{2}{y^{1/2}}\right]_{1.0}^{0.5}$$
$$= 3635\,\mathrm{s}\ \text{(or approx. 1 h)}$$

13.4 Flumes

This term is applied to devices in which the flow is locally accelerated due to:

1. a streamlined lateral contraction in the channel sides;
2. the combination of the lateral contraction, together with a streamlined hump in the invert (channel bed) (see Fig. 13.7).

The first type of flume is known as a venturi flume, and it has already been briefly introduced in section 5.9. Flumes are usually designed to achieve critical flow in the narrowest (throat) section, together with a small afflux. Flumes are especially applicable where deposition of solids must be avoided (e.g. in sewage works or in irrigation canals traversing flat terrain).

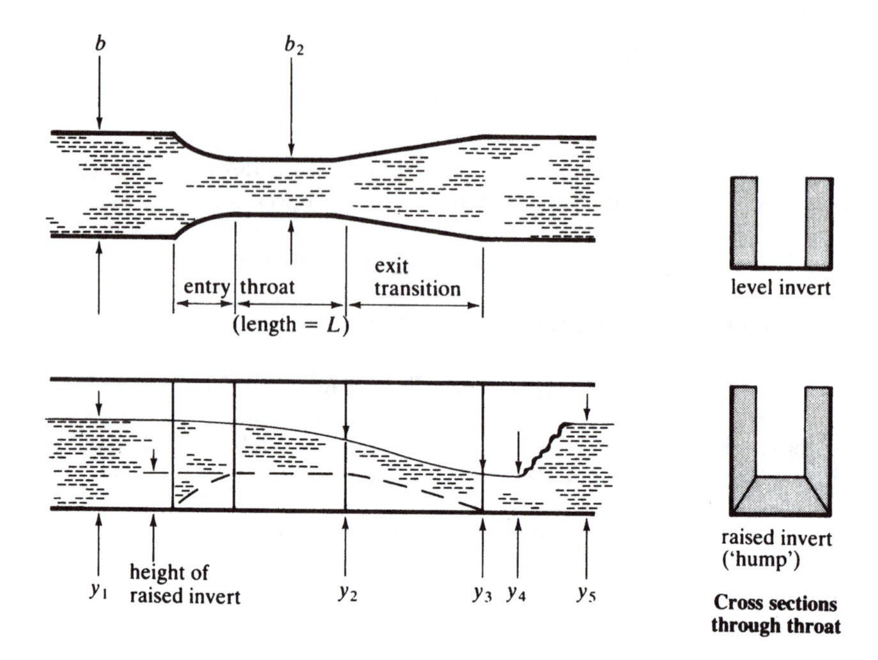

Figure 13.7 Venturi flume.

A general equation for the ideal discharge through a flume may be developed on the basis of the energy and continuity principles.

From the energy equation,

$$E_S = y_1 + \frac{V_1^2}{2g} = y_2 + \frac{V_2^2}{2g}$$

but

$$V_1 = Q/by_1 \text{ and } V_2 = Q/b_2y_2$$

Substituting for V_1 and V_2 and rearranging,

$$Q_{\text{ideal}} = \sqrt{\frac{2g(y_1 - y_2)(by_1)^2(b_2y_2)^2}{(by_1)^2 - (b_2y_2)^2}}$$

$$= by_1\sqrt{\frac{2g(y_1 - y_2)}{(by_1/b_2y_2)^2 - 1}} \qquad (13.15)$$

If critical flow is attained in the throat, then $y_2 = 2/3E_S$. If this is substituted into (13.15), then (in SI units)

$$Q_{\text{ideal}} = 1.705b_2E_S^{3/2} = 1.705b_2C_v y_1^{3/2}\ \text{m}^3/\text{s} \qquad (13.16)$$

where C_v is the velocity of approach correction factor (in foot-sec units, $Q_{ideal} = 3.09 b_2 C_v y_1^{3/2}$ ft^3/s). The actual discharge is

$$Q = 1.705 b_2 C_d C_v y_1^{3/2} \text{ m}^3/\text{s} \tag{13.17}$$

(which is the same as (5.34)).

Although the flat bed venturi flume is simpler to construct, it is sometimes necessary to raise the invert in the throat to attain critical conditions. For either the flat or raised floor, the flume throat length should ideally be sufficient to ensure that the curvature of the water surface is small, so that in the throat the water surface is parallel with the invert.

A special group of 'short' flumes are less expensive and more compact than the 'ideal' throated flume. However, the water surface profile varies rapidly, and a theoretical analysis is not really possible. Empirical relationships must therefore be developed. For this reason, compact flumes tend to be of standard designs (e.g. the Parshall Flume) for which extensive calibration experiments have been carried out (see US Bureau of Reclamation, 1967).

Example 13.3 Venturi flume

An open channel is 2 m wide and of rectangular cross-section. A venturi flume with a level invert, having a throat width of 1 m is installed at one point. Estimate the discharge:

(a) if the upstream depth is 1.2 m and critical flow occurs in the flume;
(b) if the upstream depth is 1.2 m and the depth in the throat is 1.05 m.

Take $C_v = 1$ and $C_d = 0.95$.

Solution

(a) If the flow in the throat is critical, then using (13.17),

$$Q = 1.705 \times 0.95 \times 1.0 \times 1.2^{3/2} = 2.13 \text{ m}^3/\text{s}$$

(b) If $y_1 = 1.2$ m and $y_2 = 1.05$ m, we must use (13.15):

$$Q = 0.95 \times 2.0 \times 1.2 \sqrt{\frac{2 \times 9.81(1.2 - 1.05)}{(2 \times 1.2/1 \times 1.05)^2 - 1}} = 1.903 \text{ m}^3/\text{s}$$

The apparent ease of this approach (for flumes or for long-based weirs) to some extent disguises a number of difficulties which the designer must face. These are discussed below.

Flume design methodology

(1) The principal obstacle to modular operation is submergence due to high tailwater levels. To obtain modular conditions at low discharge, it is necessary to use a narrow throat combined with a small expansion angle in the exit transition. This approach suffers from two disadvantages:

(a) it is expensive;
(b) it will cause a large afflux at high discharges.

To illustrate the effect of installing a flume, Figure 13.8(a) shows how y_1 varies with Q for two different throat widths. The appropriate values of y_n are also given. The height of the banks upstream will impose a maximum limit on the afflux, which implies a corresponding lower limit on the value of b_2. Incorporating a raised throat invert in the form of a streamlined hump often assists in deferring submergence. Indeed, it may be possible to improve the matching of the flume to the channel by raising the invert and increasing b_2.

(2) The next problem is to determine the value of C_d. The flow in the main channel upstream probably approximates to a steady fully developed uniform flow. However, the flume is normally of a different material and surface roughness from that of the channel. Consequently, as the flow passes through the flume, a boundary layer appropriate to the surface of the flume will develop. The boundary layer will influence the values of C_d and C_v. It

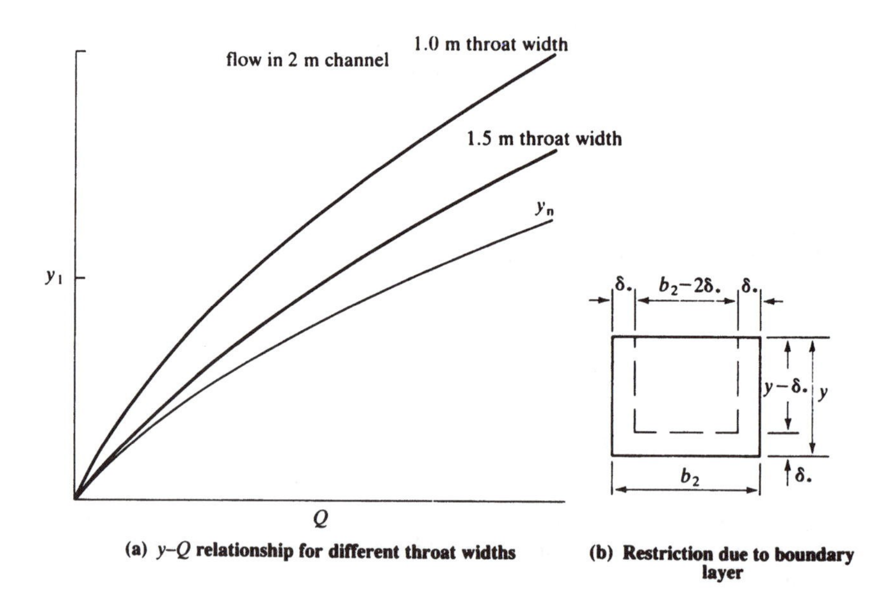

Figure 13.8 Flume design.

may be assumed (a) that boundary layer growth commences at the entry to the throat and (b) that the control is at the downstream end of the throat where the boundary layer thickness is greatest (Fig. 13.8(b)). The boundary layer has the effect of reducing the values of y_1 and b_2 to $(y_1 - \delta_*)$ and $(b_2 - 2\delta_*)$, respectively. So, for a rectangular flume and channel,

$$Q_{\text{ideal}} \propto b_2, y_1^{3/2}$$

and

$$Q \propto (b_2 - 2\delta_*), (y_1 - \delta_*)^{3/2}$$

Therefore

$$C_d = \frac{(b_2 - 2\delta_*)(y_1 - \delta_*)^{3/2}}{b_2 y_1^{3/2}} \tag{13.18}$$

and

$$C_v = \left(\frac{E_{S1} - \delta_*}{y_1 - \delta_*}\right)^{3/2} \tag{13.19}$$

(3) Matching flume geometry to a given channel must be determined by trial and error. For a flume and channel of rectangular cross-section this is quite straightforward. A limiting value of y_1 must be related to the maximum flood discharge. A corresponding initial value of b_2 may then be obtained:

$$b_2 = Q/1.705 y_1^{3/2}$$

assuming modular flow, and with b_2 in metres.

In order to estimate C_d and C_v, δ_* must first be evaluated. For a first approximation,

$$\delta_* = 0.003L \tag{13.20}$$

which is valid for $\text{Re} > 2 \times 10^5$ and $10^5 > L/k_S > 4000$ (N.B. A more detailed and accurate method of obtaining δ_* is also presented in the British Standard (BS 3680), together with illustrative worked examples.)

The value of b_2 may then be checked:

$$b_2 = Q/1.705 C_d C_v y_1^{3/2} \tag{13.21}$$

and the conditions downstream of the flume should then be assessed to ascertain whether submergence is a potential problem.

For flumes and channels which are not of rectangular section the process remains essentially the same. However, the discharge function, $Q/1.705y_1^{3/2}$, will now vary with depth.

The art of 'matching' lies in designing a flume for which the variation of discharge function with E_{S1} or y_1 approximates closely to the corresponding variation for the channel. For a more detailed treatment of this problem, Ackers *et al.* (1978) should be consulted.

(4) The designer needs to predict the range of discharge over which modular conditions apply. Two possible approaches to this will now be considered.

(a) The simplest method is that proposed in BS 3680. This suggests that, for modular flow in a flume with a full length exit transition, the total head upstream shall be at least 1.25 times the total head downstream (provided that flow downstream is subcritical). The corresponding relationship for a flume with a truncated exit transition is that total head upstream shall be at least 1.33 times that downstream.

(b) A more sophisticated method is suggested by ILRI (Bos, 1978), which is as follows. Starting with the equation for modular flow,

$$Q = \text{constant} \times C_d C_v b_2 E_{S1}^{3/2}$$

or

$$Q = \text{constant} \times C_v b_2 E_{S2}^{3/2}$$

Hence

$$E_{S1} C_d^{2/3} = E_{S2} \qquad (13.22)$$

In the exit transition, there will be losses of energy due both to friction and to 'separation'. This may be expressed as

$$E_{S2} - E_{S3} = h_{f2-3} + \frac{C_L (V_2 - V_3)^2}{2g} \qquad (13.23)$$

The frictional head loss may be deduced from an appropriate equation (e.g. the Manning equation). The separation loss coefficient C_L has been evaluated empirically (Bos, 1978): typical values are $C_L = 0.27$ for a 28° included exit angle, $C_L = 0.68$ for a 52° included exit angle.

If (13.22) is substituted into (13.23), then, with some rearrangement,

$$\frac{E_{S3}}{E_{S1}} = C_d^{2/3} - \frac{C_L (V_2 - V_3)^2}{2gE_{S1}} - \frac{h_{f2-3}}{E_{S1}}$$

This will apply as long as the hydraulic jump is downstream of the exit. Therefore, modular conditions will certainly exist if

$$\frac{E_{S3}}{E_{S1}} \leq C_d^{2/3} - \frac{C_L(V_2 - V_3)^2}{2gE_{S1}} - \frac{h_{f2-3}}{E_{S1}} \tag{13.24}$$

A similar approach may be applied to flow over long-based weirs.

It is not possible to predict the modular limit precisely:

1. because C_d and C_L cannot be accurately estimated;
2. because it is difficult to estimate the location of a hydraulic jump if it occurs in the exit transition.

In order to ensure that the streamlines in the throat are parallel, the length of the throat section should be at least twice E_{S1}. For 'short' flumes, of course, this criterion is not met.

The method outlined in this section will now be applied.

Example 13.4 Flume design using boundary layer theory

A flume is to be installed in a rectangular channel 2 m wide. The bed slope is 0.001, and Manning's n is 0.01. When the flume is installed, the increase in upstream depth must not be greater than 350 mm at the maximum discharge of 4 m^3/s. Take $C_L = 0.27$ (28° expansion).

Solution

For the maximum discharge (using Manning's equation),

$$4 = 2y_n \frac{R^{2/3}}{0.01} \times 0.001^{1/2}$$

therefore $y_n = 1.0$ m.

For a first trial, take b_2 as 1.5 m. Then

$$E_{S1}^{3/2} \simeq \frac{4}{1.705 \times 1.5}$$

so $E_{S1} = 1.35$ m. Hence $y_1 = 1.2$ m. Assume throat length $= 2 \times 1.35 = 2.7$ m. Then

$$\delta_* = 0.003 \times 2.7 = 0.008\,\text{m}$$

$$C_d = \frac{[1.5 - (2 \times 0.008)](1.2 - 0.008)^{3/2}}{1.5 \times 1.2^{3/2}} = 0.98$$

$$C_v = \left(\frac{1.35 - 0.008}{1.2 - 0.008}\right)^{3/2} = 1.195$$

Therefore

$$y_1^{3/2} = 4/(1.705 \times 0.98 \times 1.195 \times 1.5)$$

so $y_1 = 1.213\,\text{m}$. This is acceptable ($y_1 < y_n + 0.35$). If the first trial produces an unacceptable result, further trials must be undertaken.

The normal depth for the channel has been calculated above. It is necessary for the engineer to ascertain:

1. whether free discharge is attained with this depth;
2. the downstream depth which corresponds to the modular limit. Here the ILRI method (equation (13.24)) is used.

To solve the equation, V_3 must be known. For a first approximation assume $E_{S3} = E_{S1}$. Hence, $y_3 = 0.48\,\text{m}$ and $V_3 = 4.17\,\text{m/s}$. The hydraulic gradient is

$$S_{f2-3} = \frac{Q^2 n^2}{A^2 R^{4/3}}$$

In the exit transition both width and depth of flow are varying. As an approximation, the cross-sectional area and hydraulic radius are calculated for the throat. Critical depth in the throat is 0.9 m ($= \frac{2}{3} E_{S1}$). Thus

$$S_{f2-3} = \frac{4.17^2 \times 0.01^2}{(1.5 \times 0.9)^2 \left(\dfrac{1.5 \times 0.9}{1.5 + (2 \times 0.9)} \right)^{4/3}} = 0.0031$$

For a 28° exit angle, the length of the transition is 1 m for an expansion of 0.5 m, therefore $h_{f2-3} = 0.0031\,\text{m}$.

Also, for critical flow in the throat, $V_2 = 2.963\,\text{m/s}\ (= \sqrt{g y_2})$. From (13.24),

$$\frac{E_{S3}}{1.35} = 0.98^{2/3} - \frac{0.27(2.963 - 4.17)^2}{2 \times 9.81 \times 1.35} - \frac{0.0031}{1.35}$$

Therefore $E_{S3} = 1.31\,\text{m}$, and hence $y_3 = 0.5\,\text{m}$, $V_3 = 4\,\text{m/s}$, $\text{Fr}_3 = 1.806$. Assuming $y_4 \simeq y_3$, the hydraulic jump (conjugate depth) equation (equation (5.28)) is used to estimate y_5. Hence, $y_5/y_4 = 2.10$, so at the modular limit $y_5 = 2.1 \times 0.5 = 1.05\,\text{m}$.

Naturally, this calculation could be repeated for a series of discharges to cover the whole operational range.

13.5 Spillways

The majority of impounding reservoirs are formed as a result of the construction of a dam. By its very nature, the streamflow which supplies a reservoir is variable. It follows that there will be times when the reservoir is full and the streamflow exceeds the demand. The excess water must therefore be discharged safely from the reservoir. In many cases, to allow the water simply to overtop the dam would result in a catastrophic failure of the structure. For this reason, carefully designed overflow passages – known as 'spillways' (Fig. 13.9) – are incorporated as part of the dam design. The spillway capacity must be sufficient to accommodate the 'largest' flood discharge (the Probable Maximum Flood or 1 in 10 000 year flood) likely to occur in the life of the dam. Because of the high velocities of flow often attained on spillways, there is usually some form of energy dissipation and scour prevention system at the base of the spillway. This often takes the form of a stilling basin.

There are several spillway designs. The choice of design is a function of the nature of the site, the type of dam and the overall economics of the scheme. The following list gives a general outline of the various types and applications:

1. overfall and 'Ogee' spillways are by far the most widely adopted: they may be used on masonry or concrete dams which have sufficient crest length to obtain the required discharge;
2. chute and tunnel (shaft) spillways are often used on earthfill dams;

Figure 13.9 Spillway and control gates, El Chocun dam, Argentina. (Photograph courtesy of HR Wallingford Ltd.)

3. side channel and tunnel spillways are useful for dams sited in narrow gorges;
4. siphon spillways maintain an almost constant headwater level over the designed range of discharge.

Some typical designs are now considered.

Gravity ('Ogee') spillways

These are by far the most common type, being simple to construct and applicable over a wide range of conditions. They essentially comprise a steeply sloping open channel with a rounded crest at its entry. The crest profile approximates to the trajectory of the nappe from a sharp-crested weir. The nappe trajectory varies with head, so the crest can be correct only for one 'design head' H_d. Downstream of the crest region is the steeply sloping 'face', followed by the 'toe', which is curved to form a tangent to the apron or stilling basin at the base of the dam. The profile is thus in the form of an elongated 'S' (Fig. 13.10). Profiles of spillways have been developed for a wide range of dam heights and operating heads. A wealth of information is available in references published by the US Bureau of Reclamation (1964, 1977) and the US Army Waterways Experimental Station (1959). Spillways will here be considered mainly from the hydraulics standpoint.

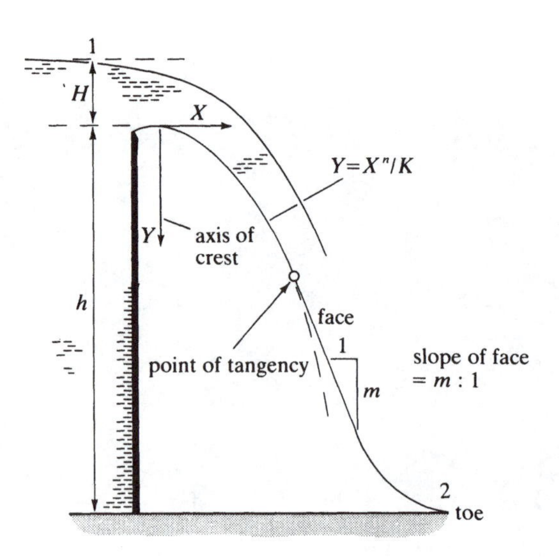

Figure 13.10 'Ogee' spillway.

The discharge relationship for a spillway is of the same form as for other weirs:

$$Q = \text{constant} \times C_d \sqrt{g} b H^{3/2}$$

In practice, this is usually written $Q = CbH^{3/2}$. C is not dimensionless, and its magnitude increases with increasing depth of flow. C usually lies within the range $1.6 < C < 2.3$ in metric units ($2.8 < C < 4.1$ in foot-sec units). The breadth, b, does not always comprise a single unbroken span. If control gates are incorporated in the scheme, the spillway crest is subdivided by piers into a number of 'bays' (Fig. 13.11). The piers form the supporting structure for the gates. The piers have the effect of inducing a lateral contraction in the

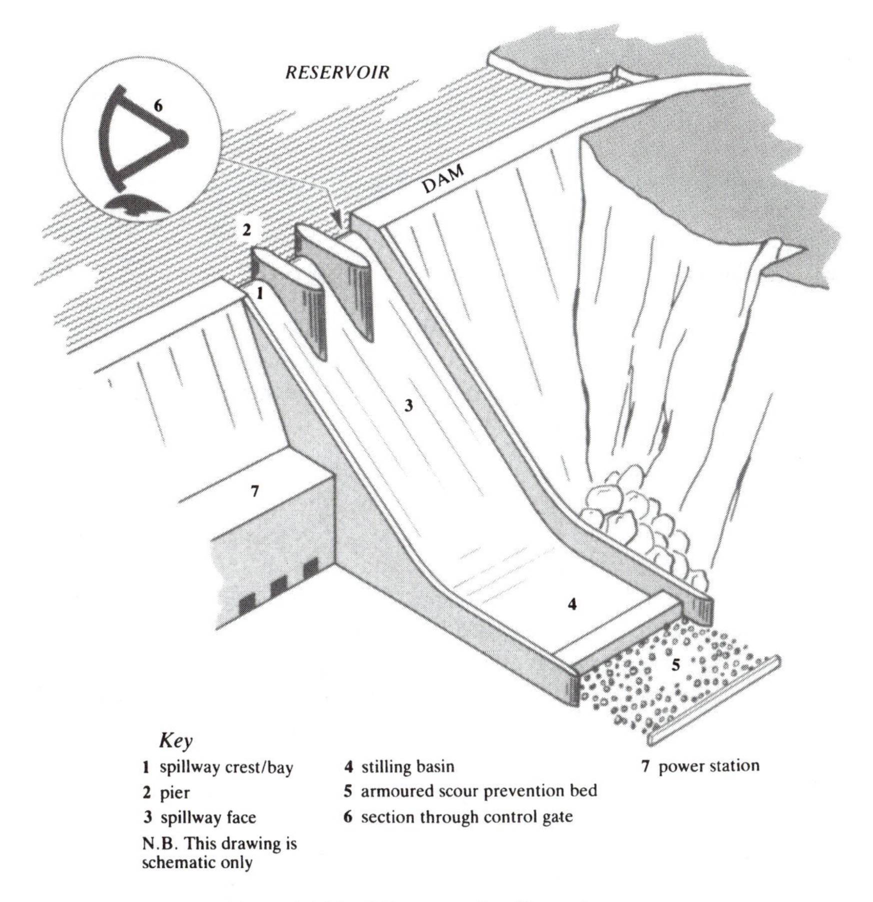

Figure 13.11 Diagram of spillway layout.

flow. In order to allow for this effect in the discharge equation, the total span, b, is replaced by b_c, the contracted width:

$$b_c = b - knH$$

where n is the number of lateral contractions and k is the contraction coefficient, which is a function of H and of the shape of the pier.

Some other important aspects of spillway hydraulics are summarized below.

(a) If $H < H_d$, the natural trajectory of the nappe falls below the profile of the spillway crest, then there will therefore be positive gauge pressures over the crest. On the other hand, if $H > H_d$, then the nappe trajectory is higher than the crest profile, so negative pressure zones tend to arise. Frictional shear will accentuate this tendency, and cavitation may occur. However, in practice, this pressure reduction is not normally a serious problem unless $H > 1.5H_d$. Indeed, recent work suggests that separation will not occur until $H \rightarrow 3H_d$.

(b) Conditions in the flow down the spillway face may be quite complex, since:

(i) the flow is accelerating rapidly, and may be 'expanding' as it leaves a bay–pier arrangement;
(ii) frictional shear promotes boundary layer growth;
(iii) the phenomenon of self-aeration of the flow may arise;
(iv) cavitation may occur.

For these reasons, the usual equations for non-uniform flow developed in Chapter 5 cannot really be applied. If it is necessary to make estimates of flow conditions on the spillway, then empirical data must be used. Each feature will now be examined in greater detail.

(i) In a region of rapidly accelerating flow, the specific energy equation (or Bernoulli's equation) is usually applied. It is possible to obtain very rough estimates of the variation of V and y down the spillway on this basis, accuracy will be slightly improved if a head loss term is incorporated. Nevertheless, in the light of (ii) and (iii), below, conditions on the spillway are far from those which underlie the energy equations.

(ii) A boundary layer will form in the spillway flow, commencing at the leading edge of the crest. The depth of the boundary layer, δ_*, will grow with distance downstream of the crest. Provided that the spillway is of sufficient length, at some point the depth, δ, will meet the free surface of the water (Fig. 13.12).

The flow at the crest is analogous to the flow round any fairly streamlined body. This may imply flow separation, eddy shedding, or both. Such conditions may be instrumental in bringing about cavitation at the spillway

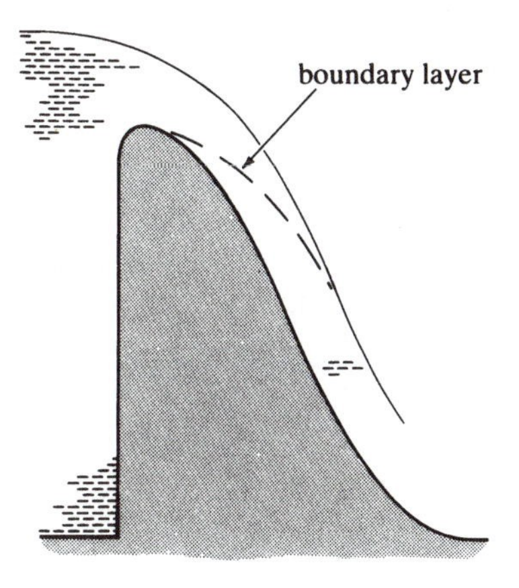

Figure 13.12 Growth of a boundary layer on spillway.

face. There have been a number of reported incidents of cavitation in major dams.

(iii) Aeration has been observed on many spillways. It entails the entrainment of substantial quantities of air into the flow, which becomes white and foamy in appearance. The additional air causes an increase in the gross volume of the flow. Observations of aeration have led to the suggestion that the point at which aeration commences coincides with the point at which the boundary layer depth meets the free surface (Henderson, 1966; Keller and Rastogi, 1975; Cain and Wood, 1981). The entrainment mechanism appears to be associated with the emergence of streamwise vortices at the free surface. Such vortices would originate in the spillway crest region. A rough estimate of the concentration of air may be made using the following equations:

$$C = 0.743 \log(S_0/q^{1/5}) + 0.876 \tag{13.25}$$

where $C =$ (volume of entrained air)/(volume of air and water) and S_0 is the bed slope, or

$$C = (x_I/y_I)^{2/3}/74 \tag{13.26}$$

where x_I and y_I are measured from the point at which entrainment commences. C may vary considerably over the width of the spillway.

(iv) Cavitation arises when the local pressure in a liquid approaches the ambient vapour pressure. Such conditions may arise on spillways for a variety of reasons, especially where the velocity of flow is high (say $V > 30\,\text{m/s}$). For example, if $H > 3H_d$, separated flow will probably arise at the crest.

Irregularities in the surface finish of the spillway may result in the generation of local regions of low pressure.

Cavitation has been observed in severely sheared flows (Kenn, 1971; Kenn and Garrod, 1981).

At the toe of the spillway, the flow will be highly supercritical. The flow must be deflected through a path curved in the vertical plane before entering the stilling zone or apron. This can give rise to very high thrust forces on the base and side walls of the spillway. A rough estimate of these forces may be made on the basis of the momentum equation.

The high velocity and energy of flow at the foot of the spillway must be dissipated, otherwise severe scouring will occur and the foundations at the toe will be undermined.

Siphon spillways

A siphon spillway is a short enclosed duct whose longitudinal section is curved as shown in Figure 13.13(a). When flowing full, the highest point in the spillway lies above the liquid level in the upstream reservoir, and the pressure at that point must therefore be sub-atmospheric, this is the essential characteristic of a siphon. A siphon spillway must be self-priming.

The way in which this type of spillway functions is best understood by considering what happens as the reservoir level gradually rises. When the water level just exceeds the crest level, the water commences to spill and flows over the downstream slope in much the same way as a simple Ogee spillway. As the water level rises further, the entrance is sealed off from the atmosphere. Air is initially trapped within the spillway, but the velocity of flow of the water tends to entrain the air (giving rise to aeration of the water) and draw it out through the exit. When all air has been expelled, the siphon is primed (i.e. running full) and is therefore acting as a simple pipe. There are thus three possible operating conditions depending on upstream depth:

1. gravity spillway flow;
2. aerated flow;
3. pipe ('blackwater') flow.

Operational problems with siphon spillways. The aerated condition is unstable and is maintained only for a short time while the siphon begins to prime, since (theoretically, at least) air cannot enter once the entry is covered. Therefore, in a simple siphon a small change in H produces a sharp increase or decrease in the discharge through the spillway. This can lead to problems if the discharge entering the reservoir is greater than the spillway flow but less than the blackwater flow, since the following cycle of events is set in train:

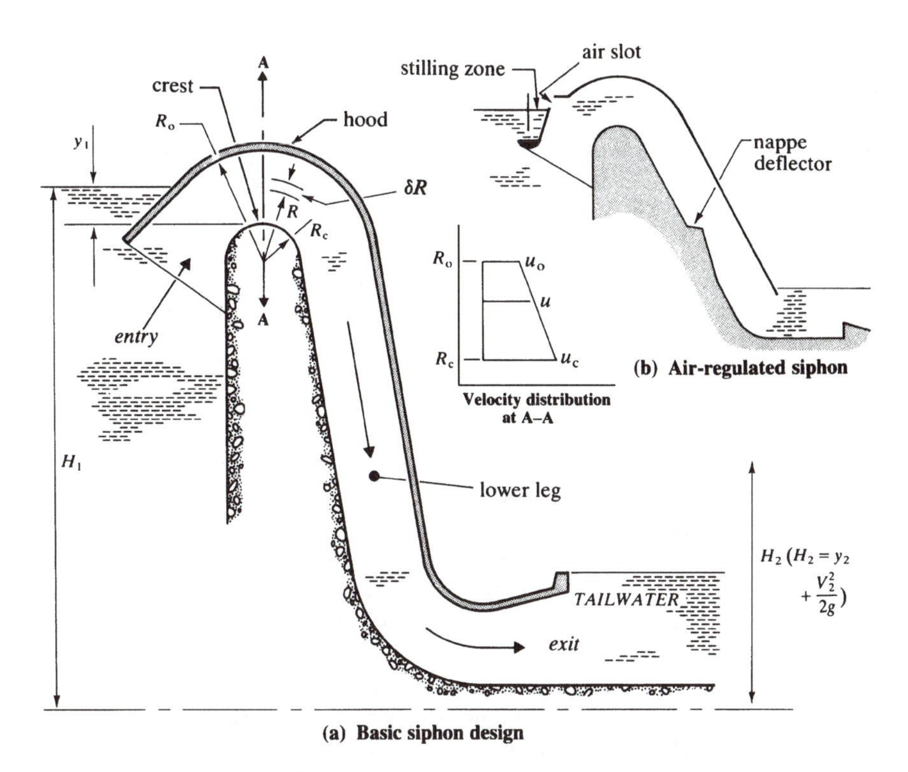

Figure 13.13 Siphon spillway.

(a) if the spillway is initially operating with gravity flow, then the upstream (reservoir) level must rise;
(b) when the upstream level has risen sufficiently, the siphon primes and the spillway discharge increases substantially;
(c) the upstream level falls until the siphon de-primes and its discharge drops.

The cycle (a) to (c) is then repeated.

Obviously, this can give rise to radical surges and stoppages in the downstream flow. This problem can be overcome by better design.

Other potential problems encountered with siphon spillways are:

1. blockage of spillway by debris (fallen trees, ice, etc.);
2. freezing of the water across the lower leg or the entry before the reservoir level rises to crest level;
3. the substantial foundations required to resist vibration;
4. waves arising in the reservoir during storms may alternately cover and uncover the entry, thus interrupting smooth siphon action.

Problem 1 may be ameliorated by installing a trash-intercepting grid in front of the intake.

Discharge through a siphon spillway. Some of the principal discharge conditions may be estimated by combining some familiar concepts, i.e. weir flow ($Q = \text{constant} \times H^{3/2}$), pipe flow ($Q = \text{constant} \times H^{1/2}$) and free vortex flow.

Thus, if H_d coincides with the onset of blackwater flow, then the hydraulic losses may be equated to the head difference as for normal pipe flow:

$$H_1 - H_2 = \left(k_1 + k_2 + k_3 + k_4 + \frac{\lambda L}{d}\right)\frac{V^2}{2g} \qquad (13.27)$$

where $k_1 - k_4$ are the loss coefficients for entry, first and second bends, and exit, respectively, and λ is the friction factor.

If the heads and coefficients are known, the corresponding velocity (or discharge) may be found. It must be emphasized that, due to the fact that the spillway is such a short pipe, the fully developed pipe flow condition is not attained and the estimates based on this equation must be regarded as rough approximations only.

For satisfactory siphon operation the pressure must nowhere approach the vapour pressure. It is therefore necessary to estimate the lowest pressure in the siphon. This will almost certainly occur in the flow around the first bend. Since the cross-section of a siphon is large compared with the length and the head, the velocity and pressure distribution will be decidedly non-uniform. As a first approximation, the flow around the bend may be assumed to resemble a free vortex. The characteristic velocity distribution is then given by the equation

$$uR = \text{constant}$$

For this flow pattern, the lowest pressures coincide with the smallest radius. Therefore, applying the Bernoulli equation,

$$H = \frac{p_c}{\rho g} + \frac{u_c^2}{2g}$$

(the subscript c refers to the spillway crest). Therefore

$$H - \frac{p_c}{\rho g} = h_c = \frac{u_c^2}{2g}$$

hence

$$u_c = \sqrt{2gh_c}$$

For the free vortex, $uR = \text{constant} = u_c R_c = \sqrt{2gh_c} R_c$, hence

$$u = \sqrt{2gh_c} R_c / R$$

Therefore the discharge through an elemental strip at radius R is

$$\delta Q = ub\, \delta R = \sqrt{2gh_c}(R_c/R) b\, \delta R$$

Thus the total discharge is

$$Q = \int_{R_c}^{R_o} \sqrt{2gh_c}(R_c/R) b\, dR = \sqrt{2gh_c} R_c b \ln(R_o/R_c) \qquad (13.28)$$

ln denoting the natural (base e) logarithm.

If the required discharge and the bend radii are known, then the corresponding value of h_c may be found. As a guide, h_c should not be more than 7 m of water below atmospheric pressure at sea level to avoid formation of vapour bubbles.

The free vortex flow pattern will be modified in practice, due to the developing boundary layer in the spillway. The existence of the shear layer may tend to increase the likelihood of cavitation, as may any irregularities in the spillway surfaces, so great care is needed both in design and construction. Because of the form of the spillway rating curves, the maximum discharge is usually assumed to coincide with the onset of blackwater flow. At flows higher than this, H increases more rapidly for a given increase in Q, with the consequent danger of overtopping the dam. Therefore, for reasons of safety, siphons are often used in conjunction with an auxiliary emergency spillway.

Design improvements to siphon spillways. It has been pointed out that uncontrolled surging can be a problem. Two methods of improving the operation of a siphon installation are (a) the use of multiple siphons with differential crest heights and (b) air regulation.

For method (a), a series of siphons are installed. One siphon has the lowest crest height, a second has a slightly higher crest and so on. The siphons thus prime sequentially as the reservoir level rises and de-prime sequentially as the level falls. The system gives a much smoother H–Q characteristic.

Method (b) involves a modification to the intake (Fig. 13.13(b)). The cowl is constructed with a carefully designed slot which admits controlled quantities of air when the head lies between the gravity flow and blackwater

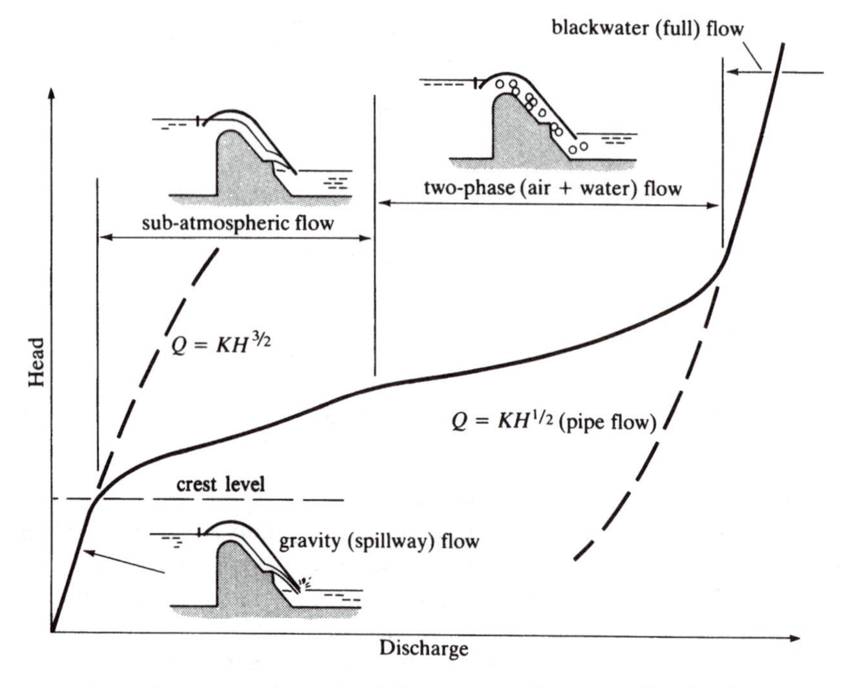

Figure 13.14 Discharge-head characteristic for air-regulated siphon.

values. The operation of such a design is therefore as follows for a rising reservoir level (Fig. 13.14):

(a) Gravity flow for low heads.
(b) As the reservoir level rises further, air trapped inside the duct tends to be entrained and the pressure falls below atmospheric pressure. However, some air is drawn in through the air-regulating slot.
(c) A slight further increase in head leads to the entrainment of air trapped under the crown. Air drawn in at the air slot now mixes with the water to form a fairly homogeneous two-phase mixture, which is foamy in appearance.
(d) Blackwater flow sets in when the reservoir level seals off the air slot.

If the air slot is well designed, a stable discharge of water can be maintained for conditions (b) and (c). Air regulation is thus a simple means of automatically matching the siphon discharge to the incoming discharge. Accurate prediction of the rate of air entrainment (and hence of the siphon discharge characteristic) is difficult. Model testing is a necessary prelude to design, though even tests do not give completely reliable results. Research is currently proceeding to clarify the reasons for the discrepancies. For an initial (and rough) estimate of entrainment, the equation of Renner may be used (see *Proceedings of the Symposium on the Design and Operation of Siphons and Siphon Spillways* (1975))

$$Q_{air}/Q = K\mathrm{Fr}^2 \tag{13.29}$$

The value of K appears to depend on the angle between the deflected nappe and the hood, but an average value is 0.002. The Froude Number refers to the flow at the toe of the deflector.

A large amount of information regarding design, modelling and operation of siphon spillways has been published. The reader may refer to the *Proceedings of the Symposium on the Design and Operation of Siphons and Siphon Spillways* (1975) and also to Head (1975), Ervine (1976), Ali and Pateman (1980) and Ervine and Oliver (1980).

Shaft (morning glory) spillways

This type of structure consists of four parts (Fig. 13.15(a)):

a circular weir at the entry;
a flared transition which conforms to the shape of a lower nappe for a sharp-crested weir;
the vertical drop shaft;
the horizontal (or gently sloping) outlet shaft.

The discharge control may be at one of three points depending on the head:

(a) When the head is low the discharge will be governed at the crest. This is analogous to weir flow and the discharge may therefore be expressed as

$$Q = CLH^{3/2}$$

where L is usually referred to the arc length at the crest. C is a function of H and the crest diameter, it is therefore not a constant. The magnitude of C is usually in the range 0.6–2.2, in SI units (1.1–4.0 in foot-sec units).

Below the crest, the flow will tend to cling to the wall of the transition and of the drop shaft. The outlet shaft will flow only partially full and is therefore, in effect, an open channel.

(b) As the head increases, so the annular nappe must increase in thickness. Eventually, the nappe expands to fill the section at the entry to the drop shaft. The discharge is now being controlled from this section, and this is often referred to as 'orifice control'. The outlet tunnel is not designed to run full at this discharge.

(c) Further increase in head will induce blackwater flow throughout the drop and outlet shafts. The Q–H relationship must now conform to that for full pipe flow, and the weir will in effect be 'submerged'. The head over the weir rises rapidly for a given increase in discharge, with a consequent danger of overtopping the dam.

$$Q = CLH^{3/2}$$

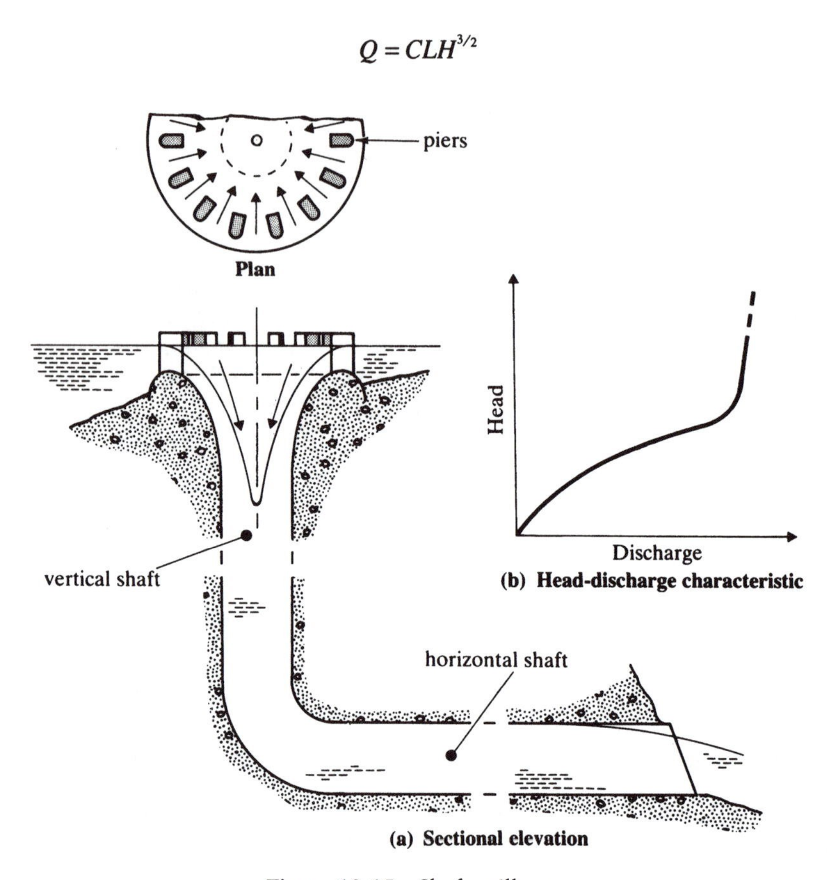

Figure 13.15 Shaft spillway.

The complete discharge–head characteristic is therefore as shown in Figure 13.15(b). The design head is usually less than the head required for blackwater flow. This is done to leave a margin of safety for exceptional floods. Even so, the discharge which can be passed by a shaft or siphon spillway is limited, so great care must be exercised at the design stage. An auxiliary emergency spillway may be necessary. It is also worth noting that as the flow enters the transition it tends to form a spiral vortex. The vortex pattern must be minimized in order to maintain a smoothly converging flow, so anti-vortex baffles or piers are often positioned around the crest. It may be undesirable for sub–atmospheric pressure to occur anywhere in the system, since this can lead to cavitation problems. To avoid such problems the system may (a) incorporate vents, (b) be designed with an outlet shaft which is large enough to ensure that the outlet end never flows full or (c) have an outlet shaft with a slight negative slope, sufficient to ensure that the outlet does not flow full (and can therefore admit air).

Some recent shaft spillways incorporate a bank of siphon spillways around the crest to reduce the rise in reservoir level for a given discharge. Trash grids must be installed around the entrance to a shaft spillway to exclude large items of debris.

13.6 Stepped spillways

Stepped spillways have found a number of applications in dams around the world. By forming the face of the spillway as a series of steps, energy is dissipated as the water flows down. The size and cost of any other energy dissipating structure (such as a stilling basin) at the foot of the spillway is therefore reduced. The flow over the steps may be in the form of a series of nappes (similar to flow over a sharp crested weir), or for larger discharges it may form a stream which 'skims' over the steps. The hydraulic behaviour and design of stepped spillways and channels is beyond the scope of a first text in hydraulics. However, readers who wish to pursue the subject further are referred to Chanson (1994).

13.7 Energy dissipators

The flow discharged from the spillway outlet is usually highly supercritical. If this flow were left uncontrolled, severe erosion at the toe of the dam would occur, especially where the stream bed is of silt or clay. Therefore, it is necessary to dissipate much of the energy, and to return the water to the normal (subcritical) depth appropriate to the stream below the dam. This is achieved by a dissipating or 'stilling' device. Typical devices of this nature are:

1. stilling basin;
2. submerged bucket; and
3. ski jump/deflector bucket.

These are now described (see Fig. 13.16).

The stilling basin

A stilling basin consists of a short, level apron at the foot of the spillway (Fig. 13.16(a)). It must be constructed of concrete to resist scour. It incorporates an integrally cast row of chute blocks at the inlet and an integral sill at the outlet. Some designs also utilize a row of baffle blocks part way along the apron.

The function of the basin is to decelerate the flow sufficiently to ensure the formation of a hydraulic jump within the basin. The jump dissipates much of the energy, and returns the flow to the subcritical state. The chute

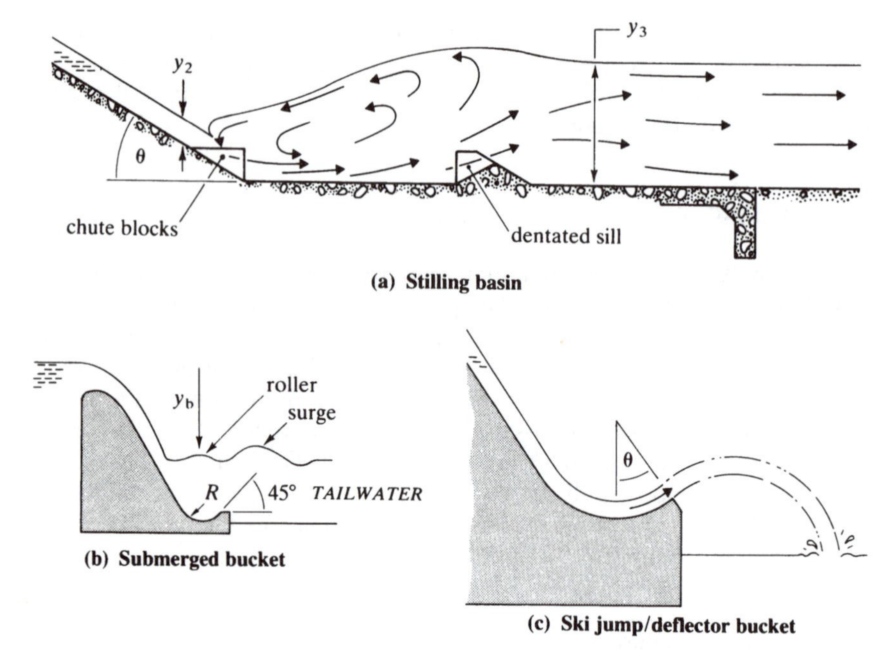

Figure 13.16 Energy dissipators.

blocks break the incoming flow into a series of jets, alternate jets being lifted from the floor as they pass over the tops of the blocks. The sill (or baffle blocks and sill) provide the resistance required to reduce energy and control the location of the jump. Baffle blocks are not usually used where the velocity of the incoming flow exceeds 20 m/s, due to the likelihood of cavitation damage.

Provided that the Froude Number of the incoming flow (Fr_2) exceeds 4.5, a stable jump can be formed. However, if $2.5 < Fr_2 < 4.5$, then the jump conditions can be less well defined and disturbances may be propagated downstream into the tailwater. Based on a combination of operational and empirical data, a number of standard stilling basin designs have been proposed. A typical set of designs were published by the US Bureau of Reclamation (1964).

US Bureau of Reclamation stilling basins

(a) Type II stilling basin (for $Fr_2 > 4.5$, $V_2 > 20\,\text{m/s}$), which has a length of $4.3y_3$;
(b) Type III stilling basin (for $Fr_2 > 4.5$, $V_2 < 20\,\text{m/s}$), which has a length of $2.7y_3$;
(c) Type IV stilling basin (for $2.5 < Fr_2 < 4.5$), which has a length of $6.1y_3$.

It is usually assumed that air entrainment makes little difference to the formation of the hydraulic jump in a stilling basin.

The stage–discharge characteristic of the tailwater is a function of the channel downstream of the dam. The stilling basin should therefore give a hydraulic jump with a downstream depth–discharge characteristic which matches that of the tailwater.

The submerged bucket

A submerged bucket is appropriate when the tailwater depth is too great for the formation of a hydraulic jump. The bucket is produced by continuing the radial arc at the foot of the spillway to provide a concave longitudinal section as shown in Figure 13.16(b). The incoming high velocity from the spillway is thus deflected upwards. The shear force generated between this flow and the tailwater leads to the formation of the 'roller' motions. The reverse roller may initially slightly scour the river bed downstream of the dam. However, the material is returned towards the toe of the dam, so the bed rapidly stabilizes.

The relationship between the depths for the incoming and tailwater flows cannot readily be derived from the momentum equation. Empirical relationships of an approximate nature may be used for initial estimates, e.g.

$$y_b \simeq 0.0664(h + E_{S1})\left(\frac{q}{g^{1/2}(h + E_{S1})^{3/2}}\right)^{0.455}$$

where h is the dam height (Fig. 13.10) and

$$\text{tailwater depth} \simeq 1.25 y_b$$

Some submerged buckets are 'slotted'. This improves energy dissipation, but may bring about excessive scour at high tailwater levels.

Detailed design data may be obtained from the appropriate publications of the US Bureau of Reclamation (1964, 1977).

The ski jump/deflector bucket

This type of dissipator has a longitudinal profile which resembles the submerged bucket (Fig. 13.16(c)). However the deflector is elevated above the tailwater level, so a jet of water is thrown clear of the dam and falls into the stream well clear of the toe of the dam. Spillways may be arranged in pairs, and it is then usual for the designer to angle the jets inwards so that they converge and collide in mid-air. This breaks up the jets, and is a very effective means of energy dissipation.

Example 13.5 Spillway and stilling basin design

Design a gravity spillway and stilling basin to pass a design flood of $2100\,\mathrm{m^3/s}$. Dam base elevation = 268 m, dam crest elevation = 300 m, reservoir design water level = 306 m and $C = 2.22\,\mathrm{m^{-1}/s}$.

Solution

$$2100 = 2.22bH_\mathrm{d}^{3/2}$$

For a first trial, assume $H_\mathrm{d} = 306 - 300 = 6\,\mathrm{m}$

$$2100 = 2.22b6^{3/2}$$

Therefore $b = 64.36$ – say 65 m. Applying the specific energy equation between the spillway crest (Station 1) and the foot (or toe) of the spillway (Station 2):

$$306 - 268 = 38 = \frac{2100^2}{2g \times 65^2 \times y_2^2} + y_2$$

so $y_2 = 1.2\,\mathrm{m}$. Therefore

$$V_2 = \frac{2100}{65 \times 1.2} = 26.9\,\mathrm{m/s}$$

and $\mathrm{Fr}_2 = 7.84$. A US Bureau of Reclamation Type II stilling basin is appropriate.
If y_3 is the depth downstream of the hydraulic jump, then

$$y_3/y_2 = \frac{1}{2}(\sqrt{1 + (8 \times 7.84^2)} - 1) = 10.6$$

therefore $y_3 = 12.72\,\mathrm{m}$. The length of the basin is $4.3 \times 12.72 = 54.7\,\mathrm{m}$ – say 55 m. The depth y_3 should be compared with the tailwater rating for this flow. It may be necessary to adjust the elevation of the spillway apron to meet tailwater requirements.

13.8 Control gates

Whenever a discharge has to be regulated, some form of variable aperture or valve is installed. These are available in a variety of forms. When gates are installed for channel or spillway regulation, they may be designed for ‘underflow’ or ‘overflow’ operation. The overflow gate is appropriate where logs or other debris must be able to pass through the control section. Some typical gate sections are shown in Figure 13.17.

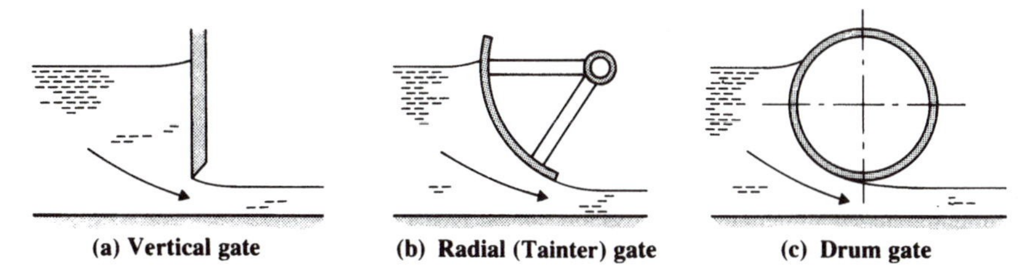

Figure 13.17 Control gates.

The choice of gate depends on the nature of the application. For example, the vertical gate (Fig. 13.17(a)) has to be supported by a pair of vertical guides. The gate often incorporates roller wheels on each vertical side, so that the gate moves as smoothly as possible in the guides. Even so, once a hydrostatic load is applied, a considerable force is needed to raise or lower the gate. Furthermore, in severe climates, icing may cause jamming of the rollers. The radial (Tainter) gate (Fig. 13.17(b)) consists of an arc-shaped face plate supported on braced radial arms. The whole structure rotates about the centre of arc on a horizontal shaft which transmits the hydrostatic load to the supporting structure. Since the vector of the resultant hydrostatic load passes through the shaft axis, no moment is applied. The hoist mechanism has therefore only to lift the mass of the gate. Tainter gates are economical to install, and are widely used in overflow and underflow formats. Other gates (such as the drum or the roller gate) are expensive, and have tended to fall out of use.

The hydraulic characteristics of gates are related to the energy equation, so for free discharge (Fig. 13.18)

$$y_1 + \frac{V_1^2}{2g} = y_2 + \frac{V_2^2}{2g}$$

which can be rearranged in terms of Q:

$$Q = by_1y_2\sqrt{\frac{2g}{y_1 + y_2}}$$

Now $y_2 = C_c y_G$. Furthermore, it is usual to express the velocity in terms of $\sqrt{2gy_1}$:

$$Q = bC_c y_G\sqrt{2gy_1\frac{y_1}{y_1 + y_2}} \tag{13.30}$$

or

$$Q = bC_d y_G\sqrt{2gy_1} \tag{13.31}$$

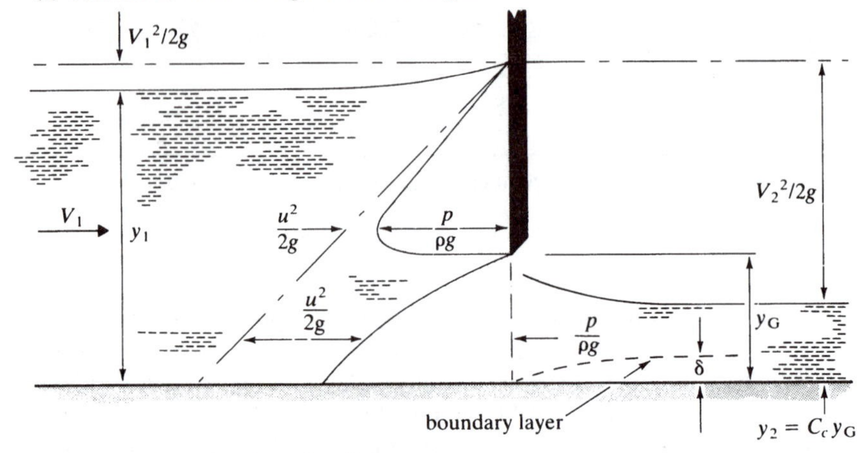

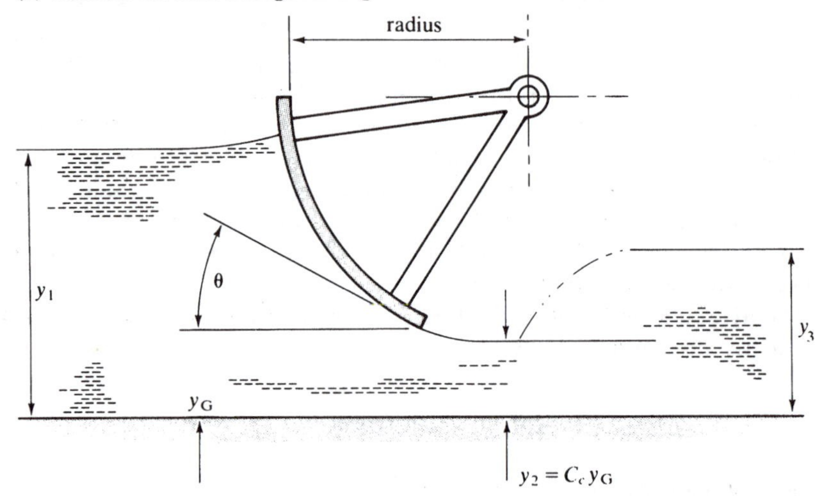

Figure 13.18 Flow past control gates.

where

$$C_d = \frac{C_c}{\sqrt{1 + (C_c y_G / y_1)}} \tag{13.32}$$

The quantity $\sqrt{2gy_1}$ should be regarded purely as a 'reference velocity', and not as an actual velocity at any point in the system.

Thus, the magnitude of C_d depends on the gate opening (y_G) and on the contraction of the jet, which in turn is a function of the gate geometry. For a vertical sluice gate under free discharge conditions, it has been found that $C_c = 0.61$ for $0 < (y_G/E_{S1}) < 0.5$. However, when a radial gate is raised or lowered, the lip angle is altered, and the value of C_c must alter correspondingly. An empirical formula for C_c has been evolved for free discharge through underflow radial gates, based on the work of a number of investigators:

$$C_c = 1 - 0.75\left(\frac{\theta}{90}\right) + 0.36\left(\frac{\theta}{90}\right)^2 \qquad (13.33)$$

where θ is measured in degrees.

Clearly, under certain conditions, the outflow from an underflow gate may be submerged (Fig. 13.19), i.e. when the downstream depth exceeds the conjugate depth to y_2. An approximate analysis may then be made as follows, assuming that all losses occur in the expanding flow downstream of the gate between Stations 2 and 3. From the energy equation,

$$y_1 + \frac{Q^2}{2gb^2y_1^2} = y + \frac{Q^2}{2gb^2y_2^2} \qquad (13.34)$$

Note that the hydrostatic term on the right-hand side is represented by the downstream depth y, not by y_2. Between Station 2 and Station 3, the momentum equation is applicable:

$$\frac{y^2}{2} + \frac{Q^2}{gb^2y_2} = \frac{y_3^2}{2} + \frac{Q^2}{gb^2y_3} \qquad (13.35)$$

In most practical situations y_1 and y_3 will be known, whereas y_2 will have to be estimated from known values of y_G and C_c.

Control gates are used for a range of applications, of which typical examples are regulation of irrigation flows and spillway flows.

Regulation of irrigation flows. It is worth pointing out that for free discharge, supercritical conditions may occur under and downstream of the gate, so a protective apron with a stilling arrangement may be required to protect the stream bed.

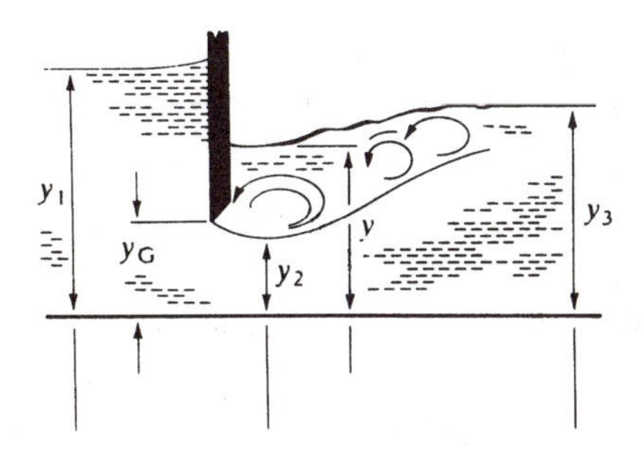

Figure 13.19 Submerged flow through an underflow sluice gate.

Spillway flows. While it is cheapest and simplest to have an unregulated spillway, it is not always the best system. This is the case where a reservoir is to be used for flood attenuation as well as storage. Consider a reservoir sited above low-lying farmland. During heavy rainstorms, the overflow down an unregulated spillway might add to flooding problems downstream. However, a system of control gates may be installed along the spillway crest. During the storm, the gates would be closed, which would increase the reservoir storage capacity and shut off the spillways. This additional storage would be available to accommodate all or part of the incoming flood flow. The excess capacity may then be released in a controlled fashion after the effects of the storm have subsided.

Example 13.6 Discharge control by sluice gate

An irrigation scheme is fed from a river by a diversion channel. The discharge into the system is controlled by an underflow vertical sluice gate. The irrigation channel is 4 m wide, and is roughly faced with cemented rubble giving an estimated value of 0.028 for Manning's n. The bed slope is 0.002. Estimate the required gate opening (y_G) and the flow condition downstream of the gate if (a) the depth of flow in the river is 2 m and the irrigation demand is $11\,m^3/s$ and (b) the depth of flow in the river is 3 m and irrigation demand has fallen to $5\,m^3/s$.

Solution

(a) $Q = 11\,m^3/s$ and y_1 (assume same as river depth) $= 2\,m$. Using Manning's equation, estimate $y_n = 1.8\,m = y_3$.

1st trial. Assume gate opening $y_G = 1.4\,m$, then $y_2 = C_c \times y_G = 0.61 \times 1.4 = 0.854\,m$. Hence

$$V_2 = 3.22\,m/s \text{ and } Fr_2 = \frac{3.22}{\sqrt{9.81 \times 0.854}} = 1.11$$

Check conjugate depth relationship:

$$\frac{y_3}{y_2} = \frac{y_3}{0.854} = \frac{1}{2}(\sqrt{1 + (8 \times 1.11^2)} - 1) = 1.147$$

Conjugate value for $y_3 = 0.98\,m$, so submerged conditions exist downstream of gate.

From the specific energy equation (13.34) (between Stations 1 and 2)

$$2 + \frac{11^2}{(4 \times 2)^2 \times 2g} = y + \frac{11^2}{(4 \times 0.854)^2 \times 2g}$$

Therefore $y = 1.568\,m$.

From the momentum equation (13.35) (between Stations 2 and 3)

$$\frac{1.568^2}{2} + \frac{11^2}{g \times 4^2 \times 0.854} = \frac{1.8^2}{2} + \frac{11^2}{g \times 4^2 \times 1.8}$$

$$2.132 = 2.048$$

2nd trial. Assume $y_G = 1.25$ m. Then

$$y_2 = C_c \times y_G = 0.61 \times 1.25 = 0.7625\,\text{m}$$

From the specific energy equation,

$$2 + \frac{11^2}{(4 \times 2)^2 \times 2g} = y + \frac{11^2}{(4 \times 0.7625)^2 \times 2g}$$

Therefore $y = 1.433$ m.

From the momentum equation,

$$\frac{1.433^2}{2} + \frac{11^2}{g \times 4^2 \times 0.7625} = \frac{1.8^2}{2} + \frac{11^2}{g \times 4^2 \times 1.8}$$
$$2.038 = 2.048$$

which is sufficiently accurate.

Therefore $y_G = 1.25$ m, and outflow is submerged.

(b) $Q = 5\,\text{m}^3/\text{s}$ and $y_1 = 3$ m. $Q = C_d b y_G \sqrt{2gy_1}$. Assume $C_d = 0.6$ for first approximation:

$$5 = 0.6 \times 4 \times y_G \sqrt{2g \times 3}$$

Therefore $y_G \simeq 0.27$ and

$$C_d = \frac{0.61}{\sqrt{1 + [(0.61 \times 0.27)/3]}} = 0.594$$

Hence,

$$Q = 0.594 \times 4 \times 0.27\sqrt{2g \times 3} = 4.92\,\text{m}^3/\text{s}$$

From Manning's formula, $y_n = 1.05\,\text{m} = y_3$.

Check whether discharge is free, using the conjugate depth relationship:

$$y_2 = C_c \times y_G = 0.61 \times 0.27 = 0.165\,\text{m}$$
$$\text{Fr}_2 = V_2/\sqrt{gy_2} = 5.86$$

Therefore

$$y_3/y_2 = \frac{1}{2}(\sqrt{1 + (8 \times 5.86^2)} - 1) = 7.8$$

so

$$y_3 = 7.8 \times 0.165 = 1.287\,\text{m}$$

which is greater than y_n, so free discharge must occur.

The hydraulic jump will be a short distance downstream of the gate. An adequate length of apron would be required to prevent scour, or a stilling arrangement could be used to control jump location.

13.9 Lateral discharge structures

Equations for lateral flow

There are a few circumstances when it is necessary to discharge water into or out of a stream (Fig. 13.20), typical instances being side weirs in sewers and side spillways. However, this presents the problem of a varying discharge in the stream. In turn, this implies that none of the equations encountered up to now is appropriate. The equations for such problems are usually based on the momentum principle, and the approach is now outlined.

The first case considered will be that of a lateral inflow of fluid (Fig. 13.20(a)). The incoming fluid is assumed to enter the channel in a direction perpendicular to the direction of flow of the stream. A force must therefore be applied to the incoming fluid to accelerate it. The total momentum in the stream must therefore be changed. The forces applied to the stream in the direction of flow are:

1. hydrostatic pressure force $= -\rho g A\,\delta y$ (where A is the cross-sectional area of the stream);
2. gravity force component $= \rho g A S_0\,\delta x$;
3. friction force $= \rho g A S_f\,\delta_x$.

The discharge, Q, in the stream will increase by δQ as it passes through a longitudinal element of length δx. Therefore, the rate of change of momentum will be

$$\begin{aligned}\frac{\delta M}{\delta t} &= \rho(Q+\delta Q)(u+\delta u) - \rho Q u \\ &= \rho Q\,\delta u + \rho u\,\delta Q\end{aligned}$$

(ignoring products of small quantities). Now $u = Q/A$ and $(u+\delta u) = (Q+\delta Q)/(A+\delta A)$, so

$$\delta u = \frac{A\,\delta Q - Q\,\delta A}{A^2}$$

and

$$\frac{\delta M}{\delta t} = \rho Q\left(\frac{2\,\delta Q}{A} - \frac{Q\,\delta A}{A^2}\right)$$

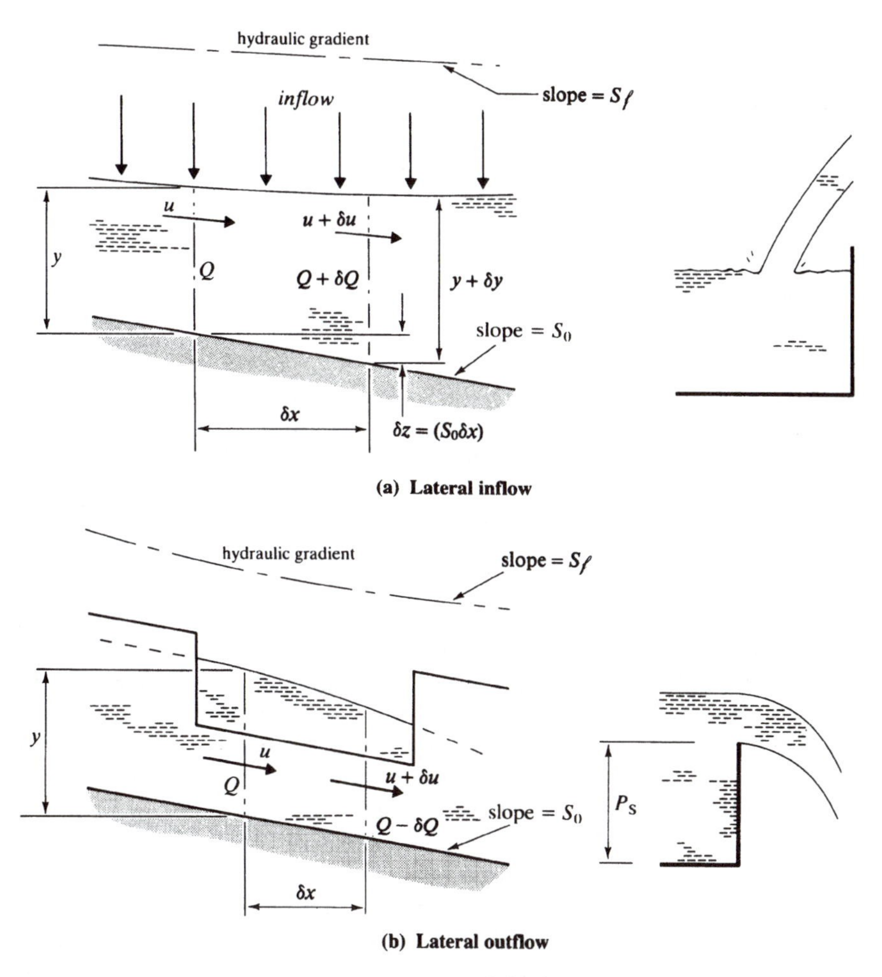

Figure 13.20 Lateral discharge.

Force equals the rate of change of momentum, so

$$-\rho gA\,\delta y + \rho gAS_0\,\delta x - \rho gAS_f\,\delta x = \rho Q\left(\frac{2\,\delta Q}{A} - \frac{Q\,\delta A}{A^2}\right)$$

Now, as $A = by$, $\delta A/A \simeq \delta y/y$. Substituting for δA in the momentum equation and dividing throughout by gA, taking the limit as $\delta x \to 0$, and rearranging,

$$\frac{\mathrm{d}y}{\mathrm{d}x} = \frac{S_0 - S_f - (2Q/gA^2)\mathrm{d}Q/\mathrm{d}x}{1 - (Q^2/gA^2y)} \tag{13.36}$$

The equation for the case of lateral outflow (Fig. 13.20(b)) may be developed in a similar manner. There is no loss of momentum in the fluid remaining in the stream, so the equation is

$$\frac{dy}{dx} = \frac{S_0 - S_f - (Q/gA^2)dQ/dx}{1 - (Q^2/gA^2y)} \tag{13.37}$$

The only difference between (13.36) and (13.37) is the coefficient of the third term of the numerator.

The reader will have noticed that the above equations bear some resemblance to the gradually varied flow equation (5.38).

If the value of the coefficient (β) departs significantly from unity, then the equation must be modified accordingly:

$$\frac{dy}{dx} = \frac{S_0 - S_f - (2Q/gA^2)dQ/dx(2\beta - 1)}{1 - (\beta Q^2/gA^2y)} \tag{13.38}$$

Producing a solution to either of these equations presents a number of problems. Both Q and y are variables. The control point is the channel section at the downstream end of the inflow or outflow structure, where the depth will be the normal depth of flow for the discharge at that point. The depth at the upstream end is initially unknown. One approach to such problems is a trial and error solution based on (13.36) or (13.37) in finite difference form (Example 13.7 illustrates how this is applied). An alternative method has been developed by Balmforth and Sarginson (1978). This makes use of a Runge–Kutta fourth-order procedure. Either of these methods may be used as the basis for a computer program.

No mention has been made of the flow profile in the channel at the weir. It is possible that the flow might, for example, be supercritical, or partly supercritical and partly subcritical with a hydraulic jump. A helpful discussion is given by Balmforth in Number 14 of the 'Wallingford Procedure' User Group Notes (see References at the end of Chapter 10). This highlights the need to include a check on the Froude Number of the flow in the channel.

Example 13.7 Side weir

A rectangular channel is 2.5 m wide, has a bed slope of 0.002 and Manning's n is 0.015. It is to incorporate a side weir. The design discharge in the channel upstream of the side weir is 5.96 m^3/s and the side weir is to draw off 0.5 m^3/s from the channel. The sill height is 0.84 m (measured above the channel bed) and the weir discharge coefficient, C_d, is 0.7. The energy and momentum coefficients are assumed to have a value of unity. Estimate the required length of the weir.

Solution

The solution procedure is as follows:

1. Calculate the normal depth of flow for the channel section at the downstream end of the weir. This is the only known depth, so it is easiest to start the calculation here and 'march' upstream.
2. Select a step length Δx.
3. The initial (downstream) value for $y(y_1)$ is known, and the value of $y(y_{i+1})$ at the upstream end of the step has to be guessed.
4. Compute the mean value of $y(\bar{y} = (y_i + y_{i+1})/2)$ for the step, hence compute the corresponding channel cross-section area and wetted perimeter $(\bar{A} = b\bar{y}, \bar{P} = b + 2\bar{y})$.
5. Compute the flow over the weir for the step using equation (13.2) $(\Delta Q = 2/3 C_d (2g)^{1/2} (\bar{y} - P_S)^{3/2} \Delta x)$. Hence calculate $Q_{i+1} (= Q_i + \Delta Q$, since the solution is proceeding upstream) and $\bar{Q} (= (Q_i + Q_{i+1})/2)$.
6. Using the Chezy–Manning equation suitably rearranged, calculate

$$S_f \left(= n^2 \bar{Q}^2 \bar{P}^{4/3} / \bar{A}^{10/3}\right)$$

7. Substitute for Q, A, y, S_f etc. in equation (13.37) and obtain y. Hence calculate $y_{i+1} (= y_i + \Delta y)$, compare the calculated value with the original (guessed) value and continue the iterative procedure until the values converge to a satisfactory degree of accuracy.
8. Proceed to next step.

Table 13.1 Solution to Example 13.7.

x_i	y_i	$\bar{y}$	$\bar{A}$	$\bar{P}$	S_f	ΔQ	Q_i
0.0	1.060						5.461
		1.053	2.630	4.604	0.00208	0.102	
−0.5	1.047						5.563
		1.041	2.601	4.581	0.00229	0.093	
−1.0	1.035						5.656
		1.030	2.575	4.560	0.00250	0.085	
−1.5	1.024						5.741
		1.019	2.548	4.548	0.00270	0.078	
−2.0	1.014						5.819
		1.010	2.525	4.520	0.00290	0.073	
−2.5	1.007						5.892
		1.004	2.51	4.508	0.00320	0.068	
−3.0	1.000						5.960
					Total weir discharge	0.499 m^3/s	

The solution to the problem has been undertaken by computer, and the results are summarized in Table 13.1. A step length of 0.5 m has been used. The result indicates that a weir length of approximately 3.0 m will draw off the required outflow.

Following on from the example, the difficulties involved in producing a solution to a practical problem may be envisaged. The incoming discharge and the side weir outflow are usually known, so the discharge in the channel downstream of the side weir can be found, and hence the corresponding normal depth calculated. However, in a real design problem, neither the length nor the height of the sill would be known (in the example the sill height has been given, to simplify matters). It would therefore be necessary to guess a sill height and produce a solution to (13.37). If the calculated outflow or the calculated weir dimensions did not conform to the client's specification, then the whole procedure would have to be repeated until an acceptable solution was found. The difficulties are compounded if a hydraulic jump is likely to occur at some point along the weir. Even if the calculation appears to indicate such an eventuality, it is not possible to estimate the flow over the weir, or the depths in the channel, with sufficient precision to predict the location of the jump with certainty. For the case considered in this example the solution is relatively straightforward as the flow is subcritical throughout. It will also be noted that the depths do not vary greatly, so in practice it would be feasible to produce an initial rough estimate of the outflow over the weir based on the assumption that the depth was constant.

13.10 Outlet structures

Outlet structures are designed to release controlled volumes of water from a reservoir into the water supply, irrigation or other system. Such structures may be conveniently divided into groups according to either the form of the structure or the function of the system. For example, a given installation could be described equally well as a 'gate controlled conduit' or as a 'river outlet', and so on. Such structures may also be used to lower the reservoir level to permit dam maintenance, etc. Some typical cases are as follows:

(a) For low heads, it is possible (and economical) to use an open channel with a control gate. The upstream end usually incorporates fish/trash grids (this is true for virtually all outlet works).
(b) For low to medium heads, a simple conduit may be used with a gate valve to control discharge. Velocities at the outlet end may be great enough to warrant a stilling system.
(c) For higher pressures, a more robust conduit is required (often steel lined). The conduit may consist of a horizontal shaft, or may incorporate

a drop shaft entry. If a drop shaft is used, it is economical to arrange two intakes (for the spillway and outlet systems) concentrically on one drop shaft. The spillway tunnel and outlet conduit are then run separately.

Where the outlet system feeds a hydroelectric or pumping scheme, it is imperative that air entrainment is avoided. Large air bubbles in the flow may cause severe damage to hydraulic machines.

13.11 Concluding remarks

A range of hydraulic structures has been considered. The range is by no means exhaustive. However, the various principles which have been applied will usually be capable of adaptation to other appropriate problems. It cannot be too strongly emphasized that our knowledge is still limited in many areas. Most of the solutions given above are approximate. In many instances, it will be important to commission model tests to confirm (or otherwise!) initial estimates.

References and further reading

Ackers, P., White, W. R., Perkins, J. A. and Harrison, A. J. M. (1978) *Weirs and Flumes for Flow Measurement*, Wiley, Chichester.

Ali, K. H. M. and Pateman, D. (1980) Theoretical and experimental investigation of air-regulated siphons. *Proc. Instn Civ. Engrs, Part* 2, **69**, 111–38.

Balmforth, D. J. and Sarginson, E. J. (1978) A comparison of methods of analysis of side weir flow. *Chartered Municipal Engr*, **105**, 273–9.

Bos, M. G. (ed.) (1978) *Discharge Measuring Structures*, 2nd edn, Int. Instn for Land Reclamation and Improvement, Wageningen.

Bradley, J. N. and Peterka, A. J. (1957) The hydraulic design of stilling basins. *Am. Soc. Civ. Engrs, J. Hydraulics Divn*, **83**(HY5), 1401–6.

Cain, P. and Wood, I. R. (1981) Measurements of self-aerated flow on a spillway. *Am. Soc. Civ. Engrs, J. Hydraulics Divn*, **107**(HY11), 1425–44.

Chanson, H. (1994) *Hydraulic Design of Stepped Cascades, Weirs and Spillways*, Pergamon, Oxford.

Chow, V. T. (1959) *Open Channel Hydraulics*, McGraw-Hill, New York.

Ellis, J. (1989) *Guide to Analysis of Open Channel Spillway flow*, Technical note 134. Construction Industries Research and Information Society (CIRIA), London.

Ervine, D. A. (1976) The design and modelling of air-regulated siphon spillways. *Proc. Instn Civ. Engrs, Part* 2, **61**, 383–400.

Ervine, D. A. and Oliver, G. S. C. (1980) The full scale behaviour of air regulated siphon spillways. *Proc. Instn. Civ. Engrs, Part* 2, **69**, 687–706.

Etheridge, M. J. (1996) The hydraulic analysis of side-channel spillways as reservoir outlets. *Water and Environmental Management*, **10**(4), 245–52.

Hager, W. H. (1991) Experiments on standard spillway flow. *Proc. Instn. Civ. Engrs, Part* 2, **91**, 399–416.

Head, C. R. (1975) Low-head air-regulated siphons. *Am. Soc. Civ. Engrs, J. Hydraulics Divn*, **101**(HY3), 329–45.
Henderson, F. M. (1966) *Open Channel Flow*, Macmillan, New York.
Keller, R. J. and Rastogi, A. K. (1975) Prediction of flow development on spillways. *Am. Soc. Civ. Engrs, J. Hydraulics Divn*, **101**(HY9), 1171–84.
Kenn, M. J. (1971) Protection of concrete from cavitation damage. *Proc. Instn. Civ. Engrs*. Tech. Note TN48 (May).
Kenn, M. J. and Garrod, A. D. (1981) Cavitation damage and the Tarbela tunnel collapse of 1974. *Proc. Instn. Civ. Engrs, Part 1*, **70**, 65–89.
Lee, W. and Hooper, J. A. (1996) Prediction of cavitation damage for spillways. *Am. Soc. Civ. Engrs. J. Hydraulic Engng.*, **122**(9), 481–8.
Novak, P., Moffat, A. I. B., Nalluri, C. and Narayanan, R. (1996) *Hydraulic Structures*, 2nd edn, E & FN Spon, London.
Proceedings of the Symposium on the Design and Operation of Siphons and Siphon Spillways (1975). British Hydromechanics Research Assoc., London.
Rao, N. S. L. (1971) Self aerated flow characteristics in wall region. *Am. Soc. Civ. Engrs, J. Hydraulics Divn*, **97**(HY9), 1285–303.
US Army Waterways Experimental Station (1959) *Hydraulic Design Criteria*. US Dept of the Army, Washington, DC.
US Bureau of Reclamation (1964) *Hydraulic Design of Spillways and Energy Dissipators*, US Dept of the Interior, Washington, DC.
US Bureau of Reclamation (1967) *Water Measurement Manual*, 2nd edn, US Dept of the Interior, Washington, DC.
US Bureau of Reclamation (1977) *Design of Small Dams*, US Dept of the Interior, Washington, DC.
Vischer, D. L. and Hager, W. H. (1997) *Dam Hydraulics*, Wiley, Chichester.

14

Computational hydraulics

14.1 Overview

Many real processes in hydraulics involve continuous variations of conditions (river or tidal flow, for example). Continuously varying processes may be represented in various ways (by scale models or electrical circuits, for example); however, because of the worldwide availability of microcomputers, computational hydraulics has assumed an increasing role in research and design. Indeed, computational hydraulics may be seen as the culmination of centuries of development in the study of fluid flows.

Computational hydraulics has been an area of rapid growth since the early 1960s, and a history of the process of development of the software and hardware may be found in Abbott (1991).

The earliest computer programs were, in essence, no more than computational versions of calculations that had previously been performed manually. The programs had to run on machines with limited memory (often less than 10 kB), slow processing speed, and cumbersome input/output devices. It was principally the demands of engineers involved in aeronautics and the space programmes that led to the developments in microelectronics, and the associated progress towards compact, reliable computer hardware. In parallel with this, software engineers were quick to respond with more sophisticated programs, so that by the early 1980s it was possible to produce good simulations of a range of real-world problems in hydraulics. Subsequent developments have been driven by market demand for user-friendly packages that can be purchased off the shelf and operated by the staff of the purchasing organization with a minimum of training. Even so, it is important that an engineer using such software should be aware of the assumptions that underly it, the methods used, and the possible shortcomings.

14.2 Mathematical models and numerical models

A computer is simply a tool for carrying out arithmetical operations at high speed, so in order to produce a computational model of a flow the problem has to be formulated in suitable terms.

The starting point is a mathematical model – a set of equations that mathematically describe the behaviour of the fluid (the equations are normally partial differential or integral equations). For many practical situations analytical solutions for the equations are not available, so an alternative solution method is required. The numerical method substitutes an algebraic form of the original equations that can be solved using simple arithmetical procedures to yield the numbers that represent the flow (such as the mean velocity, or the depth of flow) at discrete points in time and space.

The basic steps involved in setting up a numerical model of a system may be summarized as follows.

1. Define the nature of the problem.
2. Reduce the problem to some suitable mathematical form (the governing equations that form the mathematical model).
3. Make all possible simplifications consistent with adequate modelling (this implies some clear thinking about the relative importance of each aspect of the problem).
4. Replace the simplified governing equations with finite difference or other finite systems of equations.
5. Set up a representation of the domain (normally some form of 'grid') in time and space.
6. Define the boundaries of the domain and the conditions (of flow etc.) at those boundaries (the 'boundary conditions').

A numerical model is the basis for a computer program that enables a computer to undertake the repetitive calculations necessary for producing solutions to the numerical model for many points in time and space. The program encodes the exact form and order of the arithmetical operations and the input and output routines. Once the program is compiled and working it must be calibrated. Good field data are needed for this, so that the required input data can be derived. Adjustments are then made to various parameters or subroutines in the program, until the output represents the behaviour of the real system to a satisfactory standard of accuracy. The calibrated model should be verified by means of a second (independent) data set to check that it behaves correctly. The model is then ready for its task of simulating the given physical process.

Three basic approaches are widely used to discretize and solve fluid flow problems: the finite difference, finite element and finite volume methods. Because of constraints in the length of this chapter, only the finite difference method will be covered here.

As an introductory step it is necessary to revisit the basic principles governing fluid flows (conservation of mass, momentum and energy), and to form differential equations. Corresponding numerical forms of the equations may then be derived. The solutions to a number of hydraulics problems can then be investigated, and the advantages and disadvantages of computational modelling can be considered.

14.3 Derivation of conservation equations

Figure 14.1(a) shows a small volume 'cut out' from a flowing liquid. Velocities may be changing in magnitude and direction, so they are resolved into x and y components. To produce an acceleration corresponding forces must exist. Three types of force are shown in Figure 14.1(b) (only x-direction forces are shown, for simplicity): a pressure p perpendicular to a surface; a stress σ perpendicular to a surface; and a shear stress τ parallel to a surface.

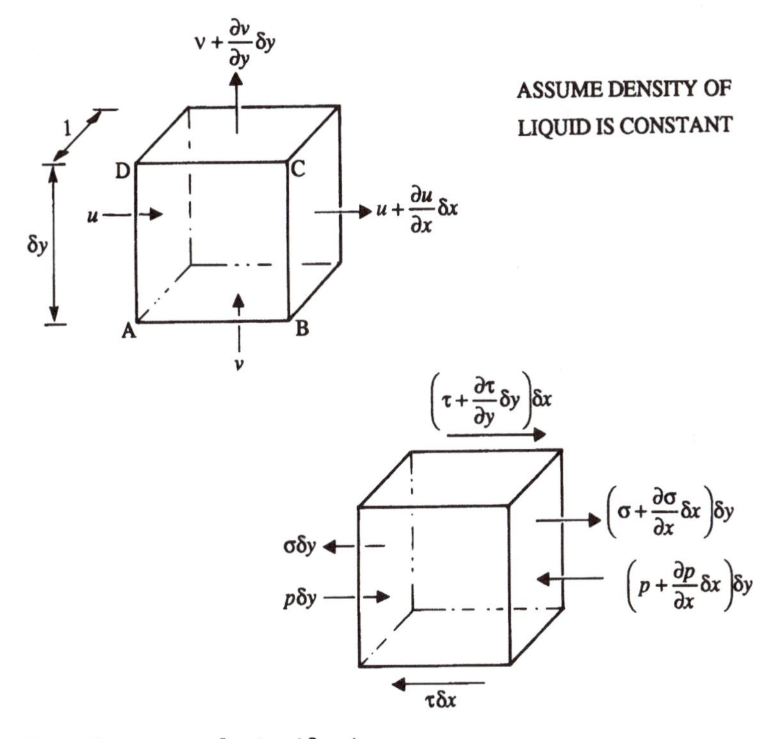

Figure 14.1 Derivation of conservation equations.

Continuity equation

Assuming that density is constant, flows into the volume must be balanced by outflows. Adopting the sign convention that inflow is negative, outflow positive:

$$-\upsilon\delta x - u\delta y + \left(\upsilon + \frac{\partial \upsilon}{\partial y}\delta y\right)\delta x + \left(u + \frac{\partial u}{\partial x}\delta x\right)\delta y = 0$$

With some rearrangement:

$$\left(\frac{\partial u}{\partial x} + \frac{\partial \upsilon}{\partial y}\right)\delta x \delta y = 0$$

Clearly it is the *sum* that must $\rightarrow 0$. That is:

$$\frac{\partial u}{\partial x} + \frac{\partial \upsilon}{\partial y} = 0 \tag{14.1}$$

This form of the continuity equation is not always the most convenient for practical use, because it may not be possible to determine the exact form of the functional relationship $u = f(x)$, $\upsilon = f(y)$. It will be seen later that the continuity equation is often integrated and depth averaged.

Equations for the acceleration of the fluid

We need expressions for the *total* change in velocities with time and distance. Applying the chain rule for partial derivatives (in the x direction of flow),

$$\delta u = \frac{\partial u}{\partial x}\delta x + \frac{\partial u}{\partial y}\delta y + \frac{\partial u}{\partial t}\delta t$$

Dividing by δt,

$$\frac{\delta u}{\delta t} = \frac{\partial u}{\partial x}\frac{\delta x}{\delta t} + \frac{\partial u}{\partial y}\frac{\delta y}{\delta t} + \frac{\partial u}{\partial t} \quad \left(\text{note that } \frac{\delta x}{\delta t} = u \text{ and } \frac{\delta y}{\delta t} = \upsilon\right)$$

Hence

$$\frac{\delta u}{\delta t} = \underbrace{u\frac{\partial u}{\partial x} + \upsilon\frac{\partial u}{\partial y}}_{\substack{\text{Advective} \\ \text{acceleration}}} + \underbrace{\frac{\partial u}{\partial t}}_{\substack{\text{Point} \\ \text{acceleration}}}$$

Total acceleration is an acceleration at a point *plus* an additional acceleration as a particle advects (moves). If acceleration is multiplied by the mass of the small volume, we obtain the rate of change of momentum for the x direction:

$$\frac{\partial M_x}{\partial t} = \rho \delta x \delta y \left(u\frac{\partial u}{\partial x} + v\frac{\partial u}{\partial y} + \frac{\partial u}{\partial t} \right) \tag{14.2}$$

Forces and momentum

Taking the x direction, there are four sets of forces acting on the element:

(a) normal stress due to the pressure of the surrounding fluid:

$$p\delta y - \left(p + \frac{\partial p}{\partial x}\delta x \right)\delta y = -\frac{\partial p}{\partial x}\delta x \delta y \tag{14.3}$$

(b) normal stresses due to velocity gradients in the fluid in the x direction:

$$\left(\sigma_{xx} + \frac{\partial \sigma_{xx}}{\partial x}\delta x \right)\delta y - \sigma_{xx}\delta y = \frac{\partial \sigma_{xx}}{\partial x}\delta x \delta y \tag{14.4a}$$

(c) shear stress due to velocity gradients in the fluid in the y direction:

$$\left(\tau_{yx} + \frac{\partial \tau_{yx}}{\partial y}\delta y \right)\delta x - \tau_{yx}\delta x = \frac{\partial \tau_{yx}}{\partial y}\delta x \delta y \tag{14.5a}$$

(d) body force (due to gravity, Coriolis force etc.) acting on the element:

$$X\,\delta x\,\delta y \tag{14.6}$$

Note that the subscripts refer to the direction of the stress: for example, τ_{yx} is in the x direction and acts on a plane perpendicular to the y axis.

Navier–Stokes equation for laminar flow

From section 3.2 τ is proportional to μ and to the rate of shear strain, for a laminar flow. The normal stress σ_{xx}, being related to a velocity gradient, is also a function of μ. In 1845, the mathematician G. Stokes showed that

$$\sigma_{xx} = \lambda_\mu \left(\frac{\partial u}{\partial x} + \frac{\partial v}{\partial y} \right) + 2\mu\frac{\partial u}{\partial x} \tag{14.4b}$$

(in which λ_μ is usually taken as equal to $-2\mu/3$) and

$$\tau_{xy} = \mu\left(\frac{\partial v}{\partial x} + \frac{\partial u}{\partial y}\right) \tag{14.5b}$$

Combining equations (14.2)–(14.5b) yields the Navier–Stokes equation for the x-direction:

$$\rho\frac{\partial u}{\partial t} + \rho u\frac{\partial u}{\partial x} + \rho v\frac{\partial u}{\partial y} = -\frac{\partial p}{\partial x} + \frac{\partial}{\partial x}\cdot\mu\left(2\frac{\partial u}{\partial x} - \tfrac{2}{3}\left(\frac{\partial u}{\partial x} + \frac{\partial v}{\partial y}\right)\right) + \frac{\partial}{\partial y}\cdot\mu\left(\frac{\partial v}{\partial x} + \frac{\partial u}{\partial y}\right) + X \tag{14.7a}$$

with a corresponding equation for the y-direction.

Continuity and Navier–Stokes equations for turbulent flows

The outline of turbulence already given (see section 3.4) indicates the difficulties in producing mathematical models of turbulent flows. The (unsteady) point velocity ($u = f(t)$) is represented as the sum of mean and fluctuating components ($u = \overline{u} + u'$). Since the time average of $u'(=\overline{u'}) = 0$, the continuity equation for turbulent flow is

$$\frac{\partial \overline{u}}{\partial x} + \frac{\partial \overline{v}}{\partial y} = 0 \tag{14.8}$$

However, in the momentum equation the additional turbulent shear stresses must be incorporated (see equation (3.3), for example). Following a similar process to that used to derive equation (14.7a), the x-momentum Navier–Stokes equation becomes

$$\rho\frac{\partial \overline{u}}{\partial t} + \rho\overline{u}\frac{\partial \overline{u}}{\partial x} + \rho\overline{v}\frac{\partial \overline{u}}{\partial y} = -\frac{\partial p}{\partial x} + \mu\left(\frac{\partial^2 u}{\partial x^2} + \frac{\partial^2 u}{\partial y^2}\right) - \rho\left(\frac{\partial \overline{u'}^2}{\partial x} + \frac{\partial \overline{u'v'}}{\partial y}\right) + X \tag{14.7b}$$

For a complete derivation see Schlichting *et al.* (1999).

Treatment of the turbulent stresses

Turbulent stresses are very complex, and no completely satisfactory treatment exists. However, it is often adequate to use 'equivalent' forms. For example:

$$\tau_T = \mu_T\frac{\partial u}{\partial y}$$

(where μ_T is often known as the 'eddy viscosity'. The magnitude of μ_T for a given flow has to be evaluated, for example, by the $k-\varepsilon$ turbulence model (see section 3.4)). The following substitutions can therefore be made for the turbulent shear terms:

$$\rho\frac{\partial\overline{u'^2}}{\partial x} \to \frac{\partial}{\partial x}\left(\mu_T\frac{\partial\overline{u}}{\partial x}\right) \quad \rho\frac{\partial\overline{u'v'}}{\partial y} \to \frac{\partial}{\partial y}\left(\mu_T\frac{\partial\overline{u}}{\partial y}\right)$$

So equation (14.7b) can be written as

$$\rho\frac{\partial\overline{u}}{\partial t} + \rho\overline{u}\frac{\partial\overline{u}}{\partial x} + \rho\overline{v}\frac{\partial\overline{u}}{\partial y} = -\frac{\partial p}{\partial x} + \frac{\partial}{\partial x}\left((\mu+\mu_T)\frac{\partial\overline{u}}{\partial x}\right) + \frac{\partial}{\partial y}\left((\mu+\mu_T)\frac{\partial\overline{u}}{\partial y}\right) + X \tag{14.7c}$$

which is the form of the Navier–Stokes equations most often encountered. *Note*: This is for the x direction, so there is another equation for the y direction if the flow is a two-dimensional flow:

$$\rho\frac{\partial\overline{v}}{\partial t}\ldots \text{etc.}$$

Also, if the only significant body force is gravity then $X = Z = 0$ and $Y = -g = \text{constant}$.

14.4 Differential equations and finite difference schemes

Some important partial differential equations

An important family of equations that is often encountered in hydraulics is based on the following equation:

$$a\frac{\partial^2 f}{\partial x^2} + b\frac{\partial^2 f}{\partial y\partial x} + c\frac{\partial^2 f}{\partial y^2} = 0 \tag{14.9}$$

(f is some variable such as velocity or the concentration of a substance).

This can take three forms depending on the value of b^2-4ac. If $b^2-4ac > 0$ then a typical form is

$$\frac{\partial^2 f}{\partial x^2} - c^2\frac{\partial^2 f}{\partial y^2} = 0 \tag{14.10}$$

This is a **hyperbolic** equation, which can be applied to unsteady flows.

If $b^2-4ac = 0$ then typically

$$\frac{\partial f}{\partial x} = c\frac{\partial^2 f}{\partial y^2} \tag{14.11}$$

This is a **parabolic** equation, which is encountered in problems involving diffusion or decay of the magnitude of a property (such as heat or concentration).

Finally if $b^2 - 4ac < 0$

$$\frac{\partial^2 f}{\partial x^2} + \frac{\partial^2 f}{\partial y^2} = 0 \tag{14.12}$$

which is the Laplace (**elliptic**) equation and may be used for equilibrium flows.

Discretization of differential equations

Because analytical solutions are not available for many of the equations we wish to solve, numerical solutions are used. A numerical solution gives numbers that represent the behaviour of a variable, but can provide such numbers only for certain discrete locations in the space–time domain. The algebraic expressions that are used to form the numerical model are based on finite differences. Finite difference equations are approximations to the equivalent differential equations. Most finite difference expressions are based on Taylor series expansions.

To illustrate this take the equation $y = f(x) = x^3 + 3x^2 + 5x + 3$. We wish to evaluate the change to $f(x)$ due to increasing x from $x = 2$ to $x = 2.1$ (that is, $\Delta x = 0.1$). By substituting $x = 2$ we find that $f(x) = 33$, and for $x = 2.1$, $f(x) = 35.991$. Hence the increase in $f(x)$ is 2.991. Now we use a finite difference method to perform the same calculation (see Fig. 14.2). First, the Taylor expansion is taken:

$$f(x + \Delta x) = f(x) + \frac{\partial f(x)}{\partial x}\Delta x + \frac{\partial^2 f(x)}{\partial x^2}\frac{\Delta x^2}{2!} + \ldots + \frac{\partial^n f(x)}{\partial x^x}\frac{\Delta x^n}{n!} \tag{14.13}$$

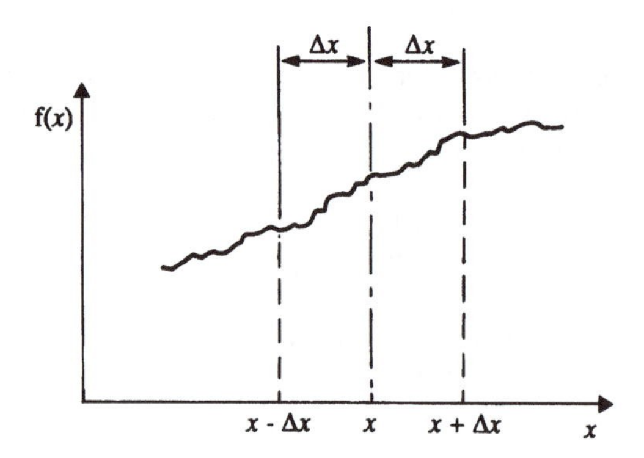

Figure 14.2 Variation of $f(x)$.

There are $n+1$ terms on the right-hand side of this expression. The simplest calculation would simply use the first term only: that is, $f(x+\Delta x)=f(x)$. This would give the result that $f(x+\Delta x)=33$. We already know that this cannot be correct. The percentage error is $(35.991-33)\times 100/35.991 = 8.31\%$.

Now try using the first *two* terms on the right-hand side of (14.13):

$$f(x+\Delta x)=f(x)+\frac{\partial f(x)}{\partial x}\Delta x$$

where

$$\frac{\partial f(x)}{\partial x}\Delta x=(3x^2+6x+5)\Delta x$$

Substituting $x=2$, $\Delta x=0.1$ yields a value of 2.9 for the second term: hence $f(x+\Delta x)=33+2.9=35.9$. This compares with the actual value of 35.991, so the percentage error is $(35.991-35.9)\times 100/35.991 = 0.25\%$.

The magnitude of the error arises from the number of terms that were omitted, so it is known as a **truncation error**. If we rearrange our Taylor expansion we obtain the following:

$$\underset{\substack{\text{partial}\\ \text{differentiation}}}{\frac{\partial f(x)}{\partial x}}=\underset{\substack{\text{finite}\\ \text{difference}}}{\frac{f(x+\Delta x)-f(x)}{\Delta x}}-\underset{\text{truncation error}}{\frac{\partial^2 f(x)}{\partial x^2}\frac{\Delta x^2}{2!}+\text{higher-order terms}} \qquad (14.14)$$

This shows how the partial differential term is being approximated. As the finite difference is obtained by subtracting $f(x)$ from $f(x+\Delta x)$ this is known as a **forward difference.**

If we wish to evaluate $f(x)$ for a reduction in x (from x to $x-\Delta x$) the corresponding Taylor expansion is

$$f(x-\Delta x)=f(x)-\frac{\partial f(x)}{\partial x}\Delta x+\frac{\partial^2 f(x)}{\partial x^2}\frac{\Delta x^2}{2!}+\text{higher-order terms} \qquad (14.15)$$

from which

$$\frac{\partial f(x)}{\partial x}=\underset{\substack{\text{finite}\\ \text{difference}}}{\frac{f(x)-f(x-\Delta x)}{\Delta x}}+\underset{\text{truncation error}}{\frac{\partial^2 f(x)\Delta x}{2!}+\text{higher-order terms}} \qquad (14.16)$$

which is a **backward difference.**

If (14.14) and (14.16) are added then

$$\frac{\partial f(x)}{\partial x} = \frac{f(x+\Delta x) - f(x-\Delta x)}{2\Delta x} + \text{truncation error} \tag{14.17}$$

If (14.16) is subtracted from (14.14):

$$\frac{\partial^2 f(x)}{\partial x^2} = \frac{f(x+\Delta x) - 2f(x) + f(x-\Delta x)}{\Delta x^2} + \text{truncation error} \tag{14.18}$$

(14.17) and (14.18) are known as **central differences.**

Discretization 'grids'. It is helpful to use a grid as a visual aid to illustrate the way in which the computation proceeds from node to node (Fig. 14.3). Locations on the x axis are indicated by subscript i, and on the y axis by subscript j. Superscript n relates to the time. Differences may be forward, backward or central as discussed above.

For example, a change in depth of water, h, with distance x at a given time t could be expressed as

$$\frac{\mathrm{d}h_i}{\mathrm{d}x} \simeq \frac{h^n_{i+1} - h^n_{i-1}}{2\Delta x} \quad \text{(central difference)} \tag{14.19}$$

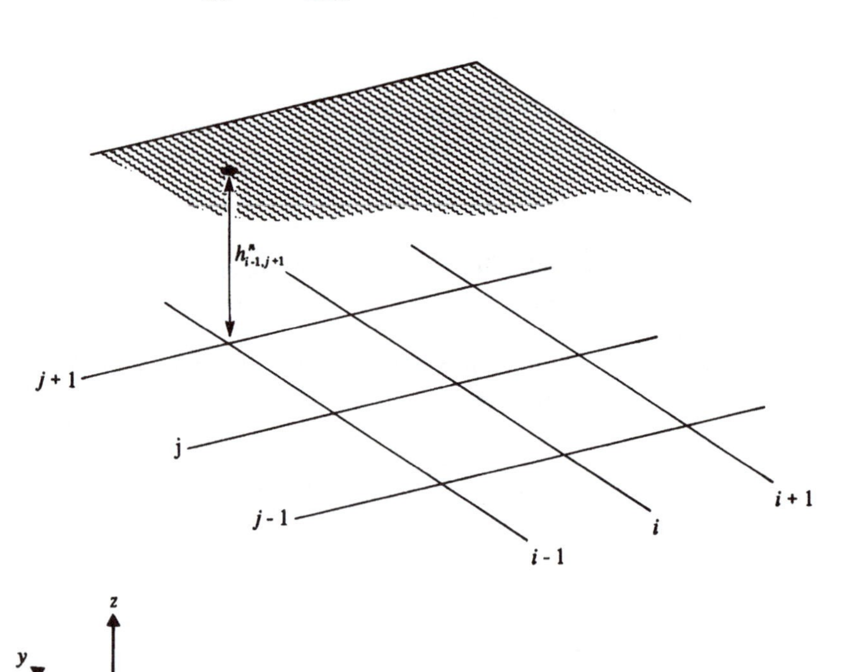

Figure 14.3 Use of grid to express location of a variable on x, y axes (time is a 'fourth' dimension).

$$\frac{\mathrm{d}h_i}{\mathrm{d}x} \simeq \frac{(h_{i+1} - h_i^n)}{\Delta x} \quad \text{(forward difference)} \tag{14.20}$$

etc.

In some cases a 'staggered grid' is used where (say) discharge is computed at $i, i+1$ etc. but depth is computed at $(i - 1/2)$, $(i + 1/2)$ etc. (that is, Δx is the same for both variables, but one grid is offset by '1/2').

Explicit and implicit solutions. Take the parabolic equation (14.11). A simple finite difference scheme may be set up, based on a forward difference for the left-hand side, and a central difference for the right-hand side. Thus if f is decaying with time, (14.11) may be written

$$\frac{\partial f}{\partial t} = c\frac{\partial^2 f}{\partial x^2}$$

To form the second-order differential take

$$\frac{\partial^2 f}{\partial x^2} = \frac{\partial(\partial f)}{\partial x} = \frac{1}{\Delta x}\left(\frac{f_1^n - f_i^n}{\Delta x} - \frac{f_i^n - f_{i-1}}{\Delta x}\right) = \frac{(f_{i+1}^n - 2f_i^n + f_{i-1}^n)}{\Delta x^2}$$

(omitting higher-order terms).

Hence the finite difference form of (14.11) is

$$\frac{f_i^{n+1} - f_i^n}{\Delta t} = c\frac{f_{i+1}^n - 2f_i^n + f_{i-1}^n}{\Delta x^2} \tag{14.21}$$

and so the solution for f_i^{n+1} is

$$f_i^{n+1} = f_i^n + c\frac{\Delta t}{\Delta x^2}(f_{i+1}^n - 2f_i^n + f_{i-1}^n) \tag{14.22}$$

This expresses the required (unknown) quantity f_i^{n+1} in terms of known quantities, and this is an example of an **explicit** solution method.

Now consider taking the solution for the spatial properties of function f based on the average of the values for the n and $(n+1)$ time levels. That is, taking the right-hand side of (14.21), f_{i+1}^n now becomes

$$f_{i+1}^{n+1/2} = \frac{f_{i+1}^{n+1} + f_{i+1}^n}{2}$$

f_i^n now becomes

$$\frac{f_i^{n+1} + f_i^n}{2}$$

f_{i-1}^n now becomes

$$f_{i-1}^{n+1/2} = \frac{f_{i-1}^{n+1} + f_{i-1}^n}{2}$$

Hence we could now write

$$\frac{\partial f}{\partial t} = c\frac{[(f_{i+1}^{n+1} + f_{i+1}^n)/2] - [f_i^{n+1} + f_i^n] + [(f_{i-1}^{n+1} + f_{i-1}^n)/2]}{\Delta x^2}$$

To solve this finite difference equation we need to know values of the function at two time levels. The values of f at time level n are known, but not those values at time level $(n+1)$. This form of finite difference equation is an example of the **implicit** approach. The solution method employed for this approach requires the simultaneous solution of the finite difference equations for all x grid points. An example of this approach is given later in this chapter.

Requirements and behaviour of finite difference schemes

There are a number of important points to be made about the nature and effects of using finite differences to represent differential equations. A satisfactory scheme must be consistent, convergent and stable. This may be summarized in two statements:

> A well-posed problem is one in which the solution to a partial differential equation exists, is unique, and is continuously dependent on the initial and boundary conditions.

and

> Given a well-posed initial value problem and a finite difference approximation to it that satisfies the consistency condition, stability is the necessary and sufficient condition for convergence (the Lax theorem).

The issues covered by these statements are of considerable importance, and are also quite complex. While it is impossible to cover them comprehensively here, it is important that they are understood.

Convergence. Convergence means that the numerical solution to a finite difference equation approaches the analytical solution as Δx, Δt etc. $\to 0$.

Consistency. Consistency means that the finite difference operators approach the differential equation as Δx, Δt etc. $\to 0$. Thus if

$$G = \frac{\partial}{\partial t} + \frac{\partial}{\partial x}$$

then

$$G(u) = \frac{\partial u}{\partial t} + \frac{\partial u}{\partial x}$$

and

$$G(u)_i = \frac{u_i^{n+1} - u_i^n}{\Delta t} + \frac{u_{i+1}^n - u_{i-1}^n}{2\Delta x} = 0$$

A Taylor series expansion for the finite difference equation yields

$$u_i^{n+1} = u_i^n + \frac{\partial u_i^n}{\partial t}\Delta t + \frac{\partial^2 u_i^n}{\partial t^2}\frac{\Delta t^2}{2} + \ldots + \text{etc.}$$

Hence

$$G(u)_i = \frac{\partial u_i^n}{\partial t} + \frac{\partial^2 u_i^n}{\partial t^2}\frac{\Delta t}{2} + \frac{\partial u_i^n}{\partial x} + \frac{\partial^3 u_i^n}{\partial x^3}\frac{\Delta x^2}{6} + \ldots + (\mathrm{O}\Delta t^2, \Delta x^4)$$

Therefore

$$|G(u)_i - G(u)| = \frac{\partial^2 u_i^n}{\partial t^2}\frac{\Delta t}{2} + \frac{\partial^3 u_i^n}{\partial x^3}\frac{\Delta x^2}{6} + \ldots + (\mathrm{O}\Delta t, \Delta x^2)$$

Therefore

$$|G(u)_i - G(u)| \rightarrow 0 \quad \text{when } \Delta t, \Delta x \rightarrow 0$$

The finite difference is therefore consistent with the differential equation. Note that O means 'of the order'.

Stability. It has already been shown that the representation of a partial differential equation by a finite difference expression may imply the existence of truncation error. In addition, computers have only a finite accuracy, so the computed solution will be taken only to a finite number of significant figures: this means that there will be a **rounding error**. The total error (e_r) is therefore the difference between the actual computer solution and the analytical solution of the partial differential equation. For the computer solution to be stable it is necessary to ensure that $|e_{ri}^{n+1}/e_{ri}^n| \leqslant 1$. If e_r grows larger as the computation progresses then the solution is unstable.

Suppose the differential equation involves some function of x. If e_r is plotted against x we shall often find that e_r varies in a fairly random fashion. If this is the case, then e_r can be represented by a Fourier series. For an unsteady flow, e_r will be a function of x and t, and its corresponding Fourier series representation can be written

$$e_\mathrm{r}(x, t) = \sum \mathrm{e}^{at}\,\mathrm{e}^{ikx}$$

where a is a constant and k is the wave number. The ratio

$$\left| \frac{e_{ri}^{n+1}}{e_{ri}^{n}} \right| = \text{amplification factor } A_{\mathrm{F}}$$

If $A_{\mathrm{F}} < 1$ then the solution will be stable. If $A_{\mathrm{F}} > 1$ it will not.

With some mathematical manipulations we can produce the **stability criterion** for a given finite difference equation. For example, the equation

$$\frac{\partial u}{\partial t} + c\frac{\partial u}{\partial x} = 0$$

may be expressed in explicit finite difference form, for which (for a given frequency)

$$A_{\mathrm{F}} = \mathrm{e}^{at} = \cos(k\,\Delta x) - i\ \mathrm{Cr}\sin(k\,\Delta x)$$

for $|\mathrm{e}^{at}| \leqslant 1$ it can be shown that $\mathrm{Cr} \leqslant 1$ where $\mathrm{Cr} = c\Delta t/\Delta x$. Cr is known as the **Courant number**, and it will be encountered again further on.

For a similar analysis of stability for parabolic equations see Anderson (1995).

Numerical dissipation. Consider again the hyperbolic equation that could be represented by the finite difference equation

$$\frac{f_i^{n+1} - f_i^n}{\Delta t} = c\frac{f_i^n - f_{i-1}^n}{\Delta x} = 0 \qquad (14.23)$$

Anderson (1995) shows that if we apply a Taylor expansion to f_i^{n+1} and f_i^n then (14.23) actually becomes

$$\frac{\partial f}{\partial t} - c\frac{\partial f}{\partial x} + c\frac{\Delta x}{2}(\nu - 1)\frac{\partial^2 f}{\partial x^2} + c\frac{(\Delta x)^2}{6}(2\nu^2 - 3\nu - 1)\frac{\partial^3 f}{\partial x^3}$$
$$+ \mathrm{O}[\Delta t^3, \Delta t^2 \Delta x, \Delta t \Delta x^2, \Delta x^3] = 0 \qquad (14.24)$$

The coefficient of the even-ordered derivative (that is, the third term in the equation) produces a form of behaviour analogous to that of the viscous terms in the Navier–Stokes equations. However, this effect has absolutely nothing to do with the actual viscosity of the fluid; it is purely a characteristic of the numerical scheme.

Numerical dispersion. In equation (14.24) there is an odd-ordered derivative (the fourth term). The coefficient produces an effect that is analogous to a dispersion. This shows up as an additional spurious oscillation superimposed on the solution.

Stability, dissipation and dispersion. The areas of numerical stability, dissipation and dispersion are extremely important aspects of the behaviour of numerical models. It has been possible to give only a very short summary here. Readers are referred to Anderson (1995) or Abbott and Basco (1989) for a more comprehensive and detailed treatment.

14.5 Boundary conditions and initial conditions

Boundary conditions

The conservation equations are general equations that, in discretized form, may be applied to any fluid flow. How then may these equations be made to provide specific numerical solutions for so many different types of flow? The answer to this lies in **boundary conditions**, which define the system limits. For example, a reservoir at the upstream end of a pipeline will imply a constant head at the entry to the pipe, which is the upstream boundary. As we shall see, there are two aspects of boundary conditions: one is the identification of the physical boundaries (such as reservoir, valve, weir); the other is the proper numerical specification of the behaviour of the fluid at the boundaries. The boundary data may be defined by:

1. the value of some function (such as the reservoir head) – this is known as a Dirichlet boundary condition;
2. the value of the derivative of some function (such as rate of change of discharge during closure of a valve in a pipeline) – this is known as the Neumann boundary condition.

Sometimes a combination of Dirichlet and Neumann conditions is required.

Initial conditions

Initial conditions are simply numbers representing the conditions of flow throughout a system at some initial time ($n = 0$). These have to be included in thc input data so that the calculation can start.

14.6 Applications of computational hydraulics

A number of problems in hydraulics will now be approached from a computational modelling viewpoint. In order to help readers who may have little or no background in this area of computing, several of the worked examples include full arithmetical calculations for the first few steps in the solution.

Gradually varying head

Problems in this category of interest to civil engineers include the time required to draw down or fill a reservoir.

Example 14.1 Outflow from a tank

A tank measures 5 m × 5 m in plan, and has a rectangular thin-plate weir, width $b = 200$ mm. If the initial head of water over the weir is 75 mm, how long will it take for the water to drain down to a head of 25 mm over the weir? Take $C_d = 0.65$ (see Fig. 14.4).

Solution

The problem is essentially an application of the continuity equation in modified form. Before proceeding, remember that the continuity equation is a general equation, which is 'tailored' to this particular problem by means of the boundary conditions and initial conditions.

The initial conditions are $h = 75$ mm and the corresponding flow over the weir.

The boundary conditions are at each of the six boundaries throughout the period during which the tank is draining down. These are:

1. the base of the tank, which is assumed impermeable and hence the boundary condition is that discharge through the base is zero (if relevant, the pressure is the hydrostatic pressure corresponding to the depth);
2. the side walls: three of these are also impermeable (discharge across these boundaries is zero), but the fourth wall contains the weir, so the boundary condition here is the discharge over the weir corresponding to h (there will also be a hydrostatic pressure force on each wall);

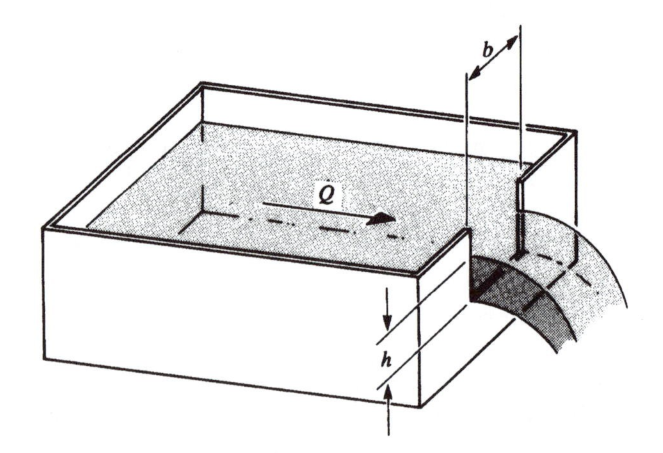

Figure 14.4 Tank with weir.

3. sixth boundary: the water surface, which is assumed to be horizontal, and can therefore be represented by the elevation at a single point.

Note that the discharge across the weir boundary will vary as the tank drains down, so this boundary condition is part of the computation.

The continuity equation for the tank may be written as $-A\delta h = Q\delta t$, where Q is the weir discharge at time t and A is the plan area of the tank ($= 5 \times 5 = 25\,\text{m}^2$). Hence

$$\frac{-A\delta h}{Q} = \delta t$$

In the limit this becomes

$$-\frac{A\text{d}h}{Q} = \text{d}t \tag{14.25}$$

Therefore

$$-\int_{0.075}^{0.025} \frac{A\text{d}h}{Q} = t$$

Since $Q = \frac{2}{3}C_d b\sqrt{2g}h^{3/2} = 0.384h^{3/2}$, an analytical solution can be found to the integral:

$$t = -\int_{0.075}^{0.025} \frac{A\text{d}h}{Q} = -\int_{0.075}^{0.025} \frac{25\,\text{d}h}{0.384h^{3/2}} = 348.1\,\text{s}$$

Now the problem is approached in terms of a finite difference solution. Rearranging (14.25) and taking $\frac{2}{3} \times \sqrt{2g} = 2.954$,

$$\frac{\Delta h}{\Delta t} = \frac{Q}{A} = \frac{2.954C_d bh^{3/2}}{A} \tag{14.26}$$

A decision has to be made about the value of h to be used to compute the flow over the weir for each step in the computation. One possibility is to use the head at the start of each time interval; however, this would give the maximum flow over the weir for that time interval. A better way is to use the mean value of h for the interval (that is, $(h^{n+1} + h^n)/2)$).

Hence the finite difference form of the differential equation is

$$\frac{h^{n+1} - h^n}{\Delta t} = -\frac{2.954C_d b[(h^{n+1} + h^n)/2]^{3/2}}{A}$$

or, in terms of Δt,

$$\Delta t = \frac{A(h^n - h^{n+1})}{2.954 C_d b[(h^n + h^{n+1})/2]^{3/2}}$$

which is the basis for the numerical scheme. To solve this a decision has to be made about the magnitude of the increment in h, from which we find the corresponding value of Δt. The total time for draining down is $\Sigma \Delta t$. Thc scheme may be encoded as a program, with input using 'data' statements, as given below.

Readers who are familiar with numerical methods may note that (14.26) may be stated in a modified form by taking $h = (0.075 - y)$, so that

$$\frac{\mathrm{d}t}{\mathrm{d}y} = \frac{A(0.075 - y)^{-3/2}}{2.954 C_d b} = f(y)$$

and

$$\Delta t = \Delta y f\left(y^n + \frac{\Delta y}{2}\right)$$

This is the standard form for a second-order Runge–Kutta solution. Analysis shows that the error for this method is of order $(\Delta y)^2$, as can easily be demonstrated by plotting $\Sigma(\Delta t)$ against $(\Delta h)^2$.

The program listing is as follows (in BASIC):

```
10  REM PROGRAM TO COMPUTE DRAIN-DOWN TIME FOR TANK
20  REM L,B ARE LENGTH AND WIDTH OF TANK, BW=WEIR WIDTH,
    CD=DISCHARGE COEFF
30  REM NI=NUMBER OF INCREMENTAL STEPS
40  DIM H(60), T(60)
100 READ L, B, BW, CD, NI, H(0), H(NI).
110 DELTAH=(H(0)-H(NI))/NI
120 T(0)=0
200 FOR N=0 TO (NI-1)
210 H(N+1)=H(N)-DELTAH
215 HMEAN=(H(N)+H(N+1))/2
220 DELTAT=(H(N)-H(N+1))*L*B/(2.954*CD*BW*(HMEAN^1.5))
230 T(N+1)=T(N)+DELTAT
240 NEXT N
250 LPRINT "TIME TO DRAIN DOWN TANK FROM" H(0)"m TO "H(NI)"m IS
    "T(N+1)"sec"
260 DATA 5,5,0.2,0.65,4,0.075,0.025
270 END
```

Notice that there are NI decrements in the head over the weir, and that the discharge for each decrement is computed for the mean head (HMEAN).

Results

To test the behaviour of the numerical scheme it has been run for different values of NI (50, 20, 10, 5), which yields the following:

NI	$\Delta H (= h^n - h^{n+1})$ (m)	Time to drain (s)
50	0.001	348.1
20	0.0025	347.9
10	0.005	347.2
5	0.01	344.5

The model is stable, consistent and convergent (note the growth of the error e_r as NI reduces and ΔH increases).

Surge tower (incompressible surge in pipelines)

The basic equations of flow for the surge tower have been developed in Chapter 12 (see equations (12.6), (12.7) and (12.8)).

A numerical description of the motion in the surge tower may be obtained by solving (12.6b) and (12.7) simultaneously. A finite difference method is often used, and for this purpose (12.7) is rearranged slightly. The flow in the pipeline may be positive (that is, in the original direction) or negative (in the reverse direction). The total resistance will always oppose motion, and accordingly the sign of each resistance term must be negative (for a 'positive' flow) or positive (for a 'negative' flow). To achieve this it is usual to write '$\pm V|V|$' instead of 'V^2'. Thus the friction loss term becomes $\pm K_f V|V|/2g$, where $|V|$ is the modulus (that is, the positive numerical value) while V may be positive or negative, depending on flow direction. Equation (12.7) now becomes

$$\Delta V = \frac{g}{L}\left(\mp z_{ST} \pm \frac{K_f V|V|}{2g} \pm \frac{K_L V|V|}{2g}\right)\Delta t \tag{14.27}$$

and (12.6b) becomes

$$\Delta z_{ST} = \frac{VA\Delta t}{A_{ST}} \tag{14.28}$$

where $K_f = \lambda L/D$ and K_L is the local loss coefficient.

Hence an explicit forward difference form may be derived (with $\Delta V = V^{n+1} - V^n$ etc.) as follows:

$$V^{n+1} = V^n + \frac{g}{L}\left[\mp z_{ST}^n \pm K_f \frac{V^n|V^n|}{2g} \pm K_L \frac{V^n|V^n|}{2g}\right]\Delta t \tag{14.29}$$

$$z_{ST}^{n+1} = z_{ST}^n + \frac{V^n A \Delta t}{A_{ST}} \tag{14.30}$$

Provided that V^n, z_{ST}^n are known for a given pipeline, these equations may be evaluated to yield V^{n+1}, z_{ST}^{n+1} for a series of increments of time. The magnitude of Δt will influence the accuracy of the solution. For good accuracy $\Delta t \ll$ 'periodic time'. While it is possible to perform such calculations on a simple hand calculator, the use of a programmable calculator or microcomputer is advisable for economy of time. The explicit technique outlined above is relatively simple to understand and to program, and thus forms a useful introduction for the newcomer. Other techniques (such as those based on the Runge–Kutta method) have been applied; also, a solution can be produced in one of the mathematics packages available for use on PC's.

A worked example follows to illustrate the technique. Three steps have been worked by hand calculator in order to show the method; however, the complete calculations are by computer.

Example 14.2 Surge tower

A hydroelectric scheme is served by a tunnel 3 m in diameter and 1.5 km long, with a reservoir at its upstream end. A surge tower of diameter 10 m is sited at the downstream end of the tunnel at the entry to the penstocks. The local loss coefficient for the entry to the surge tower is 0.785 (referred to flow velocity in the tunnel). The initial discharge is $32\,\text{m}^3/\text{s}$. The friction factor for the tunnel is 0.04. Estimate the peak upsurge height, and the periodic time, for a sudden and complete shutdown.

Solution

For $Q = 32\,\text{m}^3/\text{s}$, $V_0 = 4.53\,\text{m/s}$ and $h_{f0} = 20.92\,\text{m}$
$K_f = \lambda L/D = 20 \quad A = 7.068\,\text{m}^2 \quad A_{ST} = 78.54\,\text{m}^2$

The boundary conditions for this problem are:

1. the upstream boundary, which is the reservoir for which $h =$ constant (the datum level is taken at the top water level);
2. the downstream boundary, which is the control valve, situated just downstream of the surge tower, for which $Q = 32\,\text{m}^3/\text{s}$ before shutdown, $Q = 0$ after shutdown;
3. the water surface level in the surge tower (at which pressure is atmospheric).

The value for Δt must be selected. Its magnitude will affect the accuracy of the computations. To reinforce this point two sets of computations have been performed, one with $\Delta t = 2$ s and the other with $\Delta t = 50$ s. The first three steps of the solution (for $\Delta t = 2$ s) are given below.

Step 1 (2s after closure). The data from $t = 0$ are used to predict conditions at $\Delta t = 2$ s. From (14.29),

$$V^{n+1} = 4.53 + \frac{9.81}{1500}\left[-20.92 + \frac{20 \times 4.53|4.53|}{2 \times 9.81} + \frac{0.785 \times 4.53|4.53|}{2 \times 9.81}\right] \times 2$$
$$= 4.52\,\text{m/s}$$

and from (14.30),

$$z_{\text{ST}}^{n+1} = -20.92 + \left(4.53 \times \frac{7.068}{78.54}\right) \times 2 = -20.105\,\text{m}$$

Step 2 ($t = 4$ s)

$$V^{n+2} = 4.52 + \frac{9.81}{1500}\left[-20.105 + \frac{20 \times 4.52|4.52|}{2 \times 9.81} + \frac{0.785 \times 4.52|4.52|}{2 \times 9.81}\right] \times 2$$
$$= 4.5\,\text{m/s}$$
$$z_{\text{ST}}^{n+2} = -20.105 + \left(4.52 \times \frac{7.068}{78.54}\right) \times 2 = -19.291\,\text{m}$$

Step 3 ($t = 6$ s)

$$V^{n+3} = 4.50 + \frac{9.81}{1500}\left[-19.291 + \frac{20 \times 4.50|4.50|}{2 \times 9.81} + \frac{0.785 \times 4.50|4.50|}{2 \times 9.81}\right] \times 2$$
$$= 4.472\,\text{m/s}$$
$$z_{\text{ST}}^{n+3} = -19.291 + \left(4.50 \times \frac{7.068}{78.54}\right) \times 2 = -18.481\,\text{m}$$

The output file from the computer program has been transferred to a spreadsheet for plotting as shown (Fig. 14.5). The type of differential equation used would lead one to expect a damped sinusoidal curve to result from the calculations. This is so for $\Delta t = 2$ s. However, the points representing the results for $\Delta t = 50$ s are significantly different. Clearly a problem of instability has been produced, because Δt is 'large' compared with the periodic time of approximately 260 s. The errors generated at each successive step are clearly too great, and the results are useless. If a value $\Delta t > 50$ s were

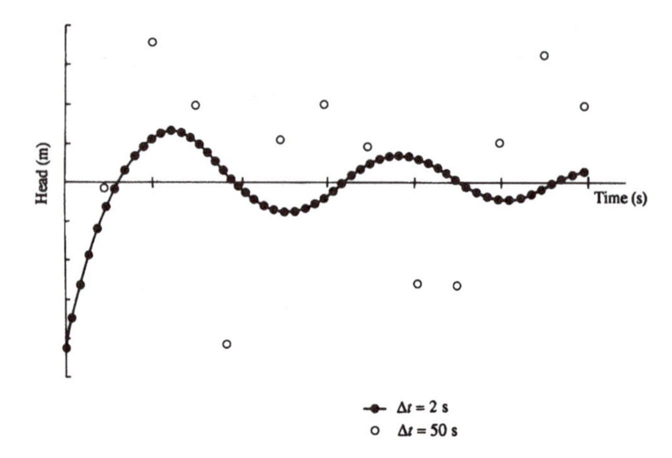

Figure 14.5 Surge tower: graph of head versus time.

used the computation would terminate because of the generation of very large numbers (for example, because of division by zero at a stage in the computation).

Unsteady flows in rivers

There are a number of open channel problems involving unsteady flows: that is, flows in which the velocity and depth of flow vary with time (examples are the passage of a flood wave along a river, or a tidal flow in an estuary). Few of these problems can be solved analytically, so they have to be solved numerically.

To illustrate the way in which a model is developed, and the assumptions and simplifications that have to be made, the case of a flood wave will be considered. For simplicity we shall assume that the river has a uniform rectangular cross-section of width B throughout its length.

The momentum (Navier–Stokes) equation for a turbulent flow in the x (longitudinal) direction is

$$\rho\frac{\partial\overline{u}}{\partial t}+\rho\overline{u}\frac{\partial\overline{u}}{\partial x}+\rho\overline{v}\frac{\partial\overline{u}}{\partial y}=-\frac{\partial p}{\partial x}+\frac{\partial}{\partial x}\left((\mu+\mu_{\mathrm{T}})\frac{\partial\overline{u}}{\partial x}\right)+\frac{\partial}{\partial y}\left((\mu+\mu_{\mathrm{T}})\frac{\partial\overline{u}}{\partial y}\right) \quad (14.31)$$

(This is the same as equation (14.7c), but omitting the body force.)

First, the viscous shear stress terms are assumed to be small compared with the turbulent shear terms. Second, it is often possible to ignore effects of vertical variations and simply to take 'depth averaged' values, which are found by integration between the bed (at height y_0) and the water surface (at height y_s) and then dividing by the depth $h = (y_s - y_0)$ thus:

$$\frac{\rho}{h}\int_{y_0}^{y_s}\left(\frac{\partial \overline{u}}{\partial t} + \overline{u}\frac{\partial \overline{u}}{\partial x} + \overline{v}\frac{\partial \overline{u}}{\partial y}\right)\mathrm{d}y = \frac{1}{h}\int_{y_0}^{y_s} -\frac{\partial p}{\partial x}\,\mathrm{d}y$$
$$+\frac{\rho}{h}\int_{y_0}^{y_s}\left(\frac{\partial}{\partial x}\mu_T\frac{\partial \overline{u}}{\partial x} + \frac{\partial}{\partial y}\mu_T\frac{\partial u}{\partial y}\right)\mathrm{d}y \quad (14.32)$$

noting that:

1. $\int u\,\mathrm{d}y = q$ (discharge per unit width) and $q/h = U$ (mean velocity);
2. $p = \rho g(y_s - y)$, therefore

$$\frac{1}{h}\int_{y_0}^{y_s}\frac{\partial p}{\partial x}\,\mathrm{d}y = \rho g\left(\frac{\partial y_s}{\partial x} - \frac{\partial y_0}{\partial x}\right) = \rho g(S_s - S_0)$$

   where S_0 bed slope and S_s water surface slope
   or

$$\frac{\mathrm{d}p}{\mathrm{d}y} = \rho g$$

3. the turbulence is assumed to be homogeneous, and shear stresses on vertical planes may be ignored, therefore the last term on the right-hand side may be simplified to

$$\frac{\rho}{h}\int_{y_0}^{y_s}\frac{\mathrm{d}}{\mathrm{d}y}\mu_T\frac{\mathrm{d}\overline{u}}{\mathrm{d}y}\,\mathrm{d}y = \frac{\tau_s - \tau_0}{h}$$

The shear stress (τ_s) at the water surface is usually neglected. Recalling (from Chapter 5, equation (5.5) that

$$\tau_0 = \rho g R S_f \quad \frac{\tau_0}{R} = \rho g S_f$$

Equation (14.32) in depth-averaged form and divided by (ρg) becomes

$$\frac{1}{g}\left(\frac{\partial U}{\partial t} + U\frac{\partial U}{\partial x} + V\frac{\partial U}{\partial y}\right) + (S_s + S_f - S_0) = 0$$

In the above equation all the terms are equivalent to gradients. The equation can be further simplified if the channel is regarded as being straight and of uniform rectangular cross-section. The channel flow can then be treated as a simple one-dimensional flow, and average values for U, S_s and S_f can be taken. Also, V (the vertical velocity term) is assumed to be zero. Hence

$$\frac{1}{g}\left(\frac{\partial U}{\partial t} + U\frac{\partial U}{\partial x}\right) + S_s + S_f - S_0 = 0 \quad (14.33a)$$

Where the channel cross-section is more complex (for example, where there is a main channel and flood plains) it is often better to use discharge rather than velocity. Equation (14.33a) then becomes

$$\frac{1}{gA}\left[\frac{\partial Q}{\partial t}+\frac{\partial}{\partial x}\left(\frac{\beta Q}{A}\right)\right]+S_s+S_f-S_0=0 \tag{14.33b}$$

The corresponding depth-integrated form for the continuity equation is based on the general concept that the difference between inflow and outflow causes any change in water surface level:

$$\frac{\partial h}{\partial t}+h\frac{\partial u}{\partial x}+u\frac{\partial h}{\partial x}=0=\frac{\partial h}{\partial t}+\frac{\partial(uh)}{\partial x} \tag{14.34a}$$

Again, this may be expressed in terms of Q and A:

$$\frac{\partial Q}{\partial x}+\frac{\partial A}{\partial t}=0 \tag{14.34b}$$

The simplest approach to computational modelling of the unsteady flow in a river is to base the solution on the triangular array of grid points shown in Figure 14.6. At $t=0$ all conditions are known. At $t=n$, there might be a *small* change in boundary conditions at $x=0$ (because of the increasing flow resulting from the rainfall), but conditions at all other points might be unchanged. At $t=n+1$, there is a further small change in the boundary conditions. The effect of prior changes to conditions at $t=n+1, x=i$

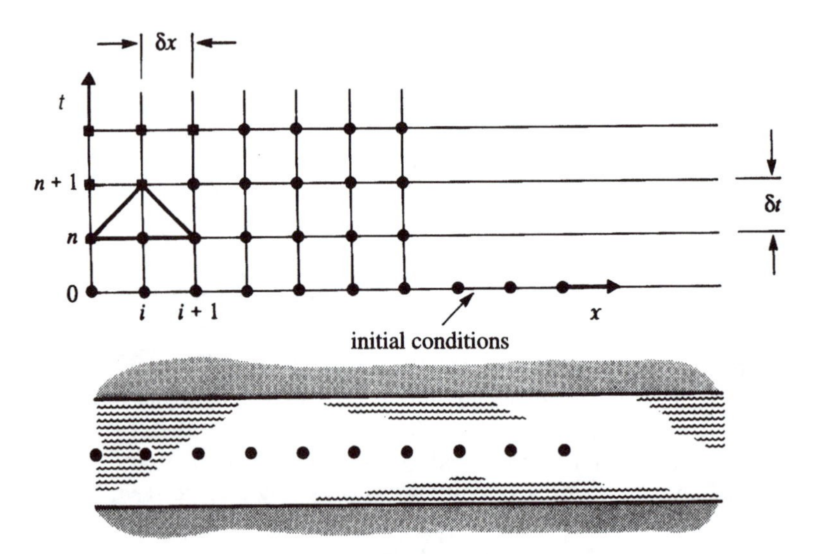

Figure 14.6 Triangular array of grid points.

etc. must be found. Taking known conditions at $t = n$ and at $x = 0, i, i+1$ we wish to compute (unknown) conditions at $t = n+1, x = i$. Taking equation (14.34a), the derivatives may be approximated:

$$\frac{\partial h}{\partial t} = \frac{h_i^{n+1} - h_i^n}{\Delta t}$$

$$h\frac{\partial U}{\partial x} = h_i^n \frac{U_{i+1}^n - U_{i-1}^n}{2\Delta x}$$

and

$$U\frac{\partial h}{\partial x} = U_i^n \frac{h_{i+1}^n - h_{i-1}^n}{2\Delta x}$$

Hence equation (14.34a) may be written

$$\left(\frac{h_i^{n+1} - h_i^n}{\Delta t}\right) + h_i^n \left(\frac{U_{i+1}^n - U_{i-1}^n}{2\Delta x}\right) + U_i^n \left(\frac{h_{i+1}^n - h_{i-1}^n}{2\Delta x}\right) = 0$$

Hence

$$h_i^{n+1} = h_i^n + \frac{1}{2}\frac{\Delta t}{\Delta x}\left(h_i^n \left(U_{i-1}^n - U_{i+1}^n\right) + U_i^n \left(h_{i-1}^n - h_{i+1}^n\right)\right) \tag{14.35}$$

If the Chezy–Manning equation is rearranged:

$$S_f = U_i^{n+1}|U_i^{n+1}|n^2/(R_i^{n+1})^{4/3}$$

Then in finite difference form equation (14.33a) becomes

$$\frac{1}{g}\left[\left(\frac{U_i^{n+1} - U_i^n}{\Delta t}\right) + U_i^n \left(\frac{U_{i+1}^n - U_{i-1}^n}{2\Delta x}\right)\right] + S_s - S_0 + U_i^{n+1}|U_i^{n+1}|n^2/(R_i^{n+1})^{4/3} = 0 \tag{14.36}$$

For convenience define

$$K_A = (R_i^{n+1})^{4/3}/n^2 g\Delta t$$

$$K_B = K_A\left[\frac{U_i^n \Delta t}{2\Delta x}(U_{i+1}^n - U_{i-1}^n) + g\Delta t S_s - g\Delta t S_0 - U_i^n\right]$$

Then equation (14.36) becomes

$$U_i^{n+1}|U_i^{n+1}| + K_A U_i^{n+1} + K_B$$

This is a quadratic expression, from which

$$U_i^{n+1} = \left[-K_A + \left(K_A^2 - 4K_B\right)^{1/2}\right]/2 \tag{14.37}$$

Equations (14.35) and (14.37) are explicit equations, from which the (unknown) depth and velocity at time step $n+1$ may be evaluated.

This type of scheme is relatively simple to derive and to program; unfortunately, its application is not so simple. There are two major drawbacks:

1. The scheme is extremely susceptible to instability (if energy dissipation and diffusion terms are omitted, the scheme is *unconditionally* unstable!)
2. The downstream boundary condition has to be calculated from conditions at the previous grid point using a numerical technique based on the 'method of characteristics' (see the section 'Compressible surge in pipelines' on p. 471 for an introduction to this method).

The above is useful simply to illustrate the processes involved in developing a scheme. It is possible to develop a more stable explicit scheme, which will now be given.

Following Koutitas (1983) and using a centred grid (that is, Q nodes are at $i, i+1, n, n+1$, but h nodes are at $i-1/2, i+1/2, n-1/2, n+1/2$, etc.) the continuity equation becomes

$$h_{i+1/2}^{n+1/2} = h_{i+1/2}^{n-1/2} - \frac{\Delta t}{B\Delta x}\left(Q_{i+1}^{n} - Q_{i}^{n}\right)$$

So $h_{i+1/2}^{n+1/2}$ may be found explicitly. Turning to the momentum equation, stability is improved by replacing local values of quantities with averages over two nodes (Lax, 1954; Lax and Wendroff, 1960). It is convenient to use the following notation for the averaged quantities:

$$\overline{h} = \frac{\left(h_{i+1/2}^{n+1/2} + h_{i+1/2}^{n-1/2} + h_{i-1/2}^{n+1/2} + h_{i-1/2}^{n-1/2}\right)}{4}$$

$$S_f = \frac{(Q_i^n)^2 n^2 (B + 2\overline{h})^{1.333}}{(B\overline{h})^{3.333}}$$

$$\overline{h}_1 = \frac{\left(h_{i-1/2}^{n+1/2} + h_{i-1/2}^{n-1/2}\right)}{2}, \quad \overline{h}_2 = \frac{\left(h_{i+1/2}^{n+1/2} + h_{i+1/2}^{n-1/2}\right)}{2},$$

$$\overline{Q}_1 = \frac{(Q_i^n + Q_{i-1}^n)}{2}, \quad \overline{Q}_2 = \frac{(Q_{i+1}^n + Q_i^n)}{2}$$

Hence equation (14.33b) becomes

$$Q_i^{n+1} = \frac{\overline{Q}_2 + \overline{Q}_1}{2} - \frac{\Delta t}{\Delta x}\left(\frac{\overline{Q}_2^2}{B\overline{h}_2} - \frac{\overline{Q}_1^2}{B\overline{h}_1}\right) - \frac{\Delta t}{\Delta x} g\overline{h}B(\overline{h}_2 - \overline{h}_2) - \Delta t g\overline{h}B(S_f - S_0)$$

Q_i^{n+1} may therefore be found explicitly (a program listing is given on the website (www.sponpress.com/supportmaterial/0415306094.html)). The above scheme is now applied to the following problem.

Example 14.4 Translatory wave in a rectangular channel

Determine the water surface profile through time, the maximum water depths and the outflow hydrograph for the following channel given the inflow hydrograph and assuming that the channel contains a step at the downstream boundary (such that critical depth is maintained).

Channel: rectangular width B = 10 m, length 10 km
Manning's $n = 0.025$
bed slope $(S_0) = 0.002$ m/m
initial depth of flow = 1.0 m

Inflow hydrograph (in addition to steady flow):

Time (h)	0	0.5	1.0	1.5	2.0	2.5
Discharge (m^3/s)	0	50	37.5	25	12.5	0

Solution

For the solution to proceed the boundary conditions must be specified. At the upstream boundary Q_i^n etc. = (inflow hydrograph + steady flow). At the downstream boundary (for critical flow), $U = \sqrt{gh}$ and $\Delta Q = \Delta A\sqrt{gh}$. The problem of instability must be avoided, and to predict the limit of stability the Courant number, Cr $(= c\Delta t/\Delta x)$, will be used. The stable limit is usually taken as Cr < 1; however, many schemes require lower values in practice. Instabilities can arise either because the values for Δt or Δx are too large, or from calculation of the boundary conditions. The solution has been produced using a program based on the above scheme, and the following investigations illustrate the behaviour of various aspects of the computational model.

Effect of boundary conditions. To ensure that calculation of the downstream boundary conditions proceeds satisfactorily, Δx must be less than 100 m. The effect of using a greater value ($\Delta x = 1000$ m), with a *constant* value for Q (no flood wave) and $\Delta t = 50$ s, is shown in Figure 14.7(a). Note the effect of the over-large Δx is to produce an instability in the computation of the boundary condition. This manifests itself in the form of artificial waves.

Effect of magnitude of Δt. Having established the magnitude of Δx at 100 m, the effect of varying Δt is investigated. Figure 14.7(b) shows the inflow hydrograph, and the corresponding outflow hydrograph for $\Delta t = 12$ s and 12.5 s. The wave speed

c in the channel may be estimated using the fact that the normal depth for the maximum discharge (65 m^3/s) is approximately 2.56 m, and therefore $c = \sqrt{gh} = \sqrt{9.81 \times 2.56} = 5.01$ m/s in still water. Because the velocity of the water in the channel is not zero, the wave speed will be $c \pm U$. Taking the maximum velocity near the downstream end as $U = 65/(10 \times 2.25) = 2.9$ m/s, then the Courant number is $\mathrm{Cr} = (5.01 + 2.9)\Delta x/\Delta t \leqslant 1$. As Δx has been fixed at 100 m we can see that the corresponding maximum value for Δt is 12.6 s. Figure 14.7(b) shows that the outflow hydrograph for $\Delta t = 12$ s is a smooth curve, and that magnitudes are reasonably in line with what might be expected. An increase in Δt to 12.5 s produces an unstable computation. Figure 14.7(c) shows the corresponding distribution of maximum water depths for the same two time intervals (12 s, 12.5 s) with the unnatural peak for the 12.5 s interval.

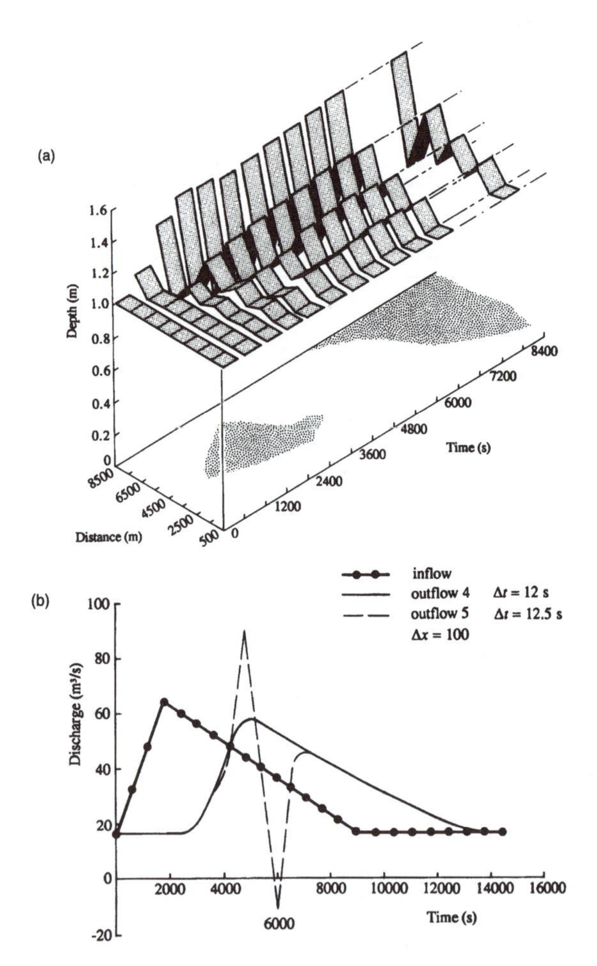

Figure 14.7 Flood wave model: (a) Unstable downstream boundary, $\Delta x = 1000$ m, $\Delta t = 50$ s; (b) inflow and outflow hydrographs.

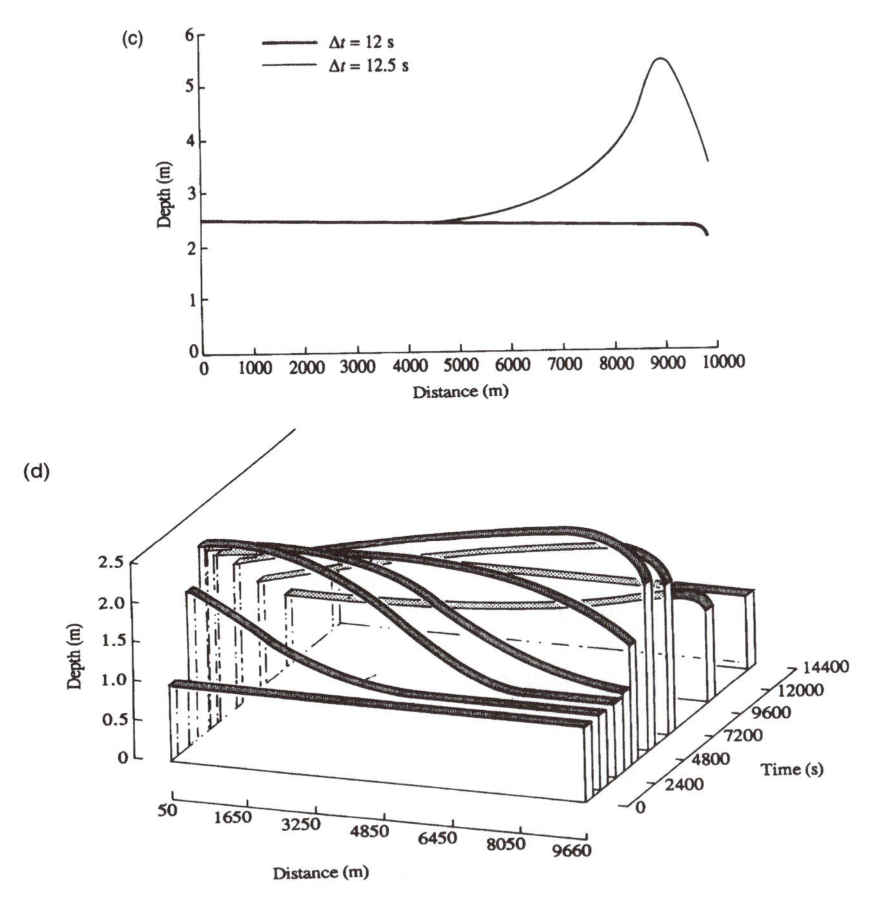

Figure 14.7 *contd:* (c) distribution of maximum water depth; (d) water surface profiles (each strip represents the computer-generated estimate of the water surface profile at a given time, due to the passage of the flood wave).

Figure 14.7(d) shows the water surface profiles for different times in the passage of the wave.

Some interesting conclusions may be gleaned from the above example. First, the attenuation of the flood peak is only 10% in 10 km (and this becomes less for steeper bed slopes). Second, the maximum depths correspond very closely to the normal depths at the peak discharges. Third, the speed of propagation is very closely estimated using the wave speed formula with the normal depth at maximum discharge.

In this case, therefore, the application of a steady flow model could have provided the necessary information to a level of accuracy acceptable for some engineering purposes. However, this is not always the case, and careful consideration needs to be given to each individual study regarding an appropriate choice of computational model.

If the various models considered up to this point are reviewed it will be seen that they are all based on the use of 'averaged' conditions (mean velocity etc.) at each point in the system. The frictional effect of turbulent (or laminar) conditions on the flow has been simulated through the Darcy or Chezy–Manning formula. This has the merit of simplicity, but any detail (such as the variation of velocity or turbulence across a section) is lost. For some cases this may not be important, but for other cases the simplifications may lead to significant errors in modelling system behaviour.

Implicit finite difference schemes

It has been seen that a major limitation in the use of explicit schemes is the onset of instability. This restricts the magnitudes of Δt and Δx, and means that quite large numbers of calculations may be involved. With modern personal computers storage is not a barrier to the solution of flows in pipes and channels of uniform cross-section. However, for flows in estuaries and along coasts (where a 2-D grid is often required) computer storage can give rise to difficulties. 'Implicit' solutions to finite difference equations can overcome these (see section 14.4). Stability analysis of implicit models shows that, provided the 'weighting coefficients' used in the equations are within specified ranges, the models are stable for any value of the Courant number.

The Preissmann four-point implicit scheme

The Preissmann scheme is an example of an implicit finite difference scheme. This scheme is commonly used in current computational models of unsteady river flows (see for example the FLUCOMP model described by Samuels and Gray (1982)).

Figure 14.8 shows four points in the x–t plane at distances x_j and x_{j+1} and times t^n and t^{n+1} at which the flow variables Q, h are to be determined. In the Preissman scheme the space and time derivatives ($\partial f/\partial x$ and $\partial f/\partial t$) of any function f are represented by a weighted average of the values of f at the four grid points divided by the space and time increments respectively. For the space derivative the weighting factor θ is given a value between 0.5 and 1, and for the time derivative the weighting factor is fixed at 0.5. Thus

$$\frac{\partial f}{\partial x} \approx \frac{\theta}{\Delta x_j}(f_{j+1}^{n+1} - f_j^{n+1}) + \frac{1-\theta}{\Delta x_j}(f_{j+1}^{n} - f_j^{n}) \tag{14.38}$$

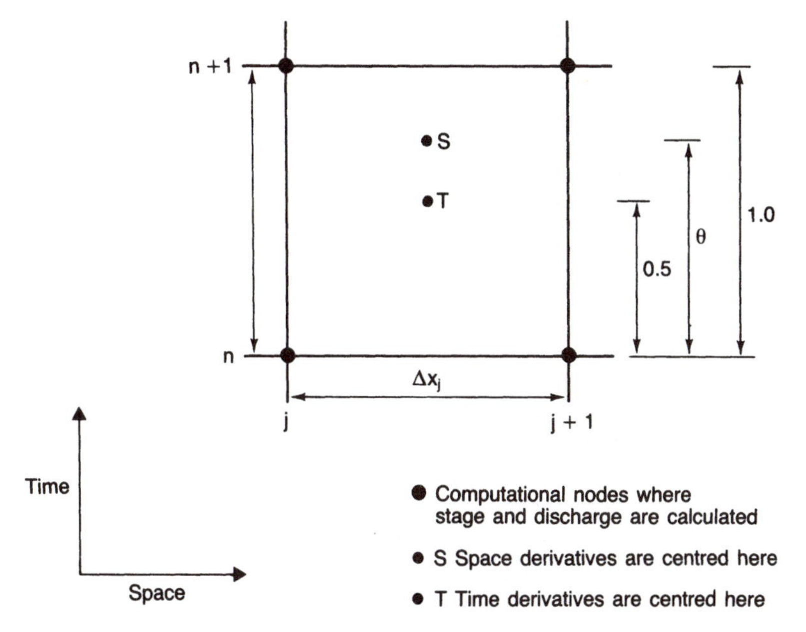

Figure 14.8 The Preismann four point implicit scheme: computational nodes where stage and discharge are calculated; S, space derivatives are centred here; T, time derivatives are centred here.

and

$$\frac{\partial f}{\partial t} \approx \frac{1}{2\Delta t}(f_{j+1}^{n+1} - f_{j+1}^{n} - f_j^n) \tag{14.39}$$

A brief outline of one such scheme follows.

Using the notation

$$\mathrm{D}_x(f_j^n) = \frac{\partial f}{\partial x} \quad \text{and} \quad \mathrm{D}_t(f_j^n) = \frac{\partial f}{\partial t}$$

where D is the difference operator, then the continuity equation may be written in finite difference form as follows (note that the discharge Q is used instead of velocity U).

$$\mathrm{D}_x(Q_j^n) + \mathrm{D}_t(A_j^n) = M(q) \tag{14.40}$$

where $M(q)$ is the weighted average value of q and in more general terms is given by:

$$M(f_j^n) = \theta(f_{j+1}^{n+1} + f_j^{n+1}) + (1-\theta)(f_{j+1}^n + f_j^n)$$

Similarly, the momentum equation (14.33b) may be written in finite difference form as

$$D_t(Q_j^n) + D_x\left(\left(\frac{\beta Q^2}{A}\right)_j^n\right) + M(gA_j^n)[D_x(h_j^n) + M(S_f)_j^n] = 0 \qquad (14.41)$$

These two equations contain four unknown quantities (the stage and discharge at the time level $n+1$ and space positions j and $j+1$). Thus they cannot be solved explicitly. However, for a given distance grid of N points, $N-1$ continuity equations and $N-1$ momentum equations may be formulated, giving $2N-2$ equations with $2N$ unknown values (that is, N values of Q and N values of h). Hence two additional equations are necessary to solve the problem. These come from the boundary conditions. For subcritical flow, the typical boundary conditions are the upstream flood hydrograph and a downstream relationship between stage and discharge (for example at a weir) or a prescribed downstream stage (as in a tidal reach, for example). Hence the system of $2N$ equations can be solved simultaneously across all grid points at each succeeding time level. Matrix methods are generally used for the efficient solution of these equations; however, the solution technique is not a trivial matter. To solve for the stage and discharge at the next time step requires a knowledge of the cross-sectional area A, wetted perimeter P and conveyance K at the next time step. This information is not known a priori, and therefore an efficient iterative solution technique must be adopted. Most implicit schemes use the current values of A, P and K as the initial estimate of their values at the next time step.

The overall solution technique may thus be summarized as follows:

1. Form the system of $2N-2$ continuity and momentum equations for h and Q in finite difference form using the Preissman four-point scheme.
2. Set up the two boundary conditions.
3. Solve the system of $2N$ equations using matrix methods and using current values of A, P, K as initial estimates of A, P, K at the next time step.
4. Using an efficient iterative technique, repeat step 3 with computed values of A, P, K until convergence to the desired level of accuracy is achieved.
5. Repeat steps 1–4 at each time step for the duration of the unsteady flow event (for example, until the complete flood hydrograph has passed down the river).

By their nature, implicit schemes do not readily lend themselves to illustration by worked example. To help the reader develop a better understanding, a program listing is given on the website (www.sponpress.com/supportmaterial/0415306094.html)

Compressible surge in pipelines: the method of characteristics

In Chapter 6 the differential equations were formed representing conservation of mass and momentum for compressible flow in pipes. Slightly rearranged and divided by g equation (6.16) can be written

$$u\frac{\partial H}{\partial x} + u\sin\theta + \frac{\partial H}{\partial t} + \frac{c^2}{g}\frac{\partial u}{\partial x} = 0 \tag{6.16a}$$

and (6.17) is

$$g\frac{\partial H}{\partial x} + \frac{\lambda u|u|}{2D} + u\frac{\partial u}{\partial x} + \frac{\partial u}{\partial t} = 0 \tag{6.17}$$

Note that equations (6.16a) and (6.17) are both equal to zero. If the momentum equation (6.17) is multiplied by some coefficient, C, and is then added to (6.16a), the resulting equation must still be equal to zero:

$$\left(u\frac{\partial H}{\partial x} + u\sin\theta + \frac{\partial H}{\partial t} + \frac{c^2}{g}\frac{\partial u}{\partial x}\right) + C\left(g\frac{\partial H}{\partial x} + \frac{\lambda u|u|}{2D} + u\frac{\partial u}{\partial x} + \frac{\partial u}{\partial t}\right) = 0$$

This may be rearranged as

$$\left[\frac{\partial H}{\partial x}(u + Cg) + \frac{\partial H}{\partial t}\right] + C\left[\frac{\partial u}{\partial x}\left(u + \frac{c^2}{Cg}\right) + \frac{\partial u}{\partial t}\right] + u\sin\theta + \frac{C\lambda u|u|}{2D} = 0 \tag{14.42}$$

Now it has been shown (see section 14.3) that the total derivative for the acceleration of a fluid may be written

$$\frac{du}{dt} = \frac{dx}{dt}\frac{\partial u}{\partial x} + \frac{\partial u}{\partial t}$$

and a similar derivative applies to H. Hence if

$$(u + Cg) = \frac{dx}{dt} \tag{14.43}$$

then the first term in square brackets in equation (14.42) is equivalent to dH/dt. Also, if

$$(u + c^2/Cg) = \frac{dx}{dt} \tag{14.44}$$

then the second term in square brackets is equivalent to du/dt. Equation (14.42) can now be written

$$\frac{dH}{dt} + C\frac{du}{dt} + u\sin\theta + \frac{C\lambda u|u|}{2D} = 0$$

or

$$dH + C\,du + u\sin\theta\,dt + \frac{C\lambda u|u|}{2D}\,dt = 0 \tag{14.45}$$

Equation (14.45) is true *only* if

$$\frac{dx}{dt} = \left(u + \frac{Cc^2}{g}\right) = \left(u + \frac{g}{C}\right) \tag{14.46}$$

for which

$$C = \pm\frac{c}{g} \tag{14.47}$$

If $u \ll c$ then

$$\frac{dx}{dt} \to \pm c$$

Equation (14.47) indicates that there are two lines representing the propagation of a disturbance, which can be plotted on a space–time plane (see Fig. 14.9). One line has a slope of $+c$, the other a slope of $-c$. Also, equation (14.45) is valid *only* if (14.47) is satisfied, and the solution of (14.45) is therefore valid only along the lines of propagation. The lines of propagation are known as **characteristic lines**, or simply as ‘characteristics’. The line with a slope of $+c$ is a positive characteristic and the line with a slope of $-c$ is a negative characteristic. The zone enclosed between the x axis and the two characteristics is the zone of determinacy. Any event occurring outside that zone will have no effect on the properties of the system within the zone.

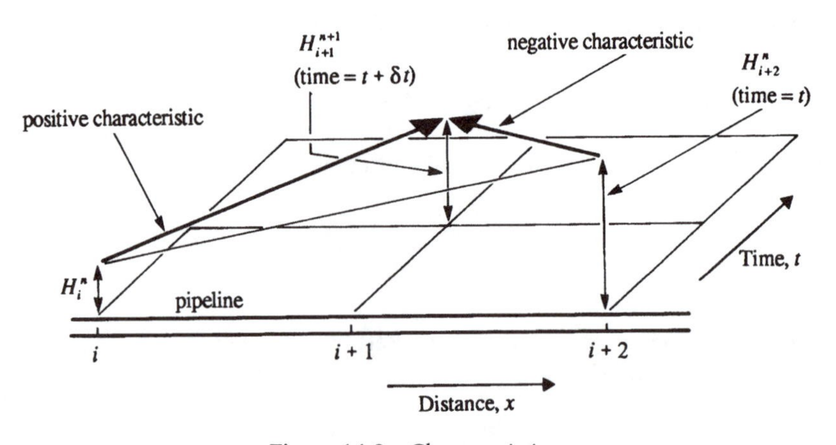

Figure 14.9 Characteristics.

To see how this may be applied, suppose that conditions at the point n, i are known: that is, we have numerical values for the location i (distance along the pipeline), the time n and the corresponding head H_i^n and velocity u_i^n. Then equation (14.45) with $C = +c/g$ is valid for any point along the positive characteristic. Similarly, if conditions are known for location $i+2$ and time n, (14.45) can be used with $C = -c/g$, along the negative characteristic. These two equations may be solved at the point $(i+1, n+1)$ where the positive and negative characteristics meet. Thus if conditions at $(i+1, n+1)$ are unknown, and there are two unknown quantities $(H_{i+1}^{n+1}, u_{i+1}^{n+1})$, then with the two equations (for the positive and the negative characteristic) these can be evaluated.

The concept of the solution of the unsteady flow equations along a line of specified slope is the essence of the method of characteristics. Furthermore, it will be shown that the equations are easily arranged into a form that is suitable for solution by computer.

Note that, for the most general case of the propagation of a shock wave through a fluid, both c and u can vary (for example, c is a function of ρ, which is not constant in a compressible flow, and u is affected by friction), so the characteristics are actually curves rather than straight lines. Even so, provided that the increments in x and t used in a numerical scheme are kept sufficiently small, it is usually possible for the solution to proceed on the assumption that $\delta x/\delta t$ is constant over one step.

We have already encountered the problem of instability. A careful look at Figure 14.9 illustrates why instabilities may arise. As shown on the figure, H_{i+1}^{n+1} lies at the junction of the two characteristics. However, the slopes of the lines are fixed (at $\pm c$), so there is a corresponding fixed relationship between δx and the maximum possible value of δt. An increase in δt beyond that maximum would therefore take the computed point $(i+1, n+1)$ outside the domain of determinacy (the calculation becomes a 'leap into the unknown'). This approach to stability only indicates a basis for determining the limiting condition; many explicit schemes require much smaller time steps to maintain stable operation, as has been seen in the previous example.

Finite difference form of the characteristic equation. The characteristic equation (14.45) may be stated in the finite difference form that is adopted for computer application:

$$\delta H + C\,\delta u + u \sin\theta\,\delta t + C\lambda u|u|\,\delta t/2D = 0 \qquad (14.48)$$

Because variations occur in space and time, the subscript and superscript notation referred to earlier is adopted. Thus H_{i+1}^{n+1} is the head at time $t + \delta t$ and location $x + \delta x$.

Thus for the positive characteristic $\delta H = (H_{i+1}^{n+1} - H_i^n)$ and $\delta u = (u_{i+1}^{n+1} - u_i^n)$, so equation (14.48) becomes

$$H_{i+1}^{n+1} - H_i^n + C(u_{i+1}^{n+1} - u_i^n) + u_i^n \sin\theta\,\delta t + \frac{C\lambda u_i^n |u_i^n|}{2D}\,\delta t = 0 \qquad (14.49)$$

Similarly for the negative characteristic, with $\delta H = (H_{i+1}^{n+1} - H_{i+2}^n)$, etc.:

$$H_{i+1}^{n+1} - H_{i+2}^n + C(u_{i+1}^{n+1} - u_{i+2}^n) + u_{i+2}^n \sin\theta\,\delta t - \frac{C\lambda u_{i+2}^n |u_{i+2}^n|}{2D}\,\delta t = 0 \quad (14.50)$$

Adding (14.49) and (14.50) and rearranging,

$$H_{i+1}^{n+1} = \frac{1}{2}\Bigg(H_i^n + H_{i+2}^n - C(u_{i+2}^n - u_i^n) - \sin\theta(u_{i+2}^n + u_i^n)\,\delta t - \frac{C\lambda}{2D}\left(u_i^n|u_i^n| - u_{i+2}^n|u_{i+2}^n|\right)\delta t\Bigg) \qquad (14.51)$$

while subtracting (14.49) from (14.50) yields

$$u_{i+1}^{n+1} = \frac{1}{2}\Bigg(\frac{H_i^n - H_{i+2}^n}{C} + u_i^n + u_{i+2}^n + \frac{\sin\theta(u_{i+2}^n - u_i^n)\,\delta t}{C} - \frac{\lambda}{2D}\left(u_i^n|u_i^n| - u_{i+2}^n|u_{i+2}^n|\right)\delta t\Bigg) \qquad (14.52)$$

Note that the above equations are now explicit expressions for H_{i+1}^{n+1} and u_{i+1}^{n+1} in terms of the known values for H, u etc. at the preceding time step.

Characteristic equations for boundary conditions

The characteristic equations for some of the important boundaries are now derived.

Valve (Fig. 14.10(a)). The valve is an orifice of variable area, and so $Q = C_d A_v \sqrt{2gH}$ (where A_v is the valve orifice area and H is the pressure head at the valve). In the following analysis, the superscript 'o' refers to the fully open setting. Thus

$$\frac{\text{discharge with valve partly open}}{\text{discharge with valve fully open}} = \frac{Q}{Q^o} = \frac{C_d A_v \sqrt{2gH}}{C_d A_v^o \sqrt{2gH^o}} = A_R\sqrt{\frac{H}{H^o}}$$

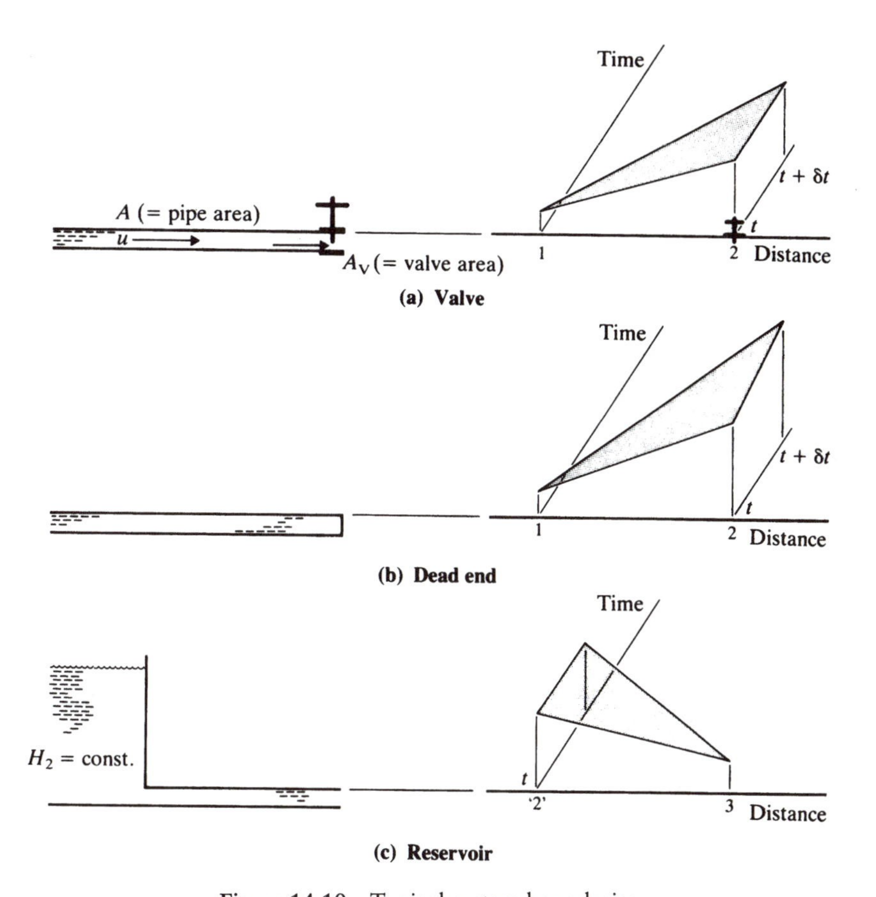

Figure 14.10 Typical system boundaries.

where $A_R = A_v/A_v^o$. Hence the pipeline velocity is

$$\frac{u}{u^o} = \frac{Q/A}{Q^o/A} = A_R\sqrt{\frac{H}{H^o}} \tag{14.53}$$

where u and u^o are the velocities in the pipe just upstream of the valve.

From Figure 14.10(a), the valve is assumed to lie at the downstream end of the pipeline, and therefore the positive characteristic equation (14.49) is applicable. Equations (14.49) and (14.53) are solved simultaneously. From (14.53),

$$\left(\frac{u_{i+1}^{n+1}}{u_{i+1}^o}\right)^2 = A_R^2\frac{H_{i+1}^{n+1}}{H_{i+1}^o}$$

$$H_{i+1}^{n+1} = \left(\frac{u_{i+1}^{n+1}}{u_{i+1}^o}\right)^2 \frac{H_{i+1}^o}{A_R^2}$$

Substituting this result in (14.49),

$$\left[\left(\frac{u_{i+1}^{n+1}}{u_{i+1}^{\mathrm{o}}}\right)^2 \frac{H_{i+1}^{\mathrm{o}}}{A_{\mathrm{R}}^2} - H_i^n\right] + C\left(u_{i+1}^{n+1} - u_i^n\right) + u_i^n \sin\theta\,\delta t + \frac{C\lambda u_i^n|u_i^n|\,\delta t}{2D} = 0$$

Multiplying through by ${u_{i+1}^{\mathrm{o}}}^2 A_{\mathrm{R}}^2 / H_{i+1}^{\mathrm{o}}$ and rearranging,

$$\left(u_{i+1}^{n+1}\right)^2 + \frac{C\,{u_{i+1}^{\mathrm{o}}}^2 A_{\mathrm{R}}^2 u_{i+1}^{n+1}}{H_{i+1}^{\mathrm{o}}} + \frac{{u_{i+1}^{\mathrm{o}}}^2 A_{\mathrm{R}}^2}{H_{i+1}^{\mathrm{o}}}\left(u_i^n \sin\theta\,\delta t + \frac{C\lambda u_i^n|u_i^n|\,\delta t}{2D} - C\,u_i^n - H_i^n\right) = 0$$

This is a quadratic equation in u_{i+1}^{n+1}, which may be solved to yield

$$u_{i+1}^{n+1} = -\frac{C\,{u_{i+1}^{\mathrm{o}}}^2 A_{\mathrm{R}}^2}{2H_{i+1}^{\mathrm{o}}} + \left[\left(\frac{C\,{u_{i+1}^{\mathrm{o}}}^2 A_{\mathrm{R}}^2}{2H_{i+1}^{\mathrm{o}}}\right)^2 - \frac{{u_{i+1}^{\mathrm{o}}}^2 A_{\mathrm{R}}^2}{H_{i+1}^{\mathrm{o}}}\left(u_i^n \sin\theta\,\delta t + \frac{C\lambda u_i^n|u_i^n|\,\delta t}{2D} - C\,u_i^n - H_i^n\right)\right]^{1/2} \quad (14.54)$$

So, conditions at time $t + \mathrm{d}t$ are computed from the (known) conditions at time t and before the commencement of closure.

Dead end (Fig. 14.10(b)) This could be a closed valve or a stopped-off pipe: therefore $u_{i+1} = 0$. Like the valve, the dead end has a positive characteristic, so equation (14.49) is used:

$$H_{i+1}^{n+1} - H_i^n + C(0 - u_i^n) + u_i^n \sin\theta\,\delta t + \frac{C\lambda u_i^n|u_i^n|}{2D}\,\delta t = 0$$

Therefore

$$H_{i+1}^{n+1} = H_i^n + C\,u_i^n - u_i^n \sin\theta\,\delta t - \frac{C\lambda u_i^n|u_i^n|}{2D}\delta t = 0 \quad (14.55)$$

Reservoir (Fig. 14.10(c)) At the pipe-reservoir interface, the head must remain constant. The reservoir is here assumed to lie at the upstream end of the pipeline. A negative characteristic solution is therefore assumed, with stations $i+1$ and $i+2$ positioned so as to correspond to (14.50). In running the procedure on a computer, conditions at station $i+2$ would already be calculated, and H_{i+1} would be a known, constant ($= H_{i+1}^{\mathrm{o}}$) magnitude, so equation (14.50) may be arranged as

$$u_{i+1}^{n+1} = \frac{H_{i+1}^{\mathrm{o}} - H_{i+2}^n}{C} + u_{i+2}^n + \frac{u_{i+2}^n \sin\theta\,\delta t}{C} - \frac{\lambda u_{i+2}^n|u_{i+2}^n|\,\delta t}{2D} = 0 \quad (14.56)$$

Other types of boundary may also be encountered in practice. Wylie and Streeter (1993) may be consulted for details.

Example 14.5 Application of the method of characteristics

A reservoir supplies water through a horizontal pipeline 1 km long and 500 mm internal diameter (Fig. 14.11). The discharge is controlled by a valve at the downstream end of the pipeline. If the valve closes in 4 s and gives a linear retardation, estimate the pressure rise at the valve and at the mid-point of the pipe over an 8 s period. The reservoir head is 100 m and the head at the valve before commencement of valve closure is 4.75 m. Friction factor $\lambda = 0.012$, and speed of sound $c = 1000$ m/s.

Solution

A time interval must first be adopted. The interval should normally be some convenient fraction of L/c. Here an interval of 0.5 s is used. This corresponds to the time for the shock wave to travel halfway along the pipe. (*Note*: $i = 1$ is reservoir boundary, $i = 2$ is pipe mid-point, $i = 3$ is valve boundary.)

Initial conditions $t = 0$:

$$h_f = 100 - 4.75 = \frac{0.012 \times 1000 \times u_3^{o2}}{2 \times 9.81 \times 0.5}$$

so

$$u_3^o = 8.824 \text{ m/s}$$

Therefore

$$H_1^o = 100 \text{ m} \quad H_2^o = 52.38 \text{ m} \quad H_3^o = 4.75 \text{ m}$$

$$C = \frac{c}{g} = \frac{1000}{9.81} \approx 100$$

Pipe is horizontal, so $\theta = 0$.

First step. $t = 0.5$ s after commencement of valve closure:

$$\text{Valve area reduction} = 0.5/4 = 0.125$$

Therefore

$$A_R = 0.875$$

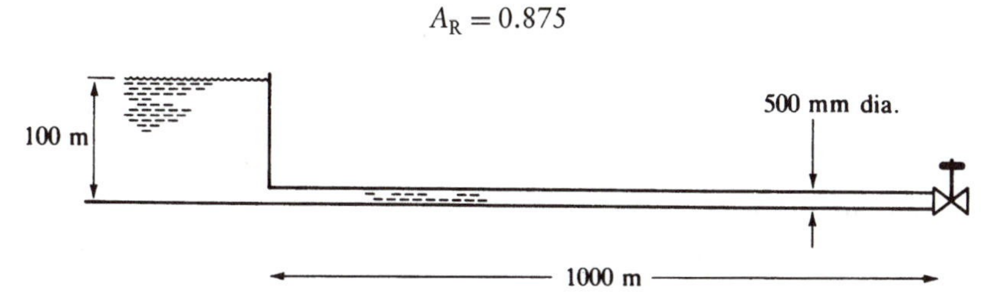

Figure 14.11 Reservoir and pipeline (Example 14.5).

Using equation (14.54), determine the velocity at the valve (for the computer solution the valve is in effect assumed to close in a series of discrete 'steps' or 'jerks', one step at each time interval):

$$u_3^{n+1} = -\frac{100 \times 8.824^2 \times 0.875^2}{2 \times 4.75} + \left[\left(\frac{100 \times 8.824^2 \times 0.875^2}{2 \times 4.75}\right)^2 - \frac{8.824^2 \times 0.875^2}{4.75}\left(\frac{100 \times 0.012 \times 8.824|8.824| \times 0.5}{2 \times 0.5} - (100 \times 8.824) - 52.38'\right)\right]^{1/2} = 8.81\,\text{m/s}$$

Notice that all the data for evaluating the equation are derived from conditions at the preceding time ($t = 0$).

The pressure head is determined from equation (14.49), because the valve is assumed to have a positive characteristic. Rearranging (14.49) into a suitable form,

$$H_3^{n+1} = H_2^n - C(u_3^{n+1} - u_2^n) - u_2^n \sin\theta\,\delta t - \frac{C\lambda u_2^n |u_2^n|}{2D}\,\delta t$$

Substituting each term,

$$H_3^{n+1} = 52.38 - 100(8.81 - 8.824) - 0 - \frac{100 \times 0.012 \times 8.824|8.824| \times 0.5}{2 \times 0.5} = 7.06\,\text{m}$$

Because of the assumption about the step closure pattern at the valve, the first shock wave has only just been generated. Conditions at Stations 1 and 2 therefore remain as at the preceding time. Summarizing,

$$H_1^{n+1} = 100\,\text{m} \qquad H_2^{n+1} = 52.38\,\text{m} \qquad H_3^{n+1} = 7.06\,\text{m}$$

$$u_1^{n+1} = 8.824\,\text{m/s} \qquad u_2^{n+1} = 8.824\,\text{m/s} \qquad u_3^{n+1} = 8.81\,\text{m/s}$$

Second step. $t = 1.0$ s:

$$\text{Valve area reduction} = 1/4 = 0.25$$

Therefore

$$A_R = 0.75$$

$$u_3^{n+2} = -\frac{100 \times 8.824^2 \times 0.75^2}{2 \times 4.75} + \left[\left(\frac{100 \times 8.824^2 \times 0.75^2}{2 \times 4.75}\right)^2 - \frac{8.824^2 \times 0.75^2}{4.75}\left(\frac{100 \times 0.012 \times 8.824|8.824| \times 0.5}{2 \times 0.5} - (100 \times 8.824) - 52.38\right)\right]^{1/2} = 8.80\,\text{m/s}$$

$$H_3^{n+2} = 52.38 - 100(8.80 - 8.824) - 0 - \frac{100 \times 0.012 \times 8.824|8.824| \times 0.5}{2 \times 0.5} = 8.06\,\text{m}$$

The shock wave that was generated at the valve at $t = 0.5$ s has just reached the mid-point of the pipe. The velocity and head at the mid-point are evaluated using (14.51) and (14.52) respectively:

$$u_2^{n+2} = \frac{1}{2}\left[\frac{100-7.06}{100} + 8.824 + 8.81 - \frac{0.012}{2\times 0.5}(8.824|8.824| + 8.81|8.81|)0.5\right]$$
$$= 8.82\,\text{m/s}$$

$$H_2^{n+2} = \frac{1}{2}\left[100 + 7.06 - 100(8.81 - 8.824) - 0 - \frac{100\times 0.012(8.824|8.824| - 8.81|8.81|)\times 0.5}{2\times 0.5}\right]$$
$$= 54.16\,\text{m}$$

Summarizing,

$$H_1^{n+2} = 100\,\text{m} \qquad H_2^{n+2} = 54.16\,\text{m} \qquad H_3^{n+2} = 8.06\,\text{m}$$
$$u_1^{n+2} = 8.824\,\text{m/s} \qquad u_2^{n+2} = 8.82\,\text{m/s} \qquad u_3^{n+2} = 8.80\,\text{m/s}$$

Notes on Example 14.5

1. This example has been set up to highlight certain aspects of surge analysis, and not as an example of good design. It would not normally be considered sensible to run a system with such high initial velocity, precisely because this would lead to surge problems.
2. The solution has been completed by computer, and the results are presented in graphical form in Figure 14.12. Note the violent fluctuation in pressure. This would almost certainly cause a rupture in the pipeline.

A flowchart for the computer program is shown in Figure 14.13.

14.7 Concluding notes

This chapter gives a basic introduction to computational hydraulics. This has been, and is likely to remain, an area of rapid development. When considering the behaviour of physical scale models (see Example 11.3 and section 11.7), we found that scale effects were produced, which limited the capability of the model to simulate the behaviour of the prototype. Numerical models based on finite difference equations do not suffer from scale effects, but they do possess their own intrinsic characteristics, which also limit their capability. Therefore numerical models, like physical models, need to be checked and calibrated carefully against real-time prototype data to ensure that they are producing results that are of acceptable accuracy.

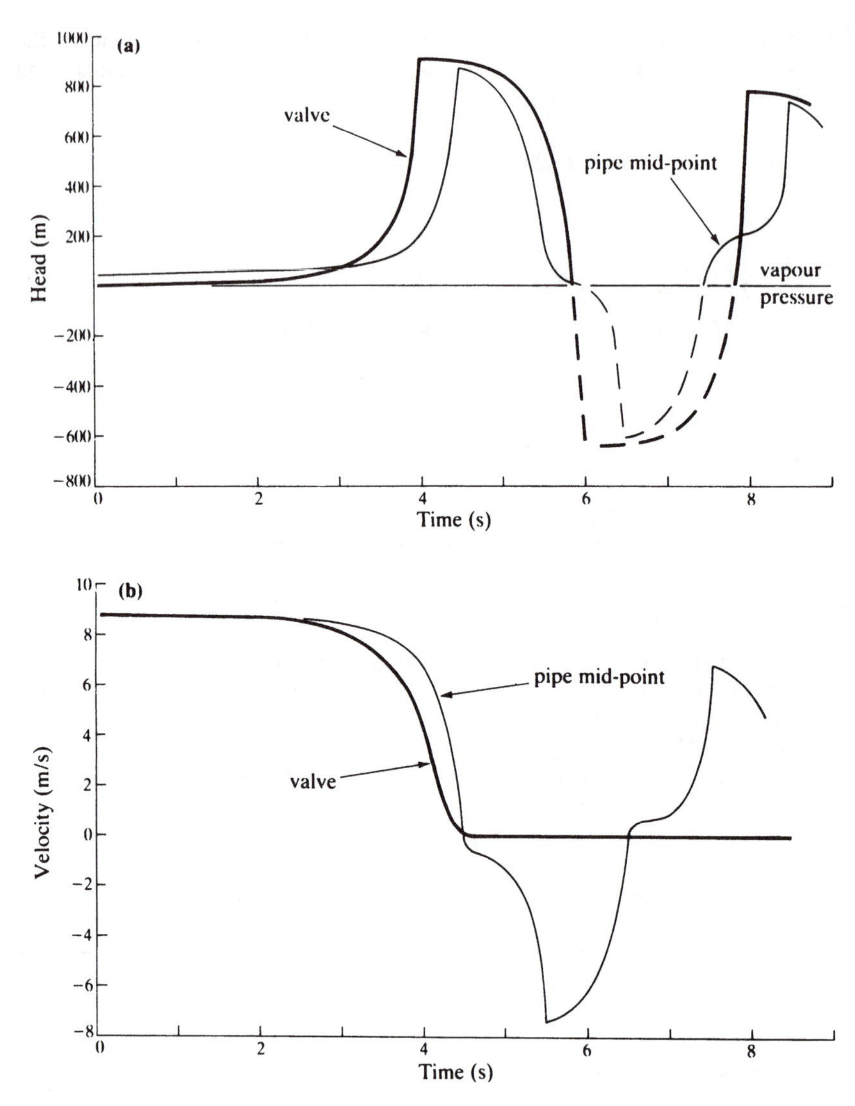

Figure 14.12 Graphical representation of solution to Example 14.5: (a) variation of head with time; (b) variation of velocity with time.

Models may be explicit, implicit or 'semi-implicit' (a hybrid). The implicit methods have the advantage that stability can be maintained over larger time steps. The disadvantages are that programming is more complicated, the time required to solve each time step is slightly greater than for the explicit approach (though this is usually outweighed by the saving due to the ability to use larger or variable time steps), and truncation errors may therefore be large. Implicit schemes have not been as extensively developed

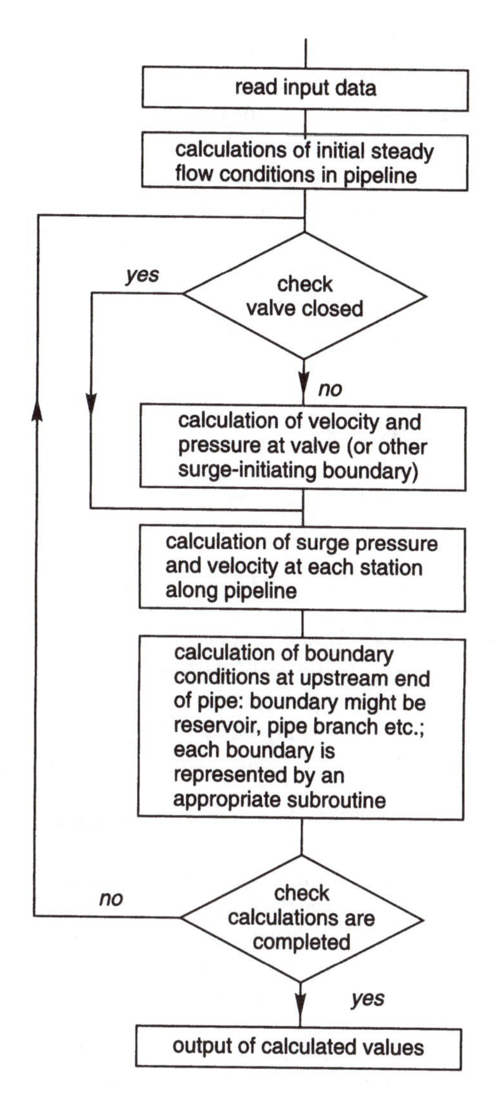

Figure 14.13 Flowchart for computer program using method of characteristics.

as explicit schemes for modelling flood waves. Explicit schemes are easier to program, but suffer from the disadvantage of stability constraints. However, a well-designed explicit scheme does tend to give good conservation of the flood wave form.

The use of implicit techniques came into favour in the 1970s, when the speed and memory size of many commercially available computers were much more limited than now. However, the inexorable advance in computer processing speed means that the engineer can exercise more freedom of choice in the selection of the technique for solving a given problem. Nevertheless, the continuously increasing demands of engineers have led to corresponding developments in both techniques.

There are now many computer packages capable of modelling a wide range of hydraulic phenomena. Commercial packages are usually far more sophisticated than the models outlined here, involving input, output and graphics routines, as well as the 'core' simulation routines.

More details of the basic modelling techniques and the general background to their application may be found in Anderson (1995) and Shaw (1992). A more advanced treatment is that of Abbott and Basco (1989).

References and further reading

Abbott, M. B. (1991) *Hydroinformatics*, Avebury Technical, Aldershot.

Abbott, M. B. and Basco, D. R. (1989) *Computational Fluid Dynamics*, Longman, Harlow.

Abbott, M. B. and Minns, A. W. (1997) *Computational Hydraulics*, 2nd edn, Ashgate, Aldershot.

Anderson, J. D. (1995) *Computational Fluid Dynamics*, McGraw-Hill, New York.

Brebbia, C. A. and Connor, J. J. (1976) *Finite Element Techniques for Fluid Flow*, Butterworth, London.

French, R. H. (1986) *Open Channel Hydraulics*, McGraw-Hill, Singapore.

Koutitas, C. (1983) *Elements of Computational Hydraulics*, Pentech Press, Plymouth.

Koutitas, C. (1988) *Mathematical Models in Coastal Engineering*, Pentech Press, Plymouth.

Lax, P. D. (1954) Weak solutions of non-linear hyperbolic equations: their numerical applications. *Communications on Pure and Applied Mathematics*, 7, 159–93.

Lax, P. D. and Wendroff, B. (1960) Systems of conservation laws. *Communications on Pure and Applied Mathematics*, **13**, 217–37.

Rodi, W. (1984) *Turbulence Models and their Applications in Hydraulics*. IAHR, Delft.

Samuels, P. G. and Gray, M. P. (1982) *The FLUCOMP River Model Package. An Engineer's Guide*, HR Ltd Report Ex 999, HR, Wallingford.

Schlichting, H., Gersten, K., Krause, E. and Oertel, H.(Jr) (trans, Mayes,C.) (1999) *Boundary Layer Theory*, 8th edn, Springer Verlag, Berlin.

Shaw, C. T. (1992) *Using Computational Fluid Dynamics*, Prentice-Hall, New York.

Vardy, A. E. and Hwang, K.-L. (1991) A characteristic model of transient friction in pipes. *Journal of Hydraulic Research*, **29**(5), 669–84.

Wylie, E. B. (1983) The microcomputer and pipeline transients. *ASCE Journal of Hydraulic Engineering*, **109**(12), 1723–39.

Wylie, E. B. and Streeter, V. L. (1993) *Fluid Transients in Systems*, Prentice-Hall, Englewood Cliffs, NJ.

15

River and canal engineering

15.1 Introduction

This chapter draws together various strands of knowledge which have been developed separately in the text together with some new concepts. Open channel flows of the steady, varying and unsteady types, hydrology and sediment transport all play a part in river engineering. The vast body of literature which has been produced over the last 100 years bears witness to the fascination, complexity and controversial nature of the subject. There is no complete theoretical approach to many of the problems, though various numerical models can help the engineer to make rough predictions, some of which are described in this chapter. Above all, the engineer concerned with river hydraulics must be a careful observer who uses his knowledge to co-operate with natural laws as far as is possible. Before any major modifications are made to the course of a river, it is usual to institute physical and/or numerical model studies as a check on the viability of the proposals.

Channels may be lined or unlined, artificial or natural (or may consist of various combinations of these). Thus, there is the need for several approaches to design. These are described in the following sections.

15.2 Optimization of a channel cross-section

In designing an artificial channel, cost is a prime consideration, so the engineer must use a channel whose geometry minimizes the cost of excavation and lining. The uniform flow equation can be used to give some indication of the optimum section shape from a hydraulic viewpoint. Take the case of a channel of prismatic section (Fig. 15.1). For uniform flow,

$$Q = \frac{1}{n}\frac{A^{5/3}}{P^{2/3}}S_0^{1/2}$$

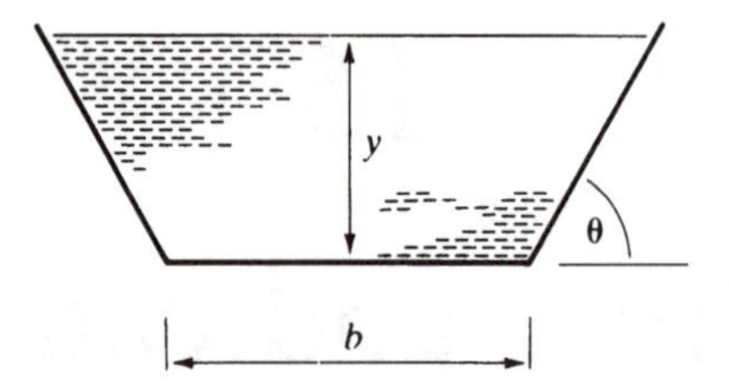

Figure 15.1 Trapezoidal channel.

(Equation 5.9). If n and S_0 can be assumed to be constant, then for an economic section the value of $A^{5/3}/P^{2/3}$ must be a maximum for a given value of A. Put another way, the requirement is for P to be the minimum value for a given value of A. From the geometry of the section,

$$A = by + y^2/\tan\theta \tag{15.1}$$

$$P = b + 2y/\sin\theta \tag{15.2}$$

From (15.1)

$$b = \frac{A}{y} - \frac{y}{\tan\theta}$$

therefore

$$P = \left(\frac{A}{y} - \frac{y}{\tan\theta}\right) + \frac{2y}{\sin\theta}$$

For the minimum P,

$$\frac{dP}{dy} = -\frac{A}{y^2} - \frac{1}{\tan\theta} + \frac{2}{\sin\theta} = 0$$

hence

$$A = -\frac{y^2}{\tan\theta} + \frac{2y^2}{\sin\theta}$$

and so

$$b = \frac{A}{y} - \frac{y}{\tan\theta} = -\frac{2y}{\tan\theta} + \frac{2y}{\sin\theta}$$

Therefore, for the optimum section,

$$R = \frac{A}{P} = \frac{-(y^2/\tan\theta) + (2y^2/\sin\theta)}{[-(2y/\tan\theta) + (2y/\sin\theta)] + 2y/\sin\theta} = \frac{y}{2} \tag{15.3}$$

Several considerations arise regarding the above:

(a) It is readily shown that the optimum trapezoidal section may be defined as that shape which approaches most nearly to an enclosed semicircle whose centre of radius lies at the free surface.
(b) The above development took into account only hydraulic considerations. Thus the area A is the area of flow, not the cross-sectional area of the excavation (the latter will naturally be greater than the former).
(c) If the channel is unlined and runs through erodible material, considerations of bank stability (section 15.3) must also enter into the calculations. Furthermore, the bed slope S_0 may not coincide with the natural stable slope in that material for the given flow.
(d) Where the channel is lined, lining methods and costs also play a determining role in deciding section shape.
(e) The slope of the local topography cannot be ignored, and this may be neither constant nor identical with the desired value for S_0.

The points above serve to emphasize the fact that estimates based on (15.3) represent an optimum section shape only from a very restricted viewpoint.

15.3 Unlined channels

Unlined channels may be natural or man-made, and may pass through rocky and/or particulate geological formations. Given a fast enough current, any material will erode (as witness the action of fast-flowing streams in gorges). Where the stream boundaries are both particulate and non-cohesive (e.g. a river traversing an alluvial plain) erosion and deposition occur. The fundamental ideas relating to sediment transport have been developed in Chapter 9. These ideas find an application here. In understanding the processes which govern channel formation, the starting point must be the current patterns which occur in channels.

Current patterns in channels

Some mention has already been made of secondary flows (section 5.4), but in view of their importance in determining the paths of natural watercourses, a further look at certain aspects is justified. In a straight watercourse, the pattern of secondary currents is usually assumed to be symmetrical. Consequently, the distribution of shear stress around the boundary should also be symmetrical and this, in turn, should theoretically give a symmetrical pattern of erosion (or deposition) and lead to a symmetrical boundary shape in an erodible bed material. However, when the flow traverses a bend, the secondary current patterns are changed. The velocity of an element of water near the channel centre at the surface will be higher than that of a second element near the channel bed (Fig. 15.2). The centrifugal force on the first element is higher than that on the second, which implies that the force pattern is not in equilibrium. The first element therefore tends to migrate outwards towards the bank, displacing other elements as it does so. This mechanism sets up a spiral vortex (Fig. 15.3) around the outside of the bend. The vortex causes an asymmetrical flow pattern which tends to erode material from the region near the outside of the bend and deposit it near the inside. This implies that the bend will tend to migrate outwards over a period of time. From observations, it would seem that secondary currents are seldom symmetrical, even in a straight channel. Furthermore, ground conditions are rarely homogeneous. Hence the problem of predicting, let alone controlling, events. Even in an apparently straight channel, the line of maximum depth may not be straight. The actual current patterns are more complicated than this outline suggests, nevertheless it gives an indication of the type of mechanism by which the alignment of an unlined watercourse changes with time. The centrifugal effects at the bend also cause a differential elevation of the water surface (superelevation) with higher levels at the outside of the bend than at the inside.

Stable channels and the 'regime' concept

An unlined channel exhibits multiple degrees of freedom, whereas the lined channel has only one (the depth). The unlined channel is subject to erosion

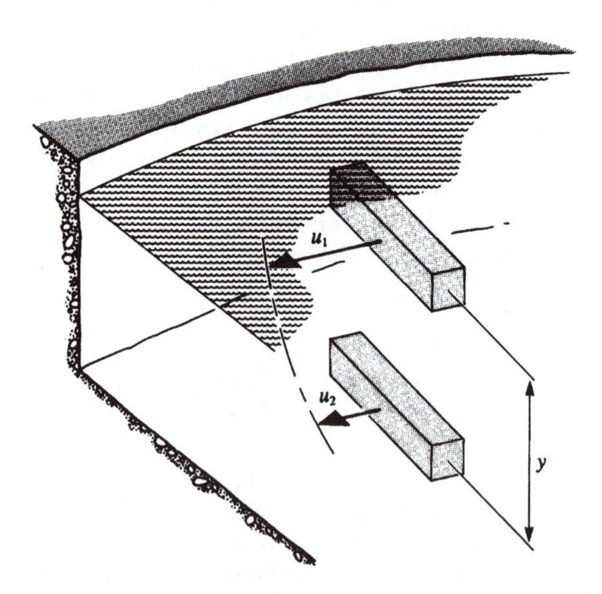

Figure 15.2 Variation of velocity with depth, in channel.

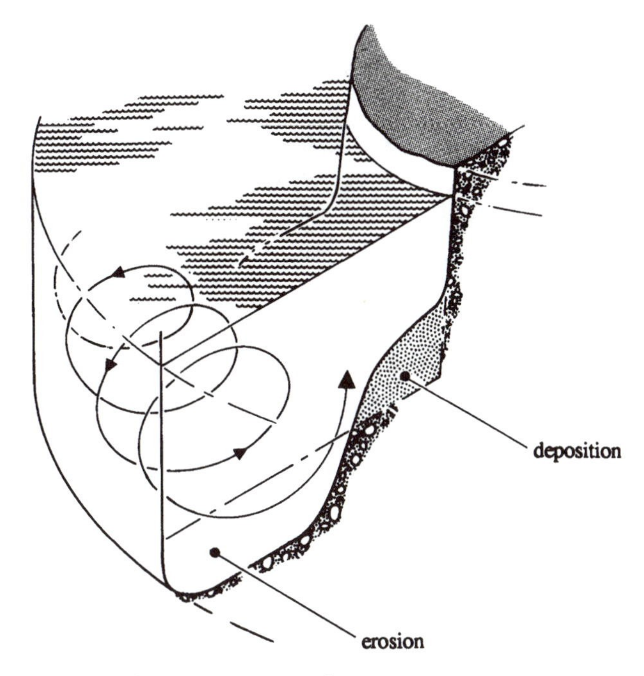

Figure 15.3 Secondary current pattern.

and deposition, which may change the bed slope and the channel cross-section and alignment. It is interesting that an artificially straight channel in an erodible bed is seldom stable. The flow will usually compel the course to 'meander' (see section 15.5). Therefore the degrees of freedom of an erodible channel include depth, width, slope and alignment. A system with one degree of freedom will rapidly stabilize in a steady state (normal depth) for a given discharge. Stabilization of a channel with multiple degrees of freedom may take a long time and, indeed, may in some cases never be quite attained. In other cases, 'armouring' (section 9.5: 'Concluding notes') will assist in the long term stabilization of the channel.

Where channels run through regions of fine-grained soils, in which irrigation may be vital, it is also vital to avoid excessive erosion of the precious topsoil. Such conditions are widespread throughout the Middle East and the Indian subcontinent, and, indeed, it was in India that some of the early work on the design of stable channels was initiated. A channel which does not exhibit long term changes in geometry and alignment, and in which scour and deposition are in equilibrium, is said to be stable (or 'in regime'). Three approaches have been developed to the design of stable channels with particulate boundaries: the 'regime' approach, the 'rational regime' approach and the 'tractive force' approach.

15.4 The design of stable alluvial channels

The 'regime' approach

This was probably originated by R. G. Kennedy towards the end of the 19th century. Kennedy studied a number of irrigation channels in the Punjab, and developed a formula for the mean velocity V_{CR} required for a stable channel:

$$V_{CR} = Ky^n \quad (\text{or } KR^n)$$

Unfortunately, n is not a constant, but varies roughly between the limits $0.5 < n < 0.73$. In common with many other 'regime' formulae, the numerical coefficient K is not dimensionless, but is purely empirical. Some engineers have criticized the regime approach on precisely the grounds that it lacks any coherent theoretical framework.

Further work has been done since Kennedy's formulae were published. Major contributors include Lacey (in a series of papers published between 1929 and 1958), Blench (in the 1950s and 1960s), Simons and Albertson (1963), Nixon (1959), Charlton, Brown and Benson (1978) and Hey (1986).

Blench suggested that the regime approach was primarily applicable to channels having the following characteristics:

1. steady discharge and with Froude Number less than unity;
2. steady sediment load (the sediment is therefore small, i.e. $D_{50} \ll y$);
3. straight alignment;
4. cross-section in which $B > 3y$ and the bank angle approximates to the 'natural' angle;
5. bed and banks are hydraulically smooth;
6. sufficiently long established to ensure that equilibrium has been attained and that the channel is stable.

In a sense, (6) begs many questions and, indeed, it is questionable whether it is possible to predict the timescale for its fulfilment when designing an artificial channel.

Simons and Albertson extended the data base for regime work using the information from India, adding data gathered from a number of North American rivers. They concluded that five channel types could be distinguished:

Type 1 – channels with sandy boundaries,
Type 2 – channels with sandy bed and cohesive banks,
Type 3 – channels with cohesive boundaries,
Type 4 – channels with coarse non-cohesive boundary material,
Type 5 – Type 2 channel with heavy load of fine silty sediment.

A selection of their equations (in metric units) is given below (a complete list may be found in Henderson (1966)):

$$B = 0.9P \quad P = \mathrm{K}_1 Q^{1/2}$$

$$R = \mathrm{K}_2 Q^{0.36} \quad S_0 = \frac{1}{R^2}\left(\frac{V}{\mathrm{K}_3}\right)^{1/n}$$

$$y_n = \begin{cases} 1.21R, & R < 2.1\,\mathrm{m} \\ 0.61 + 0.93R, & R > 2.1\,\mathrm{m} \end{cases}$$

The values for the various constants depend on the channel type. For example, a Type 2 channel would use

$$\mathrm{K}_1 = 4.71 \quad \mathrm{K}_2 = 0.484 \quad \mathrm{K}_3 = 10.81 \quad n = 0.33$$

The foregoing remarks serve mainly to make the reader aware of the concepts of regime theory. A comprehensive list of the regime equations of Blench, Simons and Albertson and Hey and Thorne (in SI units) was published in the seventh book of a series of water practice manuals by the Institution of Water and Environmental Management (Brandon (ed.), 1987). In this book clear guidance is also given on the proper application of regime equations, a topic as important as the equations themselves.

The rational approach to channel design

Stable channels are the result of quite complex interactions between fluid and sediment, some of which can currently be described by physical laws (e.g. the dynamic equation of unsteady flow) or by semi-empirical relationships (e.g. sediment transport equations). Regime equations are basically simple empirical relationships which may not take account of all the processes. Setting aside the plan and bed-form characteristics of channels, three principal variables may be considered – the width, depth and slope. If three so called **process** equations could be found then their simultaneous solution would provide the dimensions of width, depth and slope for a stable channel (e.g. a channel in regime). This is the principle of the rational approach to stable channel design.

As already intimated, two well established process equations are available, namely flow equations and the sediment transport equations. However, these two process equations require a linking equation because they are not truly independent. This arises from the fact the mean bed shear stress determines the flow resistance and contributes to the sediment transporting capacity of the channel. In turn the mean bed shear stress is a function of the grain size (and bed forms) for alluvial channels.

A third process equation might involve a relationship between the cross-sectional variation of shear stress and the other two process equations. Such an equation is not currently available. However, another concept has been developed which produces a third independent equation and hence facilitates a solution. The concept is known as the extremal hypothesis or variational principle. The hypothesis is that an alluvial channel will adjust its geometric properties such that its sediment transporting capacity is maximized or that its **stream power** is minimized.

Using these principles White, Bettess and Paris (1982) developed a rational approach to channel design. It is also described in Ackers (1983) together with other researchers' efforts, and with comments regarding its accuracy and applicability. In a later development Bettess and White (1987) review their method and the work of others. This paper also contains a very useful critique and discussion of their results by other experts in the field.

The principal details of the White, Bettess and Paris method are as follows. Firstly they used (perhaps not surprisingly!) the Ackers and White sediment transport equations (described in Chapter 9). Secondly they developed a linking equation between frictional resistance and sediment transport. Recalling equation (9.24b) for the Ackers and White particle mobility number, e.g.

$$\mathrm{F}_{\mathrm{gr}} = \frac{u_*^n}{\sqrt{gD[(\rho_S/\rho)-1]}}\left(\frac{V}{\sqrt{32}\log(10D_{\mathrm{m}}/D)}\right)^{1-n} \qquad (9.24\mathrm{b})$$

then for fine sediments n tends to 1 and a second particle mobility number F_{fg} (for fine sediments) may be deduced from the above, e.g.

$$F_{\mathrm{fg}} = \frac{u_*}{\sqrt{gD[(\rho_S/\rho)-1]}} \qquad (15.4)$$

White, Paris and Bettess (1980) found that a linear relationship existed between F_{fg} and F_{gr} with a coefficient depending on the Ackers and White dimensionless grain size D_{gr} (equation (9.25)). Following an extensive correlation exercise for a wide range of sediment sizes (from 0.04 to 10 mm) the following equation resulted:

$$1-\left(\frac{F_{\mathrm{gr}}-A}{F_{\mathrm{fg}}-A}\right) = 0.76\left(1-\exp[-(\log D_{\mathrm{gr}})^{1.7}]\right) \qquad (15.5)$$

where A is a function of D_{gr} as defined in equations (9.26a) and (9.26d).

Equation (15.5) may be solved directly for velocity (V) and hence discharge (Q) if the depth and slope are prescribed in the following way. Firstly $u_*(=\sqrt{gRS_0})$ may be calculated, then D_{gr} found for a particular grain size (and water temperature). Thus F_{fg} may be found and equation (15.5) solved for F_{gr}. Finally equation (9.24b) is invoked to solve for V and Q. Hence it may be appreciated (with a little thought) that equation (15.5) links the frictional resistance of the channel to its sediment transporting capacity through the slope and grain size.

Returning now to the application of the variational hypothesis, White, Bettess and Paris (1982) developed a computer program to solve the previously mentioned equations. For given values of discharge, sediment concentration, bed material size and water temperature they computed the width, depth, velocity and slope which maximized sediment transporting capacity at the prescribed value of sediment concentration. They then made extensive comparisons between field data and their predicted results for both sand and gravel bed channels, concluding that the method was applicable over a very wide range of conditions. Quoting from Ackers (1983) 'The method (with respect to width) is least accurate for laboratory scale models and for very large irrigation canals but otherwise shows remarkably good agreement with the bulk of the irrigation system observations'. With this favourable result, tables for the design of stable alluvial channels were produced (White, Paris and Bettess (1981)) covering discharges up to $1000\,m^3/s$, sediment concentrations from 10 to 4000 mg/l and sediment sizes from 0.06 to 100 mm.

In conclusion, the foregoing description of the rational approach to channel design represents a significant step forward on purely empirically based regime equations. Thus consideration of its use is recommended, providing its limitations are properly accounted for. Research effort in this area is ongoing and the reader should expect further developments in the future.

The 'tractive force' approach

This makes use of the tractive force equations for sediment transport (see section 9.4). Most of the early work was based on the Du Boys equation; subsequently, the Shields equations were adopted. If the Du Boys equation is used, with the shear stress evaluated by the Chézy formula ($\tau = \rho g R S_0$), then

$$\begin{aligned} q_S &= K_3 \tau_0 (\tau_0 - \tau_{CR}) \\ &= K_3 \rho g R S_0 (\rho g R S_0 - \rho g R S_{CR}) \end{aligned} \tag{15.6}$$

Note that S_{CR} is the slope estimated to give the critical shear stress, it has no necessary connection with the slope S_c which would give critical flow (i.e. a uniform flow with $Fr = 1$).

Equation (9.1b) may be used in conjunction with (15.6) to provide a simple tractive force design criterion.

Shearing action affects the banks as well as the bed. However, in Chapter 9, no explicit account was given of the effects of shear stress on the stability of channel banks. Clearly, this is an important aspect of channels with particulate boundaries, so a simple numerical model is now developed. In estimating the shear stress at the bank, account should be taken (if possible) of the secondary currents in the channel, especially in the vicinity of bends. Consideration of the forces acting on a particle resting on a bank (Fig. 15.4(a)) leads to the following:

$$\text{force perpendicular to bank} = W' \cos\theta$$

therefore

$$\text{resistance to motion} = W' \cos\theta \tan\phi$$

Forces parallel to bank are

1. due to the immersed weight of the particle ($= W' \sin\theta$) and
2. due to the drag force of the fluid on the particle ($= F_D$). The angle ϕ is the angle of repose.

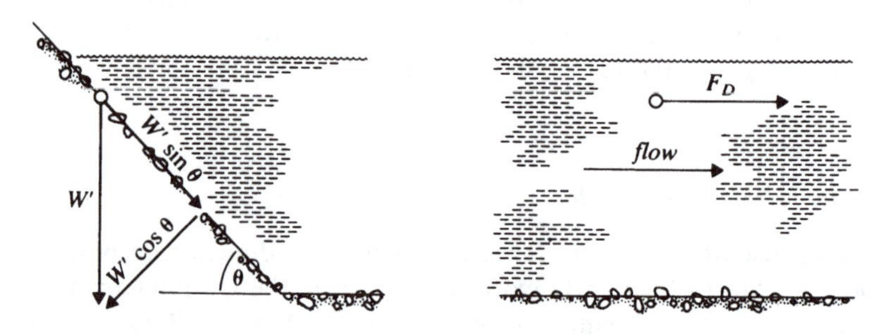

(a) Forces on particle on river bank

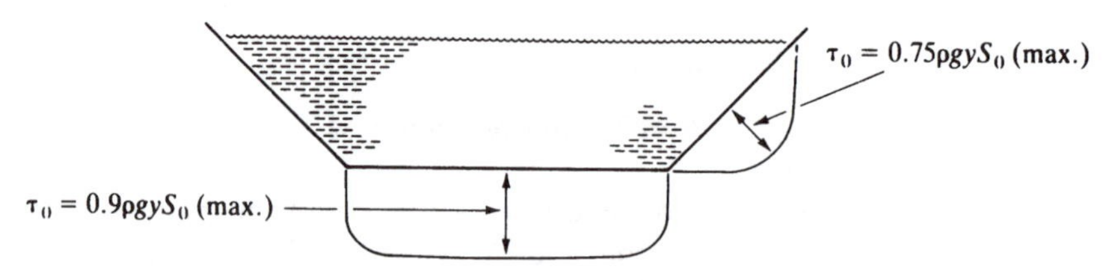

(b) Shear stress distribution

Figure 15.4 Force and shear stress in a channel.

All of the above forces can be expressed in terms of the fluid shear force. At the threshold of movement,

$$\tau_{CR} D^2 / A_p = W' \tan \phi$$

therefore

$$W' = \cos\theta \tan\phi = (\tau_{CR} D^2 / A_p) \cos\theta$$

$$W' \sin\theta = \tau_{CR} \frac{D^2}{A_p} \frac{\sin\theta}{\tan\phi}$$

$$F_D = \tau_0 D^2 / A_p$$

Therefore, at the threshold of movement (which constitutes the limit of bank stability) $\tau_0 \rightarrow \tau_{bc}$ and

$$\tau_{CR}^2 \cos^2\theta = \tau_{bc}^2 + \left(\tau_{CR} \frac{\sin\theta}{\tan\phi} \right)^2$$

therefore

$$\left(\frac{\tau_{bc}}{\tau_{CR}} \right)^2 = \cos^2\theta - \frac{\sin\theta}{\tan\phi}$$

and so

$$\frac{\tau_{bc}}{\tau_{CR}} = \cos\theta \left[1 - \left(\frac{\tan\theta}{\tan\phi} \right)^2 \right]^{1/2}$$

Since both drag (or shear) and self-weight ($W' \sin\theta$) components are combining to dislodge a grain, the shear (τ_{bc}) required to move the grain is less than τ_{CR}. In order to use the equation, it is necessary to estimate the shear stress at the bank. Figure 15.4(b) shows the shear distribution typical of a trapezoidal channel. The maximum bank shear stress is shown as $0.75\rho g y S_0$. The coefficient 0.75 is not a constant, but is a function of θ. The magnitude 0.75 departs significantly from 0.75 only when $B/y < 2$ and $45° < \theta < 90°$.

The principles outlined above have been refined and extended; for example, by Lane and the US Bureau of Reclamation (Lane, 1953) in

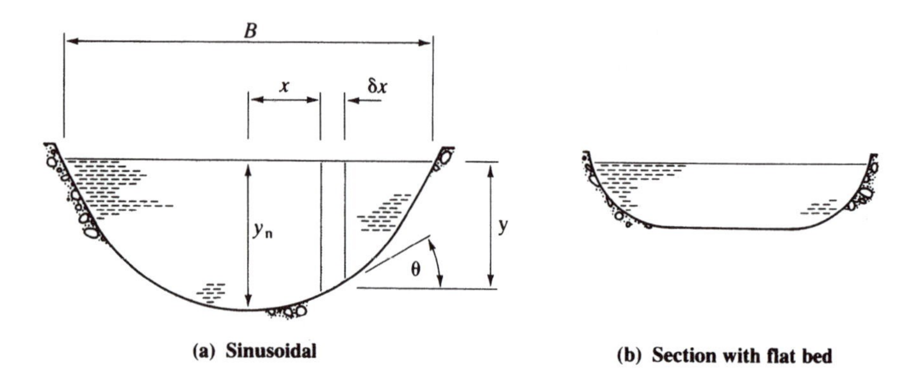

Figure 15.5 Stable section shape.

America. Using a section through a canal (Fig. 15.5(a)), the following result was derived:

$$\frac{y}{y_n} = \cos\left(\frac{x \tan\phi}{y_n}\right)$$

Thus, the stable bank profile has a sinusoidal form.

The banks may be separated by a flat base (Fig. 15.5(b)), but the shape of the bank is practically unaffected. This treatment may be further developed to yield estimates of the principal geometrical characteristics for a channel design. For example, with no flat base,

$$A = 2y_n^2/\tan\phi$$

$$P = \frac{2y_n}{\sin\phi}\int_0^{\pi/2}(1-\sin^2\phi\sin^2\theta)^{1/2}\mathrm{d}\theta$$

For channels of low to moderate width, (9.1b) can then be used to determine the bed slope appropriate to a given sediment size.

The tractive force equations tend to be used for channels through coarse non-cohesive sediments for which the threshold criterion is significant (i.e. there may be little or no transport at low flows).

In recent years a considerable amount of research has been focused on secondary currents and the distribution of boundary shear stresses in open channels. The reader's attention is drawn to Hemphill and Bramley (1989) for a review of design practice.

15.5 Morphology of natural channels

This section covers the study of the processes which are involved in the formation of natural channels. A large number of variables may be involved, not all of which may be known. The first, and most obvious, point is that the discharge in a river may be highly variable with time, being a function of the climate, geology and topography of the area. This, in turn, implies a wide variability in the nature of the sediment transport processes. Floods may denude topsoil in some areas, deposit it elsewhere, and so on. The river engineer is usually concerned with planning for the medium term, and the quality of the information available to assist in planning has improved in recent years.

The discharge

As there is no unique value of Q for a river, the question arises as to what value of Q is appropriate for making engineering estimates. A number of approaches are possible, as can be seen by the following examples.

1. The use of a numerical model for predicting the discharge hydrograph over a long period. Such an approach requires detailed information and computer analysis.
2. The use of a single value of Q, known as the 'dominant discharge'. This is defined in a number of ways by different researchers. For example, Ackers and Charlton define it as the steady discharge which would cause the same meander (bend) effect as occurs in the actual river. Alternatively, Henderson (following Leopold and Wolman) proposes that it is the discharge which would maintain a channel at its present cross-section, and which is not exceeded sufficiently often for berm build-up to occur on the banks.
3. The use of the discharge which occurs when the channel runs full–the 'bankfull' discharge.

'Braiding' and 'meandering'

The combined effect of currents and sediment transport is to modify the watercourse. Leopold and Wolman (1957) proposed that channels should be classed as straight, meandering (Fig. 15.6(a)) or braided (Fig. 15.6(b)).

For the straight and meandering cases, a single watercourse is present, whereas the braided stream is subdivided by spits. Initially straight channels with particulate boundaries will meander only if the slope and discharge exceed certain critical values, and will braid only beyond yet higher critical values. Note that the values of S_0 and Q are not independent variables here. Broadly speaking, meandering is characteristic of lowland rivers with small slopes, and braiding is characteristic of the steeper upland reaches. Braiding appears to be a natural mechanism for dissipating energy, and is associated with a degree of armouring.

It should not be assumed that braiding and meandering are always separate occurrences. A few rivers both braid and meander at certain points.

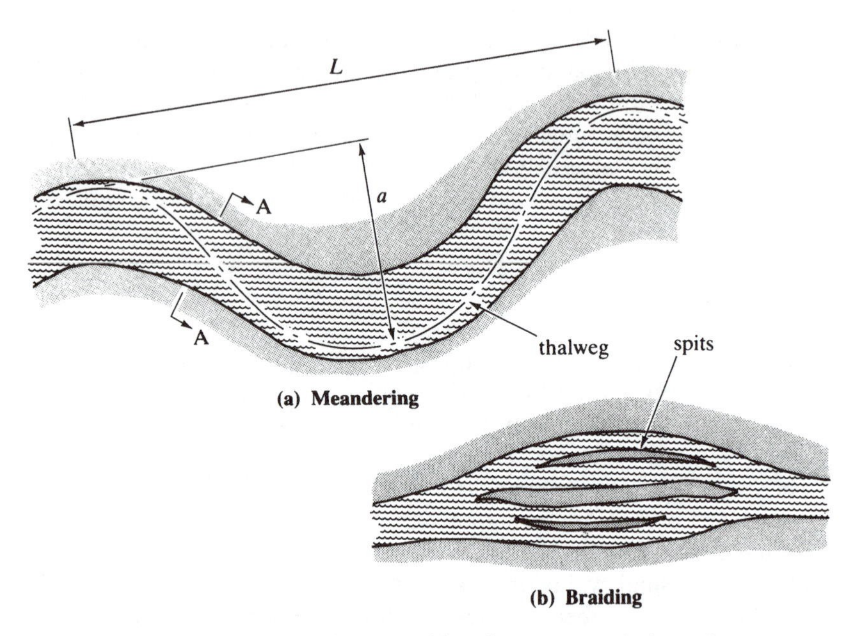

Figure 15.6 Meandering and braiding in natural channels.

Meanders may be characterized by their length, L, and amplitude, a (or radius r). They are probably brought into being (or at least strongly influenced) by the action of the asymmetrical spiral secondary current (see section 15.3) which scours the outer bank and deposits at the inner bank at a bend. The river section is therefore asymmetrical at the bend (Fig. 15.7(a)) and symmetrical at the intersection between one bend and the next (Section A–A in Fig. 15.6(a)). This means that the point of greatest depth does not lie at the centreline of the channel, but swings towards the outside of the bends. This line is known as the 'thalweg'.

This simple explanation of meanders does not command universal acceptance. Some engineers have postulated that it is the influence of small surge waves superimposed on the main flow which initiates the formation of sediment bars at certain points. The deflection of the flow by the bar then leads to meander formation.

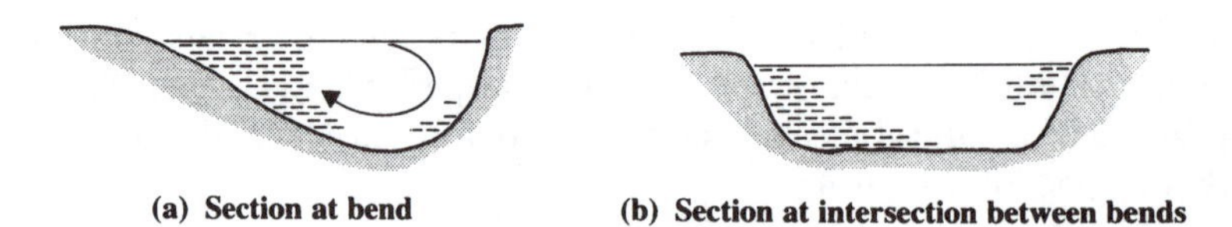

Figure 15.7 River sections.

Meander patterns are often present even in apparently straight channels, since the thalweg is often found to swing from one side of the channel to the other, thus forming a series of 'hidden bends'. Furthermore, the depth is not necessarily uniform. Measurements of depth often reveal alternating deeper and shallower zones ('pools' and 'riffles'), which exhibit much the same frequency characteristics as meanders.

Several researchers have developed relationships for the geometrical characteristics of meanders. Some examples are:

Leopold and Wolman (1960) $7 < L/B < 11, \quad 2 < r/B < 3;$
Charlton and Benson (1966) $L \propto Q^{0.515} D^{-0.285};$
Anderson (1967) $L/A^{1/2} = 72\mathrm{Fr}^{1/2}, \quad L = 46Q^{0.39}$

These equations apparently indicate a series of stable relationships (between L and Q, for example), but this disguises the complexities found in nature. Meander patterns may vary throughout the year at any one point, or they may vary from point to point due either to changes in the geological formation of the bed or to the entry of a tributary. Recent work has tended to emphasize the stochastic nature of the phenomenon (see, for example, Einstein, 1971). Other researchers have also emphasized the influence of stream power (or energy) on channel formation (for example, Lewin, 1981; Chang, 1984, pp. 106–21). Henderson (1966) proposed a criterion for the slope of a single watercourse ('straight' or meandering) in a coarse alluvial bed. In its metric form, the equation is

$$S_0 = 0.517 D_{50}^{1.14} Q^{-0.44}$$

where D_{50} is in metres and Q is in $\mathrm{m^3/s}$. Braided channels have bed slopes greatly in excess of the values derived from this equation.

Bettess and White (1983) proposed a new framework for the quantitative prediction of meandering and braiding. This emanated from their previous work on the rational approach to stable channel design. They proposed a basis for the necessary conditions for meandering and braiding to exist based on a comparison between the natural equilibrium slope of the stream and the slope of the valley in which it was situated.

Braiding is a feature of channels with steeper slopes, where flows have high energy. Therefore, the particulate banks are vulnerable to attack. It seems likely that the surplus energy is dissipated by erosion of the banks, which leads to a wider, shallower watercourse with sediment spits. Braiding may also occur where there is a heavy sediment load, or where the watercourse has (over a long timespan) varied seasonally. The latter type is observable on shingle deltas at the foot of a mountain range, where high discharges (snow melt) take a shorter route than low (summer) discharges.

Morphological computational river modelling

This constitutes a fascinating area of research which potentially has very significant practical applications. Due to the complexity of the problem only limited progress has been made to date.

In essence, the problem is to be able to predict the morphological changes in a river system due to some river engineering scheme (for example to determine the effects of dam construction on the river both upstream and downstream). Computational (flow) modelling is described in section 15.6 and, as there described, is currently limited to one-dimensional and quasi two-dimensional flows. A further complication of flood flows in compound channels, being researched at present, is discussed in section 15.7. It may thus be gleaned that one must not, therefore, expect current morphological computational models to offer more than one-dimensional predictions with any confidence.

Here, a brief description of one one-dimensional morphological computational model is presented to illustrate recent practice. The model is that of Bettess and White (1981), developed at the Hydraulics Research Station, UK, which predicts the change in bed level of a river over a number of years. They took as their flow model the one-dimensional unsteady gradually varied flow equations given in section 14.6. For sediment transport they used the Ackers and White equations given in section 9.4 and used equation (15.5) to link bed material size to frictional resistance and sediment transporting capacity (as previously described in section 15.4). A continuity equation for sediment movement (similar to the continuity equation for water flow) is also necessary (and was included). Finally they developed a method for estimating the effect of sediment sorting. This is an essential feature of their model which allows for the fact that different sediment sizes are transported downstream at different rates, which in turn affects both the frictional resistance and sediment transport at any particular point in the river.

In their 1981 paper, they described the application of the model to two test cases. The first was a study of the downstream bed level changes below a dam site. Comparison of the predicted bed level decreases below the dam with the observed decreases showed very good agreement. Additionally, the model predicted armouring of the bed immediately downstream of the reservoir. The second test case was that of accretion of the bed level upstream of a Crump weir. Here the predictions were compared with laboratory measurements. Again there was good agreement between measured and predicted values.

A more recent model of this type is that of Pender and Li (1996). This model is capable of simulating the transport, deposition and erosion of graded sediments for non-equilibrium conditions (i.e. not in regime) under unsteady flow. The model employs a so called ‘fully coupled solution method’. This means that the changing channel properties produced by the sediment transport are accounted for in the unsteady flow equations as time progresses.

One may conclude, therefore, that such one-dimensional morphological computational models can be useful tools in predicting changes in river bed levels over time due to the introduction of river engineering works.

15.6 Computational river modelling

In Chapter 14 the hydrodynamic equations for one-dimensional, unsteady, gradually varied flow were derived and their numercial solution described.

What follows in the remainder of this section is a brief discourse on the uses and limitations of some current numerical models, highlighting the various aspects of the procedures which must be carried out for the successful application of such models to the solution of river engineering problems.

Model types and applicability

Typical model types currently used in practice may be categorized as follows in increasing order of complexity:

1. simple flood routing (e.g. section 10.7)
2. steady one-dimensional flow (e.g. section 5.10)
3. unsteady one-dimensional flow (e.g. section 14.6)
4. quasi two-dimensional flow (described in this section)

Type 1 is typically used in hydrological studies to route multiple flood hydrographs through a river or storm water drainage system.

Type 2 is typically used in determining increases in river stage occasioned by the introduction of some new river engineering works (e.g. backwater curves resulting from new bridges, weirs, intake structures etc).

Types 3 and 4 are typically used when a major flood study of a river system is necessary.

Advantages and limitations

Computational models essentially are full-scale representations of the prototype, in contrast to physical models which are scaled. However, they do not model all the physical processes and the prototype topography has to be schematized and discretized. Thus computational models have certain advantages over physical scaled models and have limitations in their use.

The main advantages of computational models are:

1. they are generally less expensive than the equivalent physical scale model;
2. many alternative designs can be readily tested, quickly and cheaply;

3. the models may be stored on electronic media for future use;
4. the models do not suffer from any scaling effects.

The main limitations of computational models are:

1. They can only be applied where the main underlying physics of the flow are known and can be included in the model. (*Note*: local two- and three-dimensional effects still need to be studied using physical scale models.)
2. The minimum amount of topographical data required to obtain accurate results is difficult to quantify.
3. The presentation of the use of model results requires sophisticated graphical software.

Model accuracy

Model accuracy is dependent on a large number of factors which need to be appreciated so that undue reliance is not placed on the predictive capacity of the model. Errors in computational models arise from a number of sources including:

1. physical processes not described by the chosen equations;
2. the validity of the numerical schemes chosen to solve the equations;
3. errors in the basic topographic and hydrological data;
4. the schematization and discretization procedures;
5. the calibration procedure.

Thus the final accuracy of any computational model is very difficult to quantify. Some sources of error may be minimized but others have to be accepted and the results of the model viewed with this in mind.

Some of the physical processes not modelled in one-dimensional models are illustrated in Figures 15.8 and 15.9. Figure 15.8(a) shows a river cross-section with a rising flood in which water is transferring laterally from the main channel to the flood plains. Figure 15.8(b) shows a falling flood in which water from the flood plains is draining back into the main channel. In both circumstances the water surface is not horizontal in reality, but is assumed horizontal in a one-dimensional model. Figure 15.8(c) shows a plan view of a rising flood when overbank flow is beginning. Here the flow is seen to have components in two directions. Again this is not represented in a one-dimensional model. Figure 15.8(d) shows flow at a stage well above bank full. Under these circumstances a one-dimensional model will more accurately represent reality.

Figures 15.9(a) and (b) show circumstances in which regions of slack water and eddy currents are generated. This may even result in localized

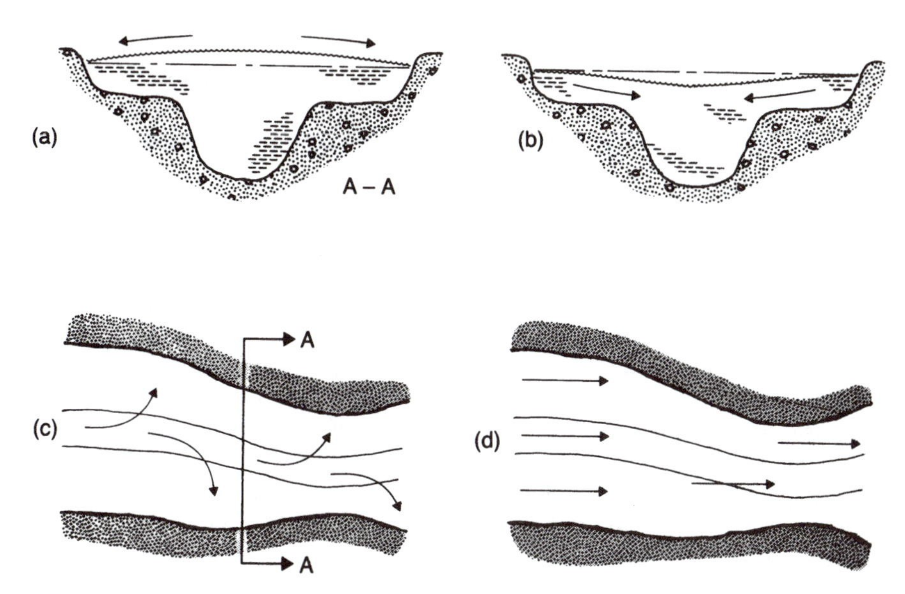

Figure 15.8 Channel and valley flow in a one-dimensional model: (a) rising flood; (b) falling flood; (c) beginning of overbank flow; (d) high flood valley flow. (After Cunge *et al.* (1980).)

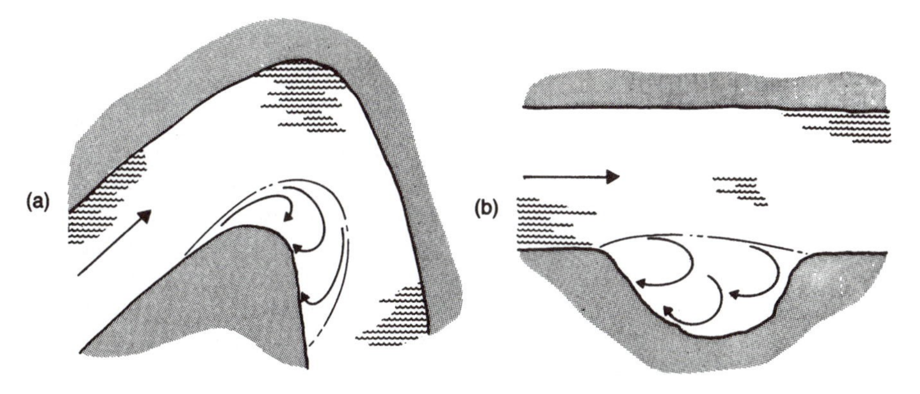

Figure 15.9 Slack water and eddy currents generated by: (a) a sharp bend; (b) an irregular channel.

reverse flow taking place. Again a one-dimensional model cannot reproduce these phenomena.

Turning next to the validity of the numerical schemes used in the model, it is not normally possible (or desirable) for a model user to adjust them. Thus it is very important to choose a computational model with an established track record and pedigree. Developments and improvements in numerical modelling schemes are still taking place and the interested reader is referred to the further reading list described at the end of this section for more details. It may be said, however, that the well-established computational models do have robust numerical schemes which perform very well under most circumstances.

Errors in the measurement of topographic data can be minimized by quality control of the survey and are generally within acceptable tolerances. However, gross errors do sometimes occur and these need to be eliminated either by quality control procedures at the data input stage (for example by plotting all the cross-sections on the computer screen) or during the model calibration procedure. Errors in the hydrological data are not so easy to eliminate and may, therefore, need to be catered for by a sensitivity analysis. For example, if say a 100-year return period flood hydrograph has been generated, it may be subject to quite wide confidence limits. Thus it would be advisable to run the computational model with several flood hydrographs to establish the sensitivity of flood stage to flood discharge.

Finally errors may be generated by an inadequate schematization and discretization of the river system and by an inadequate calibration. These two processes are key elements in the application of any computational model and are discussed separately in the next two subsections.

Schematization and discretization

Schematization refers to the process whereby all the key features of the river system are represented by some computational procedure in the model. In the case of one-dimensional models this principally involves identifying all the positions of river and flood plain cross-sections, and the location of weirs and bridges and the location of any lateral inflows.

Discretization refers to the process whereby the cross-sectional data is input to the computational model as a set of x, y coordinates (from which all the hydraulic parameters are calculated). Figure 15.10(a) is a plan view of a river valley, showing the main channel and flood plains, with the selected cross-sections marked on the plan. Figure 15.10(b) shows how this topographic information is represented in a one-dimensional river model.

Several important (if obvious) remarks concerning the discretization need to be made. Firstly the chosen cross-sections need to be representative of the river reach. Thus between cross-sections there should not be any large changes in cross-section. Secondly, the cross-sections should be drawn normal to the general flow direction. This is not a problem for the main channel, but is open to conjecture on the flood plains! Thirdly, it is very important that the discretized data preserves, as far as possible, both the river and flood plain reach lengths (and hence gradients) and the flood plain volumes.

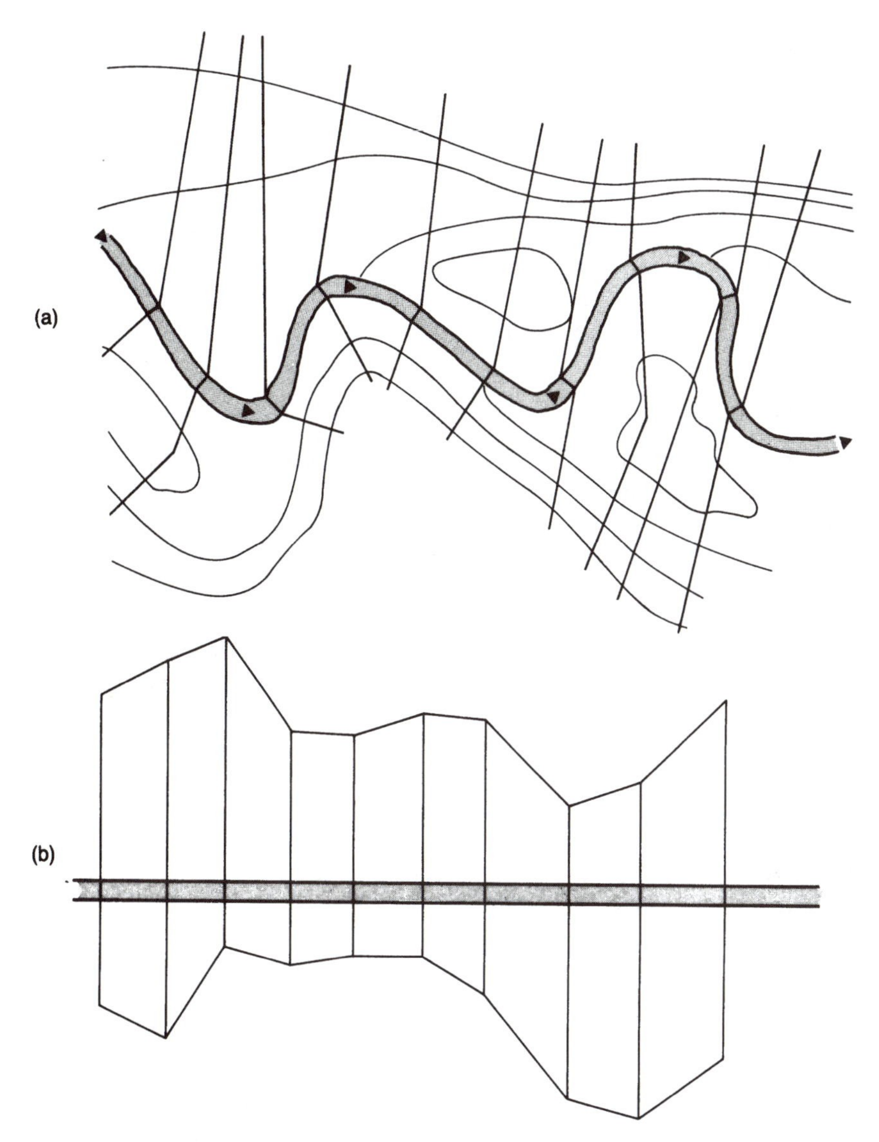

Figure 15.10 (a) Plan view of river and flood plain section lines; (b) schematic representation of (a). (After Samuels and Gray (1982) see Chapter 5 for the reference.)

Calibration and verification

Calibration (of any computational model) consists of adjusting the model parameters such that the model predictions are, as near as possible, in agreement with measured field data. In the case of river models the principal parameter to be adjusted is the friction coefficient (e.g. Manning's n or roughness height K_s) in each model river reach. To calibrate a river model, it is necessary to have recorded values of stage and discharge. Given that the friction coefficient may vary with stage it is advisable to calibrate the model for a range of stages. Where flood levels are to be predicted it is also important to obtain records of historical floods so that flood plain roughness and flooded areas can also be calibrated.

Once the model has been calibrated it should also be verified by testing the model predictions against an independent data set of recorded stages and discharges. After the model has been successfully calibrated and verified it may then be used in a predictive capacity with reasonable confidence.

Modelling bridges

In many river systems it will be necessary to model flow through bridges. Bridges come in many shapes and sizes and are often orientated at a skew angle to the main direction of the flow. Bridge crossings of rivers sometimes produce a substantial afflux defined as the increase in water level upstream due to the presence of the bridge. It is very important to be able to predict this afflux with reasonable accuracy, particularly under flood conditions. A comprehensive investigation of this problem was carried out by the US Bureau of Public Roads, and a design guide (*Hydraulics of Bridge Waterways*) was published under the authorship of Bradley (1978). In this publication calculation procedures together with a set of empirical coefficients are presented to enable the afflux to be calculated. The method incorporates procedures for allowing for the following circumstances:

1. the constriction of flood plain flow at the bridge;
2. bridge piers (size and shape);
3. flood plain eccentricity;
4. skewed crossings;
5. dual bridges;
6. abnormal stage–discharge conditions.

This method is used in commercial computational river models. More recently, new research into the afflux at arch bridges has been summarized in a report published by Hydraulics Research (1988a).

Quasi two-dimensional models

One-dimensional river models cannot be successfully applied to all river systems. In particular this applies where flow on the flood plains covers large areas, where more than one flow path is possible, where substantial interconnected flood plain storage cells are present and where the flood plains are at a lower level than the bank of the main river. Under these circumstances it is often possible to use a quasi two-dimensional model. In such models the main river can still be modelled by a one-dimensional model but the flood plains are treated separately. In essence the schematization allows for a series of interconnected 'cells'. These cells may be viewed as storage areas with inflows and outflows controlled by weir equations (or other appropriate equations). Thus storage volumes may be simply calculated using the continuity equation in a similar manner to reservoir routing. Such quasi two-dimensional commercial models are now available for general use.

Recent commercial models

Since the publication of Cunge *et al.* (1980), which is a classic text on computational river modelling, numerical techniques have continued to evolve and the major American and European hydraulic research organizations have all developed fourth generation one-and quasi two-dimensional river models. For example, ISIS (Wallingford Software Ltd and Halcrow Group Ltd, 2003), MIKE 11 (DHI Software 2003) and HEC-RAS (US Army Corps of Engineers, 2002). Details of these models can be accessed via the website (www.sponpress.com/support material/0415306094.html)

15.7 Flood discharges in compound channels

Introduction

This topic is of one of the principal concerns of river engineers. Most natural river systems consist of a main channel and one or two associated flood plains. Where man has already intervened and created artificial channels, these also often take the form of a main channel and flood channels. When the river system is in flood then the main channel is inundated and flow occurs on the flood plains. Where development has taken place on the flood plain (a very common occurrence) the consequences of flooding can be devastating both in terms of economic loss and, often tragically, loss of life. A *raison d'être* of the river engineer is to protect life and property from the consequences of flooding. Hydraulically the river engineer must be able to assess, as accurately as possible, the relationship between stage and discharge at all points in the river system, so that appropriate flood protection or alleviation measures may be undertaken.

In section 5.6, a method was described whereby the stage/discharge relationship could be evaluated for a compound channel. It was also stated that this method was only approximate and subject to wide margins of error. This is an unacceptable situation to river engineers and a considerable research effort has been devoted to this problem using the SERC flood channel facility described by Knight and Sellin (1987). What follows is a description of the problem as it is currently viewed and a summary of the research findings.

Straight compound channels

The first stage of this research was concerned with flow in straight compound channels. Figure 15.11 shows what are believed to be the main mechanisms at work in a straight compound channel. At the interface between the main channel and the flood channel, momentum is transferred across the interface by turbulent shearing action. In the main channel significant secondary flow cells develop. The whole width of the compound channel is also subjected to shearing action in both the horizontal and vertical planes. Thus it may be appreciated that any realistic method of stage/discharge assessment must account for lateral shear forces and secondary flow forces if an accurate solution is to be found.

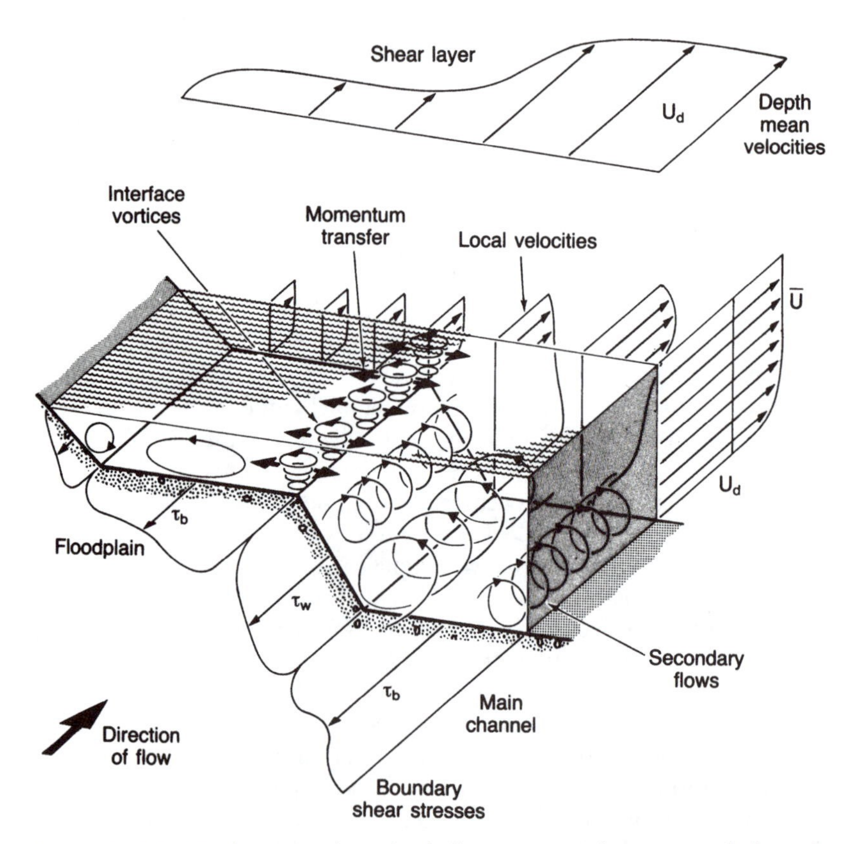

Figure 15.11 Mechanisms of overbank flow in a straight compound channel. (After Shiono and Knight, 1991.)

A fundamental approach to this problem would be to try to solve the three-dimensional Navier–Stokes equations for steady but turbulent flow. To do this requires a knowledge of the distribution of the boundary shear stresses and the adoption of a suitable turbulence model (for example the Reynolds stress model or the Prandtl eddy model described in section 3.4). Currently such an approach is subject to uncertainties regarding the necessary coefficients contained in the turbulence models. Two recent approaches to the problem are those described by Wark, Ervine and Samuels (1990) and the model of Shiono and Knight (1991). The approach of Wark, Ervine and Samuels was to solve (numerically) a two-dimensional form of the Navier–Stokes equations for steady turbulent flow which allows for lateral shear.

In their notation this is given by

$$gDS_{xf} - \frac{Bf|U|U}{8} + \frac{\partial}{\partial y}\left[v_t D\frac{\partial U}{\partial y}\right] = 0$$

or

$$gDS_{xf} - \frac{Bf|q|q}{8D^2} + \frac{\partial}{\partial y}\left[v_t \frac{\partial q}{\partial y}\right] = 0$$

gravity bed shear lateral shear

where

B = $(1+S_x^2+S_y^2)^{1/2}$: a factor relating stress on an inclined surface to stress in the horizontal plane
D = Flow depth
f = Darcy friction factor
g = Gravitational acceleration
S_x = Longitudinal slope of channel bed
S_y = Lateral slope of channel bed
x = Longitudinal coordinate direction
y = Lateral coordinate direction
q = Longitudinal unit flow (= UD)
U = Longitudinal depth averaged velocity
V_t = Lateral eddy viscosity

and in which the bed shear term may be evaluated by application of the Manning and the Darcy–Weisbach formulae in the form

$$f = \frac{8gn^2}{D^{1/3}} \quad \text{(i.e. by combining equations (5.3), (5.6) and (5.8))}$$

and the lateral shear term may be evaluated by relating the lateral eddy viscosity to bed roughness generated turbulence as in the following equation:

$$v_t = \lambda u_* D$$

where

u_* = The shear velocity = $(\tau_b/\rho)^{1/2}$
λ = The non-dimensional eddy viscosity (NEV)
ρ = Fluid density
τ_b = Bed shear stress

Thus the numerical solution provides values of the depth mean velocity and longitudinal unit flow on a grid of points across the cross-section.

The approach of Shiono and Knight has the same beginning in the Navier–Stokes equations but they have additionally allowed for secondary flows. Again, in their notation the three-dimensional Navier–Stokes equation is given by

$$\rho\left[\bar{V}\frac{\partial\bar{U}}{\partial y}+\bar{W}\frac{\partial\bar{U}}{\partial z}\right]=\rho g\sin\theta+\frac{\partial\tau_{yx}}{\partial y}+\frac{\partial\tau_{zx}}{\partial z}$$

where

$\{\overline{U},\overline{V},\overline{W}\}$ are the local mean velocities in the x (streamwise), y (lateral) and z (normal to bed) directions,
ρ is the density of the water,
g is the gravitational acceleration,
θ is the bed slope ($S_0=\sin\theta$), and
τ_{ij} is the shear stress in the j-direction on the plane perpendicular to the i-direction.

$$\left.\begin{aligned}\tau_{yx}&=-\rho\overline{u'v'}\\ \tau_{zx}&=-\rho\overline{u'w'}\end{aligned}\right\}\quad\text{are the Reynolds' stresses}$$

A depth averaged form of this equation as derived by Shiono and Knight has the following form:

$$\rho gHS_0-\rho\frac{f}{8}U_d^2\sqrt{1+\frac{1}{s^2}}+\frac{\partial}{\partial y}\left\{\rho\lambda H^2\left(\frac{f}{8}\right)^{1/2}U_d\frac{\partial U_d}{\partial y}\right\}=\frac{\partial}{\partial y}[H(\rho\overline{U}\,\overline{V})_d]$$

where f is the Darcy–Weisbach friction factor

$$\left(f=8\frac{\tau_b}{(\rho U_d^2)}\right)$$

U_d is the depth averaged streamwise velocity
τ_b is the local bed shear stress
s is the channel side slope (1 : s, vertical : horizontal) and
λ is the dimensionless eddy viscosity coefficient given by

$$\bar{\varepsilon}_{yx}=\lambda H\left(\frac{f}{8}\right)^{1/2}U_d$$

where

$\bar{\varepsilon}_{yx}$ is the depth averaged eddy viscosity

$$\bar{\tau}_{yx} = \rho\bar{\varepsilon}_{yx}\frac{\partial U_d}{\partial y}$$

This equation has been solved numerically and analytically for U_d and τ_b for a number of trial cases for which either experimental data (using the SERC flood channel facility) or field measurements had been taken. The model appears to provide very good correlation with measured stage/discharge curves, an example of which is shown in Figure 15.12.

Because this work is of such fundamental importance to hydraulic and river engineers a list of the published work (kindly supplied by Dr Donald Knight) is included in a separate appendix at the end of this chapter. Also, a first design guide, based on this work, has now been published (see Ackers, 1991, 1992, 1993).

In this design guide a new method for calculating the stage/discharge relationship in straight two-stage channels is presented, together with worked examples. The method is algebraically complex and iterative. Here, only a short introduction to the method is given. Firstly, a new parameter termed the coherence (COH) was defined, given by the ratio of the conveyance as a single channel to the sum of the conveyances of the main channel and the flood plain(s). This parameter is a good indicator of how the channel stage/discharge relationship relates to that of a single channel. At any particular flood stage the coherence can be readily calculated from the conveyances, as given in section 5.6. If this tends towards one, single-channel behaviour may be expected. If this value tends towards zero, then

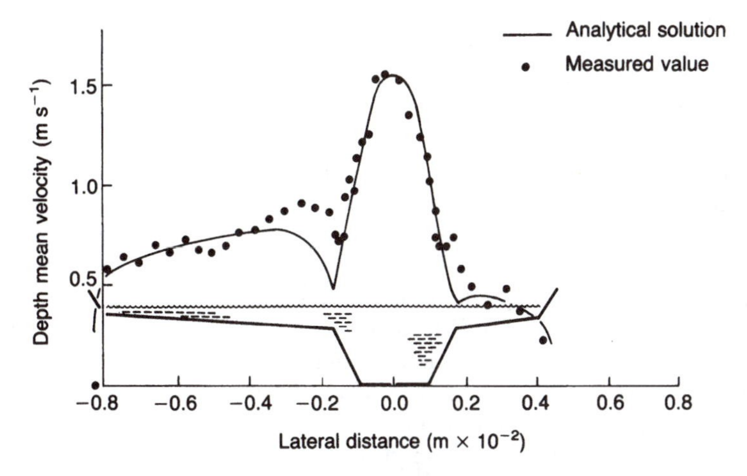

Figure 15.12 Lateral variation of depth-mean velocity at Montford. (After Knight, Shiono and Pert, 1989.)

widely different behaviour to that of a single channel may be expected. A second parameter, named the discharge adjustment factor (DISADF) was then defined by $K_D = \text{DISADF} \times K_{DB}$, where K is conveyance, D refers to the whole channel and DB to the sum of the main channel and the floodplain(s). This parameter is used to estimate the reduction in discharge at flood stages due to the interference effects between the main channel and the floodplain(s). By plotting DISADF against relative depth (ratio of difference between full depth/stage and bankfull depth/stage to bankfull depth/stage), Ackers found that four regions of flood plain flow exist. In region 1, with relatively shallow flood plain flow depths, the interference effect increases with depth. In region 2, at greater depths, interference reduces. In region 3, interference again increases. Finally in region 4, interference reduces until the channel acts as a single channel. The full procedure for calculating the actual discharge at any particular relative depth is most conveniently summarized in Appendix 1 of Ackers (1993). Suffice it to say here that the calculation involves an iterative determination of which region of flow is applicable and that different empirically derived equations are used for each region. In all cases the actual discharge calculated is less than that found by neglecting the interference effect.

Curved and meandering compound channels

The second stage of this research is concerned with flow in curved and meandering channels. Figure 15.13 illustrates flood flows in straight, curved and meandering compound channels. It can be seen that when the main channel is curved or meanders, then additional flow interactions between the main channel and the flood plains will exist, compared with the flows in a straight compound channel. The SERC flood channel facility has been used to investigate these flow interactions and some initial findings have been published by Sellin and Elliot (1990), Sellin, Ervine and Willets (1993) Ervine and Jasem (1995). Later results may also be found in the ICE journal Water and Maritime Engineering.

A design guide for curved and meandering channels has not yet appeared, but as an interim measure Greenhill and Sellin (1993) have suggested a new approach. In contrast to straight two-stage channels, where a strong vertical shear plane is present, horizontal shear planes are much more dominant in meandering channels. Their method separates the main channel and the meandering part of the floodplain horizontally, with the bed slope of the meandering floodplain taken to be equal to that of the main channel. For any remaining portion of the floodplain outside the meander belt, an inclined boundary was assumed. This method was tested against model data collected in the SERC flood channel facility and elsewhere and found to provide more accurate results than previous methods, although its accuracy is impaired for low relative depths and varies according to floodplain geometry.

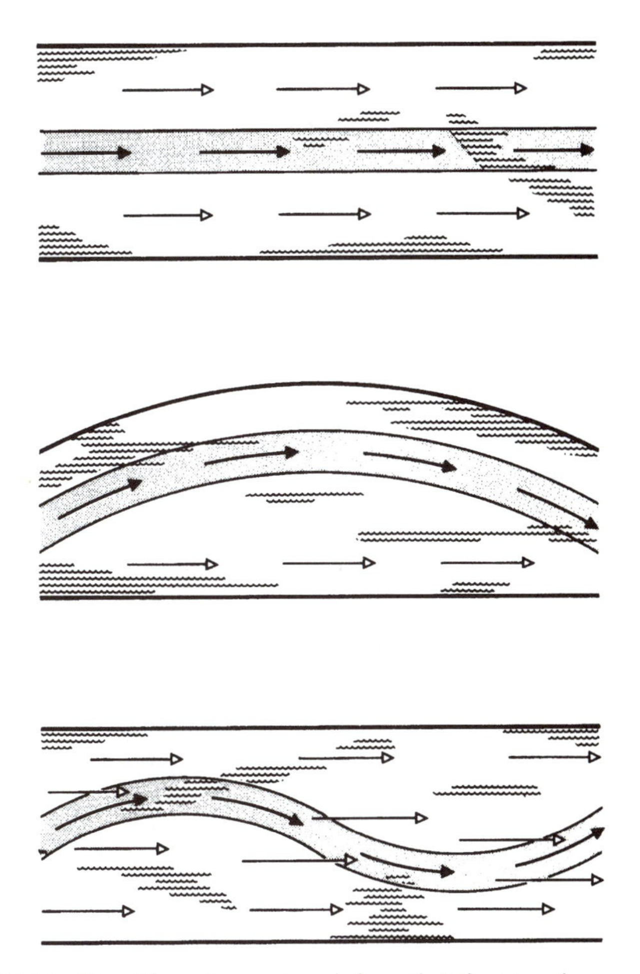

Figure 15.13 Flood flows in compound channel. (After Ramsbottom, 1989.)

15.8 River engineering

Traditional river engineering

Traditional river engineering consists of a range of constructional techniques including training, construction of embankments, docks, locks, dams, reservoirs, etc. Some of these techniques will now be reviewed.

Training

Training is the technique of confining or realigning a river to a straight and more regular course than that which occurs in nature (Fig. 15.14). In order to ensure long term stability, the banks usually have to be protected or

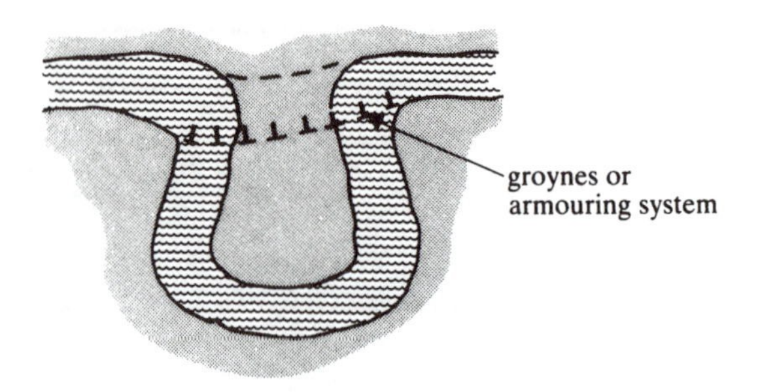

Figure 15.14 River training.

artificially armoured. A realignment usually presents a problem regarding the bed slope, since the requisite change in level now has to take place over a shorter river length. The river will to some extent readjust itself in response to the training scheme. By protecting the banks, the number of degrees of freedom has been reduced, so the adjustment process will mainly affect the slope and the depth. The shortening of the channel path often results in erosion upstream and deposition downstream, as the channel seeks to restore its natural slope. If the engineer wishes to maintain the steeper slope, it may be necessary to armour the bed artificially. Armouring often consists of the dumping of stones, though on the outer banks short groynes, plastic sheeting or sheet piling may be installed. The engineer may use regime or tractive force equations (depending on the nature of the riverbed sediments) to estimate the required geometrical characteristics for a given bankfull discharge. A number of examples of river training schemes are illustrated in Jansen *et al.* (1979).

Flood alleviation measures

Flood problems are usually most acutely felt along the lower reaches of rivers. In nature, these reaches often traverse vegetated floodplains. The floodplain provides a natural temporary storage reservoir. However, once the plain becomes populated, steps must be taken to control the extent of flooding. This may be achieved by various methods, such as:

1. the construction of embankments (Fig. 15.15);
2. the construction of embankment walls (Fig. 15.15);
3. the construction of flood storage reservoirs (on or off the river line).

The first and second of these methods increase river storage capacity, but they also raise backwater levels during the passage of a flood wave. The engineer must therefore beware of solving a problem in one place only to create another elsewhere. The use of reservoirs for routing purposes has

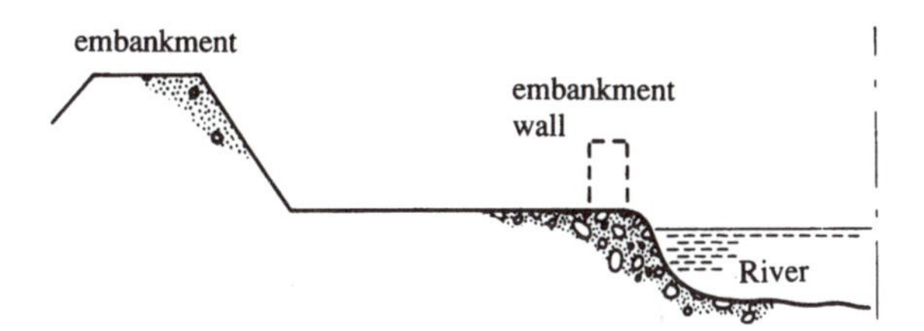

Figure 15.15 Flood protection.

been considered in section 10.7. However, great care must be taken if a dam is to be constructed across the path of the watercourse. For example, if the incoming flow is carrying a heavy sediment load, much of the sediment will settle in the reservoir (cf. Example 9.3), progressively reducing its effective capacity. This may also imply erosion of the river bed further downstream. The dam will also have considerable implications for the backwater levels. Reservoir systems often combine a water storage or hydroelectric scheme with the routing function to maximize economic viability. Design calculations for any of these (or other) systems start from hydrological data and information regarding flood waves in the existing channel.

Environmentally sound river engineering

Traditional river engineering practice may be thought of as providing a (hard) engineering solution to a site specific problem (e.g. a local flood alleviation scheme in response to a local concern about flooding). Many of the concepts introduced and discussed in this chapter have been used in this context. However, the best of modern practice in river engineering is now radically different in both approach and content. Such an approach and its form and content may be encapsulated in the term environmentally sound river engineering. The purpose of this section is to introduce the reader to some of the newly emerging philosophies and to provide references to further reading which it is essential to read if the modern philosophy is to be appreciated and understood. In a paper by Gardiner (1988) the principles of environmentally sound river engineering are propounded with a pioneering spirit. One of the principles that he expounds is that an holistic view must be taken of the whole river basin before local schemes can be properly assessed. This includes not only the traditional engineering subjects of hydrology, hydraulics and sediment transport but should also include other (environmental) topics such as river morphology and geomorphology, wildlife, landscape, amenity and recreation, fisheries, navigation, water quality, groundwater, archaeology, angling, aquatic biology and river maintenance as well as any legislative or planning requirements.

Such a holistic view is highly commendable. It is, however, very difficult to achieve without a new methodology for its implementation and the application of the techniques of cost/benefit analysis, environmental impact assessment, and sophisticated computational hydraulic modelling. Additionally, the use of geographic information management techniques is highly desirable.

Gardiner expounded his philosophy and methodology using the acronym SPIRIT representing Structure, Planning, Iterative Refinement and Interactive Teamwork. The project team needs to have specialists from all the relevant disciplines already mentioned and the emphasis of SPIRIT is on pro-active group interaction with potential solutions to specific problems being iteratively refined within the overall plan. Post-project appraisal is also seen as an essential element of the process to provide feedback. Thus long term policies, plans and schemes may be evolved holistically. Such an approach has been successfully applied by Thames Water Rivers Division, UK, as described by Gardiner, and has now become common practice in the UK through the development of catchment management plans.

Turning to more specific 'environmentally friendly' river engineering techniques, a wealth of recommendations and case study material is now available. These may be thought of as measures to 'work with nature' in a geomorphological sense and to protect and enhance the habitat of wildlife.

Brookes (1988) describes case studies of traditional river engineering schemes which have had disastrous consequences, details the effects of channelization (physical and biological) and provides some very comprehensive recommendations regarding construction, mitigation, enhancement and restoration techniques which follow the principle of 'working with nature' without precluding engineered solutions. These techniques include recognizing that river meanders, pools and riffles and bankside vegetation need to be included in engineering designs.

Turning to wildlife and nature conservancy issues, three guides have been published specifically addressing these issues by the Royal Society for Protection of Birds and the Royal Society for Nature Conservation (Lewis and Williams (1984)), the Nature Conservancy Council (Newbold, Purseglove and Holmes (1983)) and the Water Space Amenity Commission (1980).

In 1994 *The New Rivers and Wildlife Handbook* was published, produced jointly by the RSPB, NRA and the RSCN. This is an excellent source of reference, which provides details of the wildlife on, in and around British rivers and the necessary river corridor survey techniques to be employed to establish a baseline of existing habitats. It also describes a comprehensive range of river management and engineering techniques, which are designed to promote environmental enhancement while providing flood alleviation. A large number of case studies are also included, which illustrate the practical application of the techniques to a wide variety of rivers. 1999 saw the publication of the Manual of River Restoration Techniques. This manual was inspired by the success of two comprehensive restoration projects and contains a wide range of techniques and their practical application.

Finally, the role of fluvial geomorphology in river engineering is expertly reviewed by Newson (1986), and a wealth of information concerning British rivers is contained in Lewin (1981). The implications for the hydraulics of such schemes and the methods of analysis are discussed in a report published by Hydraulics Research (1988b). A new design guide for environmentally acceptable river channels was described in Ramsbottom and Fisher (1996).

References and further reading

Abbott, M. B. (1989) Reviews of recent developments in coastal modelling. See Falconer *et al.*(eds) (1989), pp. 3–19.

Ackers, P. (1983) Sediment transport problems in irrigation systems design, in P. Novak (ed.), *Developments in Hydraulic Engineering*, London, Applied Science Publishers, pp. 151–96.

Ackers, P. (1991) *Hydraulic Design of Straight Compound Channels*. Report SR 281, H. R. Wallingford, Oxfordshire.

Anderson, A. G. (1967) On the development of stream meanders. *Proc. 12th Congr. IAHR 1*, Fort Collins, Paper A46, pp. 370–8.

Bailey, R. (1991) *An Introduction to River Management*, (booklet No. 2) IWEM.

Bettess, R. and White, W. R. (1981) Mathematical simulation of sediment movement in streams. *Proc. Instn. Civ. Engrs. Part 2*, **71** (Sept), pp. 879–92.

Bettess, R. and White, W. R. (1983) Meandering and braiding of alluvial channels. *Proc. Instn. Civ. Engrs., Part 2*, **75** (Sept), pp. 525–83.

Bettess, R. and White, W. R. (1987) Extremal hypothesis applied to river regime, in Thorne, C. R., Bathhurst, J. C. and Hey, R. D. (eds), *Sediment Transport in Gravel Bed Rivers*, John Wiley, Chichester, pp. 767–89.

Blench, T. (1957) *Regime Behaviour of Canals and Rivers*, Butterworths, London.

Blench, T. (1966) *Mobile Bed Fluviology*, T Blench, Alberta.

Bradley, J. N. (1978) *The Hydraulics of Bridge Waterways*, 2nd edn, US Bureau of Public Roads, Washington, USA.

Brandon, T. W. (ed.) (1987) *River Engineering Part 1, Design Principles*, Institution of Water and Environmental Management, London.

Brookes, A. (1988) *Channelised Rivers. Perspectives for Environmental Management*, John Wiley, Chichester.

Callow, P. and Petts, G. (eds) (1992) *The Rivers Handbook: Hydrological and Ecological Principles*, Vol. 1, Blackwell, Oxford.

Callow, P. and Petts, G. (eds) (1994) *The Rivers Handbook: Hydrological and Ecological Principles*, Vol. 2, Blackwell, Oxford.

Chang, H. H. (1984) Analysis of river meanders. Am. Soc. Civ. Engrs., *Journal of Hydraulic Engineering*, **110**(1), pp. 37–50.

Charlton, F. G. and Benson, R. W. (1966) *Effect of Discharge and Sediment Charge on Meandering of Small Streams in Alluvium*. Symp. CWPRS II, Poona, pp. 285–90.

Charlton, F. G., Brown, P. M. and Benson, R. W. (1978) *The Hydraulic Geometry of some Gravel Rivers in Britain.* Report INT-180, Hydraulics Research Station, Wallingford.
Cunge, J. (1989) Review of recent developments in river modelling. See Falconer *et al.* (eds) (1989) pp. 393–410.
Cunge, J. A., Holly, F. M. (Jr) and Verwey, A. (1980) *Practical Aspects of Computational River Hydraulics*, Pitman, London.
Einstein, H. A. (1971) Probability; statistical and stochastic solutions. *Proc. First Symp. Stochastical Hydraulics*, University of Pittsburgh, Pittsburgh, USA.
Falconer, R. A., Goodwin, P. and Matthew, R. S. (eds) (1989) *Hydraulic and Environmental Modelling of Coastal, Estuarine and River Waters*, Gower Technical, Aldershot, UK.
Gardiner, J. L. (1988) Environmentally sound river engineering: examples from the Thames Catchment, *Regulated Rivers: Research and Management*, **2**, 445–69.
Gardiner, J. L. (ed.) (1991) *River Projects and Conservation: A Manual for Holistic appraisal*, John Wiley and Sons, Chichester.
Hemphill, R. W. and Bramley, M. E. (1989) *Protection of River and Canal Banks*, CIRIA, Butterworths, London.
Henderson, F. M. (1966) *Open Channel Flow*, Macmillan, New York.
Hey, R. D. (1986) River mechanics, *JIWES*, **40** (April), pp. 139–58.
Hydraulics Research (1988a) *Afflux at Bridge Arches*. HR report SR 182. Wallingford.
Hydraulics Research (1988b) *Assessing the Hydraulic Performance of Environmentally Acceptable Channels*, HR Report, EX1799. Wallingford UK.
Jansen, P., van Bendegom, L. and van der Berg, J. (1979) *Principles of River Engineering*, Pitman, London.
Knight, D. W., Yuen, K. W. H., and Alhamid, A. A. I. (1992) Boundary shear stress distributions in open channel flow. To appear in *Physical Mechanisms of Mixing and Transport in the Environment* (eds K. Bevan, P. Chatwin and J. Millbank) Wiley, Chichester. In preparation.
Lacey, G. (1953) Uniform flow in alluvial rivers and canals. *Proc. Instn. Civ. Engrs.* **237**, p. 421.
Lane, E. W. (1953) Progress report on studies on the design of stable channels by the Bureau of Reclamation. *Proc. Am. Soc. Civ. Engrs.*, No 280 (September).
Leopold, L. B. and Wolman, M. G. (1957) *River Channel Patterns: Braided, Meandering and Straight*, US Geol. Survey, Prof. Paper 282-B.
Leopold, L. B. and Wolman, M. G. (1960) River meanders. *Bull. Geol. Soc. Am.*, **71**, pp. 769–94.
Lewin, J. (ed.) (1981) *British Rivers*, Allen & Unwin, London.
Lewis, G. and Williams, G. (1984) *Rivers and Wildlife Handbook*, RSPB/RSNC.
Manual of River Restoration Techniques. RRC 1999, Silsoe, UK.
Newbold, C., Purseglove, J. and Holmes, N. (1983) *Nature Conservancy and River Engineering*, Nature Conservancy Council, Peterborough.
Newson, M. D. (1986) River basin engineering – fluvial geomorphology. *JIWES*, **40**(4), pp. 307–24.
Nixon, M. (1959) A study of the bank-full discharges of rivers in England and Wales. *Proc. Inst. Civ. Engrs.*, **12**, pp. 157–74.
Novak, P. (ed.) (1983) *Developments in Hydraulic Engineering* – 1, Applied Science Publishers, London.

Pender, G. and Ellis, J. (1990) Numerical simulation of overbank flooding in rivers. See White (ed.) (1990) pp. 403–12.

Pender, G. and Li, Q. (1996) Numerical prediction of graded sediment transport. *Proc. Instn. Civ. Engrs., Wat., Marit. & Energy*, **118** (Dec), pp. 237–45.

Petts, G. and Gurnell, A. (eds) (1995) *Changing River Channels*, Wiley, Chichester.

Purseglove, J. (1989) *Taming the Flood*, Oxford University Press, Oxford.

Ramsbottom, D. and Fisher, K. (1996) Design of environmentally acceptable river channels. *MAFF 31st Annual Conference of River and Coastal Engineers*, July, 1.3.1–1.3.16.

RSPB, NRA, RSNC (1994) *The New Rivers and Wildlife Handbook*, RSPB.

Samuels, P. G. (1985) Modelling open channel flow using Preissman's scheme. *2nd Int. Conf. Hydraulics of Floods and Flood Control*, BHRA, Paper B2.

Samuels, P. G. (1989) Some analytical aspects of depth averaged flow models. See Falconer *et al.* (eds) (1989) pp. 411–18.

Samuels, P. G. (1990) Cross-section location in 1-D models. See White (ed.) (1990) pp. 351–8.

Simons, D. B. and Albertson, M. L. (1963). Uniform water conveyance in alluvial material. *Trans. Am. Soc. Civ. Engrs*, **128**(1), pp. 65–167.

Skeels, C. P. and Samuels, P. G. (1989) Stability and accuracy analysis of numerical methods modelling open channel flow. *Proc. Hydrocomp. 89*, Dubrovnik, Yugoslavia, Elsevier, London.

Slade, J. E. and Samuels, P. G. (1990) Modelling complex river networks. See White (ed.) (1990) pp. 351–8.

Tagg, A. F. (1985) Computational modelling of the River Stour, Dorset UK. *2nd Int. Conf. Hydraulics of Floods and Flood Control*, Paper F1, BHRA.

Tagg, A. F. and Samuels, P. G. (1989) Modelling flow resistance in tidal rivers. See Falconer *et al.* (eds) (1989) pp. 441–52.

Water Space Amenity Commission (1980) *Conservation and Land Drainage Guidelines*, WSAC, London.

White, W. R. (ed.) (1990) *International Conference on River Flood Hydraulics*, Wiley, Chichester.

White, W. R., Bettess, R. and Paris, E. (1982) Analytical approach to river regime. *Journal of the Hydraulics Division, ASCE*, **108** (HY10).

White, W. R., Paris, E. and Bettess, R. (1980) The frictional characteristics of alluvial streams; a new approach. *Proc. Instn. Civ. Engrs., Part 2*, **69**, pp. 737–50.

White, W. R., Paris, E. and Bettess, R. (1981) *Tables for the Design of Stable Alluvial Channels*, Report IT208, Hydraulics Research Station, Wallingford.

Yalin, M. S. (1992) *River Mechanics*, Pergamon Press.

Appendix 15.A: Published work on the SERC Flood Channel Facility (in chronological order)

1. Hollinrake, P. G. (1987) *The Structure of Flow in Open Channels – A Literature Search*, Vol. 1, Hydraulics Research, Report No. SR96, January, pp. 1–327.
2. Knight, D. W. and Sellin, R. H. J. (1987) The SERC Flood Channel Facility. *Journal of the Institution of Water and Environmental Management*, **41**(4), August, pp. 198–204.

3. Hollinrake, P. G. (1988) *The Structure of Flow in Open Channels – A Literature Search*, Vol. 2, Hydraulics Research, Report No. SR153, January, pp. 1–126.
4. Samuels, P. G. (1988) Lateral shear layers in compound channels. *Proc. Int. Conf. on Fluvial Hydraulics 88*, Vituki, Budapest, Hungary, May.
5. Wormleaton, P. R. (1988) Determination of discharge in compound channels using the dynamic equation for lateral velocity distribution. *Proc. Int. Conf. on Fluvial Hydraulics 88*, Budapest, Hungary, May.
6. Shiono, K. and Knight, D. W. (1988) Two dimensional analytical solution for a compound channel, *3rd Int. Symp. on Refined Flow Modelling and Turbulence Measurements*, Tokyo, Japan, July, pp. 503–510.
7. Ramsbottom, D. M. (1989) Flood discharge assessment, Hydraulics Research, Report SR195, March, pp. 1–148.
8. Knight, D. W. (1989) Hydraulics of flood channels, in *Floods: hydrological, sedimentological and geomorphological implications* (eds K. Beven and P. Carling), Chapter 6, Wiley, pp. 83–105.
9. Hollinrake, P. G. (1989) *The Structure of Flow in Open Channels – A Literature Search*, Vol. 3, Hydraulics Research, Report No. 209, June, pp. 1–68.
10. Shiono, K. and Knight, D. W. (1989) Vertical and transverse measurements of Reynolds stress in a shear region of a compound channel. *7th Int. Symp. on Turbulent Shear Flows*, Stanford, August, Vol. 2, pp. 28.1.1–1.6.
11. Sellin, R. H. J., and Giles, A. (1989) Channels with spilling flood plains. *Proc. 23rd Congr, IAHR*, Ottawa, Canada, August, pp. B.499–506.
12. Myers, W. R. C. and Martin, L. A. (1989) River Main Project. *MAFF Conference of River and Coastal Engineers*, Loughborough, July.
13. Samuels, P. G. (1989) The hydraulics of two stage channels – review of current knowledge. *MAFF Conference of River and Coastal Engineers*, Loughborough, July.
14. Knight, D. W., Shiono, K. and Pirt, J. (1989) Prediction of depth mean velocity and discharge in natural rivers with overbank flow. *Proc. Int. Conf. on Hydraulics and Environmental Modelling of Coastal, Estuarine and River Waters*, University of Bradford, September. Gower, Aldershot, pp. 419–28.
15. Samuels, P. G. (1989) Some analytical aspects of depth averaged flow models. *Proc. Int. Conf. on Hydraulics and Environmental Modelling of Coastal, Estuarine and River Waters*, University of Bradford, September, Gower, Aldershot pp. 411–18.
16. Myers, W. R. C. and Brennan, E. K. (1990) Flow resistance in compound channels. *Journal of Hydraulic Research, IAHR*, **28**(2), pp. 141–55.
17. Wormleaton, P. R. and Merrett, D. (1990) An improved method of calculation for steady uniform flow in prismatic main channel/flood plain sections. *Journal of Hydraulic Research, IAHR*, **28**(2) pp. 157–74.
18. Knight, D. W. and Shiono, K. (1990) Turbulence measurements in a shear layer region of a compound channel. *Journal of Hydraulic Research, IAHR*, **28**(2), pp. 175–96.
19. Sellin, R. H. J. and Elliott, S. (1990) SERC flood channel facility: skewed flow experiments. *Journal of Hydraulic Research, IAHR*, **28**(2), pp. 197–214.
20. Hollinrake, P. G. (1990) *The Structure of Flow in Open Channels – A Literature Search*, Vol. 4, Hydraulics Research, Report No. SR227, March, pp. 1–99.

21. Knight, D. W., Samuels, P. G. and Shiono, K. (1990) River flow simulation: research and developments. *Journal of the Institution of Water and Environmental Management*, **4**(2), pp. 163–75.
22. Shiono, K. and Knight, D. W. (1990) Turbulence measurements by LDA in complex open channel flows. *Proc. 5th Int. Symp. on Applications of Laser Techniques to Fluid Mechanics*, Lisbon, Portugal, (July).
23. Guymer, I., Brockie, N. J. W. and Allen, C. M. (1990) Towards random walk models in a large scale laboratory facility. *Proc. Int. Conf. on Physical Modelling of Transport and Dispersion*, M. I. T., USA, August, pp. 1281–6.
24. Shiono, K. and Knight, D. W. (1990) Mathematical models of flow in two or multi-stage straight channels. *Proc. Int. Conf. on River Flood Hydraulics*, Wallingford, September, J. Wiley, Chichester, pp. 229–38.
25. Wark, J. B., Ervine, D. A. and Samuels, P. G. (1990) A practical method of estimating velocity and discharge in compound channels. *Proc. Int. Conf. on River Flood Hydraulics*, Wallingford, September, Wiley, Chichester, pp. 163–72.
26. Shiono, K. and Knight, D. W. (1991) Turbulent open channel flows with variable depth across the channel. *Journal of Fluid Mechanics*, **222**, pp. 617–46.
27. Ackers, P. (1992) Hydraulic design of two stage channels. *Proc. Instn. Civ. Engrs, Wat., Marit. & Energy*, **96** (Dec), pp. 247–57.
28. Ackers, P. (1993) Stage discharge functions for two- stage channels: The impact of new research. *J. IWEM*, **7** (Feb), pp. 52–61.
29. Greenhill, R. K. and Sellin, R. H. J. (1993) Development of a simple method to predict discharges in compound meandering channels. *Proc. Instn. Civ. Engrs, Wat., Marit. & Energy*, **101** (March), pp. 37–44.
30. Sellin, R. H. J., Ervine, D. A. and Willets, B. B. (1993) Behaviour of meandering two-stage channels. *Proc. Instn. Civ. Engrs, Wat., Marit. & Energy*, **101** (June), pp. 99–111.
31. Ervine, D. A. and Jasem, H. K. (1995) Observations on flows in skewed compound channels. *Proc. Instn. Civ. Engrs, Wat., Marit. & Energy*, **112** (Sept), pp. 249–59.

16

Coastal engineering

16.1 Introduction

The coastal zone has an important influence on the well-being of communities whose homes and industries lie along the coast. Civil engineers are called upon to provide protection against storm or flood damage, to maintain or improve beaches, to construct ports or marinas, or to control pollution. One of the major developments of the last decade has been the use of computers to simulate coastal processes and to model the impact of engineering works. This has occurred in parallel with a better understanding of, and concern for, the environment. Within the confines of this chapter it is possible only to summarize some aspects of wave action and sediment transport, coast protection and sea defence systems. Emphasis has also been placed on considering the environmental impacts and opportunities of coastal defence techniques.

16.2 The action of waves on beaches

This may seem a strange point from which to start discussing coastal engineering works. However, it is appropriate because an understanding of natural coastal defence mechanisms gives insights into suitable forms of engineering structures for coastal protection and sea defence.

Beaches form a natural coastal protection system. It is only when these are inadequate that further measures should be contemplated. Examples of such inadequacies include erosion of the coastline (which will result in encroachment on existing buildings and roads, etc.) and insufficient beach height (resulting in flooding by overtopping).

The action of waves on beaches depends on the type of wave and the beach material. For simplicity, wave types are generally categorized as storm waves or swell waves (refer to Chapter 8), and beach materials as sand or shingle.

As waves approach the shore they break. Where the bed slope is small, the breaking commences well offshore. The breaking process is gradual, and produces a 'surf zone' (see section 8.4), in which the wave height decreases progressively as waves approach the shore. Where the bed slope is steeper (say roughly 1 in 10) the width of the surf zone may be small or negligible, and the waves break by plunging. For very steep slopes the waves break by surging up onto the shore.

The incoming breaker will finally impact on the beach, dissipating its remaining energy in the 'uprush' of water up the beach slope. The water velocities reduce to zero and then form the 'backwash', flowing down the beach, until the next breaker arrives.

In the surf zone the seabed will be subject to a complex set of forces. The oscillatory motion due to the passage of each wave produces a corresponding frictional shear, and both incoming and reflected waves may be present. The longshore current (section 8.4) will also produce a shear stress. Finally, the bed slope itself implies the existence of a component of the gravitational force along the bed. On the beach itself forces are produced as a result of bed friction and the impact of the breaker.

16.3 Sediment transport

If the seabed and beach are of mobile material (sand or shingle), then it may be transported by the combination of forces outlined above. The 'sorting' of beach material (with larger particles deposited in one position and finer particles in another) can also be explained. For convenience, coastal sediment transport is divided into two components: perpendicular to the coastline (cross-shore transport) and parallel to the coastline (longshore transport or 'littoral drift').

Whether beaches are stable or not depends on the rates of sediment transport over long periods. The transport rates are a function of the wave, breakers and currents. Waves usually approach a shoreline at an angle α. The wave height and angle will vary with time (depending on the weather). Sediment may be transported by unbroken waves and/or currents; however, only transport due to breaking waves will be considered here.

Cross-shore transport on beaches

Under constant wave conditions, any beach will tend to form an equilibrium beach slope on which the net sediment movement is zero. The equilibrium beach slope will increase with increasing grain size. Conversely, for a given grain size, the equilibrium beach slope will reduce with increasing wave steepness.

There are several known mechanisms of on- and offshore movement, which may be explained as follows. Under swell conditions, the wave heights are small and their period long. When the waves break, material is thrown into suspension and carried up the beach (as bed and suspended load) in the direction of movement of the broken wave (the uprush). The uprush water percolates into the beach, so the volume and velocity of backwash water is reduced. Sediment is deposited by the backwash when the gravity forces predominate. The net result is an accumulation of material on the beach. In addition, the beach material is naturally sorted, with the largest particles being left highest on the beach and a gradation of smaller particles seaward. Under storm conditions, the waves are high and steep fronted, and have shorter periods. Consequently, the volumes of uprush are much larger, and the beach is quickly saturated. Under these conditions the backwash is much more severe, causing rapid removal of beach material. Also, a hydraulic jump often forms when the backwash meets the next incoming wave. This puts more material into suspension, which is then dropped seaward of the jump. The net result is depletion of the beach.

Cross-shore transport is also affected by the wave shape and by undertow. In shallow water, waves become progressively more asymmetrical in form. Under the wave crests, the velocity is directed onshore and has a higher value than that under the troughs, which is directed offshore. However, the crest velocities persist for a shorter time than the trough velocities. Thus finer sediment migrates offshore and coarser sediment onshore. A strong undertow can also be generated in the surf zone. This is an offshore-directed flow near the bed, which results from the near-surface onshore-directed flow caused by the breaking waves. These flows carry suspended sediment shorewards and bed load seawards.

On sand beaches, the material moved offshore is often deposited seaward of the breaker line as a sand bar. During storm conditions, the formation of such a bar has the effect of causing waves to break at a greater distance from the beach, thus protecting the beach head from further attack. The subsequent swell waves then progressively transport the bar material back onto the beach in readiness for the next storm attack.

Finally, the presence of long waves in the surf zone, briefly introduced in section 8.4, can have a strong influence on surf zone sediment movements, producing complex three-dimensional features, such as beach cusps and bar systems (see Huntley *et al.* (1993) for further details).

From the foregoing, it can be appreciated that beaches are an excellent means of coastal protection. Provided that sufficient beach material is available, and that the building line is kept behind the upper limit of beach movement, no further defence is necessary. However, these remarks are applicable only to stable or accreting beaches. In locations where beaches are depleting, or where no natural beach exists, other measures are necessary.

Equilibrium profiles and the depth of closure. The concept of an equilibrium profile is apparently at odds with the foregoing description of storm and swell profiles. However, if the profile is considered over a longer time period of the order of years, rather than a timescale of the order of storm events or seasons, then it has been found that many ocean-facing coastlines exhibit a concave curve, which becomes more gently sloped with distance offshore. Bruun (1954) and later Dean (1991) showed that this profile could be described by the equation

$$h = Ax^{2/3} \tag{16.1}$$

where h is the profile depth at a distance x from the shoreline, and A is a constant, which has been related to grain size by Dean ($A \equiv 0.21\,\mathrm{D}^{0.48}$ with D in mm). These equations predict that equilibrium beach slopes increase in steepness with increasing grain size. Dean also demonstrated how the profile equation could be related to physical principles, as follows.

The starting point is the assumption that an equilibrium profile will be such that a uniform energy dissipation per unit volume (D_e) in the surf zone will exist. Hence we may write

$$\frac{1}{h}\frac{\mathrm{d}P}{\mathrm{d}x} = D_e \tag{16.2}$$

In shallow water,

$$P = \frac{1}{8}\rho g H^2 (\sqrt{gh}) \tag{16.3}$$

Assuming spilling breakers, then

$$H = \gamma h \tag{16.4}$$

If equations (16.4) and (16.3) are substituted into (16.2) and then (16.2) is integrated, with $h = 0$ for $x = 0$, then the result is equation (16.1), in which

$$A = \left(\frac{24D_e}{5\rho g^{2/3}\gamma^2} \right)^{2/3}$$

This is a constant whose value can be related to grain size, as stated above.

The depth of closure, d_c, is defined as the vertical distance between the limit of wave uprush on the beach and the water depth at which waves can no longer produce any measurable change in the seabed profile. Where suitable records exist, this depth can be determined from profile data. In the absence of such records it has been shown to be of the order of $1.75H_{s12}$, where H_{s12} is the significant wave height with a frequency of occurrence of 12 h per year. Of course, the depth of closure is not really constant, but will vary with the incident wave conditions. However, when considering timescales for morphological change it is a useful parameter.

The concepts of an equilibrium profile and depth of closure have proved extremely useful in the design of beach nourishment schemes and in modelling shoreline evolution. Further details, their application in design and references to the original works may be found in the *Beach Management Manual* (Simm *et al.*, 1996).

The Bruun rule for beach erosion resulting from sea level rise. This is a simple geometric expression for shoreline recession Δx, resulting from a rise in sea level ΔS, first proposed by Bruun (1962, 1983). The principle is that an initial equilibrium profile of length l for a given depth of closure d_c will re-establish itself further landward and higher by a depth ΔS after the sea level rise (as the depth of closure remains constant). This implies that the material eroded on the upper part of the profile is deposited on the lower part of the profile. Hence

$$\Delta x d_c = l \Delta S$$

or

$$\Delta x = \frac{l}{d_c} \Delta S$$

As l is in general much larger than d_c, the shoreline recession will also be much larger than the rise in sea level.

Longshore transport ('littoral drift')

To some extent the mechanisms associated with transport of sand may be differentiated from transport of shingle. Thus for a sand seabed the oscillatory force due to the passage of a (breaking) wave will tend to stir the sediment into motion. The bed shear due to the longshore current can then transport the sand. Most shingle beaches form a bank along the shoreline. Transport is pronounced when plunging breakers form, and significant energy is being dissipated. The particles may undertake short trajectories or move as bedload. As the flow in the uprush is perpendicular to the wave crest and in the backwash is perpendicular to the beach contours, the shingle describes a 'sawtooth', or zig-zag, path along the beach.

Littoral drift does not of itself cause beach accretion or depletion. These occur only when there is an imbalance of supply of material between one point and another along the coast. For example, the construction of a large breakwater (to protect a port or marina) or the installation of a new groyne system may reduce the longshore transport locally, and in consequence beaches further down the coast may be 'starved' of material and be depleted.

Estimating longshore transport. An appreciation of the importance of beach processes is fundamental to good engineering design of coast protection works. Unfortunately, quantitative estimation of sediment transport rates is extremely difficult. Changes in beach volumes may be calculated from data derived from ground or aerial surveys. If surveys are carried out over several years a trend for accretion or depletion may be discernible. This is not necessarily a direct measure of the longshore transport rate along the coast. Rather it is an indication of any imbalances in the supply of sediment from one point to another. However, where marine structures are constructed that cut off the supply from further up the coast, comparisons of beach volumes before and after construction can give some indication of the longshore transport rates.

Direct measurement of longshore transport has been attempted using a variety of techniques, such as deposition of a tracer material (radioactive, dyed or artificial sediment) or installation of traps. Examples of field measurements may be found in Bruno *et al.* (1980), Chadwick (1989), and Kraus and Dean (1987). A comprehensive review of field data for longshore sediment transport may be found in Schoonees and Theron (1993).

Various attempts have been made to enable engineers to predict longshore transport rates. Broadly speaking, there are two approaches. The first is to estimate the total rate of transport, and the second is to model the distribution of the transport rate across the surf zone.

A well established total transport equation is found in US Army Corps of Engineers (1984):

$$Q_{LS} = \frac{1}{2}\frac{KP\sin 2\alpha}{(\rho_s - \rho)g(1-p_s)} \tag{16.5}$$

where Q_{LS} is the volumetric longshore transport (m^3/s), P is the power of the breaking wave (cf. equations (8.7)–(8.9)), p_s is the porosity, α is the angle between the wave front and the shoreline, ρ_s is the sediment density (kg/m^3), and K is a dimensionless coefficient, which is a function of particle size. For sand the value for K suggested in the *Shore Protection Manual* is 0.39 (using H_s for wave height); for shingle K would reduce to roughly 1/10th to 1/20th of that used for sand, depending on the particle size. See Van Wellen *et al.* (2000) for a recent review.

Following extensive laboratory tests and analysis of field data, Kamphuis (1990, 1991) proposed a total transport formula in the form

$$Q_{LS} = 6.4 \times 10^4 H_{sb}^2 T_p^{1.5} m^{0.75} D_{50}^{-0.25} (\sin 2\alpha)^{0.6} \tag{16.6}$$

Here D_{50} is the mean grain size, H_{sb} is the significant wave height at breaking, m is beach slope, Q_{LS} is longshore transport rate in m^3/annum, and T_p is the peak wave period.

While the transport of shingle on beaches takes place across a relatively narrow zone traversed by the plunging breakers, sand transport may take place right across the surf zone. By means of a surf zone model the hydraulic losses due to bed shear and wave breaking may be estimated for a series of points. If an appropriate transport equation is used, the corresponding transport rate at each point may be found. This approach was pioneered by Bijker (1971), who adapted a transport formula originally developed for estimating sediment transport in rivers. However, this assumes that bed friction is the principal cause of transport. In the surf zone, energy is lost because of bed friction and wave breaking. For this reason Morfett (1991) suggested that sediment transport could be related to the rate of energy dissipation, and developed a new transport formula based on this approach. Either of these two methods yields the distribution of sediment transport, and the total transport has to be found by numerical integration:

$$Q_{LS} = \int q_{LS} dx \qquad (16.7)$$

where q_{LS} is the sediment transport rate per unit width at a point.

The approaches of Bijker and Morfett use period-averaged values for the flow parameters. In effect this reduces the unsteady flow field to an equivalent steady state flow, which substantially simplifies the equations or numerical techniques used. Chadwick (1991a,b), amongst others, developed a model representing the depth-averaged unsteady flow field and the corresponding instantaneous sediment transport across the surf zone. The period-averaged distribution of sediment transport may then be found, and hence the total transport via equation (16.7).

Some of the most recent developments have used a Boussinesq model for surf zone hydrodynamics coupled to various sediment transport equations. The interested reader is referred to Fredsoe and Deigaard (1992) in the first instance and to Lawrence *et al.* (2003) for a recent example.

16.4 Shoreline evolution modelling

If a natural beach has an adequate supply of sand or shingle then it may remain in stable equilibrium over an extended period (Fig. 16.1(a)). However, if the sediment transport is intercepted (by a natural or artificial feature) then the beach will accrete on the updrift side of the feature (Fig. 16.1(b)). For a large structure, such as a breakwater, it is possible that all sediment will be trapped, and that the coastline on the downdrift side will be starved of sediment and will deplete. From a consideration of equation (16.5), transport rate is a function of angle α between the wave front and the beach contour. However, beach accretion alters the line of the

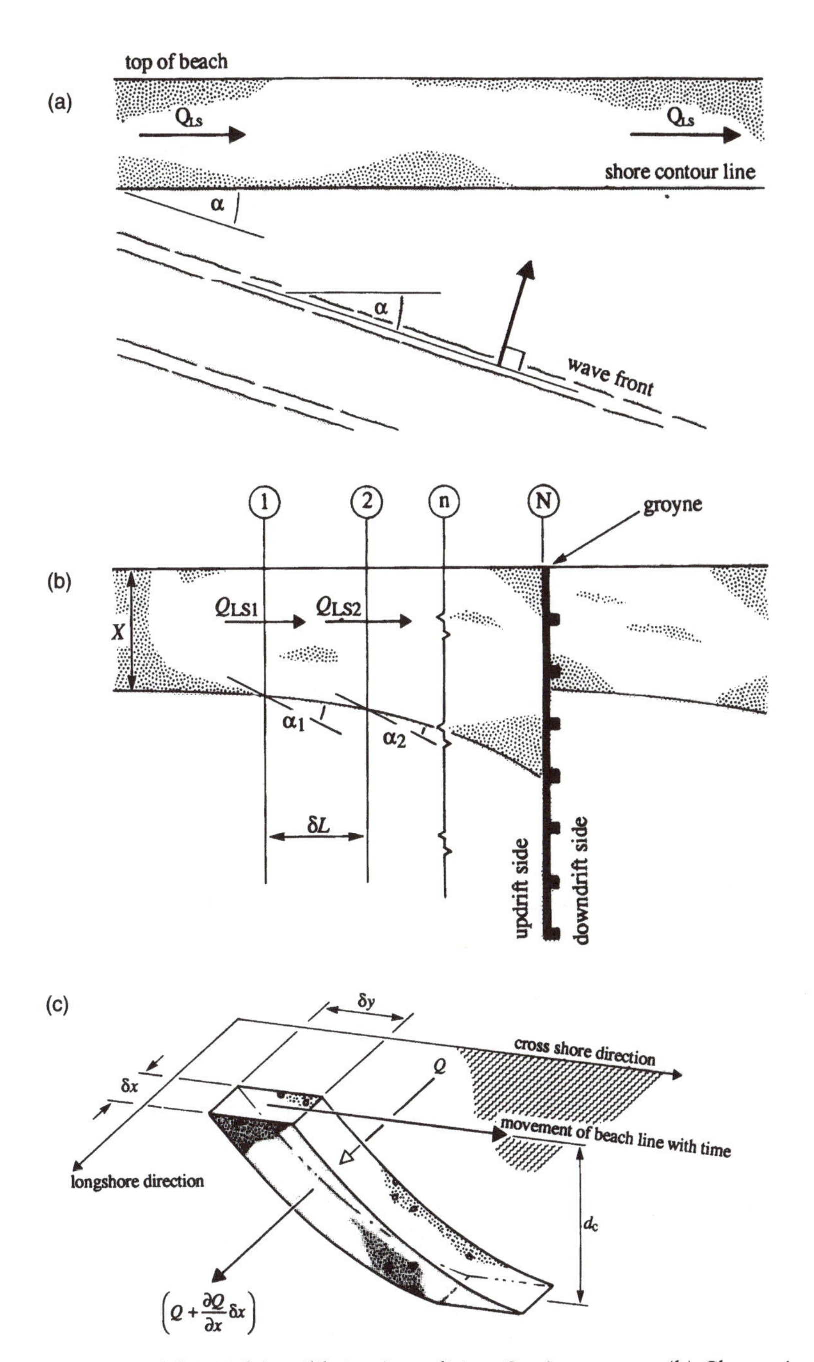

Figure 16.1 (a) Initial (equilibrium) condition Q_{LS} is constant. (b) Change in contour line at time t due to construction of groyne. (c) Definition sketch for a one-line model.

beach contour – the angle α is no longer constant – so sediment transport rate will vary with position along the shoreline.

A simple mathematical model of this situation can be developed, based on the concept of an equilibrium profile extending to the depth of closure. Consider the element of beach between boundaries 1 and 2 in Figure 16.1(b), shown in sectional elevation in Figure 16.1(c). Applying the continuity equation, in a time interval δt the change in the volume of sediment in the element is equal to the volume entering less the volume leaving. Hence

$$A\delta x - \left(A + \frac{\partial A}{\partial t}\delta t\right)\delta x = Q\delta t - \left(Q + \frac{\partial Q}{\partial x}\delta x\right)\delta t$$

where A is the cross-sectional area of the beach profile.

Simplifying,

$$\frac{\partial Q}{\partial x} = \frac{\partial A}{\partial t} \tag{16.8}$$

For an equilibrium profile any change in area must result in a horizontal movement of the profile, δy, given by $\delta A = d_c \delta y$, and substituting in (16.8) gives

$$\frac{\partial Q}{\partial x} = d_c \frac{\partial y}{\partial t} \tag{16.9}$$

Equation (16.9) can be rearranged into finite difference form in a number of ways. A simple explicit numerical scheme based on a staggered grid is

$$Y_{n+1/2,\,t+1} = \frac{1}{d_c}\left(\frac{Q_{LS_{n,t}} - Q_{LS_{n+1,t}}}{\Delta x}\right)\Delta t + Y_{n+1/2,\,t} \tag{16.10}$$

Subscript n refers to location (boundary line 1, 2, . . .), and t to time. As the values of Y alter, this will correspondingly alter α, which can be computed from the new Y values

The transport rate $Q_{LS_{n,t+1}}$ can be calculated (for example, from equation (16.5)). Equations (16.10) and (16.11) may be used as the basis for a computer program. This can be applied to a simple problem in which the waves approach the shore from one direction. The solution is started with initial values of Y and Q_{LS} at initial time t. The effect of the intercepting feature is to reduce $Q_{LS_{n,t+1}}$, say to zero. The equations are solved sequentially for all boundary lines at $t + \Delta t$, $t + 2\Delta t$ and so on.

Where the direction of the incident waves varies (as will be the case in most real situations) the above approach requires modifications. Some care is also needed in selecting the magnitudes of the distance and time differences (ΔL, Δt) to avoid problems of numerical stability.

This method gives the evolution of only one 'typical' shoreline contour, and so it is known as a one-line model. For an example of the application of a one-line model to beach evolution see Chadwick (1989). It has been assumed in the development of equations (16.9) and (16.10) that sediment transport is longshore only. There is likely to be on- and offshore transport, especially near the barrier feature. Also, in practice the selection of a value for d_c is not necessarily straightforward. To deal with the combination of longshore together with cross-shore transport multi-line models have been developed (see Fleming, 1994b), but they required detailed information about the distribution of the sediment transport rates, and this is not always available. However, for shingle beaches a one-line evolution model can be used with a parametric beach profile model (Powell, 1990) to yield more detailed predictions. The *Beach Management Manual* (Simm *et al.*, 1996) provides further information and references.

Coastal profile and coastal area models

Such models represent the most recent developments in attempting to predict shoreline evolution. Coastal profile models are restricted to the prediction of cross-shore profile development, whereas coastal area models attempt to predict the integrated effects of cross- and longshore movements. Such models generally contain one of a number of hydrodynamic models for waves and currents and a sediment transport model, based either on shear stress or energetics concepts, and allow the bathymetry to develop over time in response to the driving forces on the sediments. Such models are referred to as **morphodynamic models**. An excellent source of reference for such models may be found in a special issue of *Coastal Engineering* issued in 1993, entitled *Coastal Morphodynamics: Processes and Modelling*. Cross-shore models have been more recently reviewed by Schoonees and Theron (1995).

16.5 Natural bays, coastal cells and shoreline management planning

Natural bays

Where an erodible coastline exists between relatively stable headlands, a bay will form. The shape of such bays is determined by the wave climate and the amount of upcoast littoral drift. Silvester (1974) investigated the shape of naturally stable bays by a series of wave tank model experiments. His results indicated that, in the absence of upcoast littoral drift, a stable bay will form in the shape of a half heart, for any given offshore wave direction. These are called **crenulated bays**. The reason why crenulated bays are stable is that the breaker line is parallel to the shore along the whole bay. Littoral drift is therefore zero. Silvester also gives a method of predicting the stable bay shape and its orientation to the predominant swell directions (see section 16.7). More recent work may be found in Hsu *et al.* (1989).

These results have several significant implications. For example, the ultimate stable shape of the foreshore, for any natural bay, may be determined by drawing the appropriate crenulated bay shape on a plan of the natural bay. If the two coincide, then the bay is stable and will not recede further. If the existing bay lies seaward of the stable bay line, then either upcoast littoral drift is maintaining the bay, or the bay is receding. Also, naturally stable bays act as 'beacons' of the direction of littoral drift. Silvester has scrutinized the coastlines of the world to identify crenulated bays, and hence has determined the directions of predominant swell and littoral movements. Where such movements converge at a point, large depositions and accumulation of coastal sediments may be expected. Finally, the existence of crenulated bays suggests a method of coastal protection in sympathy with the natural processes, by the use of artificial headlands. This is discussed further in section 16.7.

Coastal cells

The concept of a coastal cell follows on quite naturally from Silvester's crenulated, stable bay. It is also of crucial importance to coastal zone management, allowing a rational basis for the planning and design of coastal defence schemes. The definition of a coastal cell is a frontage within which the longshore and cross-transport of beach material take place independently of that in adjacent cells. Such an idealized coastal cell is shown in Figure 16.2. Within such a cell, coastal defence schemes can be implemented without causing any effects in the adjacent cells. However, a more detailed

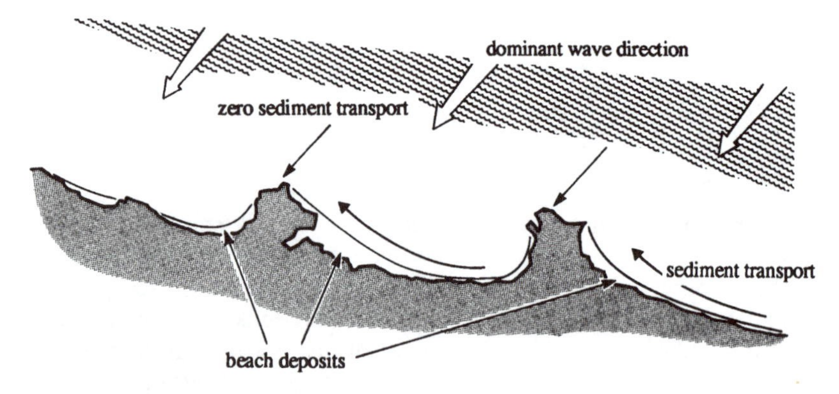

Figure 16.2 Sediment transport within coastal cells.

review of this concept reveals that a coastal cell is rather difficult to define precisely, depending on both the timescale and the sediment transport mode.

Three timescales may be considered useful: micro (for wave-by-wave events), meso (for storm events) and macro (for long-term morphological developments). For meso timescales, the local longshore drift direction can be the reverse of the macro drift direction, possibly allowing longshore transport from one cell to another. With regard to sediment transport, this may be as either bed or suspended load. Longshore transport of coarse material is predominantly by bedload across the active beach profile and largely confined to movements within the coastal cell. Conversely, longshore transport of fine material is predominantly by suspended load, which is induced by wave action but then carried by tidal as well as wave-induced currents, possibly across cell boundaries.

Despite the inherent fuzziness of the boundaries of a coastal cell, it is nevertheless a very useful concept for coastal zone management. In the UK, for example, the coastline of England has been divided into 11 primary cells with a series of subcells defined within each primary cell.

Shoreline management planning

Within the UK, the planning of new coastal defence schemes is now carried out within the context of a shoreline management plan. Many other coastal authorities throughout the world have adopted or are beginning to adopt a similar policy. The aim of a shoreline management plan is to provide the basis for sustainable coastal defence policies within a coastal cell and to set objectives for the future management of the shoreline.

To fulfil this aim, four key components and their interrelationships need to be considered. These are the coastal processes, the coastal defences, land use and the human and built environment, and finally the natural environment. An understanding of the interrelationships between coastal processes and coastal defence is fundamental to developing a sustainable defence policy. The need for coastal defence schemes arises from effects on land use, and the funding of such schemes relies on an economic assessment of whether the benefits of defence outweigh the costs of construction. Finally, the effects of defence schemes on the natural environment must be very carefully considered and an environmental assessment carried out. Environmental hazards and opportunities should be identified, and schemes should be designed to conserve or enhance the natural environment. Where conflicts arise between the needs for defence and conservation, these must be resolved by the environmental assessment. A key part of this assessment is the process of consultation with statutory bodies, local interest groups and the general public. Further details of UK practice may be found in the MAFF (1995) publication, *Shoreline Management Plans*.

16.6 Understanding coastal system behaviour

Within Coastal Engineering, there has been a strong focus upon littoral processes and this approach is frequently used as a basis for analysing coastal change and assessing future policy options and impacts. Whilst the littoral cell concept is a valid approach, it is only one aspect of coastal system behaviour and other factors also need to be taken into account when assessing future shoreline evolution. Therefore, in terms of making large-scale or longer-term predictions of coastal evolution, the cell concept can have a number of shortcomings.

Analysis of coastal dynamics and evolution is difficult due to both the range of spatial and temporal scales over which coastal changes occur, and the complex interactions that result in shoreline responses of varying, nonlinear and often unpredictable nature. There is also inter-dependence between different geomorphic features that make up the natural system, such that the evolution of one particular element of the coast is influenced by evolution in adjacent areas. Often these influences extend in a number of directions, thereby further complicating the task of assessing change.

A major project, FUTURECOAST, was commissioned by the UK government Department for the Environment Food and Rural Affairs. It was carried out by Consulting Engineers, Halcrow, with results published in 2002. A 'behavioural systems' approach, was adopted by the FUTURECOAST project. This involved the identification of the different elements that make up the coastal structure and developing an understanding of how these elements interact on a range of both temporal and spatial scales. In this work it is the interaction between the units that is central to determining the behaviour. Feedback invariably plays an important role and changes in energy/sediment inputs that affect one unit can in turn affect other units, which themselves give rise to a change in the level of energy/sediment input.

Whilst the starting point for a behavioural system is the energy and sediment pathways, it is important to identify the causative mechanism as a basis for building a robust means of predicting the response to change. This must take account of variations in sediment supply and forcing parameters, such as tide and wave energy. However, it is also important to look for situations where the system response is to switch to a different state, for example, the catastrophic failure of a spit, or the switching of channels as a consequence of episodic storm events.

Key influences upon plan-form shape, and evolution, are the underlying geology and coastal forcing, e.g. prevailing wave activity. Large-scale shoreline evolution may be broadly considered in terms of those areas that are unlikely to alter significantly, i.e. hard rock coasts, and those areas that are susceptible to change, i.e. soft coasts.

The evolution of hard rock coasts is almost exclusively a function of the resistant nature of the geology, with the influence of prevailing coastal forcing on the orientation of these shorelines only occurring over very long timescales (millennia). Differential erosion may occur along these coastlines to create indentations or narrow pockets where there is an area of softer geology, or faulting, which has been exploited by wave activity.

The evolution of softer shorelines is more strongly influenced by coastal forcing, although geology continues to play a significant role in both influencing this forcing (e.g. diffraction of waves around headlands) and dictating the rate at which change may occur. The plan-form of these shorelines will, over timescales of decades to centuries, tend towards a shape whose orientation is in balance with both the sediment supply and the capacity of the forcing parameters to transport available sediment. In general soft shorelines have already undergone considerable evolution. Some shoreline may have reached their equilibrium plan-form in response to prevailing conditions, whilst others have not and continue to change.

It is the softer shorelines that are most sensitive to changes in environmental conditions, such as climate change impacts, which may alter the coastal forcing. Such changes in conditions are not necessarily instantaneous, and can take many decades or centuries to occur. Therefore, some of the changes taking place at the shoreline over the next century may be a continuation of a response to events that occurred at some time in the past.

Further details of the FUTURECOAST project are summarised in Reeve, Chadwick and Fleming (2003). The results of the FUTURECOAST project, are available on CD and have been widely distributed in the UK.

16.7 Coastal defence principles

Coastal defence is the general term used to cover all aspects of defence against coastal hazards. Two specific terms are generally used to distinguish between different types of hazard. The term **sea defence** is normally used to describe schemes that are designed to prevent flooding of coastal regions under extremes of wave and water levels. By contrast, the term **coast protection** is normally reserved to describe schemes designed to protect an existing coastline from further erosion.

There are two approaches to the design of coastal defence schemes. The first is referred to as 'soft engineering', which aims to work in sympathy with the natural processes by mimicking natural defence mechanisms. Such an approach has the potential for achieving economies while minimizing environmental impact and creating environmental opportunities. The second is referred to as 'hard engineering', whereby structures are constructed on the coastline to resist the energy of waves and tides. Tables 16.1 and 16.2 show examples of hard and soft engineering techniques and where they are located with respect to the shoreline, for three types of coastline. These tables provide an excellent guide as to where one might expect to find

Table 16.1 Location of hard engineering schemes

Type of coast:	Open coast	Bay		Estuary
		Breakwaters, barrages		
Low shore:	Rock platform	Shingle/Sand beaches		Mud flats
		Groynes		
Upper shore:	Rocky/cliff	Shingle ridge	Marsh	Sand dune
		Sea walls		
Supra tidal:	Cliff	High backshore	Low backshore	Reclaimed marsh
		Flood embankments		
Hinterland:	(Urban Industrial Agricultural Recreational Natural)			
		Drainage		

Table 16.2 Location of hard engineering schemes

Type of coast:	Open coast	Bay		Estuary
		Stable bays		
Low shore:	Rock platform	Shingle/Sand beaches		Mud flats
		Beach feeding		
Upper shore:	Rocky/cliff	Shingle ridge	Marsh	Sand dune
		Set back		
Supra tidal:	Cliff	High backshore	Low backshore	Reclaimed marsh
		Dune building, cliff drainage		
Hinterland:	(Urban Industrial Agricultural Recreational Natural)			
		Drainage		

particular forms of coastal defence. Elements of hard and soft engineering are often used together to provide an optimal coastal defence scheme: for example, the combined use of beach feeding with groynes or breakwaters.

The project design framework

To determine an optimal coastal defence scheme, it is good practice to have a framework for the project design. Such a framework has been recommended by MAFF (1996) and is, in essence, of general applicability. It consists of six steps, which are shown in Table 16.3. Steps 2, 3 and 4 may well need to be iterative. Of particular note in the context of this section is step 2

Table 16.3 The project design framework

Step 1 Preliminary thinking
• The perceived problem
• Awareness, evaluation and consultation
Step 2 Developing and appraising the options
• Consultations
• Environmental information
• Scheme options
• Informal environmental appraisal
• Shortlist of options
Step 3 Choosing the preferred option
• Ranking of options
• Formal environmental assessment
Step 4 Design and planning
• Environmental considerations
Step 5 Operational phase
Step 6 Post-project evaluation

(developing and appraising the options). Here four stages in appraising the possible coastal defence options need to be considered. They are:

1. Do nothing: evaluate the consequences as a baseline against which other options can be measured.
2. Do minimum: adopt a lower standard of protection and subsequent maintenance requirements.
3. Sustain: maintain existing defences to sustain current defence.
4. Improve: use new soft and/or hard defences to improve the current defence or retreat the line of defence.

16.8 Coastal defence techniques

This section contains an introduction to the principal forms of coastal defence. The approach taken is to outline the theory behind the technique, provide some details of the design, and draw attention to the potential environmental impacts and opportunities. The reader's attention is again drawn to Tables 16.1 and 16.2, which show where each of these techniques is generally to be found. Further details, particularly concerning the environmental impacts and opportunities, may be found in MAFF (1993).

Artificial headlands

Silvester's experimental and field work on the stability of natural bays has already been discussed (see section 16.5). Figure 16.3 summarizes his results in graphical form. For any given angle of wave approach, β, there is a fixed ratio of bay indentation *a* to bay length *b*, which is known as the **indentation ratio,** a/b. These results suggest a method of coastal protection on a grand scale. If two artificial headlands are formed on an eroding coastline, then a new stable bay should form between them. Figure 16.4 shows how the technique can be extended by the construction of a sequence of artificial headlands. Thus, by this simple expedient, whole sections of coastline may be stabilized in one scheme. Such schemes have been constructed: for example, in Singapore (Silvester and Ho, 1972). In the UK, Muir Wood and Fleming (1981) describe another scheme at Barton on Sea, where artificial headlands have been used.

However, for a variety of reasons this method of coastal protection has not gained great popularity. Perhaps the most significant of these reasons is

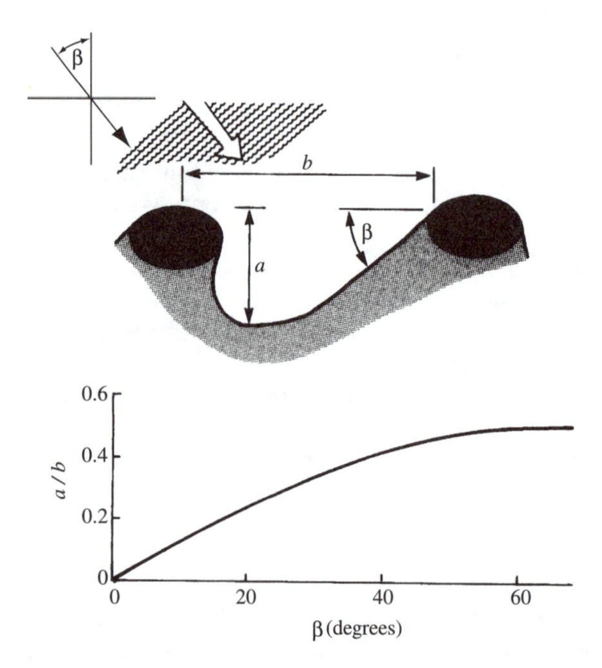

Figure 16.3 Crenulated bays and the indentation ratio.

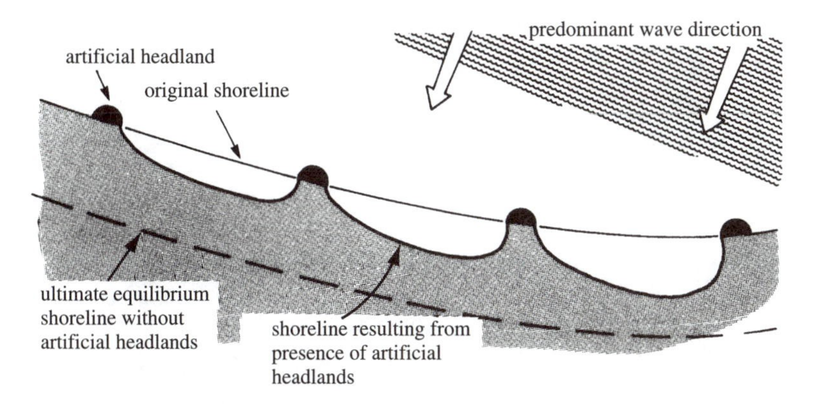

Figure 16.4 Creation of stable bays using artificial headlands.

the uncertainty of the bay stability in conditions where there is a large degree of variability in swell wave direction, and where storm attack comes from a variety of directions. Fleming (1994a) provides an excellent summary of the use of artificial headlands, and describes the key references concerning the theories developed to date.

The environmental impacts of the introduction of artificial headlands include the creation of a new shoreline, which may significantly alter existing habitats. Also, the creation of a new coastal cell will significantly alter littoral drift, possibly causing downcoast erosion. The environmental opportunities include the creation of new stable yet dynamic beaches, which may be used for recreation. Also, the artificial headlands may offer a new home to marine life.

Breakwaters and rip-rap

Breakwaters are often used in harbour works to form the primary means of protection from storm attack. They are also used offshore to protect a coastline by creating a zone of reduced wave energy at the shoreline, as shown in Figure 16.5. Wave energy is dissipated by the breakwaters themselves, and is dispersed by diffraction and refraction shoreward of the breakwaters. These processes and the wave-induced currents in the lee of the structures promote the formation of salients or tombolos. A salient will reduce longshore transport, which is often preferable to the effect of a tombolo, which will prevent longshore transport. It is therefore very important to be able to predict whether a salient or a tombolo will form. This can be achieved by the use of empirical rules, the application of a morphodynamic model or a physical model study. Such methods are summarized by Fleming (1994a). An evaluation of these techniques using

Figure 16.5 Aerial view of the Elmer offshore breakwater scheme.

field data may be found in Axe *et al.* (1997). Submerged breakwaters can also be used, which may be more cost-effective.

Many breakwaters are constructed using large blocks of rock (the 'armour units') placed randomly over suitable filter layers. More recently, rock has been replaced by numerous shapes of massive concrete blocks (for example, dolos, tetrapod and cob). Many of these shapes have been patented. A typical breakwater is shown in Figure 16.6(a), and concrete armour units in Figure 16.6(b).

The necessary size of the armour units depends on several interrelated factors. Traditionally, the Hudson formula has been used. This was derived from an analysis of a comprehensive series of physical model tests on breakwaters with relatively permeable cores and using regular waves. The formula is given by

$$W = \frac{\rho_r H^3}{K_D \Delta^3 \cot \alpha}$$

where W is the required weight of the armour unit, ρ_r, is the density of the armour unit, H is the design wave height (normally taken as H_s), Δ is the relative buoyant density $= (\rho r - \rho w)/\rho$, α is the structure slope angle, and K_D is a non-dimensional constant.

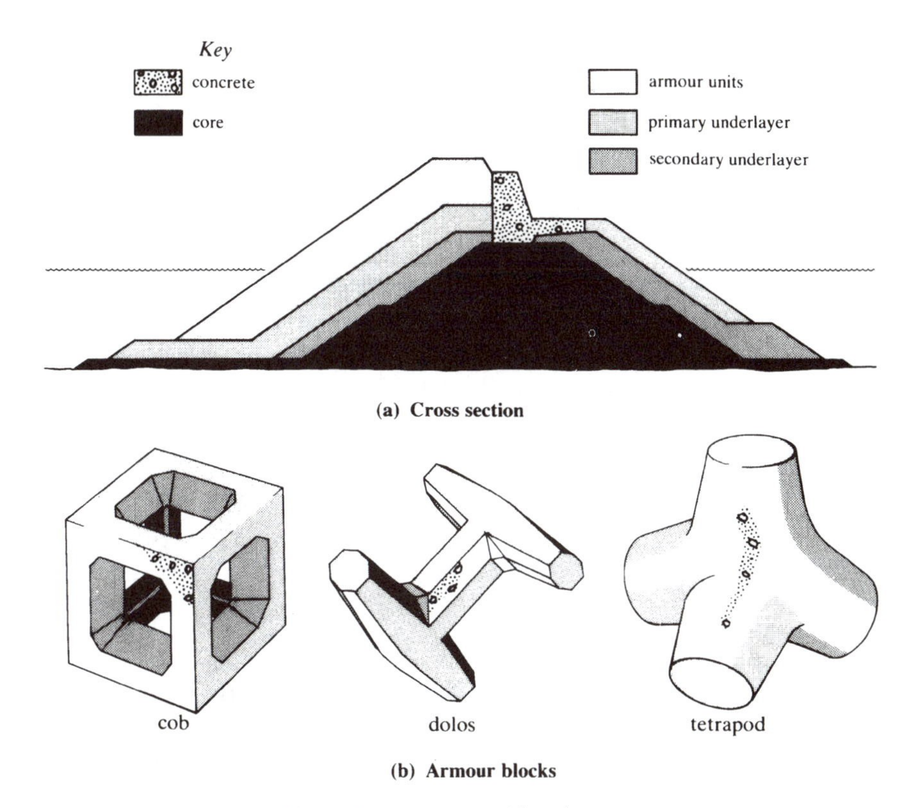

Figure 16.6 A typical breakwater.

The value of K_D is determined principally by the type of armour unit, but its value is also affected by whether the waves incident on the structure are breaking or non-breaking, and whether the structure trunk or head is being considered. For rock, its value is approximately 1–3, whereas for the 'best' concrete armour units it may be in excess of 10. These values are also based on what Hudson referred to as the 'zero damage' criterion; which allows up to 5% of the armour units to be displaced at the design wave height. Full details of the Hudson equation and its application in design may be found in the *Shore Protection Manual* (US Army corps of Engineers, 1984).

More recently (1985–88) these equations have been superseded by Van der Meer's equations. These equations were also developed from an extensive series of physical model tests. In these tests random waves were used, and the influence of wave period and number of storm waves, N, was also considered. A new damage criterion, S_d, and a notional core permeability factor, P, were developed. The equations are for use where the structure is placed in deep water with the waves either breaking on the structure or causing surging.

For plunging waves:

$$H_s/\Delta D_{n50} = 6.2P^{0.18}(S_d/\sqrt{N})^{0.2}\xi^{-0.5}$$

For surging waves:

$$H_s/\Delta D_{n50} = 1.0P^{-0.13}(S_d/\sqrt{N})^{0.2}\sqrt{\cot\alpha}\,\xi^P$$

The transition from plunging to surging waves is given by

$$\xi_c = (6.2p^{0.31}\sqrt{\tan\alpha})^{1/(P+0.5)}$$

where ξ is the Iribarren number, based on mean wave period

$$= \frac{\tan\alpha}{\sqrt{\dfrac{2\pi H_s}{gT_z^2}}}$$

and D_{n50} is the nominal rock diameter $= (W_{50}/\rho_r g)^{1/3}$.

The value of S_d corresponding to Hudson's zero damage is termed the **initial damage**, and has a value of between 2 and 3. The value of P lies between 0.1 and 0.6, corresponding to an impermeable core and no core respectively. The maximum number of waves, N, should not exceed 7500, after which the structure should have reached an equilibrium. Further details of these equations and their application to design may be found in the *Manual on the Use of Rock in Coastal Engineering* (CIRIA/CUR, 1991). A series of conferences on breakwaters (ICE, 1983, 1985, 1988, 1995) also contain much valuable information.

A rip-rap revetment consists of rock or stone placed on filter layers of finer material, and as such is similar to a breakwater. It is normally used to stabilize a shoreline where a beach does not exist. Also, rip-rap is often used as toe protection to sea walls. Details of the design (e.g. slopes, stone sizing, filter layers, wave run-up) are given in CIRIA/CUR (1991).

Environmental impacts of offshore breakwater schemes include reduced wave energy at the shoreline, the creation of a new biologically inert shoreline (due to new sediment deposition), the possible destruction of existing shoreline life, a reduction of upper and supra shore inundation affecting the local ecology, visual intrusion, limiting access to the sea, creation of a safety hazard (for navigation and recreation), and affecting longshore transport. Environmental opportunities include the creation of new beaches, the creation of new habitats for marine life, and the creation of shoreline recreation activities. Clearly, a careful environmental assessment should be undertaken to establish the merits of such a means of coastal defence!

Groynes

Groynes are shore protection structures used to control longshore transport. They act by altering the natural orientation of the beach line, and intercept the wave-induced longshore currents. This occurs because, within a groyne, bay material is transported from one end of the bay to the other by longshore transport, thus realigning the beach line to the incident wave crests and hence progressively reducing longshore transport within their zone of influence. They have little direct effect on cross-shore movements. They are most effective on shingle beaches, where they are often used to stabilize eroding beaches in conjunction with beach recharge (when there is insufficient local sediment supply). Fleming (1994a) provides some very good examples of the use of groynes in a variety of circumstances, which are shown in Figures 16.7 and 16.8.

These structures have in the past been commonly constructed of wooden posts driven into the beach, with wooden planks attached between the piles. Such groynes are termed ‘permeable’, as they allow water to flow through them. More recently rock groynes have been used.

The layout of a groyne system is governed by their length, spacing, height and orientation to the beach line. Until recently the interrelation of these parameters was largely based on empirical rules or observation. As a general guide, the following empirical rules may be used.

1. The length is normally taken as the distance between beach head and the low water line (for ease of construction and low cost).
2. The height is between 0.5 and 1.0 m, or sufficient to accommodate the difference in height between storm and swell profiles.
3. The spacing is between one and three times the length.
4. The orientation is at right angles to the beach head.

In 1990 CIRIA published a new design guide and data as to the performance of existing groyne systems (CIRIA, 1990a,b), and a very useful summary of a rational approach to the design of groyne systems is given in Fleming (1994a).

Environmental impacts of groyne systems include possibly causing downdrift erosion, modifying the position of the beach line and its orientation, visual intrusion, reducing access to shoreline, and introducing a safety hazard to beach users. Wooden groynes should also be made from sustainable forestry sources. Environmental opportunities include the recreational benefits of windshelter, sunbathing and privacy, and they also provide a habitat for marine life.

Beach nourishment

Beach nourishment may be used as an alternative to, or in conjunction with, groynes and offshore breakwaters. Its purpose is to renourish the beach

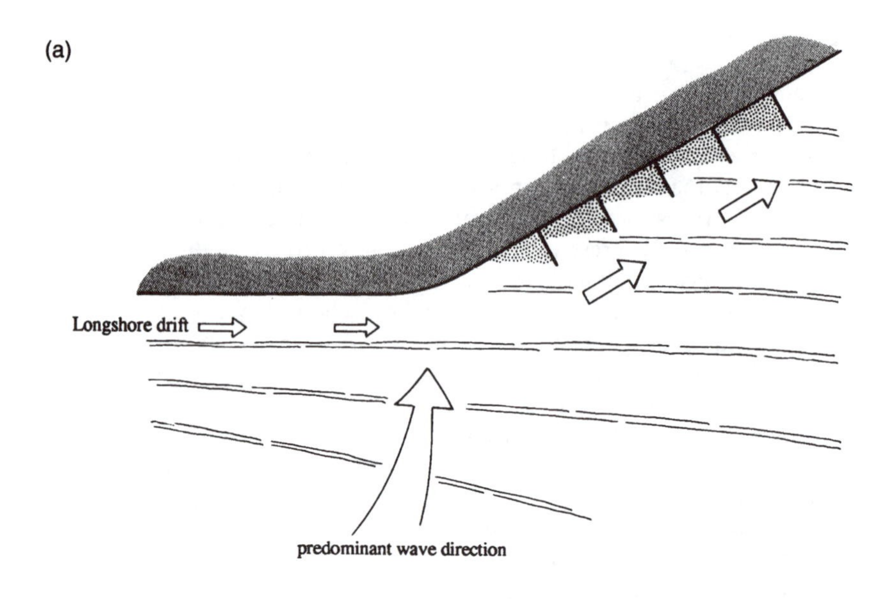

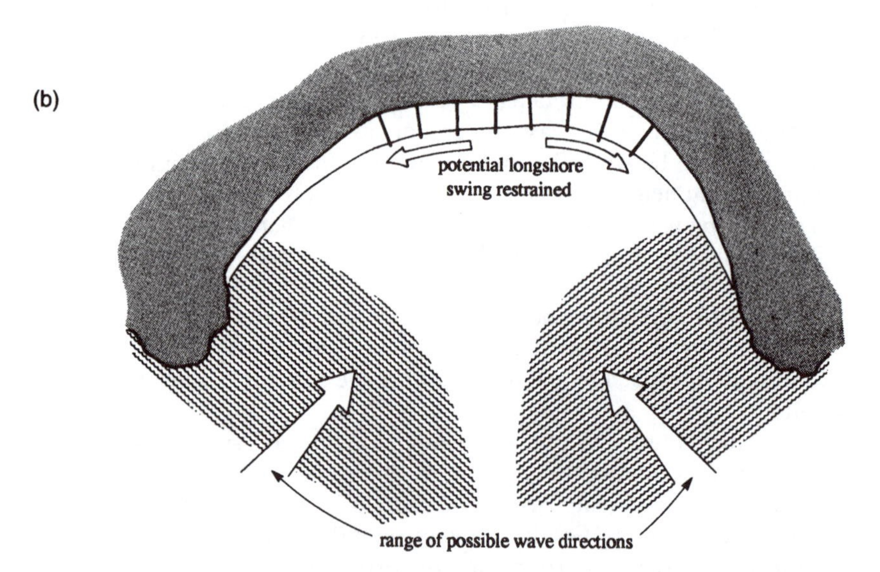

Figure 16.7 Uses of groynes: (a) arrest of alongshore drift by a groyne system; (b) use of groyne system in bay.

that forms the natural defence and protection. It is not usually necessary to sort the material, or to place it to a particular gradient, because wave action will sort and distribute it along the beach. The economics of nourishment schemes will depend on the rate of beach depletion (as distinct from the rate

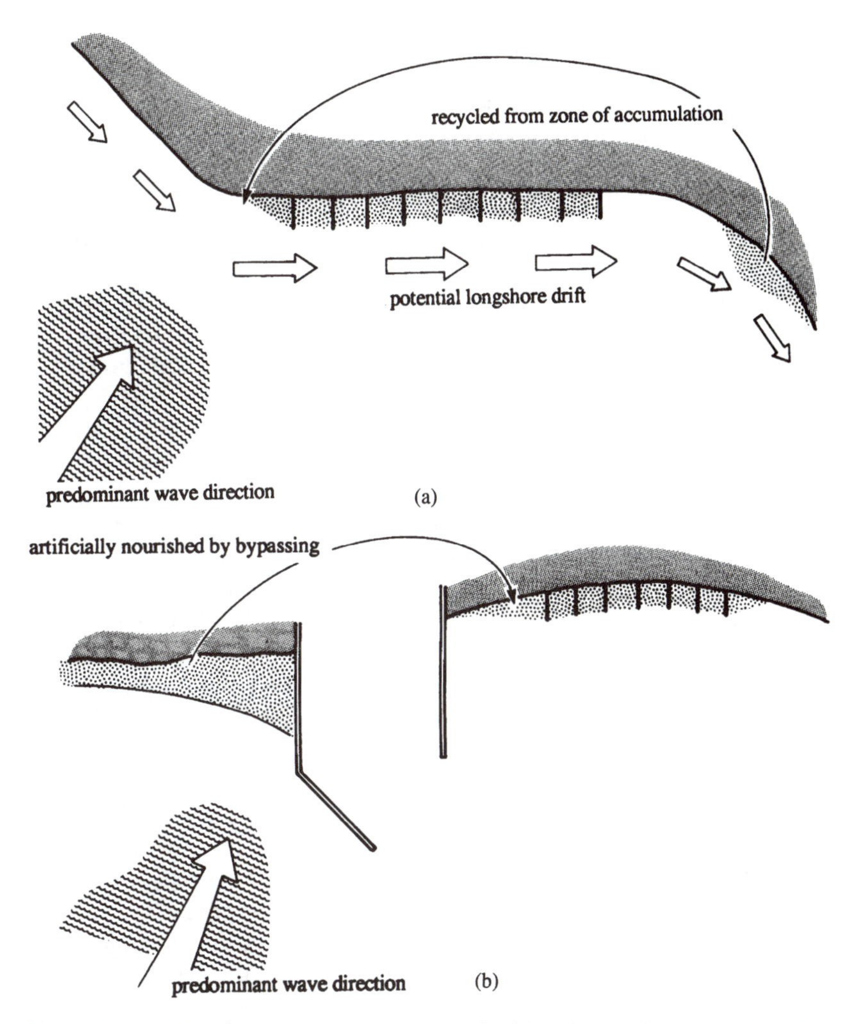

Figure 16.8 Use of groyne system on nourished beach: (a) changes in shoreline orientation; (b) arrest of downdrift erosion.

of littoral drift) and the source (and hence cost) of the supply material. In the cost–benefit study, due account must be taken of the recurrent annual costs in maintaining the beach material. The *Beach Management Manual* (Simm *et al.*, 1996) provides comprehensive advice and guidance on all these matters.

Although beach nourishment may appear at first sight to offer the most 'natural' solution to a coastal defence problem, it is not devoid of environmental impacts. During construction, there can be considerable disturbance to the shoreline, either from the use of the heavy machinery necessary to deliver and place the new material or from the large volumes of pumped water if the material is delivered in barges and pumped ashore. Washout of fine material (and the possible presence of toxic materials) from the imported beach material can have a deleterious effect. It is also necessary to consider the effects on the donor site of removing the recharge material. After construction, the beach is likely to form a new equilibrium profile, which will be steeper than the original if the source material is coarser than the existing material and vice versa if it is finer. Nevertheless, beach nourishment is regarded as one of the most environmentally friendly methods of coastal defence, if used correctly. It offers the opportunity to restore a natural shoreline at sites previously protected by hard defences, potentially providing both ecological and recreational benefits.

Sea walls

Traditionally, sea walls have been the dominant form of coastal defence to the upper shore. At first sight they appear to offer a sure means of defence and protection. However, with a little forethought, it should be evident that they can be far from satisfactory, and should be used only when all other measures have been considered.

Vertical sea walls will reflect virtually all of the incident wave energy (refer to Chapter 8). This sets up a short-crested wave system adjacent to the wall, doubling the wave heights and potentially causing severe erosive action at the sea bed. Consequently, the wall foundations can be quickly undermined unless very substantial toe protection is provided. Immediately downcoast of the sea wall, the short-crested wave system can cause further erosion, and this leads to a temptation to extend the sea wall. Any beach material in front of the sea wall can be rapidly removed downcoast under storm attack, removing any natural defence mechanism and allowing larger waves to attack the wall. However, it is also true to say that beaches in front of sea walls can also recover quickly after a storm event. Perhaps surprisingly, the effects of sea walls on beaches is still a matter of controversy and research.

The forces exerted on a sea wall by wave action can be considered to be composed of three parts: the static pressure forces, the dynamic pressure forces and the shock forces. The shock forces arise from breaking waves trapping pockets of air, which are rapidly compressed. As a result, very high localized forces will exist. Thus sea walls must have a high structural strength, and their construction materials must be able to withstand the shock forces.

Modern designs of sea walls have tended to alleviate some of the problems associated with vertical sea walls (Fig. 16.9). First, the wall is given a sloping face to reduce reflection. Second, a curved wave wall is placed at the top of the sea wall to deflect waves downwards, and thus dissipate reflected wave energy by turbulence. Finally, one of the various forms of rip-rap protection

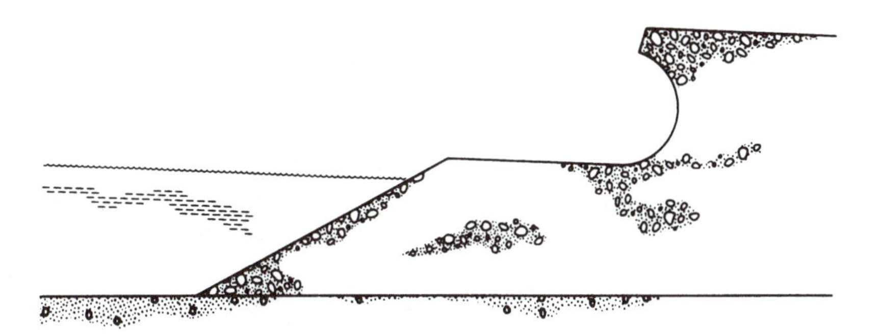

Figure 16.9 Typical form of a sea wall.

to the toe of the wall is provided. Such sea walls are expensive to construct, and the efficacy of the initial design should be tested by physical modelling before the design is finalized. A review of the hydraulic design of sea walls is given in Thomas and Hall (1992).

The environmental impacts of beach erosion in front of the structure (possibly leading to structural collapse) and terminal scour have already been described. To this can be added visual intrusion, (possibly) inhibiting access to the shoreline, the masking of supra-shore features, such as geological or archaeological sites, and reducing habitats for flora and fauna. The environmental opportunities are few, but they can provide recreational benefits by (conversely) providing access to and along the shoreline.

Managed retreat

This is a recently developed (and controversial) concept in response to concerns about global warming and the predicted sea level rise. It is most commonly associated with flood defence of inter-tidal cohesive shores (mud flats and salt marshes in estuaries), and is illustrated in Figure 16.10.

The existing flood defences are moved landwards rather than raised, resulting in a widening of the inter-tidal profile, which absorbs wave energy and tidal flood volumes and reduces the cost of relocated flood banks. Its major environmental impact is the loss of land, but this must be weighed against the environmental opportunity to increase the inter-tidal habitat.

16.9 Wave modelling

This final section is concerned with wave transformation modelling. Such modelling is often required in order that local coastal processes may be understood, that the most appropriate forms of coastal defence can be determined, and that such coastal defence schemes may be properly designed.

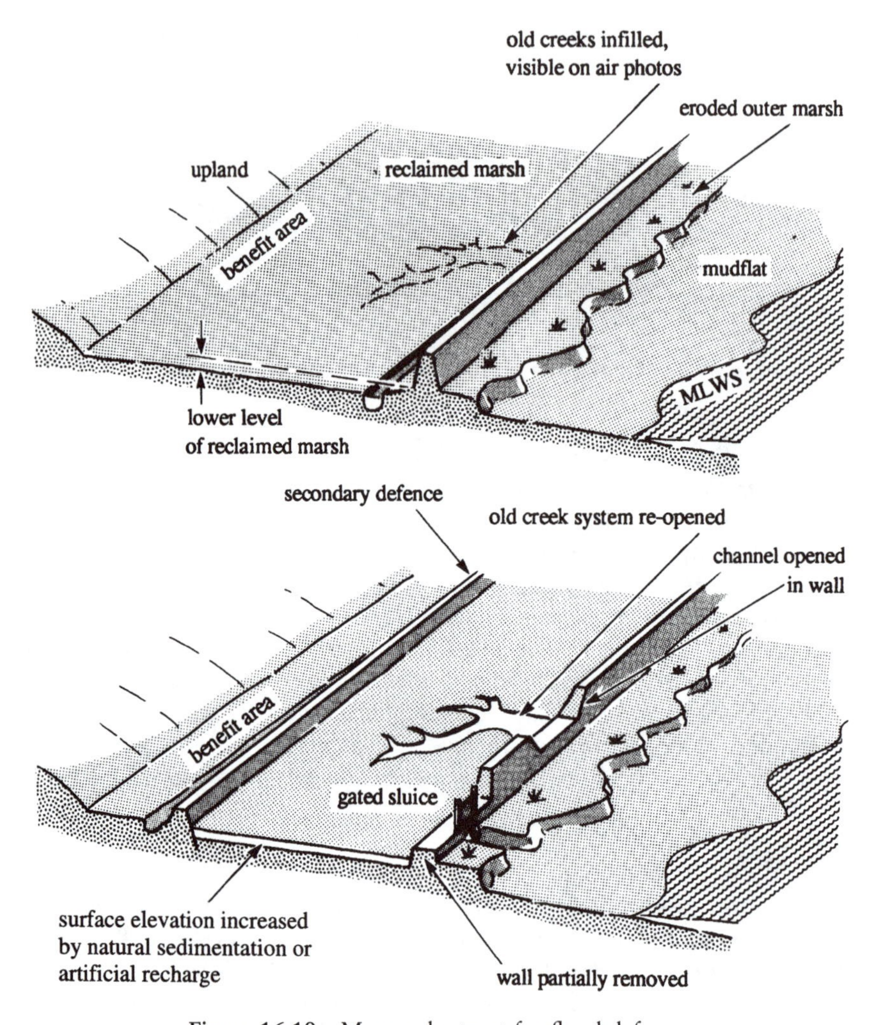

Figure 16.10 Managed retreat for flood defence.

The reader may need to refer back to Chapter 8 for details of the wave transformation processes.

The processes that need to be considered for inclusion in numerical wave transformation models are first stated, and then the main classes of model that have been developed and applied over the last 20 years or so are described. None of these models is currently capable of simulating all the processes simultaneously. The circumstances under which they may be applied is therefore pointed out and their inherent limitations highlighted.

In transitional water depths, the processes of refraction, shoaling, diffraction, reflection, wave–current interaction, set-down and frictional dissipation of wave energy need to be considered. These are most easily modelled by using a monochromatic wave model in which the sea state is represented by a representative wave height, period and direction. Such models usually rely on *linearized* representations of the processes. This implies that higher-order and cross-product terms in the governing partial differential equations are neglected. The introduction of a random (and directional) sea state complicates the matter, producing the non-linear processes of wave–wave interaction and long-wave generation due to wave groups. In shallow water, waves progressively steepen and become asymmetrical in profile, which necessitates the application of non-linear wave theory for correct prediction. Within the surf zone the additional processes of wave breaking, wave-induced currents, set-up and free long-wave generation need to be considered. Currently it has not proved possible to incorporate all these processes in a single model. Such a model would probably be undesirable in any case. One reason for this is that the available computing power is insufficient to be able to model large coastal areas. No doubt this will cease to be an issue in the near future. More significantly, our understanding of the underlying physics is still incomplete, and numerical model development lags behind some of the known physical processes.

Forward-tracking ray models

These are essentially computerized versions of the graphical ray-tracing techniques described in Chapter 8, and were the earliest models to be used in coastal engineering studies. They are based on the wave conservation equations in wave ray form, also described in Chapter 8. The model output consists of a set of wave rays across the model domain (any local bathymetry) for a particular wave period and direction, specified at the model offshore boundary. The refraction and shoaling coefficients can therefore be calculated at any point along the wave rays, and thus wave heights can also be determined at such points. The models can thus be used to give a good picture of the variation of wave height and direction along a shoreline. However, they suffer from the formation of caustics (rays crossing) and 'dead' regions (substantial divergence of wave rays) induced by the local bathymetry. Caustics and dead regions are often unrealistic, and in reality diffraction would occur, a process not included in the models. Current refraction, reflection and bottom frictional losses have been incorporated in more recent models. Offshore directional spectra can be incorporated by linear superposition of components (as described in section 8.5), but they are not very efficient for determining an inshore directional spectrum at a specific location, and non-linear processes cannot be simultaneously incorporated.

Back-tracking ray models

These were developed to overcome the problem of caustics, and to determine the inshore directional spectrum at a specific location. They are based on the principle of ray reversibility, which means that the wave ray can be determined in the reverse direction, starting the calculations from the shoreline and working back out to sea. The procedure involves backtracking fans of rays out to sea using a sequence of directional and frequency intervals. At the offshore boundary resultant ray directions are grouped. The inshore directional spectrum at the point of interest can then be calculated for any given offshore spectrum.

Finite difference refraction models

These models determine the wave height and direction at a series of grid points in which the solution marches forward towards the shoreline. They are based on the wave conservation equations in Cartesian coordinates, also described in Chapter 8. They can incorporate an offshore directional spectrum, current refraction, bottom friction and breaking. They suffer from the same drawbacks as the forward-tracking ray model, but caustics are smoothed if a directional spectrum is used.

Finite difference combined refraction and diffraction models

These models are a more modern development due to Yoo and O'Connor (1986). This uses a wave ray type of solution, but includes diffraction. It is a very powerful technique for use over large sea areas, as ray techniques require only spatial resolution of bathymetry, not wavelength. It has been extended to include wave–current interaction, bottom friction and wave breaking, but not reflection.

The mild slope equation and model systems

This is a full wave model for refraction, shoaling, diffraction and reflection over irregular bathymetry. It has also been extended to include wave–current interactions, bottom friction and wave breaking. As suggested by the name its principal restriction is to mildly sloping bottom topography (1 in 3 is acceptable). It was first derived by Berkhoff (1972) in elliptic form (see section 8.3 for the equation). The model requires a small grid size ($< 1/10$ of the wavelength) and hence has required a very large computational time for large sea areas. The equations are valid only for harmonic and steady-state solutions, and thus cannot be used to predict the development of wave fields in space and time. A simplified parabolic form of the mild slope equation derived by Radder (1979) is computationally less demanding than the elliptic form, but cannot include wave reflection and is restricted to wave directions not too oblique to the grid. It is, however, valid for transient and non-harmonic solutions.

Subsequently, solutions including wave–current interaction, bottom friction and wave breaking were derived by Booij (1981). A hyperbolic form of solution, which includes reflection but is computationally less demanding than the elliptic form, was derived by Copeland (1985). Further refinements in computational schemes continue to date. One recent development is a time-dependent evolution equation (Li, 1994), which has an extremely fast convergent speed, and allows large coastal areas to be economically modelled using a PC with a 586 processor. At the time of writing, this model is currently being evaluated from field data, with initial results reported by Ilic and Chadwick (1996). Various versions of the mild slope equation are now used by the major companies providing wave transformation modelling services.

The Boussinesq equations and model systems

This system of equations represents a solution of the time-dependent, vertically integrated equations of conservation of mass and momentum for shallow water waves ($H/L_0 < 0.22$), for which the horizontal velocity is assumed uniform with depth. They do, however, include the effect of vertical accelerations, and hence some non-linear effects are automatically included. They can reproduce the combined effects of shoaling, refraction, diffraction and reflection of directional, irregular, finite amplitude wave propagation over complex bathymetries. They are computationally very complex, requiring large computing resources, and their application is currently restricted to small areas of interest, typically harbours and small coastal areas. Most recently such models have been extended to predict surf zone waves and currents.

Some sources of further reference for coastal wave modelling

This subject is very complex, and requires a considerable period of study to gain a good understanding of the processes and models. It is an area in which a great deal of research has been carried out in the last few years, and new developments continue apace. An authoritative review of wave transformation processes in the nearshore zone is that of Hamm *et al.* (1993), which discusses the (1993) state of the art and highlights current research directions. More recently HR Wallingford have reviewed both wave transformation models and non-linear surf zone models (Dodd and Brampton, 1995; Dodd *et al.*, 1995). They concluded that wave transformation models were sufficiently advanced and suitable for engineering design, but that non-linear surf zone models required further development. In particular, Boussinesq type models still require development for their application to the prediction of sediment transport in the surf and swash zones. An excellent review of surf zone hydrodynamics and models is that of Svensen and

Putrevu (1996). These three references, although highly selective, do have the common characteristic of being up-to-date reviews containing many references to the work that has preceded them.

References and further reading

Arcilla, A. S., Stive, M. J. F. and Kraus, N. C. (eds) (1994) *Coastal Dynamics 94*, International conference on the role of large scale experiments in coastal research, Barcelona, Spain, February 1994, American Society of Civil Engineers.

Axe, P. G., Chadwick, A. J. and Ilic, S. (1997) Evaluation of beach modelling techniques behind detached breakwaters. Presented at the 25th International Conference on Coastal Engineering, September 96, Florida, USA. To be published by the American Society of Civil Engineers in 1997.

Berkhoff, J. C. W. (1972) Computation of combined refraction of surface waves using finite and infinite elements. *Int. J. Num. Meth. Eng.*, **II**, 1271–90.

Bijker, E. W. (1971). Longshore transport computations. *ASCE Journal of Waterways, Harbors and Coastal Engineering*, **97** (WW4), 687–701.

Booij, N. (1981) *Gravity Waves on Water with Non-uniform Depth and Current*, Report No. 81-1, Department of Civil Engineering, Delft University of Technology.

Briggs, M. J., Thompson, E. F. and Vincent, C. L. (1995) Wave diffraction around breakwaters. *Journal of Waterway, Port, Coastal and Ocean Engineering (ASCE)*, **121**(1), 23–35.

Bruno, R. O., Dean, R. G. and Gable, C. G. (1980) Longshore transport evaluations at a detached breakwater. *Proceedings of Conference on Coastal Engineering*, American Society of Civil Engineers, pp. 1453–75.

Bruun, P. M. (1954) *Coast Erosion and the Development of Beach Profiles*, Technical Memo No. 44, Beach Erosion Board, US Army Corps of Engineers.

Bruun, P. M. (1962) Sea-level rise as a cause of shore erosion. *Journal of Waterways, Harbors and Coastal Engineering Division (ASCE)*, **88** (WW1), 117–30.

Bruun, P. M. (1983) Review of conditions for uses of the Bruun rule of erosion. *Journal of Coastal Engineering*, 7(1), 77–89.

Chadwick, A. J. (1989) Field measurements and numerical model verification of coastal shingle transport, in *Advances in Water Modelling & Measurement BHRA 1989*, pp. 381–402.

Chadwick, A. J. (1991a) An unsteady flow bore model for sediment transport in broken waves. Part I: The development of the numerical model. *Proceedings of the Institution of Civil Engineers Part 2*, **91**, 719–37.

Chadwick, A. J. (1991b) An unsteady flow bore model for sediment transport in broken waves. Part II: The properties, calibration and testing of the numerical model. *Proceedings of the Institution of Civil Engineers Part 2*, **91**, 739–53.

CIRIA (1990a) *Guide on the Uses of Groynes in Coastal Engineering*, Report No. 119, CIRIA, London.

CIRIA (1990b) *Groynes in Coastal Engineering: Data on Performance of Existing Systems*, Technical Note No. 135, CIRIA, London.

CIRIA/CUR (1991) *Manual on the Use of Rock in Coastal and Shoreline Engineering*. CIRIA, London.

Coastal Engineering (1993) Vol 21, special issue, (1993) *Coastal Morphodynamics: Processes and Modelling.*
Copeland, G. J. M. (1985) A practical alternative to the 'mild slope' wave equation. *Coastal Engineering*, **9**, 125–49.
Dally, W. R. and Zeidler, R. B. (eds) (1995) *Coastal Dynamics 95*, international conference on the role of large scale experiments in coastal research, Gdansk, Poland, September 1995, American Society of Civil Engineers.
Dean, R. G. (1991) Equilibrium beach profiles: characteristics and applications. *Journal of Coastal Research*, 7(1), 53–84.
Dodd, N. and Brampton, A. H. (1995) *Wave Transformation Models: A Project Definition Study*, Report SR 400, HR Wallingford.
Dodd, N., Bowers, E. C. and Brampton, A. H. (1995) *Non-Linear Modelling of Surf Zone Processes: A Project Definition Study*, Report SR 398, HR Wallingford.
Fleming, C. A. (1994a) Groynes, offshore breakwaters and artificial headlands, in *Coastal, Estuarial and Harbour Engineer's Reference Book* (eds M. B. Abbott and W. A. Price), Chapman & Hall, London, pp. 311–22.
Fleming, C. A. (1994b) Beach response modelling, in *Coastal, Estuarial and Harbour Engineer's Reference Book* (eds M. B. Abbott and W. A. Price), Chapman & Hall, London, pp. 337–44.
Fredsoe, J. and Deigaard, R. (1992) *Mechanics of Coastal Sediment Transport*, World Scientific, Singapore.
Hamm, L., Madsen, P. A. and Howell, P. (1993) Wave transformation in the nearshore zone: a review. *Coastal Engineering*, **21**, 5–39.
Hsu, J. R. C., Silvester, R. and Xia, Y. M. (1989) Generalities on static equilibrium bays. *Journal of Coastal Engineering*, **12**(4), 353–69.
Huntley, D. A. and Thornton, E. B. (eds) *Coastal Dynamics 97*, international conference on the role of large scale experiments in coastal research, Plymouth, UK, June 1997, American Society of Civil Engineers.
Huntley, D. A., Davidson, M., Russell, P., Foote, Y. and Hardisty, J. (1993) Long waves and sediment movement on beaches: recent observations and implications for modelling, *Journal of Coastal Research*, Special Issue No. 15, 215–29.
ICE (1983) *Breakwaters – Design and Construction*, Thomas Telford, London.
ICE (1985) *Breakwaters*, Proceedings of 1985 Conference on Developments in Breakwaters, Thomas Telford, London.
ICE (1988) *Breakwaters*, Proceedings of 1988 Conference on Design of Breakwaters, Thomas Telford, London.
ICE (1995) *Advances in Coastal Structures and Breakwaters*, Proceedings of Conference on Coastal Structures and Breakwaters 95, Thomas Telford, London.
Ilic, S. and Chadwick, A. J. (1996) Evaluation and validation of the mild slope evolution equation model for combined refraction–diffraction using field data, in *Proceedings Coastal Dynamics '95*, Gdansk, Poland, September, American Society of Civil Engineers, pp. 149–60.
Kamphuis, J. W. (1990) Littoral transport rate, in *Proceedings of Conference on Coastal Engineering*, American Society of Civil Engineers, pp. 2402–15.
Kamphuis, J. W. (1991) Alongshore sediment transport rate. *Journal of Waterway, Port, Coastal, and Ocean Engineering (ASCE)*, **117** (WW6).
Komar, P. D. (1976) *Beach Processes and Sedimentation*, Prentice-Hall, Englewood Cliffs, NJ.

Kraus, N. C. and Dean, J. L. (1987) Longshore sediment transport rate distributions measure by trap, in *Coastal Sediments 1987*, American Society of Civil Engineers, pp. 881–96.

Lawrence, J., Karunarathna, H., Chadwick, A. J. and Fleming, C. A. (2003) Cross-shore sediment transport on mixed coarse grain sized beaches: modelling and measurements. *Proceedings of the International Conference on Coastal Engineering 2002.*

Li, B. (1994) An evolution equation for water waves. *Coastal Engineering*, **23**, 227–42.

MAFF (1993) *Coastal Defence and the Environment – A Guide to Good Practice*, Report PB 1191, Ministry of Agriculture, Fisheries and Food, London.

MAFF (1995) *Shoreline Management Plans – A Guide for Coastal Defence Authorities*, Ministry of Agriculture, Fisheries and Food, London.

MAFF (1996) *Code of Practice on Environmental Procedures for Flood Defence Operating Authorities*, Report PB 2906, Ministry of Agriculture, Fisheries and Food, London.

Morfett, J. C. (1991) Numerical model of longshore transport of sand in surf zone. *Proceedings of the Institution of Civil Engineers*, **91**, Part 2, 55–70.

Muir Wood, A. M. and Fleming, C. A. (1981) *Coastal Hydraulics*, 2nd edn, Macmillan, London.

Powell, K. A. (1990) *Predicting Short Term Profile Response for Shingle Beaches*, Report 219, HR Wallingford.

Radder, A. C. (1979) On the parabolic equation method for water wave propagation, *Journal of Fluid Mechanics*, **95**(1), 159–76.

Schaffer, H. A., Madsen, P. A. and Deigaard, R. (1993) A Boussinesq model for wave breaking in shallow water, *Coastal Engineering*, **20**, 185–202.

Schoonees, J. S. and Theron, A. K. (1993) Review of the field-data base for longshore sediment transport. *Coastal Engineering*, **19**, 1–25.

Schoonees, J. S. and Theron, A. K. (1995) Evaluation of 10 cross-shore sediment transport/morphological models. *Coastal Engineering*, **25**, 1–41.

Silvester, R. (1974) *Coastal Engineering*, Vols 1 and 2, Elsevier, Oxford.

Silvester, R. and Ho, S. K. (1972) Use of crenulate shaped bays to stabilise coasts, in *Proceedings of 13th Conference on Coastal Engineering*, **2**.

Simm, J. D., Brampton, A. H., Beech, N. W. and Brooke, J. S. (1996) *Beach Management Manual*, CIRIA, London.

Svensen, I. A. and Putrevu, U. (1996) Surf zone hydrodynamics, in *Advances in Coastal and Ocean Engineering*, Vol 2 (ed. P. L. F. Liu), World Scientific, Singapore, pp. 1–78.

Thomas, R. S. and Hall, B. (1992) *Seawall Design*, CIRIA/Butterworths, London.

US Army Corps of Engineers (1984) *Shore Protection Manual*, Coastal Engineering Research Centre, Washington.

Van der Meer, J. W. (1988) *Rock Slopes and Gravel Beaches Under Wave Attack*, Delft Hydraulics Comm., No. 396.

Van Wellen, E, Chadwick, A. J. and Mason, T. (2000) A review and assessment of longshore sediment transport equations for coarse grained beaches. *Coastal Engineering*, **40**(3), 243–75.

Yoo, D. and O'Connor, B. A. (1986) A ray model for caustic gravity waves, in *Proceedings 5th Congress Asian and Pacific Regional Division*, IAHR, August.

17

Water quality modelling

17.1 Introduction

The increased awareness of environmental/ecological issues which has surfaced over the last two or three decades is reflected in civil engineering research and development. Many major firms of consulting engineers now find that work in this field has grown considerably. In this book some environmental aspects of civil engineering hydraulics have already been encountered (e.g. hydrology, floods, sediment transport). This chapter is intended to introduce some other aspects, in particular pollutant dispersion and some problems involving density differences in fluids – the number of studies in these and cognate areas is large. A great deal of effort is being expended in improving numerical modelling techniques and rapid developments in the field of numerical models and computer packages have been a feature of recent times. These models may, for example, combine representations of the flow field and of the mechanisms which disperse pollutants, and may be used to predict the path and dilution rates of pollutants injected into a hydraulic system such as a river, lake or sea. Studies involving the modelling of these types of problem are often placed under the general heading of 'Water Quality Modelling'. This chapter will concentrate on the hydraulic aspects of three types of problem, namely dispersion, buoyant jets and intrusion due to density currents.

17.2 Dispersion processes

Consider a body of fluid (the receiving fluid) into which a small volume (or 'slug') of a second 'tracer' fluid is injected. The tracer is assumed to have a density equal to that of the receiving fluid so that it is neutrally buoyant. Even if the receiving fluid is at rest the tracer will gradually disperse as a result of *molecular diffusion*. The rate of diffusion will be extremely slow, for example a slug of dye of a few millimetres in diameter injected into water will take about 24 h to disperse through a 1 m diameter. If the

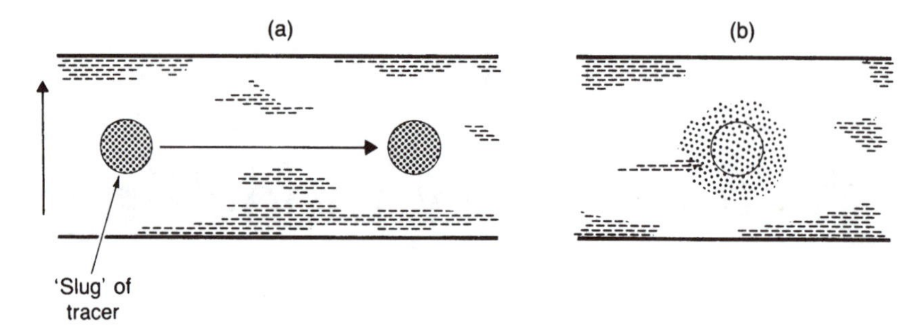

Figure 17.1 Transport processes.

receiving body of fluid is flowing, and the Reynolds' Number is sufficiently large, then the flow is turbulent and this introduces a powerful mechanism for diffusing the tracer. If a tracer slug is injected at a point in a turbulent flow (Fig. 17.1) it will be subjected to two distinct processes: (a) it will be swept along (*advected*) with a velocity comparable to that of the flow, (b) it will be mixed (*diffused*) due to turbulence, so that the tracer becomes more dilute, but its 'sphere of influence' expands. Diffusion due to turbulence is normally two or more orders of magnitude larger than molecular diffusion, so the latter will be omitted from further consideration. The combination of advection and diffusion produces dispersion.

As explained in section 3.4, a turbulent flow may be conceived as consisting of two constituents, namely a time-averaged velocity and a fluctuating velocity (u and u' say). In many real cases the velocity field is three-dimensional, with variations of velocity and turbulence in the x, y and z directions. Flow in channels is usually three-dimensional (though it is often treated as one- or two-dimensional for simplicity) and this has an effect on dispersion. Take the case of a slug of tracer injected instantaneously right across a river at a given location, so as to produce a uniform concentration over the cross-section. The tracer fluid near the centre will be advected downstream at a higher velocity, and that near the banks and bed at a slower velocity. At the same time the larger turbulent eddies near the centre will produce the more rapid diffusion. A non-uniform distribution of tracer is thus produced as it progresses downstream.

The mathematical description of turbulent diffusion is based on considerations of the conservation of mass of a tracer injected at a rate $\partial C/\partial t$ at a point in a flow, where C is the concentration of the tracer. In a turbulent flow the concentration may be split into the two components (time averaged value, c, and fluctuating component c' so that $C = c + c'$). For convenience it is conventional to use subscripts 'i' 'j' to denote direction, and superscript 'n' to denote time, in accordance with the usage in Chapter 14. The turbulent diffusion process may be represented by a differential equation

$$\frac{\partial c}{\partial t} + u_i \frac{\partial c}{\partial x_i} = \frac{\partial}{\partial x_i}\left[\mathrm{D}_{\mathrm{M}i} \frac{\partial c}{\partial x_i} - u_i' c' \right] \tag{17.1}$$

where $\mathrm{D}_{\mathrm{M}i}$ is a molecular diffusion coefficient and the last term ($u_i'c'$) represents the mixing effect of turbulence. This cannot easily be evaluated as it stands. An empirical approach based on the Prandtl eddy model yields the result $u_i'\, c_i' = (1 - \mathrm{D}_{\mathrm{T}i} \partial c / \partial x_i)$, where $\mathrm{D}_{\mathrm{T}i}$ is a coefficient of turbulent diffusion. In practice it is unusual for engineers to possess full details of the flow field in the three axes. It is therefore possible to argue that (17.1) may be simplified by averaging over the flow depth, this reduces the problem to that of a two-dimensional flow field. If the cross-sectional distribution of velocity and turbulence does not vary (as is the case for steady uniform flow in a pipe or channel) then the problem may be reduced to that of a one-dimensional flow field. Most numerical models are designed to model dispersion in one or two dimensions. As an introduction to pollution dispersion modelling, a one-dimensional flow system will be used.

17.3 Dispersion in a one-dimensional stream

For a one-dimensional model it is assumed that the flow is fully developed and uniform. Conditions are averaged over width and depth, so variations are limited to the x direction i.e. the direction of the mean velocity of flow. The theoretical basis for this model is due to Taylor (1954). It must be emphasized that the tracer (pollutant) is assumed to be fully mixed over the cross-section. The dispersion equation is then

$$\frac{\partial}{\partial t}(A\mathrm{C_a}) + \frac{\partial}{\partial x}(AV\mathrm{C_a}) = \left(\frac{\partial}{\partial x}\left(A\mathrm{K_x} \frac{\partial \mathrm{C_a}}{\partial x} \right)\right) \tag{17.2}$$

$\mathrm{C_a}$ and V are the average concentration and velocity at a cross-section, A is the cross-sectional area and $\mathrm{K_x}$ is the longitudinal mixing coefficient. Since the mixing process is being averaged over the cross-section the mixing coefficient should not be confused with $\mathrm{D}_{\mathrm{M}i}$ or $\mathrm{D}_{\mathrm{T}i}$ which relate to diffusion at a point. The equation represents both the advection and diffusion processes. For the purpose of forming a numerical model it is usually convenient to separate the two processes. Thus the advection may be expressed as

$$\frac{\partial \mathrm{C_a}}{\partial t} + V \frac{\partial \mathrm{C_a}}{\partial x} = 0 \tag{17.3}$$

and the diffusion is expressed as

$$A \frac{\partial \mathrm{C_a}}{\partial t} = \frac{\partial}{\partial x}\left(A\mathrm{K_x} \frac{\partial \mathrm{C_a}}{\partial x} \right) \tag{17.4}$$

In a typical numerical scheme derived from (17.3) and (17.4) the tracer concentration at one point and time is estimated from the concentration at some previous point and time. The preceding concentration is first advected forward and the effect of diffusion then computed.

There are several difficulties involved in producing a numerical solution to (17.3) and (17.4). One is the evaluation of K_x. To produce a value of K_x for a given site it is usually necessary to resort to systematic field measurements using a chemical tracer and a sampling system. With the growth of the available data base it may sometimes be possible to derive a value from earlier studies for similar sites. Methods for estimating K_x are given in Cunge *et al.* (1980). It should be noted that K_x is a function of velocity distribution and turbulence and consequently its magnitude may vary considerably even along one river! Another potential difficulty lies in ensuring that the solution for the numerical scheme fulfils the appropriate stability criteria. This may involve some manipulation of the space and time intervals. A difficulty which is potentially serious lies in the fact that in some models the finite difference approximation to (17.3) actually amounts to

$$\frac{\partial C_a}{\partial t} + V\frac{\partial C_a}{\partial x} = K_p\frac{\partial^2 C_a}{\partial x^2} \qquad (17.5(a))$$

It will be noted that this is actually comparable in some respects to (17.4) in that a diffusion term K_p has appeared on the right-hand side. However this is purely a function of the numerical scheme and is nothing to do with the actual physical diffusion due to turbulence. As long as the numerical diffusion is, say, one or more orders of magnitude smaller than that due to turbulence its presence may not unduly affect the final result. However if the numerical diffusion is too large then the model will produce results which are misleading.

To give an example of a numerical approximation which might give rise to numerical diffusion, consider the problem illustrated on Figure 17.2. The numerical scheme represents the diffusion process on a space–time grid. An element of tracer which arrives at station $(i+1), (n+1)$ might have departed from point i at time n (Fig. 17.2(a)). In this case the concentration at i, n will have been computed, and, for advection only, the concentration at $(i+1), (n+1)$ will be the same. This would be very convenient, however, depending on the relationship between $V, \Delta t$ and Δz. It would be more usual for the tracer arriving at $(i+1), (n+1)$ to have originated at some point 0 at time n (Fig. 17.2(b)). Because 0 is situated *between* node points $(i-1)$ and i, it is necessary to interpolate to estimate the concentration. The simplest method is to use linear interpolation, however Cunge *et al.* (1980) demonstrate that this is an approximation which does lead to numerical diffusion such that $K_p = (1 - Cr)(Cr - 2)x^2/2t$, where Cr is the Courant Number. To minimize such problems it is necessary to use a better interpolation routine. One numerical scheme for advection which claims improved

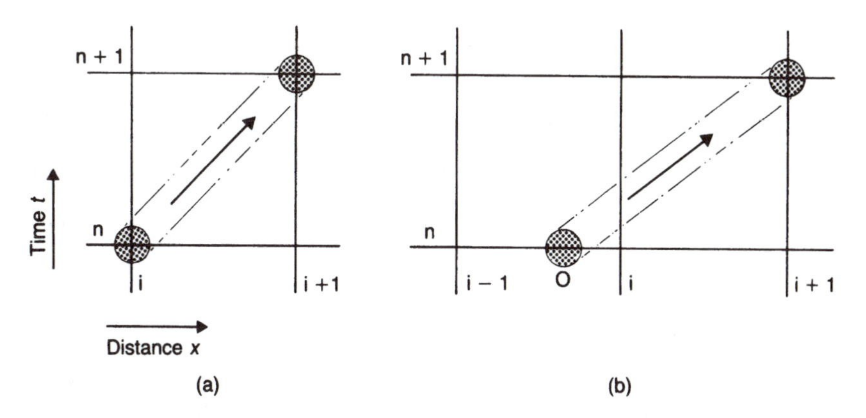

Figure 17.2 Advection path on distance–time grid.

accuracy, and which was developed with river and ocean pollution problems in mind is due to Holly and Preissmann (1977). This is now considered in detail.

Numerical scheme for advection

The scheme uses an interpolation routine based on a cubic polynomial and assuming that C_a and its derivative, $\partial C_a/\partial x$, are both known for nodes $(i-1, n)$ and (i, n). The scheme is now given in full:

$$C^{n+1}_{a\,i+1} = C^{n}_{a_0}$$

$$= k_1 C^n_{a\,i-1} + k_2 C^n_{a\,i} + k_3 \left(\frac{\partial C_a}{\partial x}\right)^n_{i+1} + k_4 \left(\frac{\partial C_a}{\partial x}\right)^n_i \tag{17.6a}$$

$$k_1 = Cr^2(3-2Cr) \tag{17.6b}$$

$$k_2 = 1 - k_1 \tag{17.6c}$$

$$k_3 = Cr^2(1-Cr)(x_i - x_{i-1}) \tag{17.6d}$$

$$k_4 = -Cr(1-Cr^2)(x_i - x_{i-1}) \tag{17.6e}$$

To solve for the derivative terms (assuming constant mean velocity of flow for the channel):

$$(\partial C_a/\partial x)^{n+1}_{i+1} = (\partial C_a/\partial x)^n_0$$

$$= k_5 C^n_{a\,i-1} + k_6 C^n_{a\,i} + k_7 \left(\frac{\partial C_a}{\partial x}\right)^n_{i-1} + k_8 \left(\frac{\partial C_a}{\partial x}\right)^n_i \tag{17.6f}$$

$$\mathrm{k}_5 = \frac{6\mathrm{Cr}(\mathrm{Cr}-1)}{(x_i - x_{i+1})} \tag{17.6g}$$

$$\mathrm{k}_6 = -\mathrm{k}_5 \tag{17.6h}$$

$$\mathrm{k}_7 = \mathrm{Cr}(3\mathrm{Cr}-2) \tag{17.6i}$$

$$\mathrm{k}_8 = (\mathrm{Cr}-1)(3\mathrm{Cr}-1) \tag{17.6j}$$

Numerical scheme for diffusion

Having computed the advected concentration, the effect of diffusion is modelled, using (17.4) as a basis. A numerical scheme due to Chevereau and Preissmann (1971) is based on the use of a centred grid (see Fig. 17.3):

$$\mathrm{C}_{\mathrm{a}\,i}^{n+1} - \mathrm{C}_{\mathrm{a}\,i}^{n} = \frac{\Delta t}{A_i(x_{i+1/2} - x_{i-1/2})}\left[A_{i-1/2}\mathrm{K}_{x\,i-1/2}\frac{\mathrm{C}_{\mathrm{a}\,i}^{n} - \mathrm{C}_{\mathrm{a}\,i-1}^{n}}{x_i - x_{i-1}} - A_{i-1/2}\mathrm{K}_{x\,i-1/2}\frac{\mathrm{C}_{\mathrm{a}\,i+1}^{n} - \mathrm{C}_{\mathrm{a}\,i}^{n}}{x_{i+1} - x_i}\right] \tag{17.7}$$

The stability criterion is

$$\frac{\mathrm{K}_{\mathrm{x}}\Delta t}{(x_{i+1} - x_i)(x_i - x_{i-1})} \leqslant 0.5 \tag{17.8}$$

The application of the numerical model to a simple pollution problem is now illustrated.

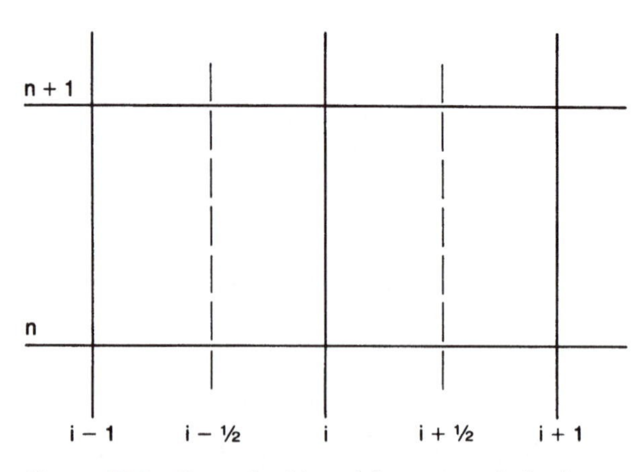

Figure 17.3 Centred grid used for a numerical scheme.

Example 17.1 Pollutant dispersion in a river

A river is 150 m wide and the depth of flow, y, is 15 m. The section may be taken as rectangular. The bed slope is 0.0001 and Manning's $n = 0.03$. An outfall from a process plant discharges pollutant at intervals into the river at one point. A typical injection pattern is shown at the upstream (left-hand) part of the x-axis in Figure 17.4. A town is situated on the river bank, 20 km downstream of the plant, and there is concern at the possible environmental consequences of the pollution on the river ecology and the dangers to people using the river for recreation. The numerical model is to be used to estimate existing levels of pollution and as the basis for determining whether a treatment system needs to be installed at the plant.

Solution

Values of Δx and Δt are 100 m and 50 s respectively (these have been chosen to conform with the suggestion of Chevereau and Preissmann that the ratio $V\partial t/\partial x = n$, where n is an integer). The value of K_x is $2.2u_*y$ where $u_* = 0.015$ m/s (the value of u_* is derived from the application of boundary layer theory). Hence $K_x = 0.5\,\text{m}^2/\text{s}$, which is a fairly weak diffusion (in very turbulent flows K_x may be 20 times greater). The value of the stability parameter for the diffusion calculation is 0.125, which is within the suggested limit (equation (17.8)).

The numerical solution has been produced by computer, and the main steps in the procedure are as follows:

1. Establish the characteristics of the channel flow. In this case the calculation is straightforward, since a uniform steady flow is specified. The mean velocity of

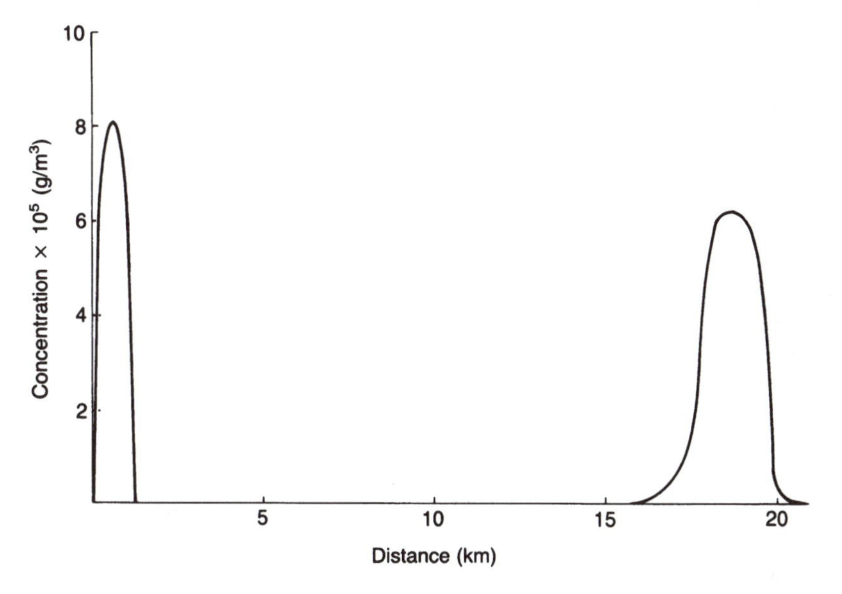

Figure 17.4 Dispersion of pollutant along a river (Example 17.1).

flow can therefore be computed using the Chezy–Manning equation. The depth of flow is constant at 15 m. The Courant Number is 1. (For a more complex problem the computer program would have to include routines for producing the velocities and depths of flow for gradually or rapidly varied flow.)
2. At each time step the pollutant concentrations at the point of discharge form the upstream boundary condition. (The discharge reduces to zero after 500 s.) At the downstream boundary it is assumed that $\partial C_a/\partial x = 0$ for the advection routine and that $\partial^2 C_a/\partial_x^2 = 0$ for the diffusion routine.
3. At each time step the advected concentrations are computed for the whole reach, then the turbulent diffusion effects are computed for the whole reach.

The resulting concentration profile at the town is shown at the downstream (right-hand) side of the x-axis in Figure 17.4. (The interested reader may like to try programming the numerical scheme and reproducing the solution, based upon the given data.) The concentrations are calculated for all x-stations at each time step, and if desired they may be printed (or plotted) to show the variation in concentration as the pollutant is transported downstream.

A major shortcoming in the solution method is that the value of K_x has been determined using data which was derived from site measurements. In applying this approach to any practical problems it would be impossible to be sure whether any errors were due to the numerical solution or the field measurements. Furthermore the results are a good approximation only for points well downstream of the pollution source. For these reasons two-dimensional schemes tend to be favoured by professionals involved in environmental modelling. In these schemes the turbulent dispersion is modelled independently, however, more extensive data are required.

17.4 Two-dimensional modelling

If it is necessary to model the transport of pollutants from their source, and if the distribution of the pollutants is not symmetrical within a simple flow field, then a two-dimensional scheme becomes imperative. A major advantage is that the mixing coefficients, D_{Tii}, D_{Tij} can be directly related to shear velocity and depth, and thus the distribution of turbulence (across a channel say) can be modelled. Two-dimensional schemes are therefore more adaptable and can be applied to river, estuarial or coastal flows.

A simple approach to two-dimensional modelling may be used for rivers. In this the river is assumed to be sub-divided into a series of streamtubes (Holly and Cunge, 1975). The discharge in each streamtube is assumed to be constant. The initial conditions are established for the starting (upstream) section (this would include the initial cross-sectional areas of the streamtubes, the corresponding discharges, longitudinal and transverse velocities and diffusion coefficients). The cross-sectional areas are not constant, but vary in such a way as to maintain constant discharge in each streamtube. The diffusion processes are broken down into

(a) the streamwise advection in each tube;
(b) the streamwise diffusion in each tube;
(c) the transverse diffusion between neighbouring streamtubes.

The corresponding differential equations are:

$$A\frac{\partial C}{\partial t} + \frac{\partial}{\partial x}[AuC] = 0 \tag{17.9}$$

$$A\frac{\partial C}{\partial t} = \frac{\partial}{\partial x}\left[A\,D_{Txx}\frac{\partial C}{\partial x}\right] \tag{17.10}$$

$$A\frac{\partial C}{\partial t} = \left[y\,D_{Txz}\frac{\partial C}{\partial z}\right]_{i-1/2} - \left[y\,D_{Txz}\frac{\partial C}{\partial z}\right]_{i+1/2} \tag{17.11}$$

Note that the index $(i+1)$ represents the right-hand boundary with a neighbouring streamtube (facing downstream) and $(i-1)$ the left-hand boundary. Hence $(i+1/2)$, $(i-1/2)$ represent mean conditions. The application of the model follows the same pattern as indicated in Example 17.1, except that an additional calculation is required (i.e. for equation (17.11)) for each node. It is assumed that the transverse velocities are small in comparison to the streamwise velocities. Holly and Cunge (1975) used the following equations for the turbulent diffusion coefficients $D_{Txx} = 5.93u_*y$, $D_{Txz} = 0.23u_*y$.

For coastal and estuarial problems the flow field is often more complex. Velocity, depth and turbulent diffusion vary from point to point and with time (due to tide and weather). Wave action becomes a significant factor. Considerable advances have been made in the numerical modelling of such problems. Falconer (e.g. Falconer, 1984, 1986) has undertaken a number of studies of pollutant and temperature distributions in estuarial and coastal environments. Reference has already been made to the surf zone (see section 8.4). A numerical surf zone model (e.g. Yoo and O'Connor, 1986) can be combined with a two-dimensional solute transport equation to simulate the flow and dispersion fields. Problems of numerical stability may become more acute in such models, and there are a number of ways in which this can be tackled. One which is in wide use is the Alternating Direction Implicit technique. This involves the subdivision of each time step into two. For one half time step the velocity and solute fields are solved by means of the implicit form of the equations for the x-direction and the explicit form for the y-direction, for the other half-step the x-direction is solved explicitly and the y-direction implicitly. Examples of this technique are given by Koutitas (1988), complete with program listings.

17.5 Differential density effects

In dealing with the problem of pollutant dispersion above, it was assumed that the density of the pollutant was the same as that of the fluid (water) in the main (receiving) stream. For some problems, however, this is not the case. Some typical examples are

1. the output of warm water pumped from a power station cooling system into an estuary;
2. the penetration of salt water (e.g. from the sea) into a freshwater river;
3. a discrete volume of sewage floating on the surface of estuarial or coastal waters.

The first example is known as a buoyant jet, the second as saline intrusion (or salt wedge), and the third as a plume. The first two cases are considered here.

17.6 Buoyant jets

Consider the situation when a fluid is pumped through a pipe and is discharged horizontally into a large, stationary body of fluid of different density and temperature. As the jet emerges from the outlet there will be a discontinuity (of temperature, density, velocity) at the boundary between the jet and the surrounding (receiving) fluid. The combination of the shear force generated at the boundary and turbulent mixing modifies the velocity profile. Two distinct processes occur here. One is diffusion (i.e. two-way mixing across the jet boundary due mainly to turbulence) and the other is entrainment (by which some of the surrounding fluid is drawn into the jet, due to the shear). The jet is thus diluted and caused to spread (Fig. 17.5(a)). Assuming that the jet fluid is lower in density than the receiving fluid, a buoyancy force will exist, causing the jet to rise to (or remain at) the surface of the surrounding fluid. After traversing some distance, Le, the discontinuity at the boundary has been eliminated, and a smooth velocity profile established. In the vicinity of the point of discharge (the nearfield) the jet flow is usually three-dimensional and there is significant entrainment. At some distance from that point (in the farfield) the jet is primarily influenced by turbulent diffusion and may usually be approximated to a two-dimensional flow, with all the attendant simplifications that this implies in numerical modelling. The jet will be advected if the receiving fluid is flowing rather than stationary. It will then be necessary to deduce the path (trajectory) traced by the jet.

The treatment of buoyant jets is usually based on a combination of dimensional analysis and experimental work. Some of the simpler models which have resulted from this work are considered below.

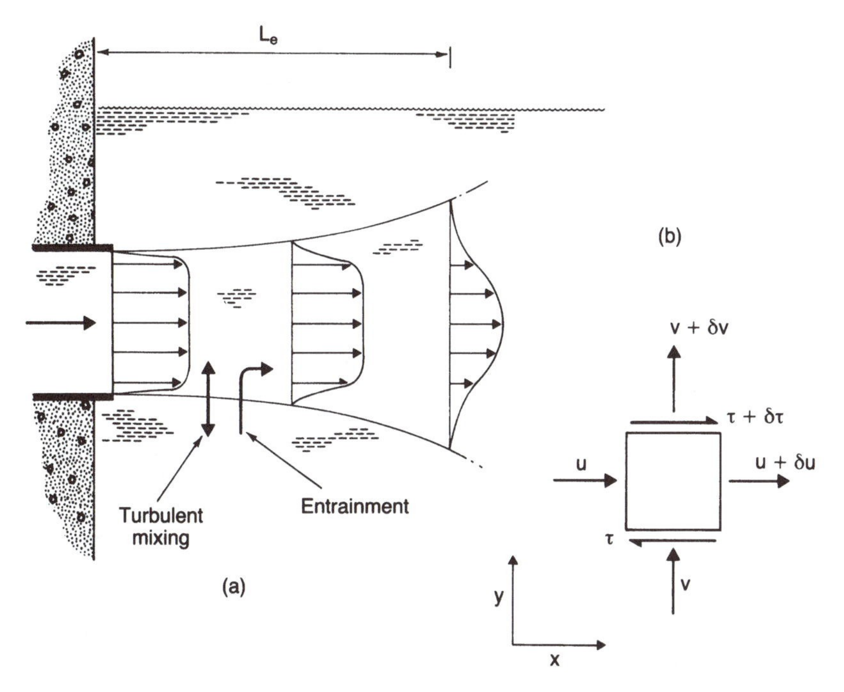

Figure 17.5 Discharge of a jet.

For the case of a neutrally buoyant jet two differential equations of flow may be established based on continuity and momentum. For simplicity only the two-dimensional case is considered (Fig. 17.5(b)).

$$u\frac{\partial u}{\partial x} + v\frac{\partial u}{\partial y} = \frac{1}{\rho}\frac{\partial \tau}{\partial y} \tag{17.12}$$

$$\frac{\partial u}{\partial x} + \frac{\partial v}{\partial y} = 0 \tag{17.13}$$

where τ is the turbulent shear stress.

It is possible to produce a solution to these equations by making assumptions about the velocity distribution. For example if a Gaussian velocity distribution is used $u/u_{max} = \exp(-y^2/2K^2x^2)$, where K is an empirical coefficient. Boundary conditions are $y \to \infty$, $u \to 0$ and $\partial v/\partial y \to 0$. Hence it is possible to obtain a value for the discharge in the jet (for details see French, 1986).

A number of investigators (e.g. Fan, 1967; Anderson *et al.*, 1973) used empirical means for estimating the length Le, between the point of discharge and the point at which jet flow may be regarded as established (i.e. where the entrainment is becoming negligible, see Fig. 17.6). This can be expressed in the form

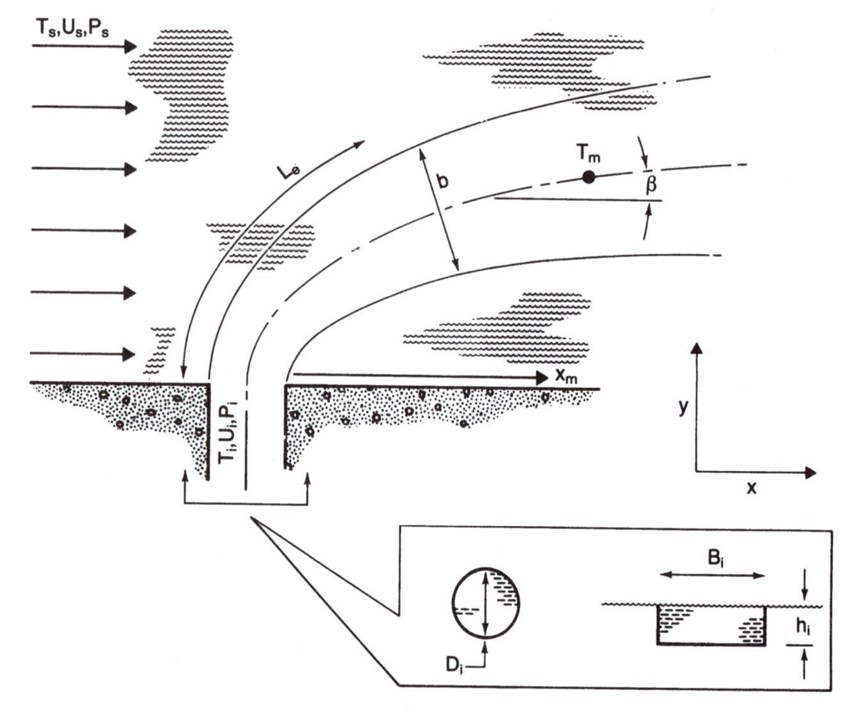

Figure 17.6 Trajectory of a jet discharging into a cross-flow.

$$Le = D_i K_1 \exp(-K_2 U_s / u_i) \quad (17.14)$$

K_1 has a value of the order of 6, while K_2 lies in the range 2.57–3.22. Another model (which relates *Le* to temperature, cross-flow velocity and density) is due to Stefan *et al.* (1975).

The effect of entrainment is to increase the discharge within the boundary of the jet. It may also produce changes to the temperature, chemical constitution, density and velocity distribution, as outlined above. If the receiving fluid is flowing, this too will influence entrainment. Fan (1967) proposed an equation for the rate of increase of discharge in a jet of circular cross-section entering a cross-flow (i.e. a flow perpendicular to the direction of the jet at the point of discharge, see Figure 17.6).

$$\frac{dQ}{dL} = 2\pi C_E b|\Delta u| \quad (17.15)$$

where C_E is an entrainment coefficient and $|\Delta u|$ is the modulus (positive value) of the vector difference between the velocities of the jet and of the receiving fluid at a point.

A simple equation representing the trajectory of a rectangular section jet discharging into a cross-flow was proposed by Jirka *et al.* (1981). The jet is B_i wide and h_i deep at the point of discharge (Fig. 17.6), and $l_0 = (h_i B_i/2)^{1/2}$.

$$\frac{yU_s}{l_0 u_i} = 2\left(\frac{xU_s}{l_0 u_i}\right)^{1/2} \quad \text{where} \quad \left(\frac{xU_s}{l_0 u_i}\right) \leqslant 2 \tag{17.16a}$$

$$\frac{yU_s}{l_0 u_i} = 2\left(\frac{xU_s}{l_0 u_i}\right)^{1/3} \quad \text{where} \quad \left(\frac{xU_s}{l_0 u_i}\right) \geqslant 2 \tag{17.16b}$$

It should be noted that where the receiving fluid is shallow (as might be the case for a stream) there is a tendency for one side of the jet to become attached to the shoreline as it is advected. Abdelwahed and Chu (1981) developed equations for the trajectory and temperature of a rectangular jet B_i wide and h_i deep. The jet is assumed to be discharging into a channel at the water surface, the cross-flow in the channel being uni-directional.

$$\frac{x_m}{l_c}\left[\mathrm{Fr}_{Di}\frac{U_s}{u_i}\right]^{-2} = \mathrm{K}_3\left[\frac{y}{l_c}\left(\mathrm{Fr}_{Di}\frac{U_s}{u_i}\right)^{-2/3}\right] \tag{17.17}$$

$$\frac{\Delta \mathrm{T}_m u_i}{\Delta \mathrm{T}_i U_s}\left[\mathrm{Fr}_{Di}\frac{U_s}{u_i}\right]^{4/3} = \mathrm{K}_4\left[\frac{x}{l_c}\left(\mathrm{Fr}_{Di}\frac{U_s}{u_i}\right)^{-2}\right] \tag{17.18}$$

The cross-flow length scale, $l_c = l_0 u_i/U_s$, x_m is the horizontal displacement, $\mathrm{K}_3, \mathrm{K}_4$ are empirically determined coefficients which are functions of the jet cross-section (shape and size) and the velocities (u_i, U_s). The densimetric Froude Number, $\mathrm{Fr}_{Di} = u_i/(l_0 g(\rho_s - \rho_i)/\rho_s)^{1/2}$. $\Delta\mathrm{T}_i = \mathrm{T}_i - \mathrm{T}_s$ and $\Delta\mathrm{T}_m = \mathrm{T}_m - \mathrm{T}_s$. The jet temperature T_m is the maximum, and corresponds to distance x_m.

A numerical model of a buoyant jet which accounts for mixing, entrainment and momentum effects is given in French (1986, pp. 527–33). From this it is possible to determine the cross-section, velocity and temperature of the jet (see Fig. 17.7).

The above treatment is restricted to modelling of horizontal jets discharging into a steady uni-directional flow. In some cases (e.g. power station cooling water discharging into an estuary), the receiving fluid will be tidal, and the flow bi-directional. Brocard (1985) gives the results of some experimental work on this area. One important case is that of the discharge of effluent from an outfall pipe at the bed of a lake, estuary or sea. The jet will often be of lower density than the ambient fluid, and will therefore rise towards the surface. However the ambient fluid near the surface may be warmer than that near the bed (due to the sun), and entrainment and mixing will affect the jet. This may cause the jet to become neutrally buoyant below the surface. Numerical modelling techniques for this type of problem are given in James (1984).

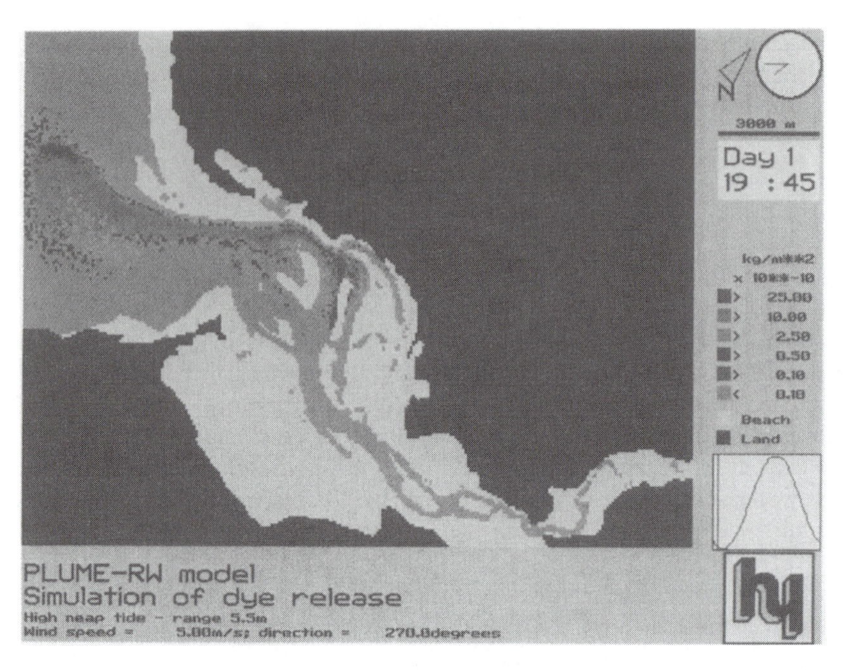

Figure 17.7 Computer screen graphics representation of a numerical model simulation of a plume in the Loughnor Estuary. (Courtesy of HR Wallingford.)

17.7 Intrusion due to density currents

Consider the situation illustrated in Figure 17.8(a), where there are two fluids of differing density in a channel. The fluids are initially separated by a vertical plate. As long as the plate remains in position, the only effect of the density difference is to produce a corresponding difference between the hydrostatic pressure forces on either side of the plate. If the plate is removed this inequilibrium of forces cannot be sustained. The heavier fluid will intrude along the channel bed, whilst the lighter fluid rises (due to buoyancy) and flows over the top of the heavier fluid (Fig. 17.8(b)).

Differences in density may be produced by differences in salinity, sediment load, temperature or chemical constituents. A typical example is that of a river mouth where a freshwater stream meets the salt water of the sea. However it will be obvious that this is a much more complex problem, since the river flow will vary throughout the year according to the rainfall etc., while the level and flow of the salt water is controlled by the tides. The extent of the saline intrusion upriver will therefore fluctuate. In some countries, where water is drawn from the regions near to the river mouth to serve the needs of the local population, a knowledge of the extent of the intrusion may be a vital concern.

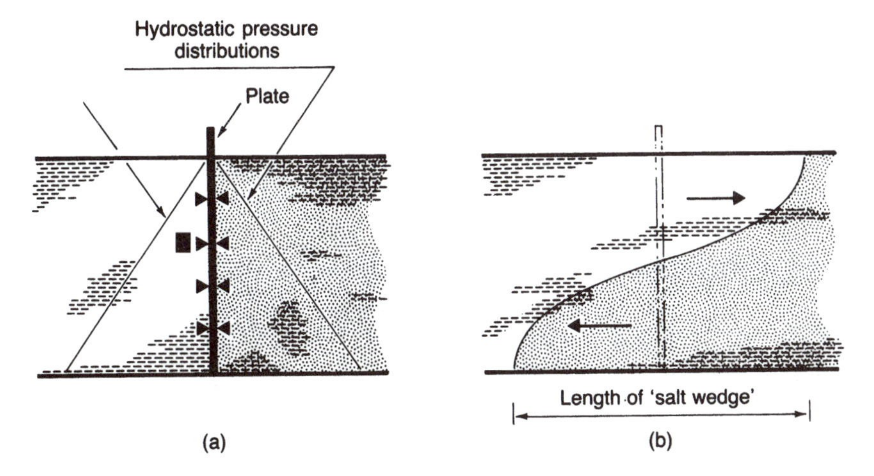

Figure 17.8 Intrusion due to density difference between fluids.

It is possible to produce a relatively simple mathematical model of salt water intrusion by taking the case of a freshwater river flowing into a non-tidal salt sea (like the Dead Sea). It is assumed that no mixing occurs at the interface between the two fluids. Figure 17.9 illustrates this problem. The denser fluid will flow upstream and beneath the less dense fluid, intruding for a distance of L_{SW}. Because the two fluids are flowing in opposite directions, a frictional shear stress will occur at the interface (bed friction is ignored). When a certain limiting value of L_{SW} is attained, the difference in hydrostatic pressure due to the density difference is balanced by the shear force and the system is in equilibrium. The 'salt wedge' is then stationary ($U_{SW} = 0$). The fresh water continues to flow over the surface of the salt wedge, forming a thin film over the salt water seaward of the wedge. For the equilibrium

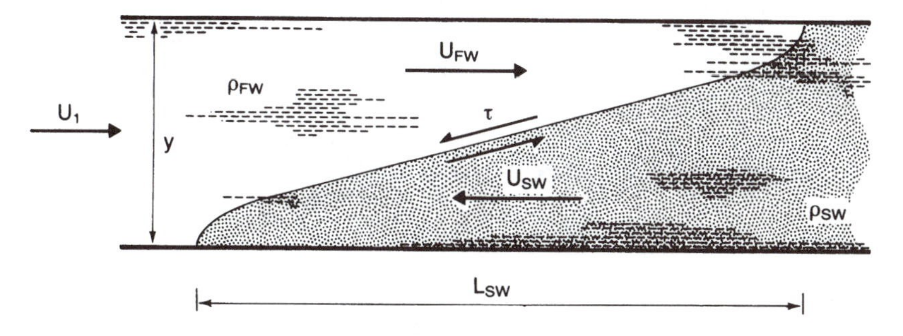

Figure 17.9 Saline intrusion.

condition L_{SW} may then be evaluated as follows, according to Schijf and Schonfeld (1953).

$$L_{SW} = \frac{2y}{f_I}\left[\frac{1}{5Fr_D^2} - 2 + 3Fr_D^{2/3} - \frac{6}{5}Fr_D^{4/3}\right] \qquad (17.19)$$

$$f_I = \frac{8\tau}{\rho_{FW}(U_{FW} - U_{SW})U_{FW} - U_{SW}}$$

$$Fr_D = \frac{U_1}{(gy(\rho_{SW} - \rho_{FW})/\rho_{FW})^{1/2}}$$

In practice a numerical model would have to represent the effects of the tide (on the water level etc.), the bed friction as well as the shear at the interface and the mixing which occurs across the interface. This is in itself a quite complicated process which is produced by a combination of turbulent mixing and entrainment. The latter is produced by the breaking of 'internal waves'. These waves occur along the interface, and where they break the salt water is entrained by the fresh water.

Density differences can produce complex effects. For example at some states of the tide the direction of the current at the water surface may be in the opposite direction to that of the underlying salt wedge. This can cause problems for the manoeuvring of ships. The salt wedge may strongly influence the siltation patterns and the flow of pollutants at river mouths and in harbours and estuaries. A more detailed discussion of these and other aspects of differential density and mixing processes can be found in McDowell and O'Connor (1977), while Koutitas (1988) gives mathematical models and computer program listings covering a number of examples.

17.8 Concluding note

This chapter has been confined to an introductory treatment of some hydraulic aspects of water quality modelling. Apart from the specific reference to saline intrusion, no consideration has been given to the nature of the substances which might be injected into a river or coastal environment, or to the ecological consequences. Readers who wish to pursue the topics of water quality or pollution further are referred to the extensive literature, of which two examples are James (1984) and Nemerow (1991).

There are a number of commercially available computer packages which simulate dispersion and differential density effects.

References and further reading

Abdelwahed, M. S. T. and Chu, V. H. (1981) *Surface Jets and Plumes in Cross Flows*, Tech. Report FML/81-1, Dept. Civil Engineering and Applied Mechanics, McGill University, Montreal.

Anderson, J., Parker, F. and Benedict, B. (1973) *Negatively Buoyant Jets in a Cross Flow*, EPA660/2-73-012, US Environmental Protection Agency, Washington.

Baddour, R. E., Lawrence, G. A. and Wallace, R. B. (1989) Two-dimensional buoyant jets in two-layer ambient fluid. *J. Hydraulic Engineering, Am. Soc. Civil Engineers*, **115**(3), 411–16.

Brocard, D. N. (1985) Surface buoyant jets in steady and reversing crossflows. *J. Hydraulic Engineering, Am. Soc. Civil Engineers*, **111**(5), 793–809.

Buhler, J., Wright, S. J. and Kim, Y. (1991) Gravity currents advancing into a co-flowing fluid. *J. Hydraulic Res.*, **29** (2), 243–57.

Chevereau, G. and Preissmann, A. (1971) Etude mathématique, program de calcul. Appendix 3, in *Modèles mathématiques de la pollution*, (S. Brebion, B. Lebrun, G. Chevereau and A. Preissmann), IRCHA, Centre de Recherche, France.

Cunge, J. A., Holly, F. M. (Jr) and Verwey, A. (1980) *Practical Aspects of Computational River Hydraulics*, Pitman, London.

Falconer, R. A. (1984) A mathematical model study of the flushing characteristics of a shallow tidal bay. *Proc. Inst. Civil Engineers*, **77**, Part 2, 311–32.

Falconer, R. A. (1985) Temperature distributions in tidal flow field. *J. Environmental Engineering, Am. Soc. Civil Engineers*, **110**(6), 1099–116.

Falconer, R. A. (1986) Water quality simulation study of a natural harbor. *J. Waterway, Port, Coastal and Ocean Engineering, Am. Soc. Civil Engineers*, **112**(1), 15–34.

Fan, L. N. (1967) *Turbulent Buoyant Jets into Stratified or Flowing Ambient Fluid*. Tech. Report KH-R-15. W M Keck Laboratory of Hydraulics and Water Resources, California Institute of Technology, Pasadena.

French, R. H. (1986) *Open-channel Hydraulics*, McGraw-Hill, New York.

Holly, F. M. (Jr) and Cunge, J. A. (1975) Time dependent mass dispersion in natural streams. *Proc. Symp. Modelling Techniques, Am. Soc. Civil Engineers*, San Francisco.

Holly, F. M. (Jr) and Preissmann, A. (1977) Accurate calculation of transport in two dimensions. *J. Hydraulics Div. Am. Soc. Civil Engineers*, **103**, HY11, 1259–78.

Hwang, B-G. and Lung, W-S. (1996) Implicit scheme for estuarine water quality models. *Am. Soc. Civil Engineers, J. Environ. Engng.*, **122**(1), 63–6.

James, A. (1984) *Introduction to Water Quality Modelling*, Wiley, Chichester.

Jin, K-R. and Raney, D. R. (1991) Horizontal salinity gradient effects in Apalachicola Bay. *J. Waterway, Port, Coastal and Ocean Engineering*, ASCE, **117**, (5), 451–70.

Jirka, G. H., Adams, E. E. and Stolzenbach, K. D. (1981) Buoyant surface jets. *J. Hydraulics Div. Am. Soc. Civil Engineers*, **107**, HY11, 1467–87.

Koutitas, C. G. (1988) *Mathematical Models in Coastal Engineering*, Pentech Press, London.

Lee, J. H. W. and Jirka, G. H. (1981) Vertical round buoyant jet in shallow water, *J. Hydraulics Div. Am. Soc. Civil Engineers*, **107**, HY12, 1651–75.

Lung, W-S. and Larson, C. E. (1996) Water quality modelling of upper Mississippi River and Lake Pepin. *Am. Soc. Civil Engineers, J. Environ. Engng.*, **121**(10), 691–9.

McDowell, D. M. and O'Connor, B. A. (1977) *Hydraulic Behaviour of Estuaries*, Macmillan, London.

Nemerow, N. L. (1991) *Stream, Lake, Estuary and Ocean Pollution*, 2nd edn, Van Nostrand Reinhold, New York.

Reichert, P. and Wanner, O. (1991) Enhanced one-dimensional modeling of transport in rivers. *J. Hydraulic Engineering, Am. Soc. Civil Engineering*, **117**(9), 1165–83.

Schijf, J. B. and Schonfeld, J. C. (1953) Theoretical considerations on the motion of salt and fresh water. *Proc. Minnesota International Hydraulics Convention*, Minneapolis.

Stefan, H., Bergstedt, L. and Mrosla, E. (1975) *Flow Establishment and Initial Entrainment of Heated Surface Water Jets*, Report EPA660/3-75-014. US Environmental Protection Agency, Corvallis.

Taylor, G. I. (1954) The dispersion of matter in turbulent flow through a pipe. *Proc. Roy. Soc.*, **233A**, 446–68.

Wallace, R. B. and Wright, S. J. (1984) Spreading layer of two-dimensional buoyant jet. *J. Hydraulic Engineering, Am. Soc. Civil Engineers*, **110**(6), 813–28.

Wright, S. J., Roberts, P. J. W., Zhongmin, Y. and Bradley, N. E. (1991) Dilution of round submerged buoyant jets. *J. Hydraulic Res.*, **29**(1), 67–89.

Wrobel, L. C. and Brebbia, C. A. (eds) (1991) *Water Pollution: modelling, measuring and prediction*, Elsevier, Barking.

Yoo, D. and O'Connor, B. A. (1986) Mathematical model of wave induced nearshore circulations. *Proc. 20th Coastal Engineering Conf., Am. Soc. Civil Engineers*, Taipei.

Postscript

In this book we have tried to juxtapose theoretical concepts and empirical results in a manner which is useful to civil engineers. We would strongly emphasize here that there are other aspects which must be considered if a knowledge of hydraulics is to be applied to the solution of engineering problems.

Engineering is essentially concerned with the provision of technological means for satisfying the physical requirements of mankind. Hence, engineers are not principally concerned with ascertaining scientific truths, *per se*, but in applying them to the benefit of mankind. To achieve this aim, the successful engineer will possess a thorough theoretical grounding, coupled with the ability to translate this into schemes which are practical, economical and environmentally sound.

A good civil engineer must observe and understand those natural processes which may be relevant to a given scheme (there is no excuse for shoddy analysis here), must appreciate the limitations of the theoretical concepts which are to be applied and must have the breadth of vision to distinguish between the general (i.e. universal principles) and the particular (that which is valid only for one case) in any scheme.

These qualities are not acquired merely by reading textbooks, or even by obtaining an engineering degree. Experience is also essential, though useful experience is not to be measured in years, but in the growth of one's ability to draw out general principles from each project, and to apply these appropriately in the future. Good engineers thus develop a sense of intuition and a judgemental capacity which enables them to weigh alternatives and reject unsuitable proposals before embarking on a detailed analysis. Such qualities are priceless when one has to break new ground (in research, design or construction), or when investigating the occasional major failure where application of current knowledge has proved to be insufficient.

It is our hope that student readers will be inspired to become good engineers, and that this book will have helped them on their way.

Problems for solution

CHAPTER 1

1. A pipe contains oil (density = $850\,kg/m^3$) at a gauge pressure of $200\,kN/m^2$. Calculate the piezometric pressure head (a) in terms of the oil and (b) in terms of water.

   [24 m, 20.4 m]

2. A sloping tube manometer has the capacity for a maximum scale reading $R_p = 150\,mm$. If it is to measure a maximum pressure of $400\,N/m^2$, what must be the angle θ at which the tube is set? Assume a fixed zero position. The gauge fluid has a density of $1800\,kg/m^3$.

   [8.69°]

3. For the manometer in Question 2, the horizontal cross-section of the tank is 40 times the tube cross-section. What will be the percentage error in the indicated pressure reading due to the fall in the level in the tank?

   [16.5%]

4. A mercury manometer is connected to a flow meter in a pipeline (Fig. P.1). The gauge pressure at Point 1 is $38\,kN/m^2$ and at Point 2 the vacuum pressure is $-50\,kN/m^2$. The fluid in the pipeline is water. Calculate the manometer reading R_p.

   [712 mm]

5. An inverted tube has its upper end sealed and the air has been evacuated to give a vacuum (Fig. P.2). The lower end is open and stands in a bath of mercury. If the air pressure is $101.5\,kN/m^2$, what will be the height of the mercury in the tube?

   [760 mm]

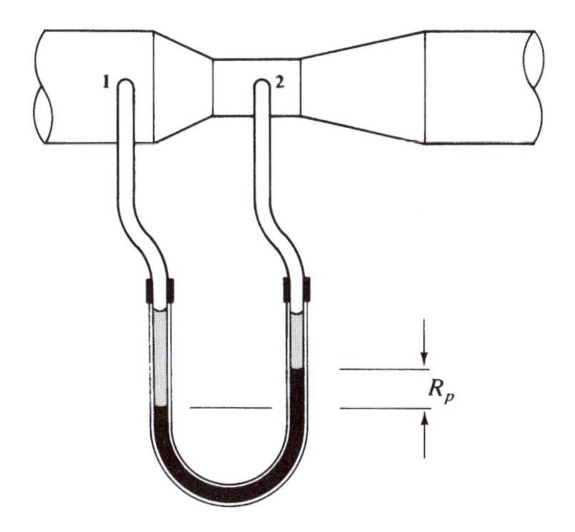

Figure P.1

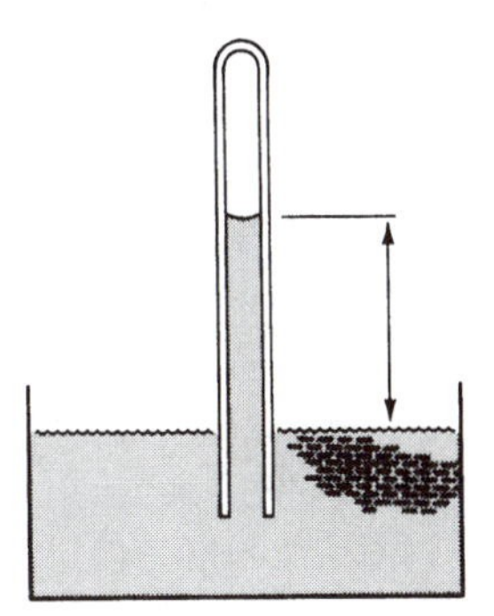

Figure P.2

6. A gas holder is sited at the foot of a hill (Fig. P.3) and contains gas at an absolute pressure of 103 000 N/m^2. If the atmospheric pressure is 101 500 N/m^2, calculate the gauge pressure in head of water. The gas holder supplies gas through a main pipeline whose highest point is 150 m above the gas holder. What is the gauge pressure at this point? Take density of air as 1.21 kg/m^3 and density of gas as 0.56 kg/m^3.

[153 mm water, 250 mm water]

7. A large drain is 1 m square in cross-section. It discharges into a sump whose wall is angled at 45° (Fig. P.4). At the outlet end of the drain there is a steel flap valve hinged along its upper edge. The mass of the flap is 100 kg, and the centre of gravity lies at its geometrical centre. The flap gate is also held shut by a weight of 250 kg on a 500 mm

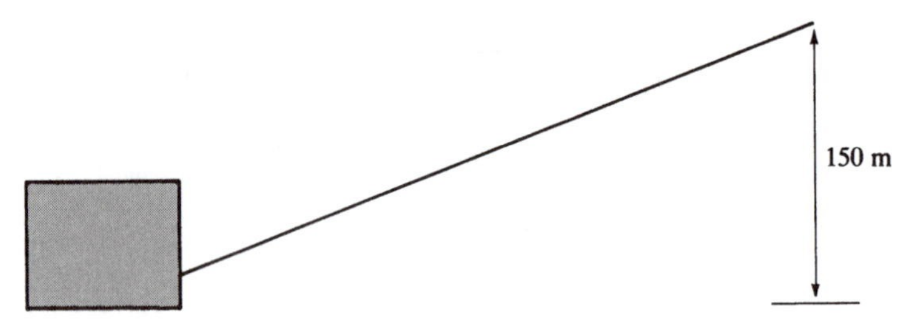

Figure P.3

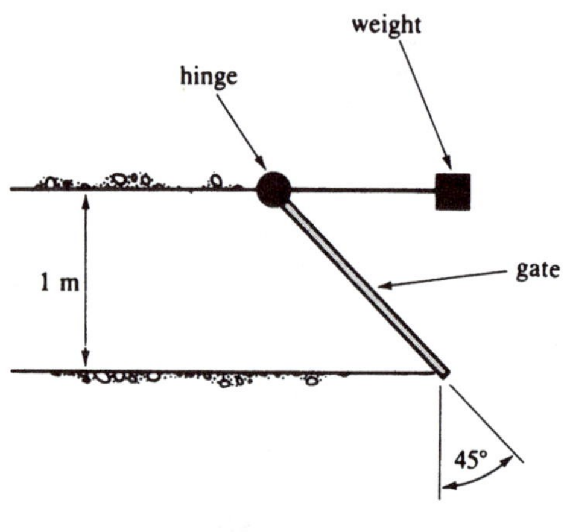

Figure P.4

cantilever arm. To what level will the water rise in the drain before the valve lifts?

[0.58 m]

8. A mass concrete dam has the section shown in Figure P.5 and spans a channel 200 m wide. Estimate the magnitude of the resultant force, its angle to the horizontal, and the point at which its line of action passes through the base line.

[630.5 MN, 13.5°, 36.6 m from *O*]

9. A radial gate is to be used to control the flow down a spillway (Fig. P.6). The gate is 10 m in radius and 12 m wide, and is supported on two shaft bearings. Calculate the load on each bearing. Prove that the resultant hydrostatic force passes through the axis of the bearings.

[0.889 MN]

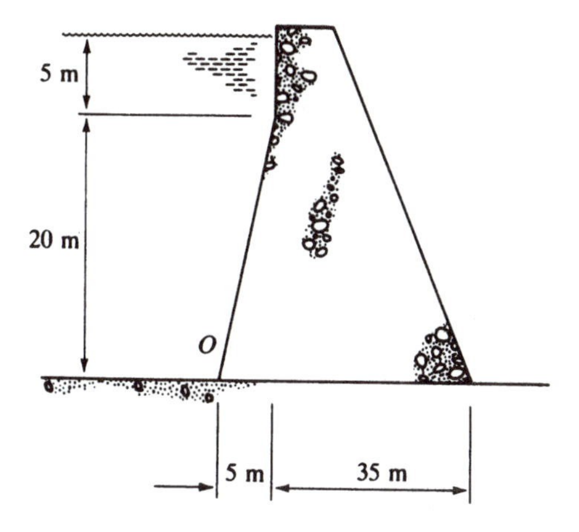

Figure P.5

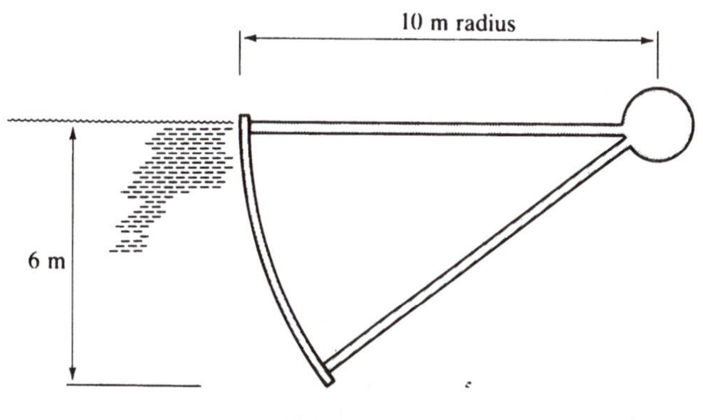

Figure P.6

10. A pontoon is to be used as a working platform for diving activities associated with a dockyard scheme. The pontoon is to be rectangular in both plan and elevation, and is to have the following specification:

    width = 6 m;
    mass = 300 000 kg;
    metacentric height $\geqslant$ 1.5 m;
    centre of gravity = 0.3 m above geometrical centre;
    freeboard (height from water level to deck) $\geqslant$ 750 mm.

    Estimate the overall length, L, and overall height, h, of the pontoon if it is floating in fresh water (density = 1000 kg/m^3).

    $$[L = 36.25\,\text{m}, h = 2.13\,\text{m}]$$

CHAPTER 2

1. For the pipeline shown in Figure P.7, estimate the velocity of flow $V(=Q/A)$ at Section 1, and the pressure p_1. The fluid is water, and the pressure head at entry is 2 m of water. Assume that there are no losses in the pipe itself. The only loss of energy is due to the dissipation of the kinetic energy at the exit (Section 2). Explain why this solution is physically impossible.

$$[24.4\,\text{m/s} - 239\,\text{kN/m}^2]$$

2. A jet of water 50 mm in diameter is directed vertically upwards. The initial (datum) velocity of the jet is 14 m/s. Determine the jet velocity and diameter 2.5 m above datum. At what height above datum would the jet come to rest?

$$[12.1\,\text{m/s}, 53.8\,\text{mm}, 10\,\text{m}]$$

3. Calculate the velocity (α) and momentum (β) coefficients for the case of laminar pipe flow if the velocity distribution is given by

$$u_r = \mathrm{K}(R^2 - r^2)$$

where u_r is the velocity at radius r, R is the pipe radius and K is a constant. (*Hint*. Figs. 4.2 & 4.3a may be of assistance.)

$$[\alpha = 2, \beta = 4/3]$$

4. A reducing pipe bend is shown in Figure P.8. If the discharge is 500 l/s and the upstream pressure is 1000 kN/m^2, find the magnitude and direction of the force on the bend.

$$[174.4\,\text{kN}, 8.26°]$$

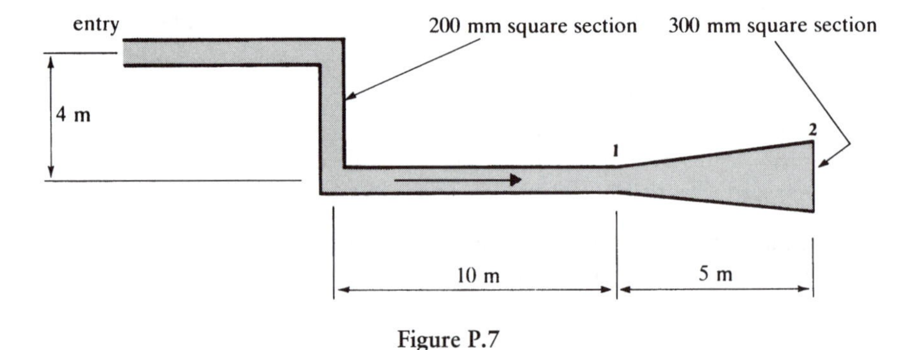

Figure P.7

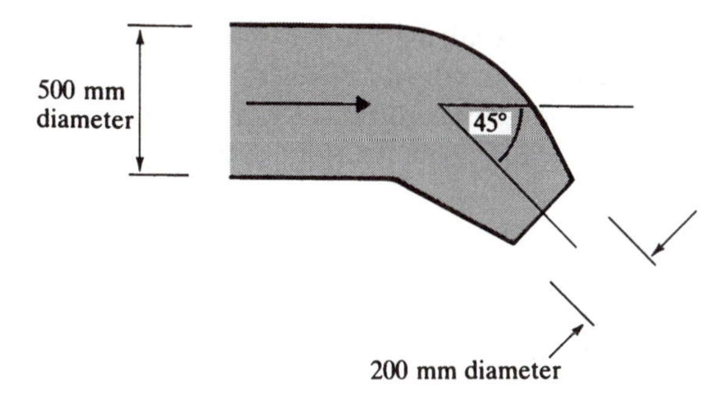

Figure P.8

5. For the pipe junction shown in Figure P.9, estimate the magnitude and direction of the required resistance force to prevent movement of the junction for the following conditions: $p_1 = 69\,\text{kN/m}^2$, $Q_1 = 570\,\text{l/s}$ and $Q_3 = 340\,\text{l/s}$.

$$[F_R = 6.93\,\text{kN}, \theta = 25.1°]$$

6. A duct of diameter 0.8 m carries gas ($\rho = 1.3\,\text{kg/m}^3$). At one point, the duct diameter reduces to 0.74 m. Starting from Bernoulli's equation, estimate the velocity of flow in the reduced section, and the mass flow rate ($= \rho Q$) through the duct if the pressure difference between a point upstream of the reduction and another point at the reduction is equivalent to 30 mm of water.

$$[41.1\,\text{m/s}, 23.0\,\text{kg/s}]$$

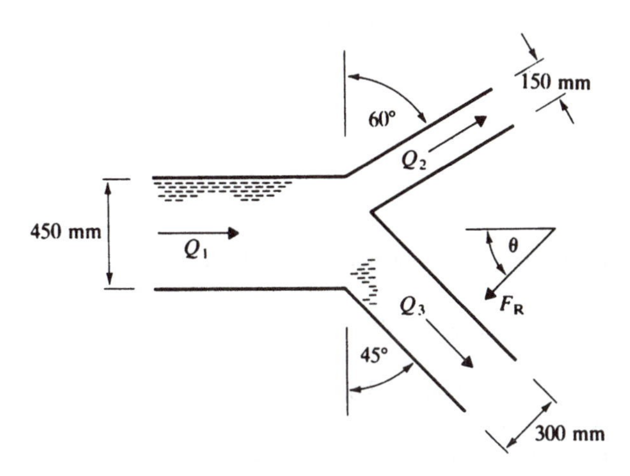

Figure P.9

7. A venturi meter is installed in a 300 mm diameter water main to measure the discharge as shown in Figure 2.9(a). If the throat diameter is 200 mm, the manometer reading is 15 mm (using mercury as a gauge fluid (density $13.6 \times 10^3\,\text{kg/m}^3$)) and $C_d = 0.98$, what is the discharge?

[66.2 l/s]

8. Starting from first principles, derive the discharge equation for a 90° V-notch weir. (*Hint*. Figure 2.11 is a good starting point.)

$$[Q_{\text{ideal}} = \frac{8}{15} \tan 45 \sqrt{2g}\, h^{5/2}]$$

9. A source, having a discharge of $4\,\text{m}^3/\text{ms}$ occurs at the origin (0, 0) of Cartesian co-ordinates. This is superimposed on a uniform rectilinear flow of 5 m/s parallel to the x-axis. Plot the flow net. Determine the co-ordinates of the stagnation point.

[127 mm, 0]

10. An ideal (potential) flow pattern is made up from a combination of:

a source with a discharge of $3.8\,\text{m}^3/\text{s}$;
a forced vortex of 300 mm diameter rotating at 100 rad/s;
a free vortex surrounding the forced vortex.

Sketch the stream function diagram which represents the above pattern. Estimate the pressure and velocity at a radius of 250 mm.

$[181.6\,\text{kN/m}^2, 9.32\,\text{m/s}]$

CHAPTER 3

1. Determine whether the following flows are laminar or turbulent:

(a) A flow of water ($\rho = 1000\,\text{kg/m}^3$, $\mu = 1.2 \times 10^{-3}\,\text{kg/m s}$) through a pipe of square cross-section. The section is 500 × 500 mm and the mean velocity of flow is 3 m/s.
(b) A flow of air ($\rho = 1.24\,\text{kg/m}^3$, $\mu = 1.7 \times 10^{-5}\,\text{kg/m s}$) through a pipe of diameter 25 mm. The mean velocity is 0.1 m/s.

2. A laminar boundary layer forms over a plate which moves at 3 m/s through water. Estimate the boundary layer thickness 5 m downstream of the leading edge. Assume $\rho = 1000\,\text{kg/m}^3$ and $\mu = 1 \times 10^{-3}\,\text{kg/m s}$. The velocity distribution is

$$\frac{u}{U_\infty} = \left(\frac{2y}{\delta} - \left(\frac{y}{\delta}\right)^2\right)$$

[7.1 mm]

3. Define the terms boundary layer, displacement thickness and momentum thickness. A rectangular surface 500 mm wide and 1 m long is immersed in water. The water is flowing at a free stream velocity of 20 m/s parallel to the surface. Evaluate the shear force of the fluid on the surface, assuming that the flow is turbulent throughout the boundary layer. For water, $\rho = 1000\,\text{kg/m}^3$ and $\mu = 1.2 \times 10^{-3}\,\text{kg/m s}$. The velocity distribution in the boundary layer $(u/U_\infty) = (y/\delta)^{1/7}$, $\tau_0 = 0.0225\rho U_\infty^2(\rho U_\infty \delta/\mu)^{-1/4}$.

[340 N]

4. A chimney for a new chemical plant is 30 m tall and 1 m in diameter, to be fabricated from stainless steel. It is to be designed for a maximum wind speed of 80 km/h. Calculate the total wind loading on the structure. If the chimney has a natural frequency of 0.5 Hz, determine whether a problem could arise due to self-induced oscillation and, if so, at what wind speed this would occur. Assume $C_D = 1.0$. The density and viscosity of air are $1.24\,\text{kg/m}^3$ and $1.7 \times 10^{-5}\,\text{kg/m s}$, respectively.

[9183 N, yes, at 9 km/h]

CHAPTER 4

1. Find the maximum discharge in a 12 mm diameter domestic plumbing system for which the flow is laminar. For this discharge find the head loss per metre run, the maximum velocity and the friction factor. Take $\mu = 1.14 \times 10^{-3}\,\text{kg/m s}$.

[$Q = 0.0215\,\text{l/s}$, $h_f = 4.9\,\text{mm/m}$, $V_{max} = 0.38\,\text{m/s}$, $\lambda = 0.032$]

2. Write an essay describing the historical development of turbulent pipe flow theory. The essay should include the contributions of all those named in Table 4.1, and should both describe their work and explain its relevance to current practice.

3. Write a computer program which will calculate the necessary data to reproduce the Moody diagram. The program calculations should be based on the Colebrook–White transition equation.

4. Write a computer program to calculate the discharge in a simple pipe system including local losses. Hence, or otherwise, find the discharge for the following two cases:

   (a) Pipe length = 10 000 m;
   pipe diameter = 500 mm;
   pipe roughness (k_S) = 0.03 mm;
   static head = 150 m;
   local loss coefficient (k_L) = 0.

   (b) Pipe length = 50 m;
   pipe diameter = 300 mm;
   pipe roughness (k_S) = 0.6 mm;
   static head = 10 m;
   local loss coefficient (k_L) = 10.

   Take $\mu = 1.14 \times 10^{-3}$ kg/m s.

   [(a) $Q = 0.68\,\text{m}^3/\text{s}$, (b) $Q = 0.265\,\text{m}^3/\text{s}$]

5. Water flows vertically down a pipe of diameter 150 mm at 2.4 m/s. The pipe suddenly enlarges to 300 mm diameter. Find the local head loss. If the flow is reversed and the coefficient of contraction is 0.62, find the new local head loss.

   [$h_L = 0.165\,\text{m}$, $h_L = 0.11\,\text{m}$]

6. A sewer of diameter 1200 mm is laid to a gradient of 1 m in 100 m. Using the HRS tables or charts with $k_S = 0.6\,\text{mm}$ find:

   (a) the full bore discharge and velocity;
   (b) the flow depth and velocity for $Q = 0.5\,\text{m}^3/\text{s}$.

   Why might this pipe size be considered unsuitable?

   [(a) $Q = 4.5\,\text{m}^3/\text{s}$, $V = 3.75\,\text{m/s}$; (b) $d = 0.28\,\text{m}$, $V = 2.5\,\text{m/s}$]

7. A concrete pipe 750 mm in diameter is laid to a gradient of 1 in 200. The estimated value of Manning's n is 0.012, and the pipe-full discharge is estimated to be $0.85\,\text{m}^3/\text{s}$. (a) Calculate the discharge for a proportional depth of 0.9 using Manning's equation. (b) Explain why the discharge in (a) is greater than the pipe-full discharge.

   [$Q = 0.91\,\text{m}^3/\text{s}$]

CHAPTER 5

1. To solve Manning's equation for depth of flow given the discharge (Q) requires an iterative procedure. Prove that for a wide rectangular channel a good estimate of the depth (y) is given by

$$y = y_1(Q/Q_1)^{0.6}$$

where y_1 and Q_1 are initial estimates of y and Q. Hence, write a computer program to solve Manning's equation for depth given discharge for any trapezoidal channel, such that the solution for y gives a discharge to within 1% of Q. Use this program to find the depth of flow for the following conditions:

(a) $b = 2\,\text{m}$, side slope 1:1, $n = 0.015$, $S_0 = 2\,\text{m/km}$, $Q = 21\,\text{m}^3/\text{s}$;
(b) $b = 10\,\text{m}$, side slope 1:2, $n = 0.04$, $S_0 = 1\,\text{m/km}$, $Q = 95\,\text{m}^3/\text{s}$.

[(a) $y = 1.846\,\text{m}$, (b) $y = 3.776\,\text{m}$]

2. Produce a graphical solution to Example 5.4, drawn accurately to scale, as shown in Figure 5.10.

3. The normal depth of flow in a rectangular channel (2 m deep and 5 m wide) is 1 m. It is laid to a slope of 1 m/km with a Manning's $n = 0.02$. Some distance downstream there is a hump of height 0.5 m on the stream bed. Determine the depth of flow (y_1) immediately upstream of the hump.

[$y_1 = 1.27\,\text{m}$]

4. Starting from first principles, show that the following equation holds true for a hydraulic jump in a trapezoidal channel:

$$\rho g\left(\frac{by^2}{2} + \frac{xy^3}{3}\right) + \frac{\rho Q^2}{(b + xy)y} = \text{constant}$$

where y is the depth of flow, b is the bottom width, Q is the discharge and x is the side slope (1 vertical to x horizontal). Hence draw the force momentum diagram for the following conditions and determine the initial depth if the sequent depth is 0.2 m: $Q = 50\,\text{l/s}$, $b = 0.46\,\text{m}$, $x = 1$.

[Initial depth $= 0.038\,\text{m}$]

5. Suppose that a stable hydraulic jump forms between the two sluice gates shown in Figure 5.15. If the flow depth downstream of a sluice gate is

61% of the gate opening (y_g), determine the downstream gate opening (y_{g2}) for the following conditions: $y_1 = 0.61\,\text{m}$, $Q = 15\,\text{m}^3/\text{s}$, $b = 5\,\text{m}$. (*Hint*. Assume no loss of energy through a sluice gate.)

$$[y_{g2} = 1.115\,\text{m}]$$

6. A rectangular concrete channel has a broad-crested weir at its downstream end as shown in Figure 5.16. The channel is 10 m wide, has a bed slope of 1 m/km and Manning's n is estimated to be 0.012. If the discharge is $150\,\text{m}^3/\text{s}$,

   (a) calculate the minimum height of the weir (Δz) to produce critical flow;
   (b) if $\Delta z = 0.5\,\text{m}$, calculate the upstream head (H_1) if $C_d = 0.88$.

   $$[\text{(a)} \quad \Delta z = 0.179\,\text{m}, \text{(b)} \quad H_1 = 5.14\,\text{m}]$$

7. A trapezoidal channel of bed width 5 m and side slopes 1:2 has a flow of $15\,\text{m}^3/\text{s}$. At a certain point in the channel the bed slope changes from 10 m/km (upstream) to 50 m/km (downstream). Taking Manning's n to be 0.035, determine the following:

   (a) the normal depth of flow upstream and downstream;
   (b) the critical depth of flow upstream and downstream;
   (c) the Froude Number upstream and downstream;
   (d) the depth at the intersection of the two slopes.

   Sketch the flow profile.

   [(a) $y_n = 0.953$ and 0.61 (m), in (b) $y_c = 0.86$ (m), in (c) $\text{Fr} = 0.84$ and 1.77, (d) $y = y_c$, sketch: see Fig. 5.18(a)]

8. A lake discharges directly into a rectangular concrete channel. If the head of water in the lake above the channel bed is 3 m and the channel is 6 m wide with Manning's $n = 0.015$, find:

   (a) the discharge for a channel bed slope of 100 m/km;
   (b) the discharge for a channel bed slope of 1 m/km.

   $$[\text{(a)} \quad Q = 53.15\,\text{m}^3/\text{s}, \text{(b)} \quad Q = 42\,\text{m}^3/\text{s}]$$

9. Using a computer program, or otherwise, verify that the three mild slope profiles given in Figure 5.20 are correct.

10. For the situation described in Problem 8(a), find the distance over which the depth reduces from critical depth to $y = 0.8$ m and the normal depth. Why is the solution unlikely to be very accurate?

    [For $Q = 53.15$ and $y_c = 2.0$ m, $y = 0.8$ m after 57.5 m, $y_n = 0.642$ m; at $y = y_c$ flow is rapidly varied and for steep slopes refer to section 13.5]

CHAPTER 6

1. A steel pipeline is 2.5 km long and 500 mm in diameter, and carries water. A control valve is sited at the downstream end. Calculate:

   (a) the celerity of the shock wave in the fluid;
   (b) the maximum discharge if the surge pressure following instantaneous shut down is limited to 2900 kN/m^2.

   Take $K = 2.11 \times 10^9$ N/m^2.

   [1450 m/s, 390 l/s]

2. The penstock to a turbine is 1 km long and 150 mm in diameter. The pipe is designed for a maximum pressure of 5600 kN/m^2. The operating discharge is 44 l/s, and the corresponding pressure at the governor valve is 2450 kN/m^2. What is the minimum time for complete shut down? Take $c = 1400$ m/s.

   [1.58 s]

3. A water supply main is 3 km long and carries water. The velocity of the water is 2.5 m/s. Calculate the surge pressure for instantanteous closure if $K = 2.13 \times 10^9$ N/m^2.

   [372 m head]

4. A steel pipeline is 750 m long and 100 mm in diameter, and has a wall thickness of 7.5 mm. It carries oil at a velocity of 1 m/s and has a control valve at the downstream end. The valve shuts in 1 s. Determine (a) the speed of sound in the medium, (b) whether the 1 s closure is 'rapid' or 'instantaneous' and (c) the surge pressure. Take $E = 205 \times 10^9$ N/m^2 for steel. For the oil, $\rho = 950$ kg/m^3 and $K = 0.67 \times 10^9$ N/m^2.

   [(a) 801 m/s, (c) 761 kN/m^2]

CHAPTER 7

1. A centrifugal pump is designed for a discharge of $1\,m^3/s$ of water. The water enters the pump casing axially, leaving the impeller with an absolute velocity of 11.5 m/s at an angle of 16.5° to the tangent at the impeller periphery. The impeller is 500 mm in diameter, and rotates at 710 rev/min. Estimate the exit vane angle and the hydraulic power delivered.

[23.4°, 205 kW]

2. A pump is to deliver 60 l/s through a 200 mm diameter pipeline. The pipe is 150 m long and rises 2 m. The friction factor $\lambda = 0.028$. Two pumps are being considered for the duty. Select the more suitable pump, and estimate the power required to drive the pump and the specific speed. Should the pump be of radial, axial or mixed flow type? Pump speed is 1385 rev/min.

Pump No. 1 performance data

Head (m)	8.6	8.35	7.56	6.35	4.95	3.7	2.3
Discharge (l/s)	0	18	39	60	75	88	100
Efficiency (%)	0	52	72	79	75	63	48

Pump No. 2 performance data

Head (m)	9.0	8.8	8.1	7.0	6.0	4.5	3.3
Discharge (l/s)	0	18	39	60	75	88	100
Efficiency (%)	0	52	75	76	67	58	46

[4.8 kW, 89, mixed flow]

3. Cavitation problems have been encountered in a mixed flow pump. The pump is sited with its intake 3 m above the water level in the reservoir and delivers $0.05\,m^3/s$. The total head at outlet is 31.7 m, and at inlet is −7 m. Atmospheric pressure $= p_A = 101.4\,kN/m^2$ and vapour pressure of water $= p_{vap} = 1.82\,kN/m^2$. Determine the Thoma cavitation number. A similar pump is to operate at the same discharge and head, but with $p_A = 93.4\,kN/m^2$ and $p_{vap} = 1.2\,kN/m^2$. What is the maximum height of the pump intake above reservoir level if cavitation must be avoided?

[0.081, 2.25 m]

CHAPTER 8

1. A deep water wave has a period of 8.5 s. Calculate the wave celerity, group wave celerity and wavelength in a transitional water depth corresponding to $d/L_0 = 0.1$.

   $[L_0 = 112.8\,\text{m}, C_0 = 13.27\,\text{m/s}, c = 9.42\,\text{m/s}, C_G = 7.63\,\text{m/s}, L = 80.1\,\text{m}]$

2. Re-solve Example 8.1 if the deep water wave is travelling at 30° to the shoreline.

   $[H_B = 4.75, c = 7.0\,\text{m/s}, d_B = 6.1, \alpha_B = 15.5°]$

3. Using Figure 8.13, determine the significant wave heights and periods corresponding to a wind speed of 20 m/s, a fetch length of 200 km and for wind durations of 1, 6 and 24 h.

   $[t = 1\,\text{h}, H_s = 2.1\,\text{m}, T_s = 4.2\,\text{s}$ (duration limited)
   $t = 6\,\text{h}, H_s = 6.0\,\text{m}, T_s = 7.4\,\text{s}$ (duration limited)
   $t = 24\,\text{h}, H_s = 6.3\,\text{m}, T_s = 7.6\,\text{s}$ (fetch limited)]

CHAPTER 9

1. A river is 100 m wide and 8 m deep. The bed slope is 1 m per 2 km. The median sediment size is 10 mm. Estimate (a) the critical shear stress, (b) the sediment bed load using the Du Boys equation and (c) the minimum stable sediment size.

   $[9.06\,\text{N/m}^2, 0.15\,\text{m}^3/\text{s}, 37.5\,\text{mm}]$

2. Further downstream, the river in Problem 1 traverses a plain where $D_{50} = 0.5\,\text{mm}$. Estimate the total sediment load. River dimensions are unchanged, Manning's $n = 0.02$.

   [$1.38\,\text{m}^3/\text{s}$ approx. using Ackers–White]

CHAPTER 10

1. 'Any quantitative assessment of flood runoff should start from an appreciation of the catchment descriptors affecting runoff.'

   (a) State whether you agree or disagree with this statement, giving your reasons, and (b) describe the effects on runoff of the catchment descriptors mentioned in section 10.3.

2. The data given below are the annual maxima floods for the River Don. Starting with linear graph paper, plot these data in a similar manner to that shown in Figures 10.2(a), (b) and (d) (i.e. starting with a histogram, converting to a pdf and finally a GL growth curve). Hence estimate the 50-year return period flood and comment on the reliability of this estimate.

Year	56	57	58	59	60	61	62
Flood (m^3/s)	38.2	58.9	120.6	57.6	159.9	40.4	66.2
Year	63	64	65	66	67	68	69
Flood (m^3/s)	73.6	66.5	206.8	84.8	97.1	142.1	82.1

$[Q_{50} = 247\,m^3/s]$

3. Calculate the peak runoff rate resulting from the probable maximum precipitation at Ardingly Reservoir from the following data:

Time (h)	0		1		2		3		4		5		6		
Rainfall (mm)		2		5		8		11		14		31			
Time (h)	6		7		8		9		10		11		12		13
Rainfall (mm)		93		31		14		11		8		5		2	

Percentage runoff = 68% (from catchment descriptors)
Baseflow = $1\,m^3/s$ (from catchment descriptors)
10 mm, 1 hour unit hydrograph time to peak = 4 h
Catchment area = $21.82\,km^2$
(*Hint.* Use a synthetic UH with $T_p = 4\,h$)

$[140\,m^3/s]$

4. Describe, in detail, what methods you would use to estimate the 50-year return period flood (both peak discharge and flood hydrograph) for the following two cases:

   (a) an ungauged catchment;
   (b) a gauged catchment with stream flow records of 15 years.

   Suggest ways in which your estimate could be improved (presuming that your improved estimates are required within 1 year!).

5. If the reservoir in Example 10.5 was not full preceding the occurrence of the PMF, but required an inflow volume of $705.6 \times 10^3\,m^3$ to bring it to spillway crest level, determine the new peak outflow.

[$115\,m^3/s$]

6. Determine the necessary pipe diameters for the first three pipes of a drainage network using the modified rational method from the data given below.

Pipe no.	Length (m)	Gradient	Area (m^2)	t_e (min)	k_s (mm)	C_V
1.0	70	0.0175	1415	4	0.6	0.9
1.1	75	0.017	3275	4	0.6	0.9
1.2	92	0.0085	4085	4	0.6	0.9

Rainfall (mm/h)	55.3	54.3	53.4	52.5	51.7	50.8	50.0	49.3	48.5	47.8	47.1
Duration (min)	4.8	5.0	5.2	5.4	5.6	5.8	6.0	6.2	6.4	6.6	6.8

[1.0, $d = 175$ mm; 1.1, $d = 225$ mm; 1.2, $d = 250$ mm]

CHAPTER 11

1. Use dimensional analysis to establish the dimensionless Π groups which are relevant to flow over a weir. Which group or groups are the most important for model analysis? Tests have been performed on a 1:50 scale model of a weir. The model discharge is 3.5 l/s. What is the corresponding discharge for the full scale weir?

[$62\,m^3/s$]

2. Show that the resistance F_R to the descent of a spherical particle through a liquid may be expressed as

$$F_R = \rho u^2 r^2 f\left(\frac{\rho u r}{\mu}\right)$$

where r is the radius of the sphere, u is the velocity of descent, and the other terms have their usual meaning. Hence prove that if F_R is proportional to u, then

$$F_R = K_{\mu} r u$$

If $K = 6\pi$, $r = 0.01$ mm, $\rho = 1000\,\text{kg/m}^3$ and $\mu = 1.1 \times 10^{-3}$ kg/m s, estimate the time taken for a particle to fall through a distance of 3 m. The density of the particle is $2500\,\text{kg/m}^3$.

[3.1 h]

3. A radial flow pump has been tested and the following results obtained:

   Speed = 2950 rev/min;
   discharge = 50 l/s;
   head = 75 m;
   efficiency = 75%

   The pump impeller is 350 mm in diameter. Determine the size and speed of a dynamically similar pump for the following duty: discharge = 450 l/s, head = 117 m.

   [940 mm, 1375 rev/min]

4. A new dock is to be constructed in an estuary. The estuary is 1.5 km wide at the seaward end and 7 km in length. The depth of water at the dock site ranges from 40 m at low tide to 50 m at high tide. The laboratory area is 20 m wide and 60 m long. Give principal model scales and the maximum flow.

   [For $\lambda_x = 1/250$, tidal period = 45.5 min, $Q'' = 4$ l/s]

5. A flood alleviation scheme is required for a stretch of river 5 km long. The river is 60 m wide, and the average depth is 7.5 m, for the passage of the probable maximum flood discharge of $1700\,\text{m}^3/\text{s}$. The laboratory for the previous question is to be used, but the model length must be accommodated across the 20 m width. The maximum laboratory discharge is 20 l/s. Estimate the model scales.

   [$\lambda_x = 1/300$, $\lambda_y = 1/50$, $Q'' = 16$ l/s]

APPENDIX A
Moments of area

One very common phenomenon in engineering is that of a pressure or stress which is distributed continuously, but not uniformly, over a surface. Typical examples are found in hydrostatics and in the bending of beams or columns. The intensity of pressure p (or stress σ) is often linearly distributed in one direction (Figs A.1(a) and (b)). That is to say, p (or σ) is proportional to distance (y, say) from a specified axis, so p (or σ) $= Ky$.

The hydrostatic force, δF, acting on a small element of area is given by

$$\delta F = p\,\delta A = Ky\,\delta A$$

The force on the whole area is obtained by integration:

$$F = \int Ky\,\mathrm{d}A = K \times (\text{first moment of area}) = KA\bar{y}$$

where $\bar{y}$ is the distance from surface of liquid to the centroid.

The moment of the force about the axis O–O is obtained by taking the product of force and distance from O–O. For the element,

$$\delta M = \delta F y = p\,\delta A y = Ky^2\,\delta A$$

Hence, the moment taken over the whole area is

$$M = \int Ky^2\,\mathrm{d}A = K \times (\text{second moment of area})$$

The second moment of area is denoted by I.

Both the first and second moments of area are simply functions of the geometry of the surface and position of the axis relative to the centroid. If the axis passes through the centroid, then the second moment has a particular value, I_0, for any given shape. Thus, for a rectangle (Fig. A.1(c))

$$\text{area of element} = \delta A = B\,\delta y$$

$$\left.\begin{array}{l}\text{second moment of area of element}\\ \text{about axis through centroid}\end{array}\right\} = \delta A y^2 = B\,\delta y y^2$$

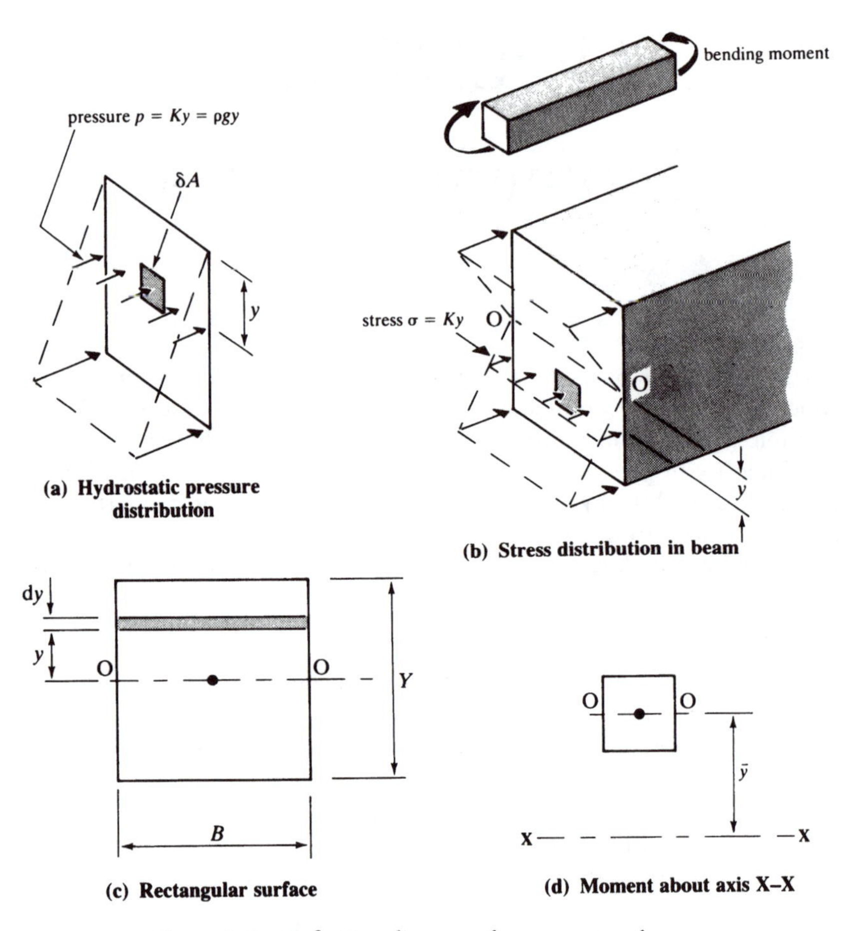

Figure A.1 Definition diagrams for moments of area.

for whole area,

$$I_0 = \int By^2\,dy = B\left[\frac{y^3}{3}\right]_{-Y/2}^{+Y/2}$$

$$= \frac{BY^3}{12} = BY\left(\frac{Y^2}{12}\right)$$

$Y^2/12$ may be regarded as the square of the length of a lever arm.

If the second moment of area about any other axis is required, then the above equations can be modified appropriately. For many common cases the arbitrary axis is parallel to O–O. Thus, for a rectangle (Fig. A.1(d)), the second moment of area of an element about axis X–X is given by

$$B\,\delta y(y+\bar{y})^2 = B\,\delta y y^2 + 2B\,\delta y y\bar{y} + B\,\delta y\bar{y}^2$$

Therefore

$$I_{\mathrm{X-X}} = \int_{-Y/2}^{+Y/2} By^2\,\mathrm{d}y + \int_{-Y/2}^{+Y/2} 2By\bar{y}\,\mathrm{d}y + \int_{-Y/2}^{+Y/2} B\bar{y}^2\,\mathrm{d}y$$

The second term → 0, so

$$I_{\mathrm{X-X}} = I_0 + A\bar{y}^2$$

where $A = BY$.

This is a statement of the Parallel Axes Theorem.

Index